CONTENTS

ADVERTISERS

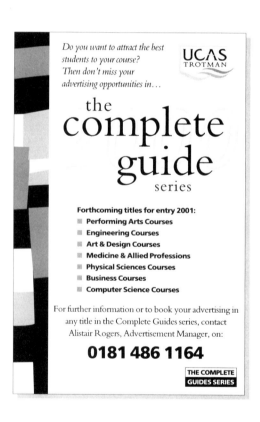

THE COMPLETE GUIDES SERIES

computer science courses 2000

UCAS TROTMAN

Computer Science Courses 2000

This edition published in the United Kingdom by UCAS and Trotman & Company Ltd.

UCAS
Rosehill
New Barn Lane
Cheltenham
Glos
GL52 3LZ

Trotman & Company Ltd
2 The Green
Richmond
Surrey
TW9 1PL

Editorial: John Crampin, Sylvia Crampin, Neil Harris, Irene Kirkman
Advertising: Alistair Rogers
Data preparation: Sylvia Crampin

Typeset at the Rosehill Headquarters of UCAS

British Library Cataloguing in Publication Data

A catalogue record of this book is available from the British Library

ISBN 0 85660 446 1

Printed and bound in Great Britain by Creative Print & Design (Wales) Ltd.

PREFACE

In the late 18th century, Thomas Newcomen invented the first practical steam engine, and further refinements by the Scottish engineer James Watt (and others) led to the Industrial Revolution in the 19th century. People's lives were irrevocably changed through the availability of a source of power that revolutionised manufacturing, transport, farming methods, and the social structure of the western world.

Future historians will undoubtedly regard the late 20th century in a similar manner because of the invention of the electronic computer and its dramatic effect on all our lives. Like the machines that replaced muscle power by steam, computers have the potential to improve the day-to-day life of the world's population in many ways. They can be used to carry out routine tasks, easing the burden of office workers in the same way that machinery eased the burden of those in the factory. They can be used to predict the effects that human activity may be having on our environment. Computers contribute to our artistic endeavours in film and music, and in the creation of fantastic patterns such as those based on 'fractals'. We use them for both leisure and for education.

The world of information technology gives us access to a wealth of information on an unprecedented scale, and has revolutionised the way we do business, both as individuals and between companies. As with all new developments since the invention of the wheel and the discovery of fire, however, there are potential dangers and ethical problems associated with the use of computer power, as foreseen, for example, in the 'Big Brother' society of George Orwell's novel *1984*.

Because of their importance to business and the rapid expansion of the industry, computer staff and information technology specialists are in great demand and can command high salaries. In many industries, these would be sufficient attractions, but working with computers is also challenging, stimulating and, most of all, fun!

The technology is developing at an accelerated pace, which means that you should not even contemplate a computing or an information technology career if you wish to study the subject only as a means to an end. You will need to update your skills constantly and, as the performance of hardware improves and prices reduce, there are more and more ways in which these skills might be used. The appeal of such a career was well summarised in the words of a programmer who once told me "I can't believe I get paid so much for doing something I enjoy anyway!"

Choose your course carefully and enjoy your career.

John Crampin
Computing Services, Universities and Colleges Admissions Service

HOW TO USE THIS BOOK

This guide is intended to help you take a realistic view of the many entry routes into a Computer Science degree. There are 4,223 detailed course entries in this book, grouped into seven areas. Because of some overlapping of subject areas, a few courses appear in more than one group. You should note that many higher education institutions use the term 'programme' rather than 'course', particularly when they are offering a subject on a modular basis. The structure of modular courses is explained in the Introduction.

When you use this book you should decide whether you wish to follow a single honours degree or a combined degree (major/minor or equal combination) which can lead to an honours degree, or the HND route, and in which specialised area. Once you have decided this you should refer to the relevant course data (starting on page 17). Within the data, sections are listed alphabetically by UCAS course code.

As well as the combined courses referred to above, there are many other subjects allied to computer sciences which can be found within the section 'Other Informatics Sciences', eg Digital Business/Entertainment, Education and IT, Environment and Computing, Food Production and Nutrition and IT, Forensic Science and IT, Industrial Information and Technology, Simulation/Virtual Environment and Technology and Virtual Reality.

A section of the book is dedicated to explaining the course characteristics and entry requirements together with information on employment opportunities and graduate destinations. There are a few case studies of undergraduates and successful graduates.

An example course entry is shown below.

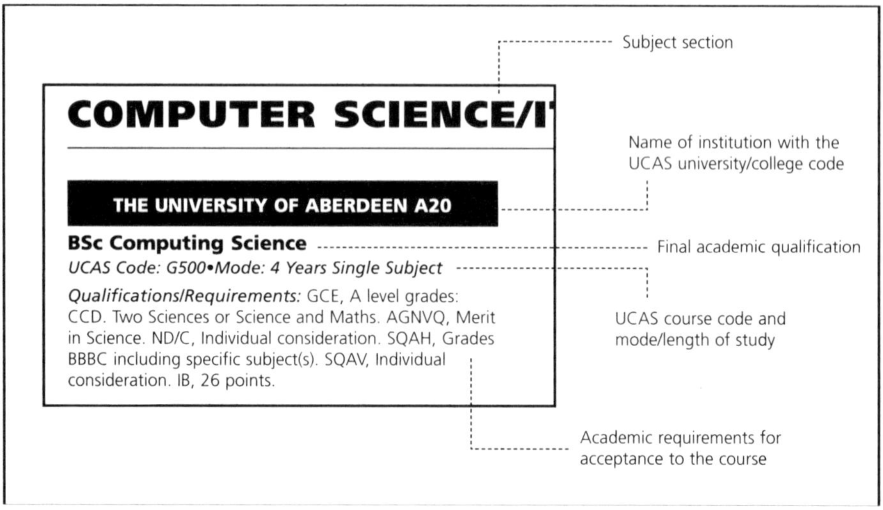

KEY TO QUALIFICATIONS

Information about individual institutional entry qualifications/requirements is as follows:

AGNVQ **Advanced General National Vocational Qualification**

Institutions often require a particular level in AGNVQ which will be shown as Pass (P), Merit (M) or Distinction (D). Some institutions may ask for 'additional units' or additional A levels or for the AGNVQ to be in a particular subject area; other institutions may consider AGNVQ on an individual basis ('Individual consideration').

GCE **General Certificate of Education Advanced Level or Advanced Supplementary (A level or AS)**

A levels/AS are shown as a points score grade A=10 for A level, 5 for AS; B=8 for A level, 4 for AS; C=6 for A level, 3 for AS etc, and A level grades are shown A–E. Some institutions may require specific A level subjects (AS may also be acceptable), eg physics, maths.

IB **International Baccalaureate**

The IB score ranges from 24 points. The data entry will show the number of points required from the diploma (eg 34 points for medicine), or the diploma will be considered on an individual basis. This is shown by the phrase 'Individual consideration'.

NC/D **National Certificate or National Diploma**

Edexcel (BTEC) offers this vocational qualification either as a National Diploma (ND) or National Certificate (NC). Specific units/courses may be required. Some institutions will consider the NC/D on an individual basis.

SQAH **Scottish Qualifications Authority (Higher)**

The data entry requirement will show grades A, B or C, or institutions will consider the Highers on an individual basis, and also if they require a Certificate of Sixth Year Studies (CSYS).

SQAV **Scottish Qualifications Authority (Vocational)**

The data entry will show if a pass is required in a National Certificate or Diploma, if a specific course or extra units are required, or if institutions will consider the Certificate/Diploma on an individual basis.

SQA **Scottish Qualifications Certificate**

The data entry will only be shown as SQA if the SQAH and SQAV are to be considered on an individual basis.

GCSE **General Certificate of Secondary Education**

This will be shown if specific GCSEs are required by the individual institution and stated as 'Specific GCSEs required'.

Most institutions will accept other qualifications and will accept mature applicants (aged 21 or over in England, Wales and Northern Ireland and 20 or over in

Scotland). For any of the courses listed in the following pages (which were correct at the time of going to press), you can check the qualification requirements further by reading the institutions' prospectuses, accessing their websites via the UCAS website (http://www.ucas.ac.uk) or speaking directly to admissions staff.

Addresses of all the institutions which offer courses are listed at the back of the book. It is important that you check the *UCAS Handbook* to see if the course you wish to study is at the main site institution or if it is located at another campus or if it is offered at an associated college. The 'Useful Sources of Information' section (see page 310) includes professional associations, government departments and other organisations which may provide information or assistance to students.

INTRODUCTION

In just over half a century, computers have developed from large specialised machines, designed to decode secret messages or to solve mathematical equations, to transform the way we live. All businesses and many individuals rely on computers to carry out their day-to-day activities. There are computers on the desks of most office workers, and many senior executives who previously relied on adminsitration and secretarial staff now use them for information gathering and may use word-processing and email themselves.

By the mid-1960s, computers were firmly established in most medium to large businesses. These 'mainframe' machines were usually located in specially designed rooms where the temperature and humidity were carefully controlled. The operating system software was designed and supplied by the hardware manufacturer, and as a result, communication between computers supplied by different manufacturers was almost impossible. Systems analysts and programmers designed and wrote the programs, and often were considered remote from other staff within the organisation for which they worked. Computer input was a specialised task, carried out by 'data preparation' staff, and the output was mostly in the form of large 'print-outs', which consisted of parts lists, account details and other relative information. Operators, who often worked shifts to enable the machine to complete its various tasks, carried out the day-to-day running of the computer.

The development of the microprocessor in the 1970s led to a rapid increase in the use of computers as the miniaturisation progressed, prices reduced and capabilities increased. This led directly to the first Personal Computers, and to the use of microprocessors (commonly referred to as microchips or chips) in other areas, eg electronic calculators, mechanical devices in manufacture, and more recently in cars, stereos, domestic appliances, telephones, and of course leisure items such as the Nintendo games console and the Sony PlayStation. Nowadays there is at least one computer chip in almost every piece of electrical equipment, every vehicle and many toys.

Miniaturisation and an increase in the speed and reliability of the storage devices to hold more and more information have paralleled the development of the microprocessor. A good 1980s PC might contain a megabyte of memory and 10 megabytes of hard disk. Now, a computer with less than 64 megabytes of memory and several gigabytes of hard disk might seem woefully inadequate. The increased capacity enables computers to store data in a wide variety of formats, including still and moving pictures and sound, as well a wealth of information in text and graphic formats.

The need for users to connect to computers from a distance, and for the interchange of information between computers, was identified in the 1960s, but it was not until a decade or so later that the first computer networks became available. Over time, and through a process of evolution, a set of common standards for exchange of information has emerged, and this has made the Internet, e-Commerce and the World Wide Web possible.

The pace of development shows no sign of slowing, which makes this an exciting and rewarding area of study.

SOME KEY FACTS

The computer area covers a very wide range of degrees and HND courses classified under the broad category of the subjects listed in the contents page, coded G or H by UCAS. There are 4,223 courses in 185 institutions covering the whole spectrum of higher education – older universities, new universities and colleges of higher education. Computer science is a very popular and expanding area for applicants. Recent UCAS statistics indicate that 103,593 applications were made during 1997/98 for a computer-related degree/HND, of which 33,153 were accepted compared to 18,129 in 1997 (see diagram below).

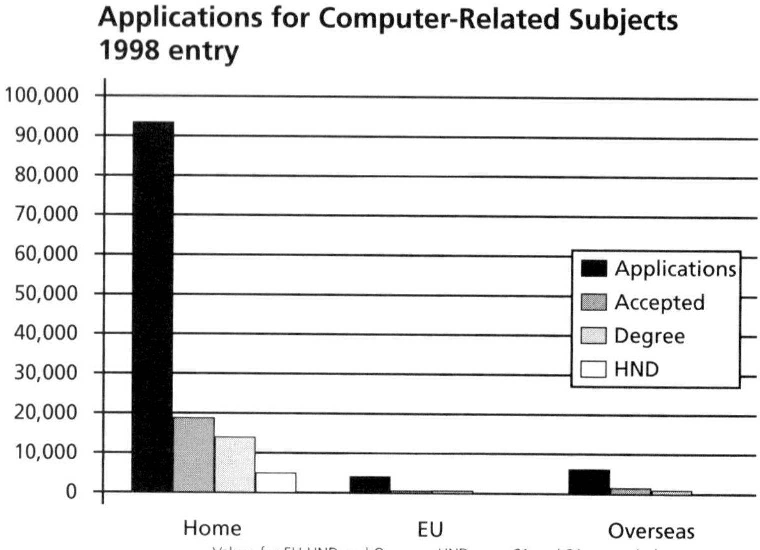

Values for EU HND and Overseas HND were 61 and 81 respectively.

Information Technology and Computing

In recent years the term 'information technology' (IT) has come into prominence when referring to various aspects of the use of computers. We speak of the need for IT skills within our workforce and within our schools. An IT department in an organisation may be responsible for setting up and maintaining a whole network of computers, the internal and external communications networks and servers, and/or the design and maintenance of a complex website and user base. Some of these activities may include programming, some a detailed knowledge of computer hardware and software, and others, for example website design, require artistic and creative skills. Taken more literally, IT also includes the design and structure of databases, and the processing and transfer of information in any form or medium.

Until recently the term 'computing' meant the operation or programming of a single computer, so that jobs in computing had titles such as Operator, Programmer or Systems Analyst. We still speak of being 'computer-literate', but this does not mean to say that someone described as such has any concept of programming, or of how the hardware or software actually operates.

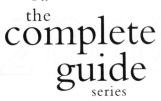

THE COURSES

The courses on offer in higher education detailed in this directory cover an enormous range of subjects and have a diverse range of titles. It is important to read prospectuses or visit websites (http://www.ucas.ac.uk) to find out precisely what the course of your choice involves, what you would learn and how the relevant professional bodies rate the course.

There are 185 institutions and 4,223 courses in the UK which offer the study of computing. Students of these courses cover the history and development of computers and the theoretical basis for their operation. A good understanding is gained of the architecture of computers, their components and how they relate to each other. Learning to use several operating systems (such as UNIX, DOS, Windows NT and MacOS) and programming languages (such as visual basic, SQL and C++) is a core activity.

Graduates from these courses are usually adept in the use of a range of software packages, especially databases and spreadsheets. They discover how to apply existing technology to a variety of different problems. Some of the courses include modules on communication by computer, the development of local and wide area computer networks (LANS and WANS), and how these are organised. After learning the basic computing knowledge and skills, students go on to apply these in a range of contexts and this usually brings them into some of the specialisms such as artificial intelligence, multimedia, and virtual reality.

Learning occurs in several ways, through lectures, tutorials, coursework and practical work in the laboratory. Good departments have large numbers of computers so that access to them is guaranteed. Departments which do not have sufficient computers for every student allocate time for priority use of the equipment. Ask the institutions you are seriously considering about access to the PC laboratories and other facilities.

Artificial Intelligence

Artificial Intelligence (AI) is a term that in its broadest sense indicates the ability of a machine to perform the same kinds of functions that characterise human thought. The possibility of developing such machines has intrigued humans since ancient times. With the growth of modern science, the search for AI has taken two major directions: psychological and physiological research into the nature of human thought, and the technological development of increasingly sophisticated computing systems.

In this sense, the term 'artificial intelligence' has been applied to computer systems and programs capable of performing tasks more complex than straightforward programming, although still far from the realm of actual thought. The most important fields of research in this area are information processing, pattern recognition, game playing, and applied fields such as medical diagnosis.

Current research in information processing deals with programs that enable a computer to understand written or spoken information and to produce summaries, answer specific questions, or redistribute information to users interested in specific areas of this information. Essential to such programs is the ability of the system to generate grammatically correct sentences and to establish links between words and ideas. Research has shown that whereas the logic of language structure – its syntax – submits to programming, the problem of meaning, or semantics, lies far deeper, in the direction of true AI.

In medicine, programs have been developed that analyse the disease symptoms, medical history and laboratory test results of a patient, and then suggest a diagnosis to the physician. The diagnostic program is an example of a so-called 'expert system' – programs designed to perform 'human' tasks in specialised areas. Expert systems take computers a step beyond straightforward programming, being based on a technique called rule-based inference, in which pre-established rule systems are used to process the data. Despite their sophistication, expert systems still do not approach the complexity of true intelligent thought.

Courses in this category may fall into the broad area of engineering, the development of 'intelligent' robots or machines, or be of a more theoretical nature, sometimes combined with psychology.

Computer Science/Information Technology

Computer science is a general term for a range of courses which cover the main uses of computers and information technology in business and industry today. In most institutions, the course will cover hardware, software and the theoretical basis of computing.

Many courses will concentrate on the production of software, from analysis and design, through programming, to testing and quality control issues. You will learn a programming language such as C++, the use of a database such as Oracle or SQL-server, and investigate the problems of implementation and delivery of the system to the user, on screen or via the Internet through a browser.

Other courses under this heading allow specialisation in different areas of computer science, such as the development of operating systems, or the theoretical and practical problems involved in networks, or the provision of complex data via the Internet.

It is vital to research thoroughly if you are contemplating a course in computer science or IT, as the content can vary widely from one institution to another. If the course concentrates on the theoretical basis for computing, you will probably need strong mathematical ability, whereas for a more business or commercially oriented course, you will not. Many computer science courses are accredited by the British Computing Society.

Computer Systems Engineering

Computer Systems Engineering refers to the design and development of computer systems, including both hardware and software, but with a greater concentration on the hardware components and the way that they interact with the software. These courses are more likely to appeal to the student who wants to understand the basis of the technology which is driving the IT revolution.

Like the computer science courses, you will study computer hardware, operating systems and software, and will learn one or more programming languages, but the emphasis will be on the interaction between the components and the course will be at a more detailed level. For this reason, there may be modules with a strong electronic engineering bias, and the Institute of Electrical Engineers accredits some of these courses.

Computing with Languages

The courses in this section will appeal to the student with a flair for language, who is considering a career in commercial computing or IT. Many commercial organisations operate in a global marketplace, and an ability in a foreign language is seen as a valuable asset in a prospective employee.

Multimedia

The recent explosion in the use of multimedia has led to a demand for experts in this growing field. Like many of the other courses, the basics of computers, operating systems and software will be included, but you will specialise in the presentation of sound coupled with still and moving pictures, and in development. To do this, you will learn the techniques involved in capturing the data, eg by using scanners, digital video cameras, sound samplers and synthesisers, to manipulating the images and sound, and compiling the required multimedia presentation using packaged software designed for the purpose.

The presentation could be recorded on CD-ROM, or stored on a server to be downloaded in compressed form to a web browser.

A multimedia course will require a good technical capability, but you will also need an artistic flair. Multimedia engineers are employed by the television and film industries, by companies requiring Computer-Based Training material (CBT courses), within education and advertising, and in the computer games industry.

Software Engineering

Software is a general term for the instructions which control a computer system and cause it to carry out the intended tasks. This includes the system software (or operating system) as well as applications software – the game, package or utility that you might buy in the shops. Some software is present in all electrical devices containing microcomputers, and this is crucial to the correct operation of the telephone, stereo, nuclear power station or aeroplane in which it resides. Most users of PCs will have had occasions when a 'bug' prevents them from continuing, or the machine crashes. Software engineering is the application of engineering principles to the development of software within electrical equipment, or is a means of enabling systems to recover from component failure without 'crashing'. The courses will share a great deal of content with pure computer science courses, in that to understand the processes of software engineering, a thorough grounding in the hardware, operating systems and a programming language is necessary.

The software engineer will also need to understand how the software might behave in all conditions, so a rigorous and structured approach to development and testing is required. Software engineers will be in demand in the telecommunications, aerospace, automotive and other industries where more and more reliance is placed on automated processes.

RELEVANCE TO PROFESSIONS

There are several professional bodies with an interest in computer science. The two most prominent ones are the British Computer Society (BCS) and the Institution of Electrical Engineers (IEE). Both of these bodies accredit courses of which they approve. Undertaking an accredited course allows students to proceed more quickly to corporate membership of an institution. Full details are given on their websites – www.bcs.org.uk and www.iee.org.uk. Some joint honours degree courses are partly accredited, which means that they help graduates to become professional members of the institutions but not as quickly as graduates from fully accredited courses. Both organisations offer student membership at reduced rates, which provides the opportunity to gain more knowledge of the profession by receiving magazines and attending local meetings of the institutions.

BCS has accredited degree and HND courses in 90% of the computer science departments in the UK and in some universities abroad. Some joint honours degrees that include the study of two subjects, not just computer science, are partly accredited, as are some ordinary degrees. Full details are given in the *BCS Yearbook*.

Both institutions provide a route to registration as a chartered engineer with The Engineering Council after a period of postgraduate training and industrial experience has been gained.

THE TYPE OF PERSON YOU NEED TO BE

If you are contemplating a career in computing and information technology after a period of study in this area, it is important to assess how deep is your interest and how relevant are your skills. There is no doubt that students with a fascination for computer systems, an inquisitiveness about how they work and the determination to master the complexity of it all are most likely to succeed.

Computers do as they are told and only what they are told. It follows that good attributes for success are a logical approach and an ability to analyse situations and problems. A facility with mathematics is a good asset although not always essential. Many of the courses include modules in mathematics, and the problem-solving techniques that mathematicians often take for granted are valuable in the world of computing.

Meticulous attention to detail is another vital quality. If you have a flair for it you will make fewer mistakes and quicker progress with computer systems than those who do not. Computers are hard taskmasters and reject all but the most accurate instructions.

Computer people are no longer isolated individuals. They usually work in teams on large projects which cannot be handled by one person. They have to listen to the needs of their clients and interpret those needs in terms of novel or upgraded systems.

They often help people who are not computer literate to understand those aspects of computing which they need to know. In short, good verbal and written communication with clients and colleagues is important.

Many employers analyse job applicants for these qualities by setting tests that measure the candidates' aptitudes in numerical, verbal and diagrammatic reasoning. If you enjoy a challenge, and are good at solving problems, computer science could be the subject for you. Vision, machine learning and translating are just three of the many areas where computers have not yet provided solutions – if you have a certain curiosity coupled with creativity and decide to work at the frontiers of knowledge, there is still plenty to explore.

THE DIFFERENT ROUTES INTO A COURSE

Entry requirements

There are now several different routes into a course in computing. Many of the courses in computer science and related subjects do not demand the previous study of any specific subjects. No previous knowledge of computers is expected. Some courses of study will expect A level, or its equivalent, in mathematics. Courses which include 'engineering' in their title, such as 'computer engineering' or 'computer systems engineering', often require qualifications in maths and a science. Of all the sciences, admissions tutors most often seek an A level or its equivalent in physics.

Qualifications in computer science, such as the Edexcel BTEC, AGNVQ in IT or an A level in computing, are welcome. Grades required vary considerably from one university to another, and applicants offering an AGNVQ in IT may still require an A level in maths. Applicants with the International Baccalaureate will probably require 5 points from the maths paper, and those with Edexcel BTEC may require a distinction.

If you are unlikely to gain the necessary grades for entry to the first year of a degree course, you might consider one of the foundation courses offered by some institutions. Mature students can be given credit (APEL) for their employment experience and may be admitted with lower grades, or have the opportunity to go straight into the second year of a three-year course if they are particularly well qualified.

Alternatively the possibility of studying the two-year computer-related HND in computing offered by institutions may be more attractive. With good grades in your HND you could go straight into employment or take an option to join the second year of a computer science degree course.

Course characteristics

Although there is considerable variation in the content of the courses offered, those with quite different titles can be very similar. Look at the details on course search on the UCAS website at www.ucas.ac.uk/ucc/index.html to find the course that interests you most. Obtain the institution prospectus or look at the institution website (which you can access from the UCAS website) for further information on each course to discover what you would actually learn.

If your future aim is to join a large IT company or a major software house, it would be wise to choose a course that gives you the full range of skills in programming, operating systems, and the use of many different software packages, together with an understanding of networks of distributed computers and how they communicate with each other. If your intention is to join the electronics industry, and your mathematics is strong, consider a course that includes software engineering.

Many of the courses offered focus on business and information systems. They help you to develop the skills and knowledge you would need to analyse business problems and systems and to use appropriate packages. Their focus is often on the application of software and computer systems to a range of business situations. Graduates of these courses do not always need to develop the entire range of skills in programming. Instead they concentrate much more on the business environment in order to prepare for a broad range of work in that area.

Computing courses usually have very close links with the industry. This has many advantages. It keeps you abreast of the latest commercial developments and is also a source of opportunities for sandwich courses, vacation work, and eventual employment.

Many higher education institutes offer attractive sandwich courses. They add a year to a normal three-year course but include a year's industrial experience. The more common type of sandwich course involves spending two years at university, then a year in industry, followed by a final year. This is known as a 'thick' sandwich. A 'thin' sandwich course consists of two semesters, or six-month periods, slotted in at different times within the four years. There are usually specialists who organise sandwich placements. The department's contacts may vary from small local businesses at one end of the spectrum to large multinational concerns at the other.

All computing departments have the opportunity to offer some of their students the chance to participate in exchanges with European universities. A few offer sandwich courses that include a year abroad. These opportunities are usually in North America or Europe. The departments may have established links with particular universities in other countries where they regularly send their students on exchanges to gain the experience of studying in another country.

The European Union operates the ERASMUS scheme of student exchanges. This provides opportunities to study in European Union and EFTA countries, plus some countries in central Europe such as the Czech Republic and Poland. Before joining a course, you should always check with the institution to find out whether it pursues this kind of opportunity with any enthusiasm. If you intend your career to start with an international firm such as Hewlett Packard, Siemens or Intel, the ability to speak and write in another European language could be important. What better way can there be to develop these abilities than to spend time immersed in the culture of another country?

The structure of each higher education institution can have an important bearing on the studies you might pursue. Some computer science courses are offered by engineering faculties and some by the electrical and electronic engineering departments. Mathematics departments run others, while many universities have independent computer science departments, schools or faculties. The flavour of the course of study will inevitably reflect the context in which it arises and the research interests of the areas in which it is offered.

Computer science departments often offer a range of different courses with a common first year, and in some there may be an option to choose the final direction of your course at a later date. This flexibility is useful because many students do not know precisely the direction of their career when they start the course of study. The first year of a computing degree (no matter which subject you decide to specialise in later) will consist of areas such as advanced programming, computer architecture, computer systems and applications, data structure, digital electronics, fundamental problem solving, software engineering, logic and program proving, computer organisation and systems software and mathematics for computer science. The precise pattern will depend on the individual institution.

CAREERS IN INFORMATION TECHNOLOGY AND COMPUTING

There is a wide range of careers available to graduates of any computing or IT course, with a bewildering variety of job titles, including programmer, software engineer, systems analyst, business analyst, webmaster, data analyst, communications specialist, network manager and database administrator. The main employers are software houses, computer manufacturers and electronic and communications equipment manufacturers, but graduates will also be found in industry and commerce, in all large and many smaller companies, as well as in the public sector.

Software houses and consultants help other organisations to improve their computer systems. They may define the system required for a specified purpose and provide the software to make it a reality. The leading software firms such as EDS, Logica and Cap Gemini are international and provide a service over the entire range of industries from finance to defence, from health-care to energy and transport. Their clients are often major companies and governments. The same is true of the consultancies such as Andersen Consulting and PricewaterhouseCoopers. Smaller software consultancies tend to specialise in niche markets such as electronic fund transfer systems for retailers or medical records systems for hospitals and surgeries.

There are numerous examples of the achievements of software houses. If the police stop you when driving your car, they can quickly look up the name and address of the owner on a vast database which was developed by a software house. The police are also developing a DNA database to help them identify all known criminals. Jobcentres will soon have access to vacancies nationwide, not just in the local area. When investment bankers get to work each morning, they know instantly how the markets have changed around the world while they were asleep. Reuters and other organisations provide the computer-based service which results in the instant availability of this information. Software systems and computer networks that can achieve such things are devised and implemented by consultancies and software houses acting on behalf of their clients.

In addition to taking on major projects, the software houses also offer to manage the computer systems of large companies. This activity, known as 'facilities management' or 'business operations management' is an area of major growth. Firms such as EDS and Andersen Consulting are continually taking on employees from other companies whose computer systems they have contracted to manage and the number of their employees is continually rising.

A third activity of these organisations is to provide staffing on an ad hoc basis for computer departments. For employees of such companies, flexibility and mobility are essential.

Telecommunications companies, including BT, Cable and Wireless and Internet service providers such as Demon and Freeserve, are a second source of jobs. The growth of the Internet is nothing less than dramatic, with Freeserve signing up hundreds of thousands of customers in its first year of operation.

Engineering is another important employer. First there are the manufacturers of computers such as Compaq, Dell, Hewlett Packard and IBM. Next are suppliers of other IT equipment including mobile phones, telephone exchanges, etc. Philips, Sony, Marconi and Nokia are typical examples of this sector. Firms which design, make and install computer control systems and robots to improve the efficiency of industrial processes are also important recruiters.

Engineering designers use computer-aided design (CAD) software that automatically prints engineering drawings, lists of parts and specifications. Additionally, many industries use simulation techniques. Examples of these are the simulation of new chemical plant, oil fields with several wells, and training systems for airline pilots.

Computer systems are ubiquitous these days. Banks rely heavily on them, not only for cash machines and keeping accounts up to date, but also for backing up trading in financial products such as derivatives and shares with sound, fast administrative systems. Modelling financial systems on computer in order to predict the future, like a financial version of a weather forecast, is another occupation that is extensively computer-centred. Meanwhile, marketing executives are collecting details of customers from credit cards, constructing databases and using them to discern trends and seek out new clients. Those responsible for logistics and transport systems are also key IT users.

In science, computers are used extensively to create model systems. From weather forecasting to modelling the development of oil wells, from the interpretation of medical images in body scanners to research into the aerodynamics of aircraft wings, computers are used extensively.

The Association of Graduate Recruiters reported in January 1999 that its members were having difficulty in finding sufficient numbers of graduates with the right skills in IT – 10–20% of vacancies remained unfilled. However, flexibility and adaptability are strong requirements for aspiring employees because the skills needed in five years' time will inevitably be different from those required now. If you decide to enter this exciting career, you will have to update your skills continually to keep in touch with rapidly advancing technology.

Employment statistics

More than 7,400 UK-domiciled students graduate each year in computer science and related subjects. Many more come from abroad to study the subject in British universities. A further 2,200 qualify with Higher National Diplomas. Only 20% of current students are women. Despite the best efforts of course organisers and employers to recruit women, men appear to be more interested in these subjects.

Computing students arguably become the most employable of graduates; in 1997 83% of them went straight into employment in a recent year when the average employment figure for all graduates was 62%; 7% stayed on at university to study for a higher degree; 5% trained as teachers and 1.4% participated in some other learning or training. Only 6% were still seeking employment six months after they had graduated.

Graduates work in a wide variety of industries. The statistics show that half took employment in software houses, consultancies and other related businesses, 13% joined manufacturing industry, while 7% were employed in telecommunications. The financial sector attracted 9% of these graduates; retailers and wholesalers employed 5%; 4% took employment in the public sector (eg civil service and local government) and a further 4% found jobs in schools, colleges and universities. The rest were spread around a broad range of enterprises including the public utilities, hotels and catering and the construction industry. These figures are based on the statistics produced by the Higher Education Statistics Agency for students who graduated in 1997.

Nayan – a work placement student at UCAS

I've been studying information technology now for the past five years, which began with a two-year Computing course at GNVQ Advanced level. From there I went on to study for an IT degree. I am actually based in Cheltenham and chose to study here for many reasons, one of which was to economise. I have spoken to friends who have studied away from home and they explained that the first few weeks of being independent was great; it's only after about a year that the excitement of being away from home seems irrelevant, including the stress of debts and bills for accommodation and welfare, all of which are on top of studying. It can be a difficult period. But this is only my point of view on student life away from home, and I'm sure that there are students out there who would disagree with me.

Enrolling on this course seemed to be the right choice as I had the option of studying two different subjects, depending on qualifications. My course is structured on a major/minor basis, Information Technology with Multimedia. Because the course is modular, you can further your options by selecting the appropriate topics/modules in each subject, so I'm working on 15 modules in IT and five in Multimedia. With the prospects of studying for at least four years, I'm sure this type of course programme will help keep my interest at a high.

You can complete your degree in three years or four years if you wish to take a one-year work placement. I was never quite sure whether to take a work placement for a year. I kept thinking that four years of education would be too long a stay, and completing the three years straight would ease the stress of any up and coming future plans. I am currently completing my work placement at UCAS, as an IT Support Technician, and I'm very much enjoying my experience. Since joining I have learnt a huge amount about networking, PC maintenance, user support and much, much more in a short time.

Comparing the two years I have spent at college with the 48-week work placement, I would say that more technical knowledge is gained through the placement, even though I completed a five-day Advanced PC Support course. I think that the IT degree should include a practical side to IT support, if not, a work placement should be compulsory for all IT students, especially as that experience is a vital element in any IT occupation.

Keith – graduate of the University of Sunderland

I attended the University of Sunderland from 1989 to 1993, studying for a BA Honours degree in Business Computing with German. The course was advertised as being a practical, hands-on approach to studying. This was achieved using a timetable of lectures, tutorials and workshops. In the first few months the pace of learning was slow, building up basic computer literacy and introducing bookkeeping, basic business German and the theory of computer systems. Our period of 'settling in' was interrupted by the introduction of new computer languages to be learned and used alongside structured system design methodologies and software testing environments. Deadlines for handing in assignments came and went rapidly, introducing the need to organise my time effectively and use the resources available at the university efficiently.

I spent the third year of the course on placement at a company in Frankfurt employed as a computer programmer. This was an immensely enjoyable time for me, working in a department that produced software for use in the commodities markets and getting the chance to sink or swim in the complexities of the German language! Regular contact was maintained with Sunderland by three visits from my placement tutor and the despatch of reports when requested.

The final year at Sunderland introduced more methodologies to be analysed, put to good practical use and repackaged in the form of presentations and assignments. German assignments and the business modules gained in complexity, or so it seemed at the time. The need to work well in teams as well as individually was emphasised by assessments being made on collective efforts in workshop situations.

Looking back now, six years after graduating, I consider the time spent at university to have been worthwhile for a number of reasons. Unlike any other situation in my life so far it was a time of self-assessment and looking at my own approach to tasks and difficulties that were encountered in an academic setting. The somewhat artificial pressure engendered within the course by self-doubts and looming deadlines was absorbing if a bit painful at the time.

I am now employed as an analyst/programmer for a high-profile company in Cheltenham.

Phil – HND and top-up degree graduate from Cheltenham & Gloucester College of Higher Education

It is now nearly 10 years since I first packed my bags in Newcastle and travelled down to Cheltenham to enrol on an HND course in Computer Studies. In the past 10 years I seem to have done so much, and yet it hardly seems any more than a few short years since I was an 18-year-old revising for my A levels. At that time I had no idea of which career I aspired to or even of any course I particularly wanted to do at college. I had applied to do a number of courses ranging from Electronic Engineering to Physiotherapy, none of which I had any great interest in.

I began the course and immediately knew I had made the right choice. Having studied maths at A level, I was able to grasp the fundamentals of computing straight away. The HND course I studied included a year out working in business. I spent that year working for an insurance company as a programmer.

I gained many things from the time I spent doing my HND. Not only did I learn a lot about computing, but having to move away from home allowed me to experience many new situations socially. The student life away from the books has got to be one of the best times of anyone's life if they enjoy socialising and meeting new people.

After finishing the HND, I decided to take a degree in another computing related course. After two more years of studying, I graduated and immediately started a job working as an analyst programmer for a company with a small development team.

I felt that by doing both an HND and a degree I was well equipped for what the real world had to offer. The HND taught me the practical skills I required, with the year's placement allowing me to reinforce and perhaps develop what I had learnt. The degree course was very much theory based and encouraged me to think more about why I was doing things in a certain way, and how to transfer what I learn in one situation to another.

ARTIFICIAL INTELLIGENCE

THE UNIVERSITY OF ABERDEEN A20

BSc Computing Science (Artificial Intelligence)

UCAS Code: G5G8 • Mode: 4 Years Major/Minor

Qualifications/Requirements: GCE, A level grades: CCD. Two Sciences or science and maths. AGNVQ, Merit in Science. ND/C, Individual consideration. SQAH, Grades BBBC including specific subject(s). SQAV, Individual consideration. IB, 26 points.

BSc Computing Science (Artificial Intelligence) with Industrial Placement

UCAS Code: G58V • Mode: 5 Years Single Subject

Qualifications/Requirements: GCE, A level grades: CCD. Two sciences or science and maths. AGNVQ, Merit in science. ND/C, Individual consideration. SQAH, Grades BBBC including specific subject(s). SQAV, Individual consideration. IB, 26 points.

THE UNIVERSITY OF WALES, ABERYSTWYTH A40

BSc Computer Science and Artificial Intelligence

UCAS Code: GG58 • Mode: 4 Years Equal Combination

Qualifications/Requirements: GCE, A/AS: 20 points. AGNVQ, Merit. ND/C, 3 Merits and 2 Distinctions. SQAH, Grades BBBCC. SQAV, Individual consideration. IB, 30 points.

THE UNIVERSITY OF BIRMINGHAM B32

BA Archaeology & Ancient History/ Artificial Intelligence

UCAS Code: GV86 • Mode: 3 Years Equal Combination

Qualifications/Requirements: GCE, A level grades: BBB. AGNVQ, Distinction with A/AS. ND/C, Individual consideration. SQAH, Grades ABBBB. SQAV, Individual consideration. IB, 32 points.

BA Artificial Intelligence/East Mediterranean History

UCAS Code: GVV1 • Mode: 3 Years Equal Combination

Qualifications/Requirements: GCE, A level grades: BBB. AGNVQ, Distinction with A/AS. ND/C, Individual consideration. SQAH, Grades ABBBB. SQAV, Individual consideration. IB, 32 points.

BA Artificial Intelligence/English

UCAS Code: GQ83 • Mode: 3 Years Equal Combination

Qualifications/Requirements: GCE, A level grades: BBB. AGNVQ, Distinction with A/AS. ND/C, Individual consideration. SQAH, Grades ABBBB. SQAV, Individual consideration. IB, 32 points.

BA Artificial Intelligence/French Studies

UCAS Code: GR81 • Mode: 4 Years Equal Combination

Qualifications/Requirements: GCE, A level grades: BBB. French at A level. AGNVQ, Distinction with A level. ND/C, Individual consideration. SQAH, Grades ABBBB. SQAV, Individual consideration. IB, 32 points.

BA Artificial Intelligence/German Studies

UCAS Code: GR82 • Mode: 4 Years Equal Combination

Qualifications/Requirements: GCE, A level grades: BBB. German. AGNVQ, Distinction with A level. ND/C, Individual consideration. SQAH, Grades ABBBB. SQAV, Individual consideration. IB, 32 points.

BA Artificial Intelligence/Hispanic Studies

UCAS Code: GR84 • Mode: 4 Years Equal Combination

Qualifications/Requirements: GCE, A level grades: BBB. AGNVQ, Distinction with A/AS. ND/C, Individual consideration. SQAH, Grades ABBBB. SQAV, Individual consideration. IB, 32 points.

BA Artificial Intelligence/Italian

UCAS Code: GR83 • Mode: 4 Years Equal Combination

Qualifications/Requirements: GCE, A level grades: BBB. AGNVQ, Distinction with A/AS. ND/C, Individual consideration. SQAH, Grades ABBBB. SQAV, Individual consideration. IB, 32 points.

BA Artificial Intelligence/Latin

UCAS Code: GQ86 • Mode: 3 Years Equal Combination

Qualifications/Requirements: GCE, A level grades: BBB. Latin. AGNVQ, Distinction with A level. ND/C, Individual consideration. SQAH, Grades ABBBB. SQAV, Individual consideration. IB, 32 points.

BA Artificial Intelligence/Modern Greek Studies

UCAS Code: GT82 • Mode: 4 Years Equal Combination

Qualifications/Requirements: GCE, A level grades: BBB. AGNVQ, Distinction with A/AS. ND/C, Individual consideration. SQAH, Grades ABBBB. SQAV, Individual consideration. IB, 32 points.

BA Artificial Intelligence/Music

UCAS Code: GW83 • Mode: 3 Years Equal Combination

Qualifications/Requirements: GCE, A level grades: AAB-ABB. Music at A level. AGNVQ, Distinction with A level. ND/C, Individual consideration. SQAH, Grades AAABB. SQAV, Individual consideration. IB, 32 points.

BA Artificial Intelligence/Philosophy

UCAS Code: GV87 • Mode: 3 Years Equal Combination

Qualifications/Requirements: GCE, A level grades: BBB. AGNVQ, Distinction with A/AS. ND/C, Individual consideration. SQAH, Grades ABBBB. SQAV, Individual consideration. IB, 32 points.

BA Artificial Intelligence/Portuguese

UCAS Code: GR85 • Mode: 4 Years Equal Combination

Qualifications/Requirements: GCE, A level grades: BBB. AGNVQ, Distinction with A/AS. ND/C, Individual consideration. SQAH, Grades ABBBB. SQAV, Individual consideration. IB, 32 points.

BA Artificial Intelligence/Russian

UCAS Code: GR88 • Mode: 4 Years Equal Combination

Qualifications/Requirements: GCE, A level grades: BBB. AGNVQ, Distinction with A/AS. ND/C, Individual consideration. SQAH, Grades ABBBB. SQAV, Individual consideration. IB, 32 points.

BA Artificial Intelligence/Theology

UCAS Code: GV88 • Mode: 3 Years Equal Combination

Qualifications/Requirements: GCE, A level grades: BBB. AGNVQ, Distinction with A/AS. ND/C, Individual consideration. SQAH, Grades ABBBB. SQAV, Individual consideration. IB, 32 points.

BSc Artificial Intelligence and Computer Science

UCAS Code: GG58 • Mode: 3 Years Equal Combination

Qualifications/Requirements: GCE, A level grades: BBC. Science at A level. AGNVQ, Individual consideration. ND/C, Individual consideration. SQAH, Individual consideration. SQAV, Individual consideration. IB, 32 points.

BSc Psychology and Artificial Intelligence

UCAS Code: CG85 • Mode: 3 Years Equal Combination

Qualifications/Requirements: GCE, A level grades: BBB. AGNVQ, Individual consideration. ND/C, Individual consideration. SQAH, Individual consideration. SQAV, Individual consideration. IB, 33 points.

UNIVERSITY OF BRISTOL B78

MEng Mathematics for Intelligent Systems

UCAS Code: J922 • Mode: 4 Years Single Subject

Qualifications/Requirements: GCE, A level grades: BBB. Maths and Physics at A level. AGNVQ, Distinction (in specific programmes) with A/AS. ND/C, Individual consideration. SQAH, CSYS required. SQAV, Higher National including specific subject(s). IB, 30 points.

COVENTRY UNIVERSITY C85

BSc Intelligent Computing

UCAS Code: G820 • Mode: 3/4 Years Single Subject

Qualifications/Requirements: GCE, A/AS: 12-16 points. Maths at A level. AGNVQ, Merit. ND/C, 3 Merits (in specific programmes). SQAH, Individual consideration. SQAV, Individual consideration. IB, Individual consideration.

UNIVERSITY OF DERBY D39

BSc Artificial Intelligence

UCAS Code: G801 • Mode: 4 Years Single Subject

Qualifications/Requirements: GCE, A/AS: 12-14 points. AGNVQ, Merit (in specific programmes). ND/C, Merit overall (in specific programmes). SQAH, Grades BCCC. SQAV, Individual consideration. IB, 26 points.

BSc Computer Studies (Artificial Intelligence)

UCAS Code: G800 • Mode: 4 Years Single Subject

Qualifications/Requirements: GCE, A/AS: 12-14 points. AGNVQ, Merit (in specific programmes). ND/C, Merit overall (in specific programmes). SQAH, Grades BCCC. SQAV, Individual consideration. IB, 26 points.

BSc Information Systems (Artificial Intelligence)

UCAS Code: G5G8 • Mode: 4 Years Major/Minor

Qualifications/Requirements: GCE, A/AS: 12-14 points. AGNVQ, Merit (in specific programmes). ND/C, Merit overall (in specific programmes). SQAH, Grades BCCC. SQAV, Individual consideration. IB, 26 points.

THE UNIVERSITY OF DURHAM D86

BSc Artificial Intelligence

UCAS Code: G800 • Mode: 3 Years Single Subject

Qualifications/Requirements: GCE, A/AS: 22 points. Two Sciences at A level. AGNVQ, Individual consideration. ND/C, Individual consideration. SQAH, Grades AABBB including specific subject(s). SQAV, Individual consideration. IB, 30 points.

THE UNIVERSITY OF EDINBURGH E56

BSc Artificial Intelligence and Computer Science

UCAS Code: GG58 • Mode: 4 Years Equal Combination

Qualifications/Requirements: GCE, A level grades: CCC. Maths at A level. AGNVQ, Pass. ND/C, Merit overall (in specific programmes). SQAH, Grades BBBB including specific subject(s). SQAV, National including specific subject(s). IB, Pass in Diploma including specific subjects.

BSc Artificial Intelligence and Psychology

UCAS Code: GL87 • Mode: 4 Years Equal Combination

Qualifications/Requirements: GCE, A level grades: BBB. Two Sciences at A level. ND/C, Merit overall (in specific programmes). SQAH, Grades ABBB. SQAV, National including specific subject(s). IB, Pass in Diploma including specific subjects.

BEng Artificial Intelligence and Software Engineering
UCAS Code: GG78 • Mode: 4 Years Equal Combination
Qualifications/Requirements: GCE, A level grades: CCC. Maths at A level. ND/C, Merit overall (in specific programmes). SQAH, Grades BBBB including specific subject(s). IB, Pass in Diploma including specific subjects.

MA Linguistics and Artificial Intelligence
UCAS Code: QG18 • Mode: 4 Years Equal Combination
Qualifications/Requirements: GCE, A level grades: BBB. AGNVQ, Individual consideration. ND/C, Individual consideration. SQAH, Grades BBBB. SQAV, Individual consideration. IB, Pass in Diploma including specific subjects.

THE UNIVERSITY OF ESSEX E70

BSc Computer Science (Artificial Intelligence)
UCAS Code: G800 • Mode: 3 Years Single Subject
Qualifications/Requirements: GCE, A/AS: 20 points. AGNVQ, Distinction. ND/C, Merit overall and 2 Distinctions. SQAH, Grades BBBB. SQAV, Individual consideration. IB, 28 points.

BSc Computer Science (Robotics and Intelligent Machines)
UCAS Code: G5G8 • Mode: 3 Years Major/Minor
Qualifications/Requirements: GCE, A/AS: 20 points. AGNVQ, Distinction. ND/C, Merit overall and 2 Distinctions. SQAH, Grades BBBB. SQAV, Individual consideration. IB, 28 points.

BSc Philosophy and Artificial Intelligence
UCAS Code: GV87 • Mode: 3 Years Equal Combination
Qualifications/Requirements: GCE, A/AS: 22 points. AGNVQ, Individual consideration. ND/C, Merit overall and 3 Distinctions. SQAH, Grades AABB. SQAV, Individual consideration. IB, 30 points.

BSc Psychology and Artificial Intelligence
UCAS Code: CG88 • Mode: 3 Years Equal Combination
Qualifications/Requirements: GCE, A/AS: 22 points. AGNVQ, Distinction. ND/C, Merit overall and 3 Distinctions. SQAH, Grades ABBB. SQAV, Individual consideration. IB, 30 points.

HERIOT-WATT UNIVERSITY, EDINBURGH H24

BSc Computer Science (Artificial Intelligence)
UCAS Code: G800 • Mode: 4 Years Single Subject
Qualifications/Requirements: GCE, A level grades: CCD. Maths. AGNVQ, Merit (in specific programmes). ND/C, Merit overall (in specific programmes). SQAH, Grades BBBC including specific subject(s). SQAV, National including specific subject(s). IB, 28 points.

UNIVERSITY OF HERTFORDSHIRE H36

BSc Intelligent Systems
UCAS Code: G800 • Mode: 4 Years Single Subject
Qualifications/Requirements: GCE, A/AS: 16 points. AGNVQ, Merit. ND/C, Merit overall. SQAH, Individual consideration. SQAV, Individual consideration. IB, 28 points.

THE UNIVERSITY OF HUDDERSFIELD H60

BA Business Computing with Artificial Intelligence
UCAS Code: G5G8 • Mode: 4 Years Major/Minor
Qualifications/Requirements: GCE, A/AS: 16 points. AGNVQ, Merit. ND/C, Merit overall. SQAH, Grades BBBB. SQAV, Individual consideration. IB, Individual consideration.

IMPERIAL COLLEGE OF SCIENCE, TECHNOLOGY AND MEDICINE (UNIVERSITY OF LONDON) I50

MEng Computing (Artificial Intelligence) (4 years) (h)
UCAS Code: G800 • Mode: 4 Years Single Subject
Qualifications/Requirements: GCE, A level grades: AAB. Maths at A level. AGNVQ, Pass. ND/C, Individual consideration. SQAH, Individual consideration. SQAV, Individual consideration. IB, Individual consideration.

UNIVERSITY OF LEEDS L23

BSc Artificial Intelligence – Philosophy
UCAS Code: GV87 • Mode: 3 Years Equal Combination
Qualifications/Requirements: GCE, A level grades: BBC. Maths at A level. AGNVQ, Individual consideration. ND/C, Individual consideration. SQAH, Individual consideration. SQAV, Individual consideration. IB, 30 points.

BSc Artificial Intelligence – Physics
UCAS Code: FG38 • Mode: 3 Years Equal Combination
Qualifications/Requirements: GCE, A level grades: BBC. Maths and physics at A level. AGNVQ, Individual consideration. ND/C, 1 Merit and 5 Distinctions (in specific programmes). SQAH, Individual consideration. SQAV, Individual consideration. IB, 30 points.

UNIVERSITY OF LUTON L93

BSc Artificial Intelligence
UCAS Code: G800 • Mode: 3 Years Single Subject
Qualifications/Requirements: GCE, A/AS: 12 points. AGNVQ, Distinction or Merit with 6 additional units or with A/AS. ND/C, 5 Merits. SQAH, Grades BBCC. SQAV, Individual consideration. IB, 32 points.

THE UNIVERSITY OF MANCHESTER M20

BSc Artificial Intelligence

UCAS Code: G800 • Mode: 3 Years Single Subject

Qualifications/Requirements: GCE, A level grades: BBB. Maths at A level. AGNVQ, Distinction in Science with A level. ND/C, Individual consideration. SQAH, Grades BBBBB including specific subject(s). SQAV, Individual consideration. IB, 30 points.

BSc Artificial Intelligence with Industrial Experience (4 years)

UCAS Code: G801 • Mode: 4 Years Single Subject

Qualifications/Requirements: GCE, A level grades: BBB. Maths at A level. AGNVQ, Distinction in Science with A level. ND/C, Individual consideration. SQAH, Grades BBBBB including specific subject(s). SQAV, Individual consideration. IB, 30 points.

THE UNIVERSITY OF MANCHESTER INSTITUTE OF SCIENCE AND TECHNOLOGY (UMIST) M25

BSc Artificial Intelligence

UCAS Code: G800 • Mode: 3 Years Single Subject

Qualifications/Requirements: GCE, A/AS: 22 points. AGNVQ, Distinction in Information Technology. ND/C, Merit overall and 5 Distinctions. SQAH, Grades ABBBB. SQAV, Individual consideration. IB, 30 points.

MNeuro Neuroscience and Artificial Intelligence

UCAS Code: BG18 • Mode: 4 Years Equal Combination

Qualifications/Requirements: GCE, A level grades: ABB.

THE MANCHESTER METROPOLITAN UNIVERSITY M40

BSc Artificial Intelligence

UCAS Code: G800 • Mode: 4 Years Single Subject

Qualifications/Requirements: GCE, A/AS: 12-18 points. AGNVQ, Distinction (in specific programmes) or Merit (in specific programmes) with 4 additional units or with A/AS. ND/C, Merits and Distinction. SQAH, Grades BBCC. SQAV, Individual consideration. IB, 24 points.

OXFORD BROOKES UNIVERSITY O66

BA/BSc Intelligent Systems/Spanish Studies

UCAS Code: GR84 • Mode: 3 Years Equal Combination

Qualifications/Requirements: Please refer to prospectus.

BSc Intelligent Systems (FT or SW)

UCAS Code: G800 • Mode: 4 Years Single Subject

Qualifications/Requirements: GCE, A level grades: CDD-BC. AGNVQ, Merit. ND/C, Individual consideration. SQAH, Individual consideration. SQAV, Individual consideration. IB, Individual consideration.

Mod Accounting and Finance/Intelligent Systems

UCAS Code: GN84 • Mode: 3 Years Equal Combination

Qualifications/Requirements: GCE, A level grades: CD-BCC. AGNVQ, Merit or Distinction with 3 additional units. ND/C, Individual consideration. SQAH, Individual consideration. SQAV, Individual consideration. IB, Individual consideration.

Mod Anthropology/Intelligent Systems

UCAS Code: GL86 • Mode: 3 Years Equal Combination

Qualifications/Requirements: GCE, A level grades: CD-BCC. AGNVQ, Merit or Distinction with 3 additional units. ND/C, Individual consideration. SQAH, Individual consideration. SQAV, Individual consideration. IB, Individual consideration.

Mod Biological Chemistry/Intelligent Systems

UCAS Code: CG78 • Mode: 3 Years Equal Combination

Qualifications/Requirements: GCE, A level grades: DD-CCC. Science. AGNVQ, Merit in Science. ND/C, Individual consideration. SQAH, Individual consideration. SQAV, Individual consideration. IB, Individual consideration.

Mod Biology/Intelligent Systems

UCAS Code: CG18 • Mode: 3 Years Equal Combination

Qualifications/Requirements: GCE, A level grades: DD-CCC. Science. AGNVQ, Merit in Science. ND/C, Individual consideration. SQAH, Individual consideration. SQAV, Individual consideration. IB, Individual consideration.

Mod Business Statistics/Intelligent Systems

UCAS Code: GG84 • Mode: 3 Years Equal Combination

Qualifications/Requirements: Please refer to prospectus.

Mod Cell Biology/Intelligent Systems

UCAS Code: CGC8 • Mode: 3 Years Equal Combination

Qualifications/Requirements: GCE, A level grades: DD-CCC. Science. AGNVQ, Merit in Science. ND/C, Individual consideration. SQAH, Individual consideration. SQAV, Individual consideration. IB, Individual consideration.

Mod Cities and Society/Intelligent Systems

UCAS Code: GL8H • Mode: 3 Years Equal Combination

Qualifications/Requirements: GCE, A level grades: DD-CCC. AGNVQ, Merit. ND/C, Individual consideration. SQAH, Individual consideration. SQAV, Individual consideration. IB, Individual consideration.

Mod Combined Studies/Intelligent Systems
UCAS Code: GY84 • Mode: 3 Years
Qualifications/Requirements: AGNVQ, Not normally sufficient. ND/C, Individual consideration. SQAH, Not normally sufficient. SQAV, Individual consideration. IB, Not normally sufficient.

Mod Complementary Therapies – Aromatherapy/Intelligent Systems
UCAS Code: GW88 • Mode: 3 Years Equal Combination
Qualifications/Requirements: Please refer to prospectus.

Mod Computing Mathematics/Intelligent Systems
UCAS Code: GG89 • Mode: 3 Years Equal Combination
Qualifications/Requirements: GCE, A level grades: CD. AGNVQ, Merit. ND/C, Individual consideration. SQAH, Individual consideration. SQAV, Individual consideration. IB, Individual consideration.

Mod Computing Science/Intelligent Systems
UCAS Code: GG85 • Mode: 3 Years Equal Combination
Qualifications/Requirements: GCE, A level grades: DD-AGNVQ, Merit. ND/C, Individual consideration. SQAH, Individual consideration. SQAV, Individual consideration. IB, Individual consideration.

Mod Computing/Intelligent Systems
UCAS Code: GG58 • Mode: 3 Years Equal Combination
Qualifications/Requirements: GCE, A level grades: BC-CDD. AGNVQ, Merit. ND/C, Individual consideration. SQAH, Individual consideration. SQAV, Individual consideration. IB, Individual consideration.

Mod Ecology/Intelligent Systems
UCAS Code: CG98 • Mode: 3 Years Equal Combination
Qualifications/Requirements: GCE, A level grades: CD. AGNVQ, Merit in Science. ND/C, Individual consideration. SQAH, Individual consideration. SQAV, Individual consideration. IB, Individual consideration.

Mod Economics/Intelligent Systems
UCAS Code: GL81 • Mode: 3 Years Equal Combination
Qualifications/Requirements: GCE, A level grades: CD-BB. AGNVQ, Merit or Merit with 3 additional units. ND/C, Individual consideration. SQAH, Individual consideration. SQAV, Individual consideration. IB, Individual consideration.

Mod Electronics/Intelligent Systems
UCAS Code: GH86 • Mode: 3 Years Equal Combination
Qualifications/Requirements: GCE, A level grades: CD-CCC. Science or Maths at A level. AGNVQ, Merit (in specific programmes). ND/C, Individual consideration. SQAH, Individual consideration. SQAV, Individual consideration. IB, Individual consideration.

Mod English Studies/Intelligent Systems
UCAS Code: GQ83 • Mode: 3 Years Equal Combination
Qualifications/Requirements: GCE, A level grades: CD-AB. AGNVQ, Merit with A/AS. ND/C, Individual consideration. SQAH, Individual consideration. SQAV, Individual consideration. IB, Individual consideration.

Mod Environmental Design and Conservation/Intelligent Systems
UCAS Code: FG98 • Mode: 3 Years Equal Combination
Qualifications/Requirements: GCE, A level grades: DD-CCC. AGNVQ, Merit. ND/C, Individual consideration. SQAH, Individual consideration. SQAV, Individual consideration. IB, Individual consideration.

Mod Environmental Chemistry/Intelligent Systems
UCAS Code: GF81 • Mode: 3 Years Equal Combination
Qualifications/Requirements: GCE, A level grades: DD-CCC. Science. AGNVQ, Merit in Science. ND/C, Individual consideration. SQAH, Individual consideration. SQAV, Individual consideration. IB, Individual consideration.

Mod Environmental Policy/Intelligent Systems
UCAS Code: KG38 • Mode: 3 Years Equal Combination
Qualifications/Requirements: GCE, A level grades: DD-CCC. Science. AGNVQ, Merit in Science. ND/C, Individual consideration. SQAH, Individual consideration. SQAV, Individual consideration. IB, Individual consideration.

Mod Environmental Sciences/Intelligent Systems
UCAS Code: FGX8 • Mode: 3 Years Equal Combination
Qualifications/Requirements: GCE, A level grades: CD. Science. AGNVQ, Merit or Distinction in Science. ND/C, Individual consideration. SQAH, Individual consideration. SQAV, Individual consideration. IB, Individual consideration.

Mod Fine Art/Intelligent Systems
UCAS Code: GW81 • Mode: 3 Years Equal Combination
Qualifications/Requirements: GCE, A level grades: CC-BC. Art. A portfolio of work is required. AGNVQ, Merit in Art & Design with A/AS. ND/C, Individual consideration. SQAH, Individual consideration. SQAV, Individual consideration. IB, Individual consideration.

Mod French Studies/Intelligent Systems
UCAS Code: RG18 • Mode: 3 Years Equal Combination
Qualifications/Requirements: Please refer to prospectus.

Mod Geographic Information Science/Intelligent Systems
UCAS Code: GF88 • Mode: 3 Years Equal Combination
Qualifications/Requirements: Please refer to prospectus.

Mod Geography/Intelligent Systems

UCAS Code: GL88 • Mode: 3 Years Equal Combination

Qualifications/Requirements: GCE, A level grades: CD-BB. AGNVQ, Merit. ND/C, Individual consideration. SQAH, Individual consideration. SQAV, Individual consideration. IB, Individual consideration.

Mod Geology/Intelligent Systems

UCAS Code: FG68 • Mode: 3 Years Equal Combination

Qualifications/Requirements: GCE, A level grades: DD-CD. Science or Maths. AGNVQ, Merit or Pass in Science. ND/C, Individual consideration. SQAH, Individual consideration. SQAV, Individual consideration. IB, Individual consideration.

Mod Geotechnics/Intelligent Systems

UCAS Code: GH82 • Mode: 3 Years Equal Combination

Qualifications/Requirements: GCE, A level grades: DD-CCC. Science, Maths, Design & Technology or Electronics. AGNVQ, Merit (in specific programmes). ND/C, Individual consideration. SQAH, Individual consideration. SQAV, Individual consideration. IB, Individual consideration.

Mod German Studies/Intelligent Systems

UCAS Code: GR8G • Mode: 4 Years Equal Combination

Qualifications/Requirements: GCE, A level grades: CD-DDD. German. AGNVQ, Merit with A level. ND/C, Individual consideration. SQAH, Individual consideration. SQAV, Individual consideration. IB, Individual consideration.

Mod History of Art/Intelligent Systems

UCAS Code: GV84 • Mode: 3 Years Equal Combination

Qualifications/Requirements: GCE, A level grades: BCC-CD. AGNVQ, Merit with A/AS. ND/C, Individual consideration. SQAH, Individual consideration. SQAV, Individual consideration. IB, Individual consideration.

Mod History/Intelligent Systems

UCAS Code: GV81 • Mode: 3 Years Equal Combination

Qualifications/Requirements: GCE, A level grades: BB-CD. AGNVQ, Merit with A/AS. ND/C, Individual consideration. SQAH, Individual consideration. SQAV, Individual consideration. IB, Individual consideration.

Mod Hospitality Management Studies/ Intelligent Systems

UCAS Code: GN87 • Mode: 3 Years Equal Combination

Qualifications/Requirements: GCE, A level grades: CC-DDD. AGNVQ, Merit or Merit with 3 additional units. ND/C, Individual consideration. SQAH, Individual consideration. SQAV, Individual consideration. IB, Individual consideration.

Mod Human Biology/Intelligent Systems

UCAS Code: BG18 • Mode: 3 Years Equal Combination

Qualifications/Requirements: GCE, A level grades: DD-CCC. Science. AGNVQ, Merit in Science. ND/C, Individual consideration. SQAH, Individual consideration. SQAV, Individual consideration. IB, Individual consideration.

Mod Information Systems/Intelligent Systems

UCAS Code: GGM8 • Mode: 3 Years Equal Combination

Qualifications/Requirements: GCE, A level grades: CDD-BC. AGNVQ, Merit. ND/C, Individual consideration. SQAH, Individual consideration. SQAV, Individual consideration. IB, Individual consideration.

Mod Intelligent Systems/Japanese Studies

UCAS Code: GT84 • Mode: 3 Years Equal Combination

Qualifications/Requirements: Please refer to prospectus.

Mod Intelligent Systems/Law

UCAS Code: GM83 • Mode: 3 Years Equal Combination

Qualifications/Requirements: GCE, A level grades: CCC-ABB. AGNVQ, Merit or Distinction with 3 additional units. ND/C, Individual consideration. SQAH, Individual consideration. SQAV, Individual consideration. IB, Individual consideration.

Mod Intelligent Systems/Leisure Planning

UCAS Code: KGH8 • Mode: 3 Years Equal Combination

Qualifications/Requirements: GCE, A level grades: DDD-BC. AGNVQ, Merit. ND/C, Individual consideration. SQAH, Individual consideration. SQAV, Individual consideration. IB, Individual consideration.

Mod Intelligent Systems/Mapping and Cartography

UCAS Code: FG88 • Mode: 3 Years Equal Combination

Qualifications/Requirements: GCE, A level grades: CC-DDD. AGNVQ, Merit. ND/C, Individual consideration. SQAH, Individual consideration. SQAV, Individual consideration. IB, Individual consideration.

Mod Intelligent Systems/Marketing Management

UCAS Code: GN8N • Mode: 3 Years Equal Combination

Qualifications/Requirements: GCE, A level grades: CD-BCC. AGNVQ, Merit or Distinction with 3 additional units. ND/C, Individual consideration. SQAH, Individual consideration. SQAV, Individual consideration. IB, Individual consideration.

Mod Intelligent Systems/Mathematics

UCAS Code: GG18 • Mode: 3 Years Equal Combination

Qualifications/Requirements: GCE, A level grades: DD-BC. Maths. AGNVQ, Merit with A level. ND/C, Individual consideration. SQAH, Individual consideration. SQAV, Individual consideration. IB, Individual consideration.

Mod Intelligent Systems/Multimedia Systems
UCAS Code: GG8P • Mode: 3 Years Equal Combination
Qualifications/Requirements: Please refer to prospectus.

Mod Intelligent Systems/Music
UCAS Code: GW83 • Mode: 3 Years Equal Combination
Qualifications/Requirements: GCE, A level grades: DD-CD. Music. AGNVQ, Merit. ND/C, Individual consideration. SQAH, Individual consideration. SQAV, Individual consideration. IB, Individual consideration.

Mod Intelligent Systems/Palliative Care
UCAS Code: BGR8 • Mode: 3 Years Equal Combination
Qualifications/Requirements: Please refer to prospectus.

Mod Intelligent Systems/Physical Geography
UCAS Code: FGV8 • Mode: 3 Years Equal Combination
Qualifications/Requirements: GCE, A level grades: CC-BBC. AGNVQ, Merit. ND/C, Individual consideration. SQAH, Individual consideration. SQAV, Individual consideration. IB, Individual consideration.

Mod Intelligent Systems/Planning Studies
UCAS Code: GK84 • Mode: 3 Years Equal Combination
Qualifications/Requirements: GCE, A level grades: DDD-BC. AGNVQ, Merit. ND/C, Individual consideration. SQAH, Individual consideration. SQAV, Individual consideration. IB, Individual consideration.

Mod Intelligent Systems/Politics
UCAS Code: MG18 • Mode: 3 Years Equal Combination
Qualifications/Requirements: GCE, A level grades: CD-AB. AGNVQ, Merit with A/AS. ND/C, Individual consideration. SQAH, Individual consideration. SQAV, Individual consideration. IB, Individual consideration.

Mod Intelligent Systems/Psychology
UCAS Code: CG88 • Mode: 3 Years Equal Combination
Qualifications/Requirements: GCE, A level grades: CCC-ABB. AGNVQ, Merit with A/AS. ND/C, Individual consideration. SQAH, Individual consideration. SQAV, Individual consideration. IB, Individual consideration.

Mod Intelligent Systems/Publishing
UCAS Code: GP85 • Mode: 3 Years Equal Combination
Qualifications/Requirements: GCE, A level grades: CDD-BB. AGNVQ, Merit or Merit (in specific programmes) with 3 additional units. ND/C, Individual consideration. SQAH, Individual consideration. SQAV, Individual consideration. IB, Individual consideration.

Mod Intelligent Systems/Rehabilitation
UCAS Code: BGT8 • Mode: 3 Years Equal Combination
Qualifications/Requirements: Please refer to prospectus.

Mod Intelligent Systems/Retail Management
UCAS Code: GN85 • Mode: 3 Years Equal Combination
Qualifications/Requirements: GCE, A level grades: CD-BCC. AGNVQ, Merit with A/AS or Merit in Business with 4 additional units. ND/C, Individual consideration. SQAH, Individual consideration. SQAV, Individual consideration. IB, Individual consideration.

Mod Intelligent Systems/Sociology
UCAS Code: GL83 • Mode: 3 Years Equal Combination
Qualifications/Requirements: GCE, A level grades: CDD-BCC. AGNVQ, Merit with A/AS. ND/C, Individual consideration. SQAH, Individual consideration. SQAV, Individual consideration. IB, Individual consideration.

Mod Intelligent Systems/Software Engineering
UCAS Code: GG78 • Mode: 3 Years Equal Combination
Qualifications/Requirements: GCE, A level grades: CDD-BC. AGNVQ, Merit. ND/C, Individual consideration. SQAH, Individual consideration. SQAV, Individual consideration. IB, Individual consideration.

Mod Intelligent Systems/Statistics
UCAS Code: GG48 • Mode: 3 Years Equal Combination
Qualifications/Requirements: GCE, A level grades: DD-CD. AGNVQ, Merit. ND/C, Individual consideration. SQAH, Individual consideration. SQAV, Individual consideration. IB, Individual consideration.

Mod Intelligent Systems/ Telecommunications
UCAS Code: GH8P • Mode: 3 Years Equal Combination
Qualifications/Requirements: GCE, A level grades: DD-CCC. Science or Maths. AGNVQ, Merit (in specific programmes). ND/C, Individual consideration. SQAH, Individual consideration. SQAV, Individual consideration. IB, Individual consideration.

Mod Intelligent Systems/Tourism
UCAS Code: GP87 • Mode: 3 Years Equal Combination
Qualifications/Requirements: GCE, A level grades: CD-BC. AGNVQ, Merit or Merit with 3 additional units. ND/C, Individual consideration. SQAH, Individual consideration. SQAV, Individual consideration. IB, Individual consideration.

Mod Intelligent Systems/Transport and Travel
UCAS Code: GN89 • Mode: 3 Years Equal Combination
Qualifications/Requirements: GCE, A level grades: CD-CC. AGNVQ, Merit. ND/C, Individual consideration. SQAH, Individual consideration. SQAV, Individual consideration. IB, Individual consideration.

Mod Intelligent Systems/Water Resources

UCAS Code: GH8F • Mode: 3 Years Equal Combination

Qualifications/Requirements: GCE, A level grades: DD-CCC. AGNVQ, Merit. ND/C, Individual consideration. SQAH, Individual consideration. SQAV, Individual consideration. IB, Individual consideration.

THE UNIVERSITY OF READING R12

BSc Applied Intelligent Systems

UCAS Code: GC8V • Mode: 4 Years Equal Combination

Qualifications/Requirements: GCE, A/AS: 20 points. Biology, Chemistry, Maths or Physics. AGNVQ, Distinction in Science with 6 additional units or with A level. ND/C, 3 Merits and 2 Distinctions (in specific programmes). SQAH, Grades BBBB including specific subject(s). SQAV, Individual consideration.

BSc Intelligent Systems

UCAS Code: GC88 • Mode: 3 Years Equal Combination

Qualifications/Requirements: GCE, A/AS: 20 points. Biology, Chemistry, Maths or Physics. AGNVQ, Distinction in Science with 6 additional units or with A level. ND/C, 3 Merits and 2 Distinctions (in specific programmes). SQAH, Grades BBBB including specific subject(s). SQAV, Individual consideration.

ROYAL HOLLOWAY, UNIVERSITY OF LONDON R72

BSc Computer Science with Artificial Intelligence

UCAS Code: G5G8 • Mode: 3 Years Major/Minor

Qualifications/Requirements: GCE, A level grades: BBB.

THE UNIVERSITY OF SHEFFIELD S18

BSc Artificial Intelligence and Computer Science

UCAS Code: GG85 • Mode: 3 Years Equal Combination

Qualifications/Requirements: GCE, A/AS: 24 points. Maths at A level. AGNVQ, Distinction with A level. ND/C, 2 Merits and 4 Distinctions (in specific programmes). SQAH, Grades AABB including specific subject(s). IB, 32 points.

UNIVERSITY OF SOUTHAMPTON S27

BSc Computer Science with Artificial Intelligence

UCAS Code: G5G8 • Mode: 3 Years Major/Minor

Qualifications/Requirements: GCE, A/AS: 24 points. Maths at A level. AGNVQ, Distinction with A level. ND/C, Individual consideration. SQAH, CSYS required. SQAV, Individual consideration. IB, 32 points.

STAFFORDSHIRE UNIVERSITY S72

BSc Computer Graphics, Imaging and Visualisation

UCAS Code: GW5F • Mode: 4 Years Equal Combination

Qualifications/Requirements: GCE, A/AS: 12 points. AGNVQ, Merit. ND/C, Individual consideration. SQAH, Grades CCC. IB, 27 points.

BSc Foundation Intelligent Systems

UCAS Code: G808 • Mode: 5 Years Single Subject

Qualifications/Requirements: Please refer to prospectus.

BSc Intelligent Systems

UCAS Code: G801 • Mode: 4 Years Single Subject

Qualifications/Requirements: GCE, A/AS: 12 points. AGNVQ, Merit. ND/C, Individual consideration. SQAH, Grades CCC. IB, 27 points.

BEng Foundation Intelligent Systems

UCAS Code: G807 • Mode: 5 Years Single Subject

Qualifications/Requirements: Please refer to prospectus.

UNIVERSITY OF SUSSEX S90

BA Artificial Intelligence: School of Cognitive and Computer Science

UCAS Code: G800 • Mode: 3 Years Single Subject

Qualifications/Requirements: GCE, A level grades: BBB. AGNVQ, Merit with 6 additional units. ND/C, Merit overall. SQAH, Individual consideration. SQAV, Individual consideration. IB, Individual consideration.

BSc Artificial Intelligence with European Studies (4 years)

UCAS Code: G8T2 • Mode: 4 Years Major/Minor

Qualifications/Requirements: GCE, A level grades: BBB. AGNVQ, Merit with 6 additional units. ND/C, Merit overall. SQAH, Individual consideration. SQAV, Individual consideration. IB, Individual consideration.

BSc Artificial Intelligence with Management Studies

UCAS Code: G8N1 • Mode: 3 Years Major/Minor

Qualifications/Requirements: GCE, A level grades: BBB. AGNVQ, Merit with 6 additional units. ND/C, Merit overall. SQAH, Individual consideration. SQAV, Individual consideration. IB, Individual consideration.

BSc Computer Science and Artificial Intelligence

UCAS Code: G575 • Mode: 3 Years Single Subject

Qualifications/Requirements: GCE, A level grades: BBB. AGNVQ, Merit with 6 additional units. ND/C, Merit overall (in specific programmes). SQAH, Individual consideration. SQAV, Individual consideration. IB, Individual consideration.

BSc Mathematics and Artificial Intelligence

UCAS Code: GG51 • Mode: 3 Years Equal Combination

Qualifications/Requirements: GCE, A level grades: BBC. Maths at A level. AGNVQ, Merit in Science with A level. ND/C, Merit overall (in specific programmes). SQAH, Individual consideration. SQAV, Individual consideration. IB, Individual consideration.

UNIVERSITY OF WESTMINSTER W50

BSc Artificial Intelligence

UCAS Code: G800 • Mode: 3 Years Single Subject

Qualifications/Requirements: GCE, A/AS: 12-14 points. AGNVQ, Distinction. ND/C, 5 Merits. SQAV, Individual consideration.

COMPUTER SCIENCE/INFORMATION TECHNOLOGY

THE UNIVERSITY OF ABERDEEN A20

BSc Computing Science

UCAS Code: G500 • Mode: 4 Years Single Subject

Qualifications/Requirements: GCE, A level grades: CCD. Two Sciences or Science and Maths. AGNVQ, Merit in Science. ND/C, Individual consideration. SQAH, Grades BBBC including specific subject(s). SQAV, Individual consideration. IB, 26 points.

BSc Computing Science (Artificial Intelligence)

UCAS Code: G5G8 • Mode: 4 Years Major/Minor

Qualifications/Requirements: GCE, A level grades: CCD. Two Sciences or Science and Maths. AGNVQ, Merit in Science. ND/C, Individual consideration. SQAH, Grades BBBC including specific subject(s). SQAV, Individual consideration. IB, 26 points.

BSc Computing Science (Artificial Intelligence) with Industrial Placement

UCAS Code: G58V • Mode: 5 Years Single Subject

Qualifications/Requirements: GCE, A level grades: CCD. Two Sciences or Science and Maths. AGNVQ, Merit in Science. ND/C, Individual consideration. SQAH, Grades BBBC including specific subject(s). SQAV, Individual consideration. IB, 26 points.

BSc Computing Science (Business Computing)

UCAS Code: G520 • Mode: 4 Years Single Subject

Qualifications/Requirements: GCE, A level grades: CCD. Two Sciences or Science and Maths. AGNVQ, Merit in Science. ND/C, Individual consideration. SQAH, Grades BBBC including specific subject(s). SQAV, Individual consideration. IB, 26 points.

BSc Computing Science (Business Computing) with Industrial Placement

UCAS Code: G524 • Mode: 5 Years Single Subject

Qualifications/Requirements: GCE, A level grades: CCD. Two Sciences or Science and Maths. AGNVQ, Merit in Science. ND/C, Individual consideration. SQAH, Grades BBBC including specific subject(s). SQAV, Individual consideration. IB, 26 points.

BSc Computing Science with French

UCAS Code: G5R1 • Mode: 4 Years Major/Minor

Qualifications/Requirements: GCE, A level grades: CCD. Two Sciences or Science and Maths. AGNVQ, Merit in Science. ND/C, Individual consideration. SQAH, Grades BBBC including specific subject(s). SQAV, Individual consideration. IB, 26 points.

BSc Computing Science with French with Industrial Placement

UCAS Code: G5RC • Mode: 5 Years Major/Minor

Qualifications/Requirements: GCE, A level grades: CCD. Two Sciences or Science and Maths. AGNVQ, Merit in Science. ND/C, Individual consideration. SQAH, Grades BBBC including specific subject(s). SQAV, Individual consideration. IB, 26 points.

BSc Computing Science with Industrial Placement

UCAS Code: G501 • Mode: 4 Years Single Subject

Qualifications/Requirements: GCE, A level grades: CCD. Two Sciences or Science and Maths. AGNVQ, Merit in Science. ND/C, Individual consideration. SQAH, Grades BBBC including specific subject(s). SQAV, Individual consideration. IB, 26 points.

BSc Computing Science with Spanish

UCAS Code: G5R4 • Mode: 4 Years Major/Minor

Qualifications/Requirements: GCE, A level grades: CCD. Two Sciences or Science and Maths. AGNVQ, Merit in Science. ND/C, Individual consideration. SQAH, Grades BBBC including specific subject(s). SQAV, Individual consideration. IB, 26 points.

BSc Computing Science with Spanish with Industrial Placement

UCAS Code: G5RK • Mode: 5 Years Major/Minor

Qualifications/Requirements: GCE, A level grades: CCD. Two Sciences or Science and Maths. AGNVQ, Merit in Science. ND/C, Individual consideration. SQAH, Grades BBBC including specific subject(s). SQAV, Individual consideration. IB, 26 points.

BSc Computing Science/Management Studies (3 or 4 years)

UCAS Code: GN5C • Mode: 4 Years Equal Combination

Qualifications/Requirements: GCE, A level grades: CCD. Two Sciences or Science and Maths. AGNVQ, Merit in Science. ND/C, Individual consideration. SQAH, Grades BBBC including specific subject(s). SQAV, Individual consideration. IB, 26 points.

BSc Computing Science/Mathematics

UCAS Code: GGMC • Mode: 4 Years Equal Combination

Qualifications/Requirements: GCE, A level grades: CCD. Two Sciences or Science and Maths. AGNVQ, Merit in Science. ND/C, Individual consideration. SQAH, Grades BBBC including specific subject(s). SQAV, Individual consideration. IB, 26 points.

BSc Computing Science/Psychology

UCAS Code: GC58 • Mode: 4 Years Equal Combination

Qualifications/Requirements: GCE, A level grades: CCD. Two Sciences or Science and Maths. AGNVQ, Merit in Science. ND/C, Individual consideration. SQAH, Grades BBBC including specific subject(s). SQAV, Individual consideration. IB, 26 points.

BSc Computing Science/Statistics

UCAS Code: GG54 • Mode: 4 Years Equal Combination

Qualifications/Requirements: GCE, A level grades: CCD. Maths and Science. AGNVQ, Merit in Science. ND/C, Individual consideration. SQAH, Grades BBBC including specific subject(s). SQAV, Individual consideration. IB, 26 points.

BSc Information Systems and Computing
UCAS Code: G521 • Mode: 4 Years Single Subject

Qualifications/Requirements: GCE, A level grades: CCD. Two Sciences or Science and Maths. AGNVQ, Merit in Science. ND/C, Individual consideration. SQAH, Grades BBBC including specific subject(s). SQAV, Individual consideration. IB, 26 points.

MA Computing
UCAS Code: G502 • Mode: 4 Years Single Subject

Qualifications/Requirements: GCE, A level grades: CCC. Maths. AGNVQ, Merit (in specific programmes). ND/C, Individual consideration. SQAH, Grades BBBB including specific subject(s). SQAV, Individual consideration. IB, 30 points.

MA Computing with Industrial Placement
UCAS Code: G505 • Mode: 5 Years Single Subject

Qualifications/Requirements: GCE, A level grades: CCC. Maths. AGNVQ, Merit (in specific programmes). ND/C, Individual consideration. SQAH, Grades BBBB including specific subject(s). SQAV, Individual consideration. IB, 30 points.

MA Computing/Entrepreneurship
UCAS Code: NGC5 • Mode: 4 Years Equal Combination

Qualifications/Requirements: GCE, A level grades: CCC. Maths. AGNVQ, Merit (in specific programmes). ND/C, Individual consideration. SQAH, Grades BBBB including specific subject(s). SQAV, Individual consideration. IB, 30 points including specific subjects.

MA Computing/Finance
UCAS Code: GN53 • Mode: 4 Years Equal Combination

Qualifications/Requirements: GCE, A level grades: CCC. Maths. AGNVQ, Merit (in specific programmes). ND/C, Individual consideration. SQAH, Grades BBBB including specific subject(s). SQAV, Individual consideration. IB, 30 points.

MA Computing/Mathematics
UCAS Code: GG51 • Mode: 4 Years Equal Combination

Qualifications/Requirements: GCE, A level grades: CCC. Maths. AGNVQ, Merit (in specific programmes). ND/C, Individual consideration. SQAH, Grades BBBB including specific subject(s). SQAV, Individual consideration. IB, 30 points.

MA Information Systems and Computing
UCAS Code: G523 • Mode: 3/4 Years Single Subject

Qualifications/Requirements: GCE, A level grades: CCC. AGNVQ, Merit (in specific programmes). ND/C, Individual consideration. SQAH, Grades BBBB including specific subject(s). SQAV, Individual consideration. IB, 30 points.

MA Information Systems and Computing with Industrial Placement
UCAS Code: G525 • Mode: 5 Years Single Subject

Qualifications/Requirements: GCE, A level grades: CCC. AGNVQ, Merit (in specific programmes). ND/C, Individual consideration. SQAH, Grades BBBB including specific subject(s). SQAV, Individual consideration. IB, 30 points.

MA Information Systems and Management
UCAS Code: GN51 • Mode: 4 Years Equal Combination

Qualifications/Requirements: GCE, A level grades: CCC. Maths. AGNVQ, Merit (in specific programmes). ND/C, Individual consideration. SQAH, Grades BBBB including specific subject(s). SQAV, Individual consideration. IB, 30 points.

UNIVERSITY OF ABERTAY DUNDEE A30

BA Computer Arts
UCAS Code: GW52 • Mode: 4/5 Years Equal Combination

Qualifications/Requirements: Please refer to prospectus.

BSc Business Computing
UCAS Code: G561 • Mode: 4/5 Years Single Subject

Qualifications/Requirements: GCE, A level grades: CD. AGNVQ, Individual consideration. ND/C, Individual consideration. SQAH, Grades BCC. SQAV, Individual consideration. IB, Individual consideration.

BSc Computer Games Technology
UCAS Code: G570 • Mode: 4/5 Years Single Subject

Qualifications/Requirements: GCE, A level grades: BCC. Maths at A level. AGNVQ, Individual consideration. ND/C, Individual consideration. SQAH, Grades BBBBB including specific subject(s). SQAV, Individual consideration. IB, Individual consideration.

BSc Computing
UCAS Code: G500 • Mode: 4/5 Years Single Subject

Qualifications/Requirements: GCE, A level grades: CD. AGNVQ, Individual consideration. ND/C, Individual consideration. SQAH, Grades BCC. SQAV, Individual consideration. IB, Individual consideration.

BSc Computing (Applications Development)
UCAS Code: G501 • Mode: 1 Year Single Subject

Qualifications/Requirements: ND/C, Higher National.

BSc Information Technology
UCAS Code: G560 • Mode: 4/5 Years Single Subject

Qualifications/Requirements: Please refer to prospectus.

HND Computing: Software Development
UCAS Code: 005G • Mode: 2 Years Single Subject

Qualifications/Requirements: GCE, A level grades: D. AGNVQ, Individual consideration. ND/C, Individual consideration. SQAH, Grades CC. SQAV, Individual consideration. IB, Individual consideration.

THE UNIVERSITY OF WALES, ABERYSTWYTH A40

BSc Accounting/Computer Science

UCAS Code: GN54 • Mode: 3 Years Equal Combination

Qualifications/Requirements: GCE, A/AS: 20 points. AGNVQ, Distinction with 3 additional units or Merit with 6 additional units. ND/C, 3 Merits and 3 Distinctions. SQAH, Grades BBBCC. SQAV, Individual consideration. IB, 30 points.

BSc Computer Science

UCAS Code: G500 • Mode: 3 Years Single Subject

Qualifications/Requirements: GCE, A/AS: 20 points. AGNVQ, Merit. ND/C, 3 Merits and 2 Distinctions. SQAH, Grades BBBCC. SQAV, Individual consideration. IB, 30 points.

BSc Computer Science and Artificial Intelligence

UCAS Code: GG58 • Mode: 4 Years Equal Combination

Qualifications/Requirements: GCE, A/AS: 20 points. AGNVQ, Merit. ND/C, 3 Merits and 2 Distinctions. SQAH, Grades BBBCC. SQAV, Individual consideration. IB, 30 points.

BSc Computer Science with Business Studies

UCAS Code: G5N1 • Mode: 3 Years Major/Minor

Qualifications/Requirements: GCE, A/AS: 20 points. AGNVQ, Merit. ND/C, 3 Merits and 2 Distinctions. SQAH, Grades BBBCC. SQAV, Individual consideration. IB, 30 points.

BSc Computer Science with German (4 years)

UCAS Code: G5R2 • Mode: 4 Years Major/Minor

Qualifications/Requirements: GCE, A/AS: 20 points. German at A level. AGNVQ, Merit with A level. ND/C, Not normally sufficient. SQAH, Grades BBBCC including specific subject(s). SQAV, Individual consideration. IB, 30 points.

BSc Computer Science with Italian (4 years)

UCAS Code: G5R3 • Mode: 4 Years Major/Minor

Qualifications/Requirements: GCE, A/AS: 20 points. A modern foreign language at A level. AGNVQ, Merit with A level. ND/C, Not normally sufficient. SQAH, Grades BBBCC including specific subject(s). SQAV, Individual consideration. IB, 30 points.

BSc Computer Science with Spanish (4 years)

UCAS Code: G5R4 • Mode: 4 Years Major/Minor

Qualifications/Requirements: GCE, A/AS: 20 points. Science at A level. AGNVQ, Merit with A level. ND/C, Not normally sufficient. SQAH, Grades BBBCC including specific subject(s). SQAV, Individual consideration. IB, 30 points.

BSc Computer Science/Geography

UCAS Code: FG85 • Mode: 3 Years Equal Combination

Qualifications/Requirements: GCE, A/AS: 20-22 points. Geography at A level. AGNVQ, Merit with A level. ND/C, Not normally sufficient. SQAH, Grades BBBBC including specific subject(s). SQAV, Individual consideration. IB, 30 points.

BSc Computer Science/Physics

UCAS Code: FG35 • Mode: 3 Years Equal Combination

Qualifications/Requirements: GCE, A/AS: 20 points. Physics and Maths at A level. AGNVQ, Merit with A level. ND/C, Not normally sufficient. SQAH, Grades BBBCC including specific subject(s). SQAV, Individual consideration. IB, 30 points.

BSc Cymraeg gyda Ffrangeg (4 blynedd)

UCAS Code: G5R1 • Mode: 4 Years Major/Minor

Qualifications/Requirements: GCE, A/AS: 20 points. French at A level. AGNVQ, Merit with A level. ND/C, Not normally sufficient. SQAH, Grades BBBCC including specific subject(s). SQAV, Individual consideration. IB, 30 points.

BEng Interactive Multimedia Engineering (including integrated individual and professional training)

UCAS Code: GG65 • Mode: 4 Years Equal Combination

Qualifications/Requirements: GCE, A/AS: 20 points. AGNVQ, Merit. ND/C, 3 Merits and 2 Distinctions. SQAH, Grades BBBCC. SQAV, Individual consideration. IB, 30 points.

ANGLIA POLYTECHNIC UNIVERSITY A60

BA/BSc Business and Internet Technology

UCAS Code: GN6C • Mode: 3 Years Equal Combination

Qualifications/Requirements: GCE, A/AS: 12 points. Maths or Physics at A level. AGNVQ, Merit. ND/C, 4 Merits. SQAH, Grades BBCC. SQAV, Individual consideration. IB, Individual consideration.

BA/BSc Graphic Design and Internet Technology

UCAS Code: GW62 • Mode: 3 Years Equal Combination

Qualifications/Requirements: GCE, A/AS: 14 points. Art and Maths or Physics at A level. AGNVQ, Merit in Art & Design. ND/C, 6 Merits. SQAH, Grades BBBC. SQAV, Individual consideration. IB, Pass in Diploma.

BSc Audiotechnology and Computer Science

UCAS Code: HGPM • Mode: 3 Years Equal Combination

Qualifications/Requirements: GCE, A/AS: 12 points. Maths or Physics at A level. AGNVQ, Merit (in specific programmes). ND/C, 4 Merits. SQAH, Grades BBCC. SQAV, National. IB, Pass in Diploma including specific subjects.

BSc Audiotechnology and Real Time Computer Systems

UCAS Code: HG6M • Mode: 3 Years Equal Combination

Qualifications/Requirements: GCE, A/AS: 12 points. Maths or Physics at A level. AGNVQ, Merit (in specific programmes). ND/C, 4 Merits. SQAH, Grades BBCC. SQAV, National. IB, Pass in Diploma including specific subjects.

BSc Biomedical Science and Computer Science

UCAS Code: BG95 • Mode: 3 Years Equal Combination

Qualifications/Requirements: GCE, A/AS: 10 points. AGNVQ, Merit in Science. ND/C, 3 Merits. SQAH, Grades BCCC. SQAV, National. IB, Pass in Diploma including specific subjects.

BSc Business Information Systems

UCAS Code: G520 • Mode: 3 Years Single Subject

Qualifications/Requirements: GCE, A/AS: 14 points. AGNVQ, Pass. ND/C, 6 Merits. SQAH, Grades BBCC. SQAV, Individual consideration. IB, Pass in Diploma.

BSc Cell and Molecular Biology and Computer Science

UCAS Code: CG65 • Mode: 3 Years Equal Combination

Qualifications/Requirements: GCE, A/AS: 10 points. AGNVQ, Merit in Science. ND/C, 3 Merits. SQAH, Grades BCCC. SQAV, National. IB, Pass in Diploma including specific subjects.

BSc Chemistry and Computer Science

UCAS Code: FG15 • Mode: 3 Years Equal Combination

Qualifications/Requirements: GCE, A/AS: 10 points. AGNVQ, Merit in Science. ND/C, 3 Merits. SQAH, Grades BCCC. SQAV, National. IB, Pass in Diploma including specific subjects.

BSc Computer Science

UCAS Code: G500 • Mode: 3 Years Single Subject

Qualifications/Requirements: GCE, A/AS: 10 points. AGNVQ, Merit (in specific programmes). ND/C, 3 Merits. SQAH, Grades BCCC. SQAV, National. IB, Pass in Diploma.

BSc Computer Science and Ecology and Conservation

UCAS Code: DG25 • Mode: 3 Years Equal Combination

Qualifications/Requirements: GCE, A/AS: 10 points. AGNVQ, Pass. ND/C, 3 Merits. SQAH, Grades BCCC. SQAV, National. IB, Pass in Diploma.

BSc Computer Science and Electronics

UCAS Code: GH56 • Mode: 3 Years Equal Combination

Qualifications/Requirements: GCE, A/AS: 10 points. Maths or Physics at A level. AGNVQ, Merit (in specific programmes). ND/C, 3 Merits. SQAH, Grades BCCC. SQAV, National. IB, Pass in Diploma.

BSc Computer Science and Environmental Biology

UCAS Code: GC59 • Mode: 3 Years Equal Combination

Qualifications/Requirements: GCE, A/AS: 12 points. AGNVQ, Merit. ND/C, 4 Merits. SQAH, Grades BBCC. SQAV, National. IB, Pass in Diploma.

BSc Computer Science and Forensic Science

UCAS Code: BG15 • Mode: 3 Years Equal Combination

Qualifications/Requirements: GCE, A/AS: 14 points. AGNVQ, Merit (in specific programmes). ND/C, 6 Merits. SQAH, Grades BBBC. IB, Pass in Diploma.

BSc Computer Science and Imaging Science

UCAS Code: BG85 • Mode: 3 Years Equal Combination

Qualifications/Requirements: GCE, A/AS: 10 points. AGNVQ, Merit in Science. ND/C, 3 Merits. SQAH, Grades BCCC. SQAV, National. IB, Pass in Diploma including specific subjects.

BSc Computer Science and Internet Technology

UCAS Code: GG56 • Mode: 3 Years Equal Combination

Qualifications/Requirements: GCE, A/AS: 12 points. AGNVQ, Merit in Science. ND/C, 4 Merits. SQAH, Grades BBCC. SQAV, National. IB, Pass in Diploma.

BSc Computer Science and Mathematics or Statistics/Statistical Modelling

UCAS Code: GG51 • Mode: 3 Years Equal Combination

Qualifications/Requirements: GCE, A/AS: 10 points. AGNVQ, Merit (in specific programmes). ND/C, 3 Merits. SQAH, Grades BCCC. SQAV, National. IB, Pass in Diploma.

BSc Computer Science and Microbiology

UCAS Code: GC55 • Mode: 3 Years Equal Combination

Qualifications/Requirements: GCE, A/AS: 10 points. AGNVQ, Merit. ND/C, 4 Merits. SQAH, Grades CCCC. SQAV, Individual consideration. IB, Pass in Diploma.

BSc Computer Science and Natural History

UCAS Code: CGD5 • Mode: 3 Years Equal Combination

Qualifications/Requirements: GCE, A/AS: 12 points. AGNVQ, Merit. ND/C, 4 Merits. SQAH, Grades BBCC. SQAV, National. IB, Pass in Diploma.

BSc Computer Science and Ophthalmic Dispensing

UCAS Code: BG55 • Mode: 3 Years Equal Combination

Qualifications/Requirements: GCE, A/AS: 10 points. AGNVQ, Merit (in specific programmes). ND/C, 3 Merits. SQAH, Grades BCCC. SQAV, National. IB, Pass in Diploma.

BSc Computer Science and Psychology

UCAS Code: CG85 • Mode: 3 Years Equal Combination

Qualifications/Requirements: GCE, A/AS: 16 points. Science at A level. AGNVQ, Distinction in Science. ND/C, Distinction. SQAH, Grades BBBCC. SQAV, National. IB, Pass in Diploma including specific subjects.

BSc Construction Management and Information Technology
UCAS Code: GK5F • Mode: 3 Years Equal Combination
Qualifications/Requirements: GCE, A/AS: 14 points. AGNVQ, Merit. ND/C, 6 Merits. SQAH, Grades BBBC. SQAV, National. IB, Pass in Diploma.

BSc Ecology and Conservation and Real Time Computer Systems
UCAS Code: DG2M • Mode: 3 Years Equal Combination
Qualifications/Requirements: GCE, A/AS: 10 points. AGNVQ, Merit (in specific programmes). ND/C, 3 Merits. SQAH, Grades BCCC. SQAV, National. IB, Pass in Diploma.

BSc Electronics and Real Time Computer Systems
UCAS Code: HG65 • Mode: 3 Years Equal Combination
Qualifications/Requirements: GCE, A/AS: 10 points. AGNVQ, Merit (in specific programmes). ND/C, 3 Merits. SQAH, Grades BCCC. SQAV, National. IB, Pass in Diploma including specific subjects.

BSc Information Systems
UCAS Code: G521 • Mode: 1/2 Years Single Subject
Qualifications/Requirements: GCE, A/AS: 10 points. AGNVQ, Merit in Science. ND/C, 3 Merits. SQAH, Grades BCCC. SQAV, National. IB, Pass in Diploma.

BSc Information Systems and Multimedia
UCAS Code: G524 • Mode: 3 Years Single Subject
Qualifications/Requirements: GCE, A/AS: 12 points. AGNVQ, Merit. ND/C, 4 Merits. SQAH, Grades CCCC. SQAV, National. IB, Pass in Diploma.

BSc Information Systems and Product Design
UCAS Code: GH57 • Mode: 3 Years Equal Combination
Qualifications/Requirements: GCE, A/AS: 10 points. AGNVQ, Merit in Science. ND/C, 3 Merits. SQAH, Grades BCCC. SQAV, Individual consideration. IB, Individual consideration.

BSc Internet Technology and Maths with Statistics
UCAS Code: GG16 • Mode: 3 Years Equal Combination
Qualifications/Requirements: GCE, A/AS: 12 points. AGNVQ, Merit (in specific programmes). ND/C, 4 Merits. SQAH, Grades BCCC. SQAV, National. IB, Individual consideration.

BSc Psychology and Real Time Computer Systems
UCAS Code: CG8M • Mode: 3 Years Equal Combination
Qualifications/Requirements: GCE, A/AS: 16 points. AGNVQ, Distinction in Science. ND/C, Distinction. SQAH, Grades BBBC. SQAV, National. IB, Pass in Diploma including specific subjects.

BSc Surveying and Information Technology
UCAS Code: GK5G • Mode: 3 Years Equal Combination
Qualifications/Requirements: GCE, A/AS: 14 points. AGNVQ, Merit. ND/C, 6 Merits. SQAH, Grades BBCC. SQAV, National. IB, Pass in Diploma.

Mod Business Administration and Information Systems
UCAS Code: GN51 • Mode: 3 Years Equal Combination
Qualifications/Requirements: GCE, A/AS: 12 points. AGNVQ, Merit. ND/C, 4 Merits. SQAH, Grades BBCC. SQAV, Individual consideration. IB, Pass in Diploma.

Mod Business and Computer Science
UCAS Code: NG15 • Mode: 3 Years Equal Combination
Qualifications/Requirements: GCE, A/AS: 12 points. AGNVQ, Merit (in specific programmes). ND/C, 4 Merits. SQAH, Grades BBCC. SQAV, Individual consideration. IB, Pass in Diploma.

Mod Communication Studies and Computer Science
UCAS Code: GP53 • Mode: 3 Years Equal Combination
Qualifications/Requirements: GCE, A/AS: 14 points. An approved subject from restricted list. AGNVQ, Merit (in specific programmes). ND/C, 6 Merits. SQAH, Grades BBBC. SQAV, Individual consideration. IB, Pass in Diploma including specific subjects.

Mod Communication Studies and Real Time Computer Systems
UCAS Code: PG35 • Mode: 3 Years Equal Combination
Qualifications/Requirements: GCE, A/AS: 14 points. An approved subject from restricted list. AGNVQ, Merit (in specific programmes). ND/C, 6 Merits. SQAH, Grades BBBC. SQAV, Individual consideration. IB, Pass in Diploma including specific subjects.

Mod Computer Science and French
UCAS Code: GR51 • Mode: 3/4 Years Equal Combination
Qualifications/Requirements: GCE, A/AS: 12 points. AGNVQ, Merit (in specific programmes). ND/C, 4 Merits. SQAH, Grades BBCC. SQAV, Individual consideration. IB, Pass in Diploma.

Mod Computer Science and German
UCAS Code: GR52 • Mode: 3/4 Years Equal Combination
Qualifications/Requirements: GCE, A/AS: 12 points. AGNVQ, Merit (in specific programmes). ND/C, 4 Merits. SQAH, Grades BBCC. SQAV, National. IB, Pass in Diploma.

Mod Computer Science and Heritage
UCAS Code: GV59 • Mode: 3 Years Equal Combination
Qualifications/Requirements: Please refer to prospectus.

Mod Computer Science and Italian
UCAS Code: GR53 • Mode: 3/4 Years Equal Combination
Qualifications/Requirements: GCE, A/AS: 12 points. AGNVQ, Merit (in specific programmes). ND/C, 4 Merits. SQAH, Grades BBCC. SQAV, National. IB, Pass in Diploma.

Mod Computer Science and Spanish

UCAS Code: GR54 • Mode: 3/4 Years Equal Combination

Qualifications/Requirements: GCE, A/AS: 10 points. AGNVQ, Merit (in specific programmes). ND/C, 3 Merits. SQAH, Grades BBCC. SQAV, National. IB, Pass in Diploma.

HND Business Information Technology

UCAS Code: 065G • Mode: 2 Years Single Subject

Qualifications/Requirements: GCE, A/AS: 6 points. AGNVQ, Pass. ND/C, National. SQAH, Grades CC. SQAV, Individual consideration. IB, Pass in Diploma.

HND Computing

UCAS Code: 105G • Mode: 2 Years Single Subject

Qualifications/Requirements: GCE, A/AS: 10 points. AGNVQ, Pass. ND/C, 3 Merits. SQAH, Grades BCCC. SQAV, Individual consideration. IB, Pass in Diploma.

ASTON UNIVERSITY A80

BSc Biology/Computer Science

UCAS Code: CG15 • Mode: 3/4 Years Equal Combination

Qualifications/Requirements: GCE, A/AS: 18 points. Biology. AGNVQ, Distinction (in specific programmes) with 6 additional units or with A level. ND/C, 5 Merits and 5 Distinctions (in specific programmes). SQAH, Grades BBBBC including specific subject(s). SQAV, Individual consideration. IB, 29 points.

BSc Biology/Computer Science (Year Zero)

UCAS Code: CG1M • Mode: 4/5 Years Equal Combination

Qualifications/Requirements: Please refer to prospectus.

BSc Business Administration/Computer Science

UCAS Code: GN51 • Mode: 3/4 Years Equal Combination

Qualifications/Requirements: GCE, A/AS: 18 points. AGNVQ, Distinction (in specific programmes) with 6 additional units or with A/AS. ND/C, 3 Merits and 7 Distinctions. SQAH, Grades BBBBB. SQAV, Individual consideration. IB, 30 points.

BSc Business Administration/Computer Science (Year Zero)

UCAS Code: GN5C • Mode: 4/5 Years Equal Combination

Qualifications/Requirements: Please refer to prospectus.

BSc Business Administration/ Geographical Information Systems

UCAS Code: FN81 • Mode: 3/4 Years Equal Combination

Qualifications/Requirements: GCE, A/AS: 18 points. Maths or Science at A level. AGNVQ, Distinction in Business with A level. ND/C, 5 Merits and 5 Distinctions. SQAH, Grades BBBBC including specific subject(s). IB, 29 points.

BSc Business Administration/ Geographical Information Systems (Year Zero)

UCAS Code: FN8C • Mode: 4/5 Years Equal Combination

Qualifications/Requirements: Please refer to prospectus.

BSc Business Computing & IT

UCAS Code: NG45 • Mode: 3/4 Years Equal Combination

Qualifications/Requirements: GCE, A level grades: BBB. AGNVQ, Distinction in Business with 6 additional units or with A/AS. ND/C, 3 Merits and 7 Distinctions. SQAH, Grades AABBB. SQAV, Individual consideration. IB, 31 points.

BSc Computer Science/Environmental Science & Technology

UCAS Code: GF59 • Mode: 3/4 Years Equal Combination

Qualifications/Requirements: GCE, A/AS: 18 points. Science. AGNVQ, Distinction (in specific programmes) with 6 additional units or with A level. ND/C, 5 Merits and 5 Distinctions (in specific programmes). SQAH, Grades BBBBC including specific subject(s). SQAV, Individual consideration. IB, 29 points.

BSc Computer Science/Environmental Science & Technology (Year Zero)

UCAS Code: GF5X • Mode: 4/5 Years Equal Combination

Qualifications/Requirements: Please refer to prospectus.

BSc Computer Science/European Studies

UCAS Code: GT52 • Mode: 3/4 Years Equal Combination

Qualifications/Requirements: GCE, A/AS: 18 points. AGNVQ, Distinction (in specific programmes) with 6 additional units or with A/AS. ND/C, 5 Merits and 5 Distinctions. SQAH, Grades BBBBC. SQAV, Individual consideration. IB, 29 points.

BSc Computer Science/European Studies (Year Zero)

UCAS Code: GT5F • Mode: 4/5 Years Equal Combination

Qualifications/Requirements: Please refer to prospectus.

BSc Computer Science/French

UCAS Code: GR51 • Mode: 4 Years Equal Combination

Qualifications/Requirements: GCE, A/AS: 20 points. French. AGNVQ, Distinction (in specific programmes) with 6 additional units or with A level. ND/C, Not normally sufficient. SQAH, Grades BBBBB including specific subject(s). SQAV, Individual consideration. IB, 30 points.

BSc Computer Science/Geographical Information Systems

UCAS Code: FG85 • Mode: 3/4 Years Equal Combination

Qualifications/Requirements: GCE, A/AS: 18 points. Science or Maths at A level. AGNVQ, Distinction (in specific programmes) with A level. ND/C, 5 Merits and 5 Distinctions. SQAH, Grades BBBBC including specific subject(s). IB, 29 points.

BSc Computer Science/Geographical Information Systems (Year Zero)
UCAS Code: FG8M • Mode: 4/5 Years Equal Combination
Qualifications/Requirements: Please refer to prospectus.

BSc Computer Science/Health & Safety Management
UCAS Code: GJ59 • Mode: 3/4 Years Equal Combination
Qualifications/Requirements: GCE, A/AS: 18 points. AGNVQ, Distinction (in specific programmes) with 6 additional units or with A/AS. ND/C, 5 Merits and 5 Distinctions. SQAH, Grades BBBBC. SQAV, Individual consideration. IB, 29 points.

BSc Computer Science/Health & Safety Management (Year Zero)
UCAS Code: GJ5Y • Mode: 4/5 Years Equal Combination
Qualifications/Requirements: Please refer to prospectus.

BSc Computer Science/Human Psychology
UCAS Code: LG75 • Mode: 3/4 Years Equal Combination
Qualifications/Requirements: GCE, A/AS: 20 points. AGNVQ, Distinction (in specific programmes) with 6 additional units or with A/AS. ND/C, 3 Merits and 7 Distinctions. SQAH, Grades BBBBB. SQAV, Individual consideration. IB, 30 points.

BSc Computer Science/Human Psychology (Year Zero)
UCAS Code: LG7M • Mode: 4/5 Years Equal Combination
Qualifications/Requirements: Please refer to prospectus.

BSc Computer Science/Politics
UCAS Code: MG15 • Mode: 3/4 Years Equal Combination
Qualifications/Requirements: GCE, A/AS: 18 points. AGNVQ, Distinction (in specific programmes) with 6 additional units or with A level. ND/C, 5 Merits and 5 Distinctions (in specific programmes). SQAH, Grades BBBBC including specific subject(s). SQAV, Individual consideration. IB, 29 points.

BSc Computer Science/Politics (Year Zero)
UCAS Code: MG1M • Mode: 4/5 Years Equal Combination
Qualifications/Requirements: Please refer to prospectus.

BSc Computer Science/Public Policy & Management
UCAS Code: GM51 • Mode: 3/4 Years Equal Combination
Qualifications/Requirements: GCE, A/AS: 18 points. AGNVQ, Distinction (in specific programmes) with 6 additional units or with A/AS. ND/C, 5 Merits and 5 Distinctions. SQAH, Grades BBBBB. SQAV, Individual consideration. IB, 30 points.

BSc Computer Science/Public Policy & Management (Year Zero)
UCAS Code: GM5C • Mode: 4/5 Years Equal Combination
Qualifications/Requirements: Please refer to prospectus.

BSc Computer Science/Telecommunications
UCAS Code: HG65 • Mode: 3/4 Years Equal Combination
Qualifications/Requirements: GCE, A/AS: 18 points. Maths or Physics. AGNVQ, Distinction (in specific programmes) with 6 additional units or with A level. ND/C, 5 Merits and 5 Distinctions (in specific programmes). SQAH, Grades BBBBC including specific subject(s). SQAV, Individual consideration. IB, 29 points.

BSc Computing Science
UCAS Code: G500 • Mode: 3/4 Years Single Subject
Qualifications/Requirements: GCE, A/AS: 20 points. AGNVQ, Distinction (in specific programmes) with 6 additional units or with A/AS. ND/C, 2 Merits and 4 Distinctions. SQAH, Grades BBBBB. SQAV, Individual consideration. IB, 30 points.

BSc Geographical Information Systems/Mathematics
UCAS Code: FG81 • Mode: 3/4 Years Equal Combination
Qualifications/Requirements: GCE, A/AS: 20 points. Maths at A level. AGNVQ, Distinction (in specific programmes) with A level. SQAH, Grades BBBBB including specific subject(s). IB, 30 points.

BSc Geographical Information Systems/Mathematics (Year Zero)
UCAS Code: FGVC • Mode: 4/5 Years Equal Combination
Qualifications/Requirements: Please refer to prospectus.

BSc Information Technology for Business
UCAS Code: G560 • Mode: 3/4 Years Single Subject
Qualifications/Requirements: GCE, A/AS: 20 points. AGNVQ, Distinction (in specific programmes) with 6 additional units or with A/AS. ND/C, 2 Merits and 4 Distinctions. SQAH, Grades BBBBC. SQAV, Individual consideration. IB, 30 points.

BEng Electronic Engineering and Computer Science
UCAS Code: GH56 • Mode: 3/4 Years Equal Combination
Qualifications/Requirements: GCE, A/AS: 20 points. Maths and any physical science, Computing or Electronics. AGNVQ, Merit (in specific programmes) with 6 additional units or with A level. ND/C, 2 Merits and 4 Distinctions (in specific programmes). SQAH, Grades BBBBB including specific subject(s). SQAV, Individual consideration. IB, 30 points.

UNIVERSITY OF WALES, BANGOR B06

BA Marketing/Computer Studies
UCAS Code: GN51 • Mode: 3 Years Equal Combination
Qualifications/Requirements: GCE, A level grades: CCC. AGNVQ, Distinction with A/AS. ND/C, Individual consideration. SQAH, Grades BBCC including specific subject(s). SQAV, Individual consideration. IB, 26 points.

BSc Computer Science

UCAS Code: G500 • Mode: 3 Years Single Subject

Qualifications/Requirements: GCE, A level grades: CCC. Maths and Physics, Electronics or Computing at A level. AGNVQ, Merit (in specific programmes) with 6 additional units or with A level. ND/C, 3 Merits (in specific programmes). SQAH, Grades BBCC including specific subject(s). SQAV, Individual consideration. IB, 26 points.

BSc Information Technology

UCAS Code: G560 • Mode: 3 Years Single Subject

Qualifications/Requirements: GCE, A/AS: 16-18 points. Two sciences at A level. AGNVQ, Merit (in specific programmes) with 6 additional units or with A level. ND/C, 3 Merits (in specific programmes). SQAH, Grades BBCC including specific subject(s). SQAV, Individual consideration. IB, 26 points.

BARNSLEY COLLEGE B13

HND Computing

UCAS Code: 005G • Mode: 2 Years Single Subject

Qualifications/Requirements: GCE, A/AS: 2 points. AGNVQ, Pass (in specific programmes). ND/C, Individual consideration. SQAH, Individual consideration. SQAV, Individual consideration. IB, Individual consideration.

DipHE Business and Management with Information Technology

UCAS Code: N1G5 • Mode: 2 Years Major/Minor

Qualifications/Requirements: GCE, A/AS: 2 points. AGNVQ, Pass (in specific programmes). ND/C, Individual consideration. SQAH, Individual consideration. SQAV, Individual consideration. IB, Individual consideration.

UNIVERSITY OF BATH B16

BSc Computer Information Systems (3 years)

UCAS Code: G520 • Mode: 3 Years Single Subject

Qualifications/Requirements: GCE, A/AS: 22-24 points. Maths. AGNVQ, Individual consideration. ND/C, Individual consideration. SQAH, CSYS required. SQAV, Individual consideration. IB, 32 points.

BSc Computer Information Systems (4 year sandwich)

UCAS Code: G521 • Mode: 4 Years Single Subject

Qualifications/Requirements: GCE, A/AS: 22-24 points. Maths. AGNVQ, Individual consideration. ND/C, Individual consideration. SQAH, CSYS required. SQAV, Individual consideration. IB, 32 points.

BSc Computer Science

UCAS Code: G500 • Mode: 3 Years Single Subject

Qualifications/Requirements: GCE, A/AS: 22-24 points. Maths. AGNVQ, Individual consideration. ND/C, Individual consideration. SQAH, CSYS required. SQAV, Individual consideration. IB, 32 points.

BSc Computer Science (4 year sandwich)

UCAS Code: G501 • Mode: 4 Years Single Subject

Qualifications/Requirements: GCE, A/AS: 22-24 points. Maths. AGNVQ, Individual consideration. ND/C, Individual consideration. SQAH, CSYS required. SQAV, Individual consideration. IB, 32 points.

BSc Mathematics and Computing (3 years)

UCAS Code: G5GC • Mode: 3 Years Major/Minor

Qualifications/Requirements: GCE, A/AS: 22-24 points. Maths at A level. AGNVQ, Individual consideration. ND/C, Individual consideration. SQAH, CSYS required. SQAV, Individual consideration. IB, 32 points.

BSc Mathematics and Computing (4 years SW)

UCAS Code: G5G1 • Mode: 4 Years Major/Minor

Qualifications/Requirements: GCE, A/AS: 22-24 points. Maths at A level. AGNVQ, Individual consideration. ND/C, Individual consideration. SQAH, CSYS required. SQAV, Individual consideration. IB, 32 points.

BEng Networks & Information Engineering (3 years)

UCAS Code: GG56 • Mode: 3 Years Equal Combination

Qualifications/Requirements: GCE, A/AS: 20 points. Maths and Physics or Electronics at A level. AGNVQ, Individual consideration. ND/C, Individual consideration. SQAH, CSYS required. SQAV, Individual consideration. IB, 30 points.

BEng Networks & Information Engineering (4 years)

UCAS Code: GG5P • Mode: 4 Years Equal Combination

Qualifications/Requirements: GCE, A/AS: 20 points. Maths and Physics or Electronics at A level. AGNVQ, Individual consideration. ND/C, Individual consideration. SQAH, CSYS required. SQAV, Individual consideration. IB, 30 points.

MEng Networks & Information Engineering (4 years)

UCAS Code: GG6M • Mode: 4 Years Equal Combination

Qualifications/Requirements: GCE, A/AS: 24 points. Maths and Physics or Electronics at A level. AGNVQ, Individual consideration. ND/C, Individual consideration. SQAH, CSYS required. SQAV, Individual consideration. IB, 32 points.

MEng Networks & Information Engineering (5 years)

UCAS Code: GG6N • Mode: 5 Years Equal Combination

Qualifications/Requirements: GCE, A/AS: 24 points. Maths and Physics or Electronics at A level. AGNVQ, Individual consideration. ND/C, Individual consideration. SQAH, CSYS required. SQAV, Individual consideration. IB, 32 points.

HND Computing and Multimedia

UCAS Code: 035G • Mode: 2 Years Single Subject

Qualifications/Requirements: Please refer to prospectus.

BELL COLLEGE OF TECHNOLOGY B26

BA Business Information Management (1 year)

UCAS Code: GN51 • Mode: 1 Year Equal Combination

Qualifications/Requirements: SQAV, Higher National including specific subject(s).

HND Computing (Support)

UCAS Code: 105G • Mode: 2 Years Single Subject

Qualifications/Requirements: GCE, A level grades: EE. Maths, English or Computing. AGNVQ, Individual consideration. ND/C, Individual consideration. SQAH, Grades CC including specific subject(s). SQAV, Individual consideration. IB, Individual consideration.

THE UNIVERSITY OF BIRMINGHAM B32

BA Computer Studies/East Mediterranean History

UCAS Code: GVM1 • Mode: 3 Years Equal Combination

Qualifications/Requirements: GCE, A level grades: BBB. AGNVQ, Distinction with A/AS. ND/C, Individual consideration. SQAH, Grades ABBBB. SQAV, Individual consideration. IB, 32 points.

BA Computer Studies/English

UCAS Code: GQ53 • Mode: 3 Years Equal Combination

Qualifications/Requirements: GCE, A level grades: BBB. AGNVQ, Distinction with A/AS. ND/C, Individual consideration. SQAH, Grades ABBBB. SQAV, Individual consideration. IB, 32 points.

BA Computer Studies/French Studies

UCAS Code: GR51 • Mode: 4 Years Equal Combination

Qualifications/Requirements: GCE, A level grades: BBB. French at A level. AGNVQ, Distinction with A level. ND/C, Individual consideration. SQAH, Grades ABBBB. SQAV, Individual consideration. IB, 32 points.

BA Computer Studies/German Studies

UCAS Code: GR52 • Mode: 4 Years Equal Combination

Qualifications/Requirements: GCE, A level grades: BBB. German. AGNVQ, Distinction with A level. ND/C, Individual consideration. SQAH, Grades ABBBB. SQAV, Individual consideration. IB, 32 points.

BA Computer Studies/Hispanic Studies

UCAS Code: GR54 • Mode: 4 Years Equal Combination

Qualifications/Requirements: GCE, A level grades: BBB. AGNVQ, Distinction with A/AS. ND/C, Individual consideration. SQAH, Grades ABBBB. SQAV, Individual consideration. IB, 32 points.

BA Computer Studies/Italian

UCAS Code: GR53 • Mode: 4 Years Equal Combination

Qualifications/Requirements: GCE, A level grades: BBB. AGNVQ, Distinction with A/AS. ND/C, Individual consideration. SQAH, Grades ABBBB. SQAV, Individual consideration. IB, 32 points.

BA Computer Studies/Latin

UCAS Code: GQ56 • Mode: 3 Years Equal Combination

Qualifications/Requirements: GCE, A level grades: BBB. Latin. AGNVQ, Distinction with A level. ND/C, Individual consideration. SQAH, Grades ABBBB. SQAV, Individual consideration. IB, 32 points.

BA Computer Studies/Modern Greek Studies

UCAS Code: GT52 • Mode: 4 Years Equal Combination

Qualifications/Requirements: GCE, A level grades: BBB. AGNVQ, Distinction with A/AS. ND/C, Individual consideration. SQAH, Grades ABBBB. SQAV, Individual consideration. IB, 32 points.

BA Computer Studies/Music

UCAS Code: GW53 • Mode: 3 Years Equal Combination

Qualifications/Requirements: GCE, A level grades: AAB-ABB. Music at A level. AGNVQ, Distinction with A level. ND/C, Individual consideration. SQAH, Grades AAABB. SQAV, Individual consideration. IB, 32 points.

BA Computer Studies/Philosophy

UCAS Code: GV57 • Mode: 3 Years Equal Combination

Qualifications/Requirements: GCE, A level grades: BBB. AGNVQ, Distinction with A/AS. ND/C, Individual consideration. SQAH, Grades ABBBB. SQAV, Individual consideration. IB, 32 points.

BA Computer Studies/Portuguese

UCAS Code: GR55 • Mode: 4 Years Equal Combination

Qualifications/Requirements: GCE, A level grades: BBB. AGNVQ, Distinction with A/AS. ND/C, Individual consideration. SQAH, Grades ABBBB. SQAV, Individual consideration. IB, 32 points.

BA Computer Studies/Russian

UCAS Code: GR58 • Mode: 4 Years Equal Combination

Qualifications/Requirements: GCE, A level grades: BBB. AGNVQ, Distinction with A/AS. ND/C, Individual consideration. SQAH, Grades ABBBB. SQAV, Individual consideration. IB, 32 points.

BA Computer Studies/Theology

UCAS Code: GV58 • Mode: 3 Years Equal Combination

Qualifications/Requirements: GCE, A level grades: BBB. AGNVQ, Distinction with A/AS. ND/C, Individual consideration. SQAH, Grades ABBBB. SQAV, Individual consideration. IB, 32 points.

BSc Computer Science/Software Engineering

UCAS Code: GG57 • Mode: 3 Years Equal Combination

Qualifications/Requirements: GCE, A level grades: BBB. Science at A level. AGNVQ, Individual consideration. ND/C, Individual consideration. SQAH, Individual consideration. SQAV, Individual consideration. IB, 32 points.

BSc Computer Science/Software Engineering with Business Studies

UCAS Code: G5N1 • Mode: 3 Years Major/Minor

Qualifications/Requirements: GCE, A level grades: BBB. Science at A level. AGNVQ, Individual consideration. ND/C, Individual consideration. SQAH, Individual consideration. SQAV, Individual consideration. IB, 32 points.

BSc Mathematics and Computer Science

UCAS Code: GG15 • Mode: 3 Years Equal Combination

Qualifications/Requirements: GCE, A level grades: ACC. Maths at A level. AGNVQ, Individual consideration. ND/C, Individual consideration. SQAH, Individual consideration. SQAV, Individual consideration. IB, 30 points.

BLACKBURN COLLEGE B40

BSc Computing

UCAS Code: G501 • Mode: 3 Years Single Subject

Qualifications/Requirements: Please refer to prospectus.

BSc Computing (1 year)

UCAS Code: G500 • Mode: 1 Year Single Subject

Qualifications/Requirements: ND/C, Higher National (in specific programmes). SQAH, Individual consideration. SQAV, Individual consideration. IB, Individual consideration.

HND Computing (with Business options)

UCAS Code: 005G • Mode: 2 Years Single Subject

Qualifications/Requirements: AGNVQ, Pass (in specific programmes). SQAH, Individual consideration. SQAV, Individual consideration. IB, Individual consideration.

BLACKPOOL AND THE FYLDE COLLEGE B41

BSc Computing (Conversion to degree)

UCAS Code: G500 • Mode: 2 Years Single Subject

Qualifications/Requirements: ND/C, Higher National.

BSc Information Technology

UCAS Code: G562 • Mode: 2 Years Single Subject

Qualifications/Requirements: ND/C, Higher National.

HND Computing

UCAS Code: 065G • Mode: 2 Years Single Subject

Qualifications/Requirements: GCE, A/AS: 2 points. AGNVQ, Pass (in specific programmes). ND/C, National. SQAH, Individual consideration. SQAV, Individual consideration. IB, Individual consideration.

BOLTON INSTITUTE OF HIGHER EDUCATION B44

BSc Business Information Systems

UCAS Code: G520 • Mode: 3 Years Single Subject

Qualifications/Requirements: GCE, A level grades: CD. AGNVQ, Merit. ND/C, Merit overall. SQAH, Grades BBCC. SQAV, Individual consideration. IB, 24 points.

BSc Computing

UCAS Code: G500 • Mode: 3 Years Single Subject

Qualifications/Requirements: GCE, A level grades: CD. AGNVQ, Merit. ND/C, Merit overall. SQAH, Grades BBCC. SQAV, Individual consideration. IB, 24 points.

BSc Computing Technology

UCAS Code: G600 • Mode: 3 Years Single Subject

Qualifications/Requirements: GCE, A/AS: 10 points. AGNVQ, Merit (in specific programmes). ND/C, 3 Merits. SQAH, Individual consideration. SQAV, Individual consideration. IB, Individual consideration.

BSc Computing Technology (3.5 years)

UCAS Code: G601 • Mode: 3/4 Years Single Subject

Qualifications/Requirements: GCE, A/AS: 4 points. AGNVQ, Pass. ND/C, National. SQAH, Individual consideration. SQAV, Individual consideration. IB, Individual consideration.

Mod Accountancy and Business Information Systems

UCAS Code: NG45 • Mode: 3 Years Equal Combination

Qualifications/Requirements: GCE, A level grades: CD. AGNVQ, Merit. ND/C, Merit overall. SQAH, Grades BBCC. SQAV, Individual consideration. IB, 24 points.

Mod Accountancy and Leisure Computing Technology

UCAS Code: GN74 • Mode: 3 Years Equal Combination

Qualifications/Requirements: GCE, A level grades: CD. AGNVQ, Merit. ND/C, National. SQAH, Individual consideration. SQAV, Individual consideration. IB, Individual consideration.

Mod Art & Design History and Computing

UCAS Code: VG45 • Mode: 3 Years Equal Combination

Qualifications/Requirements: GCE, A level grades: CD. AGNVQ, Merit. ND/C, Merit overall. SQAH, Grades BBCC. SQAV, Individual consideration. IB, 24 points.

Mod Biology and Business Information Systems

UCAS Code: GC51 • Mode: 3 Years Equal Combination

Qualifications/Requirements: GCE, A level grades: CD. AGNVQ, Merit. ND/C, Merit overall. SQAH, Grades BBCC. SQAV, Individual consideration. IB, 24 points.

Mod Biology and Computing
UCAS Code: CG15 • Mode: 3 Years Equal Combination
Qualifications/Requirements: GCE, A level grades: CD. AGNVQ, Merit. ND/C, Merit overall. SQAH, Grades BBCC. SQAV, Individual consideration. IB, 24 points.

Mod Business Information Systems and Creative Writing
UCAS Code: WG95 • Mode: 3 Years Equal Combination
Qualifications/Requirements: GCE, A level grades: CD. AGNVQ, Merit. ND/C, Merit overall. SQAH, Grades BBCC. SQAV, Individual consideration. IB, 24 points.

Mod Business Information Systems and English
UCAS Code: GQM3 • Mode: 3 Years Equal Combination
Qualifications/Requirements: GCE, A level grades: CD. AGNVQ, Merit. ND/C, Merit overall. SQAH, Grades BBCC. SQAV, Individual consideration. IB, 24 points.

Mod Business Information Systems and Enterprise Development
UCAS Code: NGD5 • Mode: 3 Years Equal Combination
Qualifications/Requirements: GCE, A level grades: CD. AGNVQ, Merit. ND/C, Merit overall. SQAH, Grades BBCC. SQAV, Individual consideration. IB, 24 points.

Mod Business Information Systems and Environmental Studies
UCAS Code: GF59 • Mode: 3 Years Equal Combination
Qualifications/Requirements: GCE, A level grades: CD. AGNVQ, Merit. ND/C, Merit overall. SQAH, Grades BBCC. SQAV, Individual consideration. IB, 24 points.

Mod Business Information Systems and European Studies
UCAS Code: GT52 • Mode: 3 Years Equal Combination
Qualifications/Requirements: GCE, A level grades: CD. AGNVQ, Merit. ND/C, Merit overall. SQAH, Grades BBCC. SQAV, Individual consideration. IB, 24 points.

Mod Business Information Systems and Film Studies
UCAS Code: GW55 • Mode: 3 Years Equal Combination
Qualifications/Requirements: GCE, A level grades: CD. Media Studies or Theatre Studies at A level. AGNVQ, Individual consideration. ND/C, Individual consideration. SQAH, Grades BBCC. SQAV, Individual consideration. IB, 24 points.

Mod Business Information Systems and French
UCAS Code: GR5C • Mode: 3 Years Equal Combination
Qualifications/Requirements: GCE, A level grades: CD. French at A level. AGNVQ, Individual consideration. ND/C, Individual consideration. SQAH, Grades BBCC. SQAV, Individual consideration. IB, 24 points.

Mod Business Information Systems and German
UCAS Code: GR5F • Mode: 3 Years Equal Combination
Qualifications/Requirements: GCE, A level grades: CD. German at A level. AGNVQ, Individual consideration. ND/C, Individual consideration. SQAH, Grades BBCC. SQAV, Individual consideration. IB, 24 points.

Mod Business Information Systems and History
UCAS Code: GVM1 • Mode: 3 Years Equal Combination
Qualifications/Requirements: GCE, A level grades: CD. AGNVQ, Merit. ND/C, Merit overall. SQAH, Grades BBCC. SQAV, Individual consideration. IB, 24 points.

Mod Business Information Systems and Human Resource Management
UCAS Code: GN5C • Mode: 3 Years Equal Combination
Qualifications/Requirements: GCE, A level grades: CD. AGNVQ, Merit. ND/C, Merit overall. SQAH, Grades BBCC. SQAV, Individual consideration. IB, 24 points.

Mod Business Information Systems and Human Sciences
UCAS Code: GCM9 • Mode: 3 Years Equal Combination
Qualifications/Requirements: GCE, A level grades: CD. AGNVQ, Merit. ND/C, Merit overall. SQAH, Grades BBCC. SQAV, Individual consideration. IB, 24 points.

Mod Business Information Systems and Language Studies
UCAS Code: GQMG • Mode: 3 Years Equal Combination
Qualifications/Requirements: GCE, A level grades: CD. AGNVQ, Merit. ND/C, Merit overall. SQAH, Grades BBCC. SQAV, Individual consideration. IB, 24 points.

Mod Business Information Systems and Law
UCAS Code: GM5H • Mode: 3 Years Equal Combination
Qualifications/Requirements: GCE, A level grades: CD. AGNVQ, Merit. ND/C, Merit overall. SQAH, Grades BBCC. SQAV, Individual consideration. IB, 24 points.

Mod Business Information Systems and Leisure Computing Technology
UCAS Code: GG5R • Mode: 3 Years Equal Combination
Qualifications/Requirements: GCE, A level grades: CD. AGNVQ, Merit. ND/C, Merit overall. SQAH, Grades BBCC. SQAV, Individual consideration. IB, 24 points.

Mod Business Information Systems and Leisure Studies
UCAS Code: GLMH • Mode: 3 Years Equal Combination
Qualifications/Requirements: GCE, A level grades: CD. AGNVQ, Merit. ND/C, Merit overall. SQAH, Grades BBCC. SQAV, Individual consideration. IB, 24 points.

Mod Business Information Systems and Literature
UCAS Code: GQ5F • Mode: 3 Years Equal Combination
Qualifications/Requirements: GCE, A level grades: CD. AGNVQ, Merit. ND/C, Merit overall. SQAH, Grades BBCC. SQAV, Individual consideration. IB, 24 points.

Mod Business Information Systems and Marketing
UCAS Code: GN55 • Mode: 3 Years Equal Combination
Qualifications/Requirements: GCE, A level grades: CD. AGNVQ, Merit. ND/C, Merit overall. SQAH, Grades BBCC. SQAV, Individual consideration. IB, 24 points.

Mod Business Information Systems and Mathematics
UCAS Code: GG51 • Mode: 3 Years Equal Combination
Qualifications/Requirements: GCE, A level grades: DD. Maths at A level. AGNVQ, Individual consideration. ND/C, Individual consideration. SQAH, Grades BBCC. SQAV, Individual consideration. IB, 24 points.

Mod Business Information Systems and Music Technology
UCAS Code: GWMH • Mode: 3 Years Equal Combination
Qualifications/Requirements: GCE, A/AS: 10 points. AGNVQ, Merit. ND/C, 3 Merits. SQAH, Individual consideration. SQAV, Individual consideration. IB, Individual consideration.

Mod Business Information Systems and Peace & War Studies
UCAS Code: GVMC • Mode: 3 Years Equal Combination
Qualifications/Requirements: GCE, A level grades: CD. AGNVQ, Merit. ND/C, Merit overall. SQAH, Grades BBCC. SQAV, Individual consideration. IB, 24 points.

Mod Business Information Systems and Philosophy
UCAS Code: GV5R • Mode: 3 Years Equal Combination
Qualifications/Requirements: GCE, A level grades: CD. AGNVQ, Merit. ND/C, Merit overall. SQAH, Grades BBCC. SQAV, Individual consideration. IB, 24 points.

Mod Business Information Systems and Psychology
UCAS Code: GL5R • Mode: 3 Years Equal Combination
Qualifications/Requirements: GCE, A/AS: 12 points. AGNVQ, Distinction. ND/C, Merit overall. SQAH, Grades BBCC. SQAV, Individual consideration. IB, 24 points.

Mod Business Information Systems and Sociology
UCAS Code: LG35 • Mode: 3 Years Equal Combination
Qualifications/Requirements: GCE, A level grades: CD. AGNVQ, Merit. ND/C, Merit overall. SQAH, Individual consideration. SQAV, Individual consideration. IB, 24 points.

Mod Business Information Systems and Sport and Exercise Science
UCAS Code: GBM6 • Mode: 3 Years Equal Combination
Qualifications/Requirements: GCE, A level grades: CD. Science or Physical Education. AGNVQ, Merit. ND/C, Merit overall. SQAH, Individual consideration. SQAV, Individual consideration. IB, 24 points.

Mod Business Information Systems and Statistics
UCAS Code: GG54 • Mode: 3 Years Equal Combination
Qualifications/Requirements: GCE, A level grades: CD. AGNVQ, Merit. ND/C, Merit overall. SQAH, Individual consideration. SQAV, Individual consideration. IB, 24 points.

Mod Business Information Systems and Textiles
UCAS Code: GJ54 • Mode: 3 Years Equal Combination
Qualifications/Requirements: GCE, A level grades: CD. AGNVQ, Merit. ND/C, Merit overall. SQAH, Grades BBCC. SQAV, Individual consideration. IB, 24 points.

Mod Business Information Systems and Theatre Studies
UCAS Code: GW54 • Mode: 3 Years Equal Combination
Qualifications/Requirements: GCE, A level grades: CD. Media Studies or Theatre Studies at A level. AGNVQ, Individual consideration. ND/C, Individual consideration. SQAH, Grades BBCC. SQAV, Individual consideration. IB, 24 points.

Mod Business Information Systems and Tourism Studies
UCAS Code: GP5R • Mode: 3 Years Equal Combination
Qualifications/Requirements: GCE, A level grades: CD. AGNVQ, Merit. ND/C, Merit overall. SQAH, Grades BBCC. SQAV, Individual consideration. IB, 24 points.

Mod Business Information Systems and Urban & Cultural Studies
UCAS Code: GL5J • Mode: 3 Years Equal Combination
Qualifications/Requirements: GCE, A level grades: CD. AGNVQ, Merit. ND/C, Merit overall. SQAH, Grades BBCC. SQAV, Individual consideration. IB, 24 points.

Mod Business Studies and Computing
UCAS Code: GN51 • Mode: 3 Years Equal Combination
Qualifications/Requirements: GCE, A level grades: CD. AGNVQ, Merit. ND/C, Merit overall. SQAH, Grades BBCC. SQAV, Individual consideration. IB, 24 points.

Mod Community Studies and Computing
UCAS Code: LG55 • Mode: 3 Years Equal Combination
Qualifications/Requirements: GCE, A level grades: CD. AGNVQ, Merit. ND/C, Merit overall. SQAH, Grades BBCC. SQAV, Individual consideration. IB, 24 points.

Mod Computing and Creative Writing
UCAS Code: GW5X • Mode: 3 Years Equal Combination
Qualifications/Requirements: GCE, A level grades: CD. AGNVQ, Merit. ND/C, Merit overall. SQAH, Grades BBCC. SQAV, Individual consideration. IB, 24 points.

Mod Computing and English
UCAS Code: GQ53 • Mode: 3 Years Equal Combination
Qualifications/Requirements: GCE, A level grades: CD. AGNVQ, Merit. ND/C, Merit overall. SQAH, Grades BBCC. SQAV, Individual consideration. IB, 24 points.

Mod Computing and Enterprise Development
UCAS Code: NGDM • Mode: 3 Years Equal Combination
Qualifications/Requirements: GCE, A level grades: CD. AGNVQ, Merit. ND/C, Merit overall. SQAH, Grades BBCC. SQAV, Individual consideration. IB, 24 points.

Mod Computing and Environmental Studies
UCAS Code: FG95 • Mode: 3 Years Equal Combination
Qualifications/Requirements: GCE, A level grades: CD. AGNVQ, Merit. ND/C, Merit overall. SQAH, Grades BBCC. SQAV, Individual consideration. IB, 24 points.

Mod Computing and European Studies
UCAS Code: GLM3 • Mode: 3 Years Equal Combination
Qualifications/Requirements: GCE, A level grades: CD. AGNVQ, Merit. ND/C, Merit overall. SQAH, Grades BBCC. SQAV, Individual consideration. IB, 24 points.

Mod Computing and Film Studies
UCAS Code: GW5M • Mode: 3 Years Equal Combination
Qualifications/Requirements: GCE, A level grades: CD. Media Studies or Theatre Studies at A level. AGNVQ, Individual consideration. ND/C, Individual consideration. SQAH, Grades BBCC. SQAV, Individual consideration. IB, 24 points.

Mod Computing and French
UCAS Code: GR51 • Mode: 3 Years Equal Combination
Qualifications/Requirements: GCE, A level grades: CD. French at A level. AGNVQ, Individual consideration. ND/C, Individual consideration. SQAH, Grades BBCC. SQAV, Individual consideration. IB, 24 points.

Mod Computing and German
UCAS Code: GR52 • Mode: 3 Years Equal Combination
Qualifications/Requirements: GCE, A level grades: CD. German at A level. AGNVQ, Individual consideration. ND/C, Individual consideration. SQAH, Grades BBCC. SQAV, Individual consideration. IB, 24 points.

Mod Computing and History
UCAS Code: GV5C • Mode: 3 Years Equal Combination
Qualifications/Requirements: GCE, A level grades: CD. AGNVQ, Merit. ND/C, Merit overall. SQAH, Grades BBCC. SQAV, Individual consideration. IB, 24 points.

Mod Computing and Human Resource Management
UCAS Code: GNM1 • Mode: 3 Years Equal Combination
Qualifications/Requirements: GCE, A level grades: CD. AGNVQ, Merit. ND/C, Merit overall. SQAH, Grades BBCC. SQAV, Individual consideration. IB, 24 points.

Mod Computing and Human Sciences
UCAS Code: GC59 • Mode: 3 Years Equal Combination
Qualifications/Requirements: GCE, A level grades: CD. AGNVQ, Merit. ND/C, Merit overall. SQAH, Grades BBCC. SQAV, Individual consideration. IB, 24 points.

Mod Computing and Language Studies
UCAS Code: GQ51 • Mode: 3 Years Equal Combination
Qualifications/Requirements: GCE, A level grades: CD. AGNVQ, Merit. ND/C, Merit overall. SQAH, Grades BBCC. SQAV, Individual consideration. IB, 24 points.

Mod Computing and Law
UCAS Code: GM53 • Mode: 3 Years Equal Combination
Qualifications/Requirements: GCE, A level grades: CD. AGNVQ, Merit. ND/C, Merit overall. SQAH, Grades BBCC. SQAV, Individual consideration. IB, 24 points.

Mod Computing and Leisure Studies
UCAS Code: GL5H • Mode: 3 Years Equal Combination
Qualifications/Requirements: GCE, A level grades: CD. AGNVQ, Merit. ND/C, Merit overall. SQAH, Grades BBCC. SQAV, Individual consideration. IB, 24 points.

Mod Computing and Literature
UCAS Code: GQ52 • Mode: 3 Years Equal Combination
Qualifications/Requirements: GCE, A level grades: CD. AGNVQ, Merit. ND/C, Merit overall. SQAH, Grades BBCC. SQAV, Individual consideration. IB, 24 points.

Mod Computing and Manufacturing Systems
UCAS Code: GH57 • Mode: 3 Years Equal Combination
Qualifications/Requirements: GCE, A level grades: CD. AGNVQ, Merit. ND/C, Merit overall. SQAH, Individual consideration. SQAV, Individual consideration. IB, Individual consideration.

Mod Computing and Marketing
UCAS Code: GN5M • Mode: 3 Years Equal Combination
Qualifications/Requirements: GCE, A level grades: CD. AGNVQ, Merit. ND/C, Merit overall. SQAH, Grades BBCC. SQAV, Individual consideration. IB, 24 points.

Mod Computing and Motor Vehicle Studies
UCAS Code: GH5J • Mode: 3 Years Equal Combination
Qualifications/Requirements: GCE, A level grades: CD. Maths or Science. AGNVQ, Merit (in specific programmes). ND/C, 3 Merits. SQAH, Individual consideration. SQAV, Individual consideration. IB, Individual consideration.

Mod Computing and Music Technology
UCAS Code: GW5H • Mode: 3 Years Equal Combination

Qualifications/Requirements: GCE, A level grades: CD. AGNVQ, Merit. ND/C, 3 Merits. SQAH, Individual consideration. SQAV, Individual consideration. IB, Individual consideration.

Mod Computing and Peace & War Studies
UCAS Code: GV5D • Mode: 3 Years Equal Combination

Qualifications/Requirements: GCE, A level grades: CD. AGNVQ, Merit. ND/C, Merit overall. SQAH, Grades BBCC. SQAV, Individual consideration. IB, 24 points.

Mod Computing and Philosophy
UCAS Code: GV57 • Mode: 3 Years Equal Combination

Qualifications/Requirements: GCE, A level grades: CD. AGNVQ, Merit. ND/C, Merit overall. SQAH, Grades BBCC. SQAV, Individual consideration. IB, 24 points.

Mod Computing and Psychology
UCAS Code: GL57 • Mode: 3 Years Equal Combination

Qualifications/Requirements: GCE, A/AS: 12 points. AGNVQ, Distinction. ND/C, Merit overall. SQAH, Grades BBCC. SQAV, Individual consideration. IB, 24 points.

Mod Computing and Simulation/Virtual Environment
UCAS Code: GG57 • Mode: 3 Years Equal Combination

Qualifications/Requirements: GCE, A level grades: CD. AGNVQ, Merit. ND/C, Merit overall. SQAH, Grades BBCC. SQAV, Individual consideration. IB, 24 points.

Mod Computing and Sociology
UCAS Code: LG3M • Mode: 3 Years Equal Combination

Qualifications/Requirements: GCE, A level grades: CD. AGNVQ, Merit. ND/C, Merit overall. SQAH, Individual consideration. SQAV, Individual consideration. IB, 24 points.

Mod Computing and Sport and Exercise Science
UCAS Code: GB56 • Mode: 3 Years Equal Combination

Qualifications/Requirements: GCE, A level grades: CD. Science or Physical Education. AGNVQ, Merit. ND/C, Merit overall. SQAH, Individual consideration. SQAV, Individual consideration. IB, 24 points.

Mod Computing and Statistics
UCAS Code: GG5K • Mode: 3 Years Equal Combination

Qualifications/Requirements: GCE, A level grades: CD. AGNVQ, Merit. ND/C, Merit overall. SQAH, Individual consideration. SQAV, Individual consideration. IB, 24 points.

Mod Computing and Theatre Studies
UCAS Code: WG45 • Mode: 3 Years Equal Combination

Qualifications/Requirements: GCE, A level grades: CD. Media Studies or Theatre Studies at A level. AGNVQ, Individual consideration. ND/C, Individual consideration. SQAH, Grades BBCC. SQAV, Individual consideration. IB, 24 points.

Mod Computing and Tourism Studies
UCAS Code: GP57 • Mode: 3 Years Equal Combination

Qualifications/Requirements: GCE, A level grades: CD. AGNVQ, Merit. ND/C, Merit overall. SQAH, Grades BBCC. SQAV, Individual consideration. IB, 24 points.

Mod Computing and Transport Studies
UCAS Code: GJ59 • Mode: 3 Years Equal Combination

Qualifications/Requirements: GCE, A level grades: CD. Maths or Science. AGNVQ, Merit. ND/C, Merit overall. SQAH, Grades BBCC. SQAV, Individual consideration. IB, 24 points.

Mod Computing and Urban and Cultural Studies
UCAS Code: GL53 • Mode: 3 Years Equal Combination

Qualifications/Requirements: GCE, A level grades: CD. AGNVQ, Merit. ND/C, Merit overall. SQAH, Grades BBCC. SQAV, Individual consideration. IB, 24 points.

Mod Computing Technology (foundation)
UCAS Code: G608 • Mode: 4/5 Years Single Subject

Qualifications/Requirements: GCE, A/AS: 4 points. AGNVQ, Pass. ND/C, National. SQAH, Individual consideration. SQAV, Individual consideration. IB, Individual consideration.

Mod Design and Business Information Systems
UCAS Code: GW52 • Mode: 3 Years Equal Combination

Qualifications/Requirements: GCE, A level grades: CD. AGNVQ, Merit. ND/C, Merit overall. SQAH, Grades BBCC. SQAV, Individual consideration. IB, 24 points.

Mod Music Technology and Simulation/ Virtual Environment
UCAS Code: GWTH • Mode: 3 Years Equal Combination

Qualifications/Requirements: GCE, A/AS: 10 points. AGNVQ, Merit. ND/C, 3 Merits. SQAH, Individual consideration. SQAV, Individual consideration. IB, Individual consideration.

HND Business Information Technology
UCAS Code: 065G • Mode: 2 Years Single Subject

Qualifications/Requirements: GCE, A/AS: 2 points. AGNVQ, Pass. ND/C, National. SQAH, Individual consideration. SQAV, National. IB, Pass in Diploma.

HND Computing
UCAS Code: 105G • Mode: 2 Years Single Subject

Qualifications/Requirements: GCE, A/AS: 2 points. AGNVQ, Pass. ND/C, National. SQAH, Individual consideration. SQAV, National. IB, Pass in Diploma.

HND Computing Technology
UCAS Code: 006G • Mode: 2 Years Single Subject

Qualifications/Requirements: GCE, A/AS: 4 points. AGNVQ, Pass. ND/C, National. SQAH, Individual consideration. SQAV, Individual consideration. IB, Individual consideration.

BOURNEMOUTH UNIVERSITY B50

BSc Business Decision Management
UCAS Code: G522 • Mode: 4 Years Single Subject

Qualifications/Requirements: GCE, A/AS: 14 points. AGNVQ, Merit. ND/C, Merit overall. SQAH, Individual consideration. SQAV, Individual consideration. IB, Individual consideration.

BSc Business Information Systems (HND top-up)
UCAS Code: G521 • Mode: 1 Year Single Subject

Qualifications/Requirements: ND/C, Higher National.

BSc Business Information Systems Management
UCAS Code: G520 • Mode: 4 Years Single Subject

Qualifications/Requirements: GCE, A/AS: 14 points. AGNVQ, Merit (in specific programmes). ND/C, Merit overall. SQAH, Individual consideration. SQAV, Individual consideration. IB, Individual consideration.

BSc Business Information Technology
UCAS Code: G560 • Mode: 4 Years Single Subject

Qualifications/Requirements: GCE, A/AS: 16-18 points. AGNVQ, Merit. ND/C, Merit overall. SQAH, Individual consideration. SQAV, Individual consideration. IB, Individual consideration.

BSc Computing
UCAS Code: G710 • Mode: 4 Years Single Subject

Qualifications/Requirements: GCE, A/AS: 16-18 points. AGNVQ, Distinction (in specific programmes). ND/C, Merit overall and Distinction. SQAH, Grades CCCC. SQAV, Individual consideration. IB, 30 points.

HND Business Information Technology
UCAS Code: 265G • Mode: 2 Years Single Subject

Qualifications/Requirements: GCE, A/AS: 6 points. AGNVQ, Pass. ND/C, 3 Merits. SQAH, Individual consideration. SQAV, Individual consideration. IB, Individual consideration.

THE ARTS INSTITUTE AT BOURNEMOUTH B53

HND Interactive Information Design
UCAS Code: 52GE • Mode: 2 Years Equal Combination

Qualifications/Requirements: Please refer to prospectus.

HND Interactive Information Design
UCAS Code: 52GW • Mode: 2 Years Equal Combination

Qualifications/Requirements: Please refer to prospectus.

THE UNIVERSITY OF BRADFORD B56

BSc Computer Animation and Special Effects
UCAS Code: G543 • Mode: 3 Years Single Subject

Qualifications/Requirements: GCE, A level grades: BCC. AGNVQ, Merit with A level. ND/C, 5 Merits and 1 Distinction. SQAH, Individual consideration. SQAV, Individual consideration. IB, Individual consideration.

BSc Computer Science (3 years)
UCAS Code: G500 • Mode: 3 Years Single Subject

Qualifications/Requirements: GCE, A/AS: 18 points. AGNVQ, Distinction (in specific programmes) with 4 additional units or with A/AS. ND/C, 3 Merits and 2 Distinctions. SQAH, Individual consideration. SQAV, Individual consideration. IB, Individual consideration.

BSc Computer Science (4 years)
UCAS Code: G501 • Mode: 4 Years Single Subject

Qualifications/Requirements: GCE, A/AS: 18 points. AGNVQ, Distinction (in specific programmes) with 4 additional units or with A/AS. ND/C, 3 Merits and 2 Distinctions. SQAH, Individual consideration. SQAV, Individual consideration. IB, Individual consideration.

BSc Computing and Information Systems (3 years)
UCAS Code: G520 • Mode: 3 Years Single Subject

Qualifications/Requirements: GCE, A/AS: 18 points. AGNVQ, Distinction (in specific programmes) with 4 additional units or with A/AS. ND/C, 3 Merits and 2 Distinctions. SQAH, Individual consideration. SQAV, Individual consideration. IB, Individual consideration.

BSc Computing and Information Systems (4 years)
UCAS Code: G521 • Mode: 4 Years Single Subject

Qualifications/Requirements: GCE, A/AS: 18 points. AGNVQ, Distinction (in specific programmes) with 4 additional units or with A/AS. ND/C, 3 Merits and 2 Distinctions. SQAH, Individual consideration. SQAV, Individual consideration. IB, Individual consideration.

BSc Computing and Mathematics
UCAS Code: GG51 • Mode: 3 Years Equal Combination

Qualifications/Requirements: AGNVQ, Individual consideration. ND/C, Individual consideration. SQAH, Individual consideration. SQAV, Individual consideration. IB, Individual consideration.

BSc Computing and Mathematics
UCAS Code: GG5C • Mode: 4 Years Equal Combination

Qualifications/Requirements: AGNVQ, Individual consideration. ND/C, Individual consideration. SQAH, Individual consideration. SQAV, Individual consideration. IB, Individual consideration.

BSc Computing with Management (3 years)
UCAS Code: G5N1 • Mode: 3 Years Major/Minor
Qualifications/Requirements: GCE, A/AS: 20 points. AGNVQ, Distinction (in specific programmes) with 4 additional units or with A/AS. ND/C, 3 Merits and 2 Distinctions. SQAH, Individual consideration. SQAV, Individual consideration. IB, Individual consideration.

BSc Computing with Management (4 years)
UCAS Code: G5NC • Mode: 4 Years Major/Minor
Qualifications/Requirements: GCE, A/AS: 20 points. AGNVQ, Distinction (in specific programmes) with 4 additional units or with A/AS. ND/C, 3 Merits and 2 Distinctions. SQAH, Individual consideration. SQAV, Individual consideration. IB, Individual consideration.

BSc Interactive Systems and Video Games Design
UCAS Code: G573 • Mode: 3 Years Single Subject
Qualifications/Requirements: GCE, A level grades: BCC. AGNVQ, Merit with A level. ND/C, 5 Merits and 1 Distinction. SQAH, Individual consideration. SQAV, Individual consideration. IB, Individual consideration.

BSc Internet, WWW, and the Information Society
UCAS Code: G563 • Mode: 3 Years Single Subject
Qualifications/Requirements: GCE, A level grades: BCC. AGNVQ, Merit with A level. ND/C, 5 Merits and 1 Distinction. SQAH, Individual consideration. SQAV, Individual consideration. IB, Individual consideration.

BSc Networks Information Management (3 years)
UCAS Code: G530 • Mode: 3 Years Single Subject
Qualifications/Requirements: GCE, A/AS: 18 points. AGNVQ, Distinction (in specific programmes) with 4 additional units or with A/AS. ND/C, 3 Merits and 2 Distinctions. SQAH, Individual consideration. SQAV, Individual consideration. IB, Individual consideration.

BSc Networks Information Management (4 years)
UCAS Code: G531 • Mode: 4 Years Single Subject
Qualifications/Requirements: GCE, A/AS: 18 points. AGNVQ, Distinction (in specific programmes) with 4 additional units or with A/AS. ND/C, 3 Merits and 2 Distinctions. SQAH, Individual consideration. SQAV, Individual consideration. IB, Individual consideration.

BRADFORD COLLEGE B60

BSc Business Information Technology (top-up)
UCAS Code: G560 • Mode: 1 Year Single Subject
Qualifications/Requirements: AGNVQ, Not normally sufficient. ND/C, Higher National. SQAH, Not normally sufficient. SQAV, Not normally sufficient. IB, Not normally sufficient.

HND Computing
UCAS Code: 005G • Mode: 2 Years Single Subject
Qualifications/Requirements: Please refer to prospectus.

UNIVERSITY OF BRIGHTON B72

BA Computing and Information Systems (4-year sandwich)
UCAS Code: G560 • Mode: 4 Years Single Subject
Qualifications/Requirements: GCE, A/AS: 18 points. AGNVQ, Merit with 6 additional units or with A/AS. ND/C, Distinction Overall. SQAH, Grades BBBC. SQAV, Individual consideration. IB, 27 points.

BSc Computer Science (4-year sandwich)
UCAS Code: G501 • Mode: 4 Years Single Subject
Qualifications/Requirements: GCE, A/AS: 18 points. AGNVQ, Merit (in specific programmes) with 6 additional units or with A level. ND/C, Distinction Overall (in specific programmes). SQAH, Grades BBBC including specific subject(s). SQAV, Individual consideration. IB, 27 points.

BSc Computer Studies
UCAS Code: G500 • Mode: 3/4 Years Single Subject
Qualifications/Requirements: GCE, A/AS: 18 points. AGNVQ, Merit with 6 additional units or with A/AS. ND/C, Distinction Overall. SQAH, Grades BBBC. SQAV, Individual consideration. IB, 27 points.

BSc Computing and Mathematics
UCAS Code: GG51 • Mode: 3/4 Years Equal Combination
Qualifications/Requirements: GCE, A/AS: 14 points. Maths. AGNVQ, Not normally sufficient. ND/C, Not normally sufficient. SQAH, Grades BBCC including specific subject(s). SQAV, Not normally sufficient. IB, Pass in Diploma including specific subjects.

BSc Computing and Media
UCAS Code: GP54 • Mode: 3/4 Years Equal Combination
Qualifications/Requirements: GCE, A/AS: 18 points. AGNVQ, Merit with 6 additional units or with A/AS. ND/C, Distinction Overall. SQAH, Grades BBBC. SQAV, Individual consideration. IB, 27 points.

BSc Computing and Operational Research
UCAS Code: GN52 • Mode: 3/4 Years Equal Combination
Qualifications/Requirements: GCE, A/AS: 14 points. Maths. AGNVQ, Not normally sufficient. ND/C, Not normally sufficient. SQAH, Grades BBCC including specific subject(s). SQAV, Not normally sufficient. IB, Pass in Diploma including specific subjects.

BSc Computing and Statistics
UCAS Code: GG54 • Mode: 3/4 Years Equal Combination
Qualifications/Requirements: GCE, A/AS: 14 points. Maths or Statistics. AGNVQ, Not normally sufficient. ND/C, Not normally sufficient. SQAH, Grades BBCC including specific subject(s). SQAV, Not normally sufficient. IB, Pass in Diploma including specific subjects.

BSc Mathematics for Computing
UCAS Code: G170 • Mode: 3/4 Years Single Subject

Qualifications/Requirements: GCE, A/AS: 14 points. Maths at A level. AGNVQ, Not normally sufficient. ND/C, Not normally sufficient. SQAH, Grades BBCC including specific subject(s). SQAV, Not normally sufficient. IB, Pass in Diploma including specific subjects.

HND Computing (Client Server Systems)
UCAS Code: 125G • Mode: 2 Years Single Subject

Qualifications/Requirements: GCE, A/AS: 4 points. AGNVQ, Merit. ND/C, Merit overall. SQAH, Grades CC. SQAV, Individual consideration. IB, Pass in Diploma.

HND Computing (Information Systems)
UCAS Code: 025G • Mode: 2 Years Single Subject

Qualifications/Requirements: GCE, A/AS: 4 points. AGNVQ, Merit. ND/C, Merit overall. SQAH, Grades CC. SQAV, Individual consideration. IB, Pass in Diploma.

HND Computing (Real Time Systems)
UCAS Code: 005G • Mode: 2 Years Single Subject

Qualifications/Requirements: GCE, A/AS: 4 points. AGNVQ, Merit. ND/C, Merit overall. SQAH, Grades CC. SQAV, Individual consideration. IB, Pass in Diploma.

UNIVERSITY OF BRISTOL B78

BSc Computer Science
UCAS Code: G500 • Mode: 3 Years Single Subject

Qualifications/Requirements: GCE, A level grades: ABB. Maths. AGNVQ, Distinction (in specific programmes) with A/AS. ND/C, Higher National (in specific programmes). SQAH, Grades AAABB. SQAV, Higher National including specific subject(s). IB, 32 points.

BSc Computer Science with Mathematics
UCAS Code: G5G1 • Mode: 3 Years Major/Minor

Qualifications/Requirements: GCE, A level grades: ABB. Maths at A level. AGNVQ, Distinction (in specific programmes) with A/AS. ND/C, Higher National (in specific programmes). SQAH, Grades AAABB. SQAV, Higher National including specific subject(s). IB, 32 points.

MEng Computer Science
UCAS Code: G503 • Mode: 4 Years Single Subject

Qualifications/Requirements: GCE, A level grades: ABB. Maths. AGNVQ, Distinction (in specific programmes) with A/AS. ND/C, Higher National (in specific programmes). SQAH, Grades AAABB. SQAV, Higher National including specific subject(s). IB, 32 points.

MEng Computer Science with Study in Continental Europe
UCAS Code: G501 • Mode: 4 Years Single Subject

Qualifications/Requirements: GCE, A level grades: ABB. Maths. AGNVQ, Distinction (in specific programmes) with A/AS. ND/C, Higher National (in specific programmes). SQAH, Grades AAABB. SQAV, Higher National including specific subject(s). IB, 32 points.

UNIVERSITY OF THE WEST OF ENGLAND, BRISTOL B80

BA Business Decision Analysis
UCAS Code: G520 • Mode: 3/4 Years Single Subject

Qualifications/Requirements: GCE, A/AS: 16 points. AGNVQ, Merit (in specific programmes). ND/C, 4 Merits (in specific programmes). SQAH, Grades BBB including specific subject(s). SQAV, Individual consideration. IB, 24 points.

BA Business Information Systems
UCAS Code: G562 • Mode: 4 Years Single Subject

Qualifications/Requirements: GCE, A/AS: 16 points. AGNVQ, Merit (in specific programmes). ND/C, Merit overall. SQAH, Grades BBB. SQAV, Individual consideration. IB, 24 points.

BA English and Information Systems
UCAS Code: QG37 • Mode: 3 Years Equal Combination

Qualifications/Requirements: GCE, A/AS: 18-20 points. English Literature. AGNVQ, Merit (in specific programmes) with A/AS. ND/C, 4 Merits and 2 Distinctions (in specific programmes). SQAH, Grades BBB including specific subject(s). SQAV, Individual consideration. IB, 28 points.

BA Geography and Information Systems
UCAS Code: LG87 • Mode: 3 Years Equal Combination

Qualifications/Requirements: GCE, A/AS: 18 points. Geography. AGNVQ, Merit with A/AS or Distinction. ND/C, 5 Merits and 1 Distinction. SQAH, Grades BBBC. SQAV, Individual consideration. IB, 26 points.

BA Information Systems Analysis
UCAS Code: G710 • Mode: 4 Years Single Subject

Qualifications/Requirements: GCE, A/AS: 16 points. AGNVQ, Merit. ND/C, Merit overall. SQAH, Grades BBC including specific subject(s). SQAV, Individual consideration. IB, 24 points.

BA Information Systems and Drama
UCAS Code: GW74 • Mode: 3 Years Equal Combination

Qualifications/Requirements: GCE, A/AS: 20 points. ND/C, 4 Merits and 2 Distinctions. SQAH, Grades BBBB. SQAV, Individual consideration. IB, 28 points.

BA Information Systems and Linguistics
UCAS Code: GQ71 • Mode: 3 Years Equal Combination

Qualifications/Requirements: GCE, A/AS: 16-18 points. AGNVQ, Merit with A/AS or Distinction. ND/C, 6 Merits. SQAH, Grades BBB. IB, 24 points.

BA Information Systems and Social Science
UCAS Code: G5L3 • Mode: 3 Years Major/Minor

Qualifications/Requirements: GCE, A/AS: 14-16 points. AGNVQ, Merit. ND/C, 5 to 6 Merits. SQAH, Grades BBC. SQAV, Individual consideration. IB, 24 points.

BA Information Systems, EFL and French
UCAS Code: G5Q3 • Mode: 3 Years Major/Minor

Qualifications/Requirements: GCE, A/AS: 18-22 points. AGNVQ, Merit with A level. ND/C, 4 to 5 Merits. SQAH, Grades BBBC. SQAV, Individual consideration. IB, 28 points.

BA Information Systems, EFL and German
UCAS Code: G5QH • Mode: 3 Years Major/Minor

Qualifications/Requirements: GCE, A/AS: 18-22 points. AGNVQ, Merit with A level. ND/C, 4 to 5 Merits. SQAH, Grades BBBC including specific subject(s). SQAV, Individual consideration. IB, 28 points.

BA Information Systems, EFL and Spanish
UCAS Code: G5QJ • Mode: 3 Years Major/Minor

Qualifications/Requirements: GCE, A/AS: 18-22 points. AGNVQ, Merit with A level. ND/C, 4 to 5 Merits. SQAH, Grades BBBC. SQAV, Individual consideration. IB, 28 points.

BA Information Systems, French and German
UCAS Code: G5T9 • Mode: 3 Years Major/Minor

Qualifications/Requirements: GCE, A/AS: 18-22 points. French or German at A level. AGNVQ, Merit with A level. ND/C, 4 to 5 Merits. SQAH, Grades BBBC. SQAV, Individual consideration. IB, 28 points.

BA Information Systems, French and Spanish
UCAS Code: G5TX • Mode: 3 Years Major/Minor

Qualifications/Requirements: GCE, A/AS: 18-22 points. French or Spanish at A level. AGNVQ, Merit with A level. ND/C, 4 to 5 Merits. SQAH, Grades BBBC. SQAV, Individual consideration. IB, 28 points.

BA Information Systems, German and Spanish
UCAS Code: G5TY • Mode: 3 Years Major/Minor

Qualifications/Requirements: GCE, A/AS: 18-22 points. German or Spanish at A level. AGNVQ, Merit with A level. ND/C, 4 to 5 Merits. SQAH, Grades BBBC including specific subject(s). SQAV, Individual consideration. IB, 28 points.

BA Law and Information Systems
UCAS Code: MG35 • Mode: 3 Years Equal Combination

Qualifications/Requirements: GCE, A/AS: 18-20 points. AGNVQ, Merit with A/AS Distinction. ND/C, 5 Merits and 1 Distinction. SQAH, Grades BBBC. SQAV, Individual consideration. IB, 26 points.

BA Marketing and Information Systems
UCAS Code: NG55 • Mode: 3 Years Equal Combination

Qualifications/Requirements: GCE, A/AS: 16 points. AGNVQ, Merit. ND/C, 6 Merits. SQAH, Grades BBB. SQAV, Individual consideration. IB, 24 points.

BSc Biology and Information Technology in Science
UCAS Code: CG15 • Mode: 3/4 Years Equal Combination

Qualifications/Requirements: GCE, A/AS: 12 points. Biology and Chemistry. AGNVQ, Merit. ND/C, 4 Merits (in specific programmes). SQAH, Grades BCC including specific subject(s). SQAV, Individual consideration. IB, 24 points.

BSc Chemistry and Information Technology in Science
UCAS Code: FG15 • Mode: 3/4 Years Equal Combination

Qualifications/Requirements: GCE, A/AS: 10 points. Chemistry. AGNVQ, Merit. ND/C, 3 Merits. SQAH, Grades BB including specific subject(s). SQAV, National including specific subject(s). IB, 24 points.

BSc Computer Science
UCAS Code: G500 • Mode: 4 Years Single Subject

Qualifications/Requirements: GCE, A/AS: 16 points. Computing, Maths or Science. AGNVQ, Merit (in specific programmes). ND/C, Merit overall. SQAH, Grades BBB. SQAV, National including specific subject(s). IB, 24 points.

BSc Computing (Foundation)
UCAS Code: GG6R • Mode: 4/5 Years Equal Combination

Qualifications/Requirements: GCE, A/AS: 2 points. AGNVQ, Pass (in specific programmes). ND/C, National (in specific programmes). SQAH, Grades C. SQAV, Individual consideration. IB, 24 points.

BSc Computing and Information Systems
UCAS Code: G501 • Mode: 4 Years Single Subject

Qualifications/Requirements: GCE, A/AS: 16 points. AGNVQ, Merit (in specific programmes). ND/C, Merit overall. SQAH, Grades BBB. SQAV, Individual consideration. IB, 24 points.

BSc Computing for Real Time Systems
UCAS Code: GG67 • Mode: 4 Years Equal Combination

Qualifications/Requirements: GCE, A/AS: 16 points. Maths, Computing or Science. AGNVQ, Merit (in specific programmes). ND/C, Merit overall (in specific programmes). SQAH, Grades BBB including specific subject(s). SQAV, National including specific subject(s). IB, 24 points.

BSc Environmental Science and Information Technology in Science
UCAS Code: FG95 • Mode: 3/4 Years Equal Combination

Qualifications/Requirements: GCE, A/AS: 6 points. Science. AGNVQ, Pass in Science. ND/C, National (in specific programmes). SQAH, Grades CCC including specific subject(s). SQAV, Individual consideration. IB, 24 points.

BSc Information Systems and Psychology
UCAS Code: GL77 • Mode: 3 Years Equal Combination

Qualifications/Requirements: GCE, A/AS: 20 points. AGNVQ, Distinction. ND/C, 4 Merits and 2 Distinctions. SQAH, Grades BBBB. SQAV, Individual consideration. IB, 28 points.

BSc Information Technology (1 year top-up post Dip)
UCAS Code: G560 • Mode: 1 Year Single Subject

Qualifications/Requirements: Please refer to prospectus.

BSc Psychology and Information Technology in Science
UCAS Code: CG85 • Mode: 3/4 Years Equal Combination

Qualifications/Requirements: GCE, A/AS: 10 points. Science. AGNVQ, Merit. ND/C, 3 Merits (in specific programmes). SQAH, Grades BB including specific subject(s). SQAV, Individual consideration. IB, 24 points.

BSc Statistics and Information Systems
UCAS Code: GG45 • Mode: 3/4 Years Equal Combination

Qualifications/Requirements: GCE, A/AS: 16 points. Maths or Statistics. AGNVQ, Merit with A level. ND/C, Merit overall. SQAH, Grades BBB including specific subject(s). SQAV, National including specific subject(s). IB, 24 points.

HND Business Decision Analysis
UCAS Code: 025G • Mode: 2 Years Single Subject

Qualifications/Requirements: GCE, A/AS: 6 points. AGNVQ, Pass (in specific programmes). ND/C, 3 Merits (in specific programmes). SQAH, Grades CC including specific subject(s). SQAV, Individual consideration. IB, 24 points.

HND Computing
UCAS Code: 105G • Mode: 2 Years Single Subject

Qualifications/Requirements: GCE, A/AS: 6 points. Computing, Maths or Science. AGNVQ, Pass (in specific programmes). ND/C, 3 Merits. SQAH, Grades CCC. SQAV, National including specific subject(s). IB, 24 points.

HND Computing Studies (Bridgwater College)
UCAS Code: 005G • Mode: 2 Years Single Subject

Qualifications/Requirements: GCE, A/AS: 4 points. AGNVQ, Pass with 6 additional units or with A level. ND/C, 2 Merits. SQAH, Grades CC. SQAV, National including specific subject(s). IB, 24 points.

HND Information Systems
UCAS Code: 125G • Mode: 2 Years Single Subject

Qualifications/Requirements: GCE, A/AS: 6 points. AGNVQ, Pass (in specific programmes). ND/C, 3 Merits. SQAH, Grades CC. SQAV, Individual consideration. IB, 24 points.

BRUNEL UNIVERSITY B84

BSc Computer Science
UCAS Code: G502 • Mode: 3 Years Single Subject

Qualifications/Requirements: GCE, A level grades: BBB-BBC. Science or Maths. AGNVQ, Distinction in Information Technology with 4 additional units or with A level. ND/C, Merit overall and 3 Distinctions (in specific programmes). SQAH, Grades BBBCC including specific subject(s). SQAV, Individual consideration. IB, 30 points.

BSc Computer Science (4 years thick SW)
UCAS Code: G507 • Mode: 4 Years Single Subject

Qualifications/Requirements: GCE, A level grades: BBB-BBC. Science or Maths. AGNVQ, Distinction in Information Technology with 4 additional units or with A level. ND/C, Merit overall and 3 Distinctions (in specific programmes). SQAH, Grades BBBCC including specific subject(s). SQAV, Individual consideration. IB, 30 points.

BSc Computer Science (4 years thin SW)
UCAS Code: G500 • Mode: 4 Years Single Subject

Qualifications/Requirements: GCE, A level grades: BBB-BBC. Science or Maths. AGNVQ, Distinction in Information Technology with 4 additional units or with A level. ND/C, Merit overall and 3 Distinctions (in specific programmes). SQAH, Grades BBBCC including specific subject(s). SQAV, Individual consideration. IB, 30 points.

BSc Information Systems
UCAS Code: G523 • Mode: 3 Years Single Subject

Qualifications/Requirements: GCE, A level grades: BBB-BBC. AGNVQ, Distinction (in specific programmes) with 4 additional units or with A level. ND/C, Merit overall and 3 Distinctions (in specific programmes). SQAH, Grades BBBCC. SQAV, Individual consideration. IB, 30 points.

BSc Information Systems (4 years thick SW)
UCAS Code: G524 • Mode: 4 Years Single Subject

Qualifications/Requirements: GCE, A level grades: BBB-BBC. AGNVQ, Distinction (in specific programmes) with 4 additional units or with A level. ND/C, Merit overall and 3 Distinctions (in specific programmes). SQAH, Grades BBBCC. SQAV, Individual consideration. IB, 30 points.

BSc Information Systems (4 years thin SW)
UCAS Code: G522 • Mode: 4 Years Single Subject

Qualifications/Requirements: GCE, A level grades: BBB-BBC. AGNVQ, Distinction (in specific programmes) with 4 additional units or with A level. ND/C, Merit overall and 3 Distinctions (in specific programmes). SQAH, Grades BBBCC. SQAV, Individual consideration. IB, 30 points.

Mod Computer Studies/Accounting
UCAS Code: G5N4 • Mode: 3 Years Major/Minor

Qualifications/Requirements: GCE, A/AS: 18 points. AGNVQ, Distinction. ND/C, Merit overall (in specific programmes). SQAH, Grades BCCC including specific subject(s). SQAV, Individual consideration. IB, 26 points.

Mod Computer Studies/Business Studies
UCAS Code: N1G5 • Mode: 3 Years Major/Minor

Qualifications/Requirements: GCE, A/AS: 18 points. Economics. AGNVQ, Distinction. ND/C, Merit overall (in specific programmes). SQAH, Grades BCCC including specific subject(s). SQAV, Individual consideration. IB, 26 points.

Mod Sport Sciences/Computer Studies
UCAS Code: B6GM • Mode: 3 Years Major/Minor

Qualifications/Requirements: GCE, A/AS: 20 points. AGNVQ, Distinction. ND/C, 6 Distinctions. SQAH, Grades BBCCC. SQAV, Individual consideration. IB, 28 points.

HND Business Information Technology
UCAS Code: 265G • Mode: 2 Years Single Subject

Qualifications/Requirements: GCE, A/AS: 8 points. Business Studies or Computing. AGNVQ, Merit in Business. ND/C, Merit overall (in specific programmes). SQAH, Grades CCCC. SQAV, Individual consideration. IB, 26 points.

HND Computing (Multi-Media Production)
UCAS Code: 45PG • Mode: 2 Years Equal Combination

Qualifications/Requirements: GCE, A/AS: 8 points. AGNVQ, Merit. ND/C, Merit overall (in specific programmes). SQAH, Grades BCCC including specific subject(s). SQAV, Individual consideration. IB, 24 points.

THE UNIVERSITY OF BUCKINGHAM B90

BSc Information Systems
UCAS Code: G500 • Mode: 2 Years Single Subject

Qualifications/Requirements: GCE, A/AS: 12 points. AGNVQ, Merit. ND/C, 5 Merits. SQAH, Grades CCCC. SQAV, Individual consideration. IB, 24 points.

BSc Information Systems with Accounting
UCAS Code: G5N4 • Mode: 2 Years Major/Minor

Qualifications/Requirements: GCE, A/AS: 12 points. AGNVQ, Merit. ND/C, 3 Merits and 2 Distinctions. SQAH, Grades CCCC. SQAV, Individual consideration. IB, 24 points.

BSc Information Systems with Business Studies
UCAS Code: G5N1 • Mode: 2 Years Major/Minor

Qualifications/Requirements: GCE, A/AS: 12 points. AGNVQ, Merit. ND/C, 5 Merits. SQAH, Grades CCCC. SQAV, Individual consideration. IB, 24 points.

BSc Information Systems with French
UCAS Code: G5R1 • Mode: 2 Years Major/Minor

Qualifications/Requirements: GCE, A/AS: 12 points. AGNVQ, Merit. ND/C, 5 Merits. SQAH, Grades CCCC. SQAV, Individual consideration. IB, 24 points.

BSc Information Systems with Operations Management
UCAS Code: G5N2 • Mode: 2 Years Major/Minor

Qualifications/Requirements: GCE, A/AS: 12 points. AGNVQ, Merit. ND/C, 5 Merits. SQAH, Grades BCCC. SQAV, Individual consideration. IB, 24 points.

BSc Information Systems with Spanish
UCAS Code: G5R4 • Mode: 2 Years Major/Minor

Qualifications/Requirements: GCE, A/AS: 12 points. AGNVQ, Merit. ND/C, 5 Merits. SQAH, Grades CCCC. SQAV, Individual consideration. IB, 24 points.

BUCKINGHAMSHIRE CHILTERNS UNIVERSITY COLLEGE B94

BSc Computer Engineering
UCAS Code: G501 • Mode: 3 Years Single Subject

Qualifications/Requirements: GCE, A/AS: 8-10 points. AGNVQ, Merit. ND/C, Merit overall. SQAH, Grades CCCC. SQAV, Individual consideration. IB, Individual consideration.

BSc Computer Engineering (1 or 2 years conv to Degree)
UCAS Code: G502 • Mode: 1/2 Years Single Subject

Qualifications/Requirements: ND/C, Higher National. SQAV, Individual consideration.

BSc Computer Engineering with Image Processing
UCAS Code: G5W2 • Mode: 3 Years Major/Minor

Qualifications/Requirements: GCE, A/AS: 8-10 points. AGNVQ, Merit. ND/C, Merit overall. SQAH, Grades CCCC. SQAV, Individual consideration. IB, Individual consideration.

BSc Computer Engineering with Management

UCAS Code: G5N1 • Mode: 3 Years Major/Minor

Qualifications/Requirements: GCE, A/AS: 8-10 points. AGNVQ, Merit. ND/C, Merit overall. SQAH, Grades CCCC. SQAV, Individual consideration. IB, Individual consideration.

BSc Computer Engineering with Marketing

UCAS Code: G5N5 • Mode: 3 Years Major/Minor

Qualifications/Requirements: GCE, A/AS: 8-10 points. AGNVQ, Merit. ND/C, Merit overall. SQAH, Grades CCCC. SQAV, Individual consideration. IB, Individual consideration.

BSc Computing

UCAS Code: G500 • Mode: 3 Years Single Subject

Qualifications/Requirements: GCE, A/AS: 8-10 points. AGNVQ, Merit. ND/C, Merit overall. SQAH, Grades CCCC. SQAV, Individual consideration. IB, Individual consideration.

BSc Computing (1/2 years)

UCAS Code: G503 • Mode: 1/2 Years Single Subject

Qualifications/Requirements: ND/C, Higher National. SQAV, Higher National.

BSc Computing and Psychology

UCAS Code: GL57 • Mode: 3 Years Equal Combination

Qualifications/Requirements: GCE, A/AS: 8-10 points. AGNVQ, Merit. ND/C, Merit overall. SQAH, Grades CCCC. SQAV, Individual consideration. IB, 27 points.

BSc Computing with Image Processing

UCAS Code: G5WF • Mode: 3 Years Major/Minor

Qualifications/Requirements: GCE, A/AS: 8-10 points. AGNVQ, Merit. ND/C, Merit overall. SQAH, Grades CCCC. SQAV, Individual consideration. IB, Individual consideration.

BSc Computing with Management

UCAS Code: G5NC • Mode: 3 Years Major/Minor

Qualifications/Requirements: GCE, A/AS: 8-10 points. AGNVQ, Merit. ND/C, Merit overall. SQAH, Grades CCCC. SQAV, Individual consideration. IB, Individual consideration.

BSc Computing with Marketing

UCAS Code: G5NM • Mode: 3 Years Major/Minor

Qualifications/Requirements: GCE, A/AS: 8-10 points. AGNVQ, Merit. ND/C, Merit overall. SQAH, Grades CCCC. SQAV, Individual consideration. IB, Individual consideration.

BSc Computing with Multimedia

UCAS Code: G5P4 • Mode: 3 Years Major/Minor

Qualifications/Requirements: GCE, A/AS: 8-10 points. AGNVQ, Merit. ND/C, Merit overall. SQAH, Grades CCCC. SQAV, Individual consideration. IB, Individual consideration.

HND Computing

UCAS Code: 105G • Mode: 2 Years Single Subject

Qualifications/Requirements: GCE, A/AS: 4-6 points. AGNVQ, Pass. ND/C, National. SQAH, Grades CCC. SQAV, Individual consideration. IB, Individual consideration.

CAMBRIDGE UNIVERSITY C05

BA Computer Science

UCAS Code: G500 • Mode: 3 Years Single Subject

Qualifications/Requirements: GCE, A level grades: AAA-AAB. Maths at A level. AGNVQ, Pass. ND/C, Individual consideration. SQAH, CSYS required. SQAV, Individual consideration. IB, Individual consideration.

CANTERBURY CHRIST CHURCH UNIVERSITY COLLEGE C10

BSc Business Information Management

UCAS Code: G520 • Mode: 3 Years Single Subject

Qualifications/Requirements: GCE, A level grades: CC. AGNVQ, Merit. ND/C, Merit overall. SQAH, Individual consideration. SQAV, Individual consideration. IB, 24 points.

BSc Computing

UCAS Code: G500 • Mode: 3 Years Single Subject

Qualifications/Requirements: GCE, A level grades: DD. AGNVQ, Merit. ND/C, Merit overall. SQAH, Individual consideration. SQAV, Individual consideration. IB, 24 points.

Mod Art with Information Technology

UCAS Code: W1G5 • Mode: 3 Years Major/Minor

Qualifications/Requirements: GCE, A level grades: DD. Art at A level. AGNVQ, Merit in Art & Design. ND/C, Merit overall. SQAH, Individual consideration. SQAV, Individual consideration. IB, 24 points.

Mod Information Technology and Marketing

UCAS Code: GN55 • Mode: 3 Years Equal Combination

Qualifications/Requirements: GCE, A level grades: CC. AGNVQ, Merit. ND/C, Merit overall. SQAH, Individual consideration. SQAV, Individual consideration. IB, 24 points.

Mod Information Technology with American Studies

UCAS Code: G5Q4 • Mode: 3 Years Major/Minor

Qualifications/Requirements: GCE, A level grades: CC. AGNVQ, Merit. ND/C, Merit overall. SQAH, Individual consideration. SQAV, Individual consideration. IB, 24 points.

Mod Information Technology with Art

UCAS Code: G5W1 • Mode: 3 Years Major/Minor

Qualifications/Requirements: GCE, A level grades: DD. Art at A level. AGNVQ, Merit in Art & Design. ND/C, Merit overall. SQAH, Individual consideration. SQAV, Individual consideration. IB, 24 points.

Mod Information Technology with Business Studies

UCAS Code: G5N1 • Mode: 3 Years Major/Minor

Qualifications/Requirements: GCE, A level grades: CC. AGNVQ, Merit. ND/C, Merit overall. SQAH, Individual consideration. SQAV, Individual consideration. IB, 24 points.

Mod Information Technology with Early Childhood Studies

UCAS Code: G5X9 • Mode: 3 Years Major/Minor

Qualifications/Requirements: GCE, A level grades: CC. AGNVQ, Merit. ND/C, Merit overall. SQAH, Individual consideration. SQAV, Individual consideration. IB, 24 points.

Mod Information Technology with English

UCAS Code: G5Q3 • Mode: 3 Years Major/Minor

Qualifications/Requirements: GCE, A level grades: CC. English at A level. AGNVQ, Merit with A level. ND/C, Merit overall. SQAH, Individual consideration. SQAV, Individual consideration. IB, 24 points.

Mod Information Technology with French

UCAS Code: G5R1 • Mode: 3 Years Major/Minor

Qualifications/Requirements: GCE, A level grades: DD. French at A level. AGNVQ, Merit with A level. ND/C, Merit overall. SQAH, Individual consideration. SQAV, Individual consideration. IB, 24 points.

Mod Information Technology with Geography

UCAS Code: G5L8 • Mode: 3 Years Major/Minor

Qualifications/Requirements: GCE, A level grades: DD. AGNVQ, Merit. ND/C, Merit overall. SQAH, Individual consideration. SQAV, Individual consideration. IB, 24 points.

Mod Information Technology with History

UCAS Code: G5V1 • Mode: 3 Years Major/Minor

Qualifications/Requirements: GCE, A level grades: DD. History at A level. AGNVQ, Merit with A level. ND/C, Merit overall. SQAH, Individual consideration. SQAV, Individual consideration. IB, 24 points.

Mod Information Technology with Marketing

UCAS Code: G5N5 • Mode: 3 Years Major/Minor

Qualifications/Requirements: GCE, A level grades: CC. AGNVQ, Merit. ND/C, Merit overall. SQAH, Individual consideration. SQAV, Individual consideration. IB, 24 points.

Mod Information Technology with Mathematics

UCAS Code: G5G1 • Mode: 3 Years Major/Minor

Qualifications/Requirements: GCE, A level grades: DD. Maths at A level. AGNVQ, Individual consideration. ND/C, Individual consideration. SQAH, Individual consideration. SQAV, Individual consideration. IB, 24 points.

Mod Information Technology with Media & Cultural Studies

UCAS Code: G5P4 • Mode: 3 Years Major/Minor

Qualifications/Requirements: GCE, A level grades: CC. AGNVQ, Merit. ND/C, Merit overall. SQAH, Individual consideration. SQAV, Individual consideration. IB, 24 points.

Mod Information Technology with Music

UCAS Code: G5W3 • Mode: 3 Years Major/Minor

Qualifications/Requirements: GCE, A level grades: CC. Music at A level. AGNVQ, Merit with A level. ND/C, Merit overall. SQAH, Individual consideration. SQAV, Individual consideration. IB, 24 points.

Mod Information Technology with Religious Studies

UCAS Code: G5V8 • Mode: 3 Years Major/Minor

Qualifications/Requirements: GCE, A level grades: DD. AGNVQ, Merit. ND/C, Merit overall. SQAH, Individual consideration. SQAV, Individual consideration. IB, 24 points.

Mod Information Technology with Science (Natural)

UCAS Code: G5YC • Mode: 3 Years

Qualifications/Requirements: GCE, A level grades: DD. Science at A level. AGNVQ, Merit in Science. ND/C, Merit overall. SQAH, Individual consideration. SQAV, Individual consideration. IB, 24 points.

Mod Information Technology with Social Science

UCAS Code: G5L3 • Mode: 3 Years Major/Minor

Qualifications/Requirements: GCE, A level grades: CC. AGNVQ, Merit. ND/C, Merit overall. SQAH, Individual consideration. SQAV, Individual consideration. IB, 24 points.

Mod Information Technology with Statistics & Operational Research

UCAS Code: G5G4 • Mode: 3 Years Major/Minor

Qualifications/Requirements: GCE, A level grades: DD. Maths at A level. AGNVQ, Individual consideration. ND/C, Individual consideration. SQAH, Individual consideration. SQAV, Individual consideration. IB, 24 points.

Mod Information Technology with Tourism & Leisure Studies
UCAS Code: G5P7 • Mode: 3 Years Major/Minor
Qualifications/Requirements: GCE, A level grades: CC. AGNVQ, Merit. ND/C, Merit overall. SQAH, Individual consideration. SQAV, Individual consideration. IB, 24 points.

Mod IT with Science (Natural) (including foundation)
UCAS Code: G5YD • Mode: 4 Years
Qualifications/Requirements: Please refer to prospectus.

Mix American Studies and Information Technology
UCAS Code: GQ54 • Mode: 3 Years Equal Combination
Qualifications/Requirements: GCE, A level grades: CC. AGNVQ, Merit. ND/C, Merit overall. SQAH, Individual consideration. SQAV, Individual consideration. IB, 24 points.

Mix Art and Information Technology
UCAS Code: WG15 • Mode: 3 Years Equal Combination
Qualifications/Requirements: GCE, A level grades: DD. Art at A level. AGNVQ, Merit in Art & Design. ND/C, Merit overall. SQAH, Individual consideration. SQAV, Individual consideration. IB, 24 points.

Mix Business Studies and Information Technology
UCAS Code: NG15 • Mode: 3 Years Equal Combination
Qualifications/Requirements: GCE, A level grades: CC. AGNVQ, Merit. ND/C, Merit overall. SQAH, Individual consideration. SQAV, Individual consideration. IB, 24 points.

Mix English and Information Technology
UCAS Code: QG35 • Mode: 3 Years Equal Combination
Qualifications/Requirements: GCE, A level grades: CC. English at A level. AGNVQ, Merit with A level. ND/C, Merit overall. SQAH, Individual consideration. SQAV, Individual consideration. IB, 24 points.

Mix Geography and Information Technology
UCAS Code: GL58 • Mode: 3 Years Equal Combination
Qualifications/Requirements: GCE, A level grades: DD. AGNVQ, Merit. ND/C, Merit overall. SQAH, Individual consideration. SQAV, Individual consideration. IB, 24 points.

Mix Information Technology and Mathematics
UCAS Code: GG15 • Mode: 3 Years Equal Combination
Qualifications/Requirements: GCE, A level grades: DD. Maths at A level. AGNVQ, Individual consideration. ND/C, Individual consideration. SQAH, Individual consideration. SQAV, Individual consideration. IB, 24 points.

Mix Information Technology and Media & Cultural Studies
UCAS Code: GP54 • Mode: 3 Years Equal Combination
Qualifications/Requirements: GCE, A level grades: CC. AGNVQ, Merit. ND/C, Merit overall. SQAH, Individual consideration. SQAV, Individual consideration. IB, 24 points.

Mix Information Technology and Music
UCAS Code: WG35 • Mode: 3 Years Equal Combination
Qualifications/Requirements: GCE, A level grades: CC. Music at A level. AGNVQ, Merit with A level. ND/C, Merit overall. SQAH, Individual consideration. SQAV, Individual consideration. IB, 24 points.

Mix Information Technology and Religious Studies
UCAS Code: GV58 • Mode: 3 Years Equal Combination
Qualifications/Requirements: GCE, A level grades: DD. AGNVQ, Merit. ND/C, Merit overall. SQAH, Individual consideration. SQAV, Individual consideration. IB, 24 points.

Mix Information Technology and Science (Natural)
UCAS Code: GY51 • Mode: 3 Years
Qualifications/Requirements: GCE, A level grades: DD. Science at A level. AGNVQ, Individual consideration. ND/C, Individual consideration. SQAH, Individual consideration. SQAV, Individual consideration. IB, 24 points.

Mix Information Technology and Social Science
UCAS Code: GL53 • Mode: 3 Years Equal Combination
Qualifications/Requirements: GCE, A level grades: CC. AGNVQ, Merit. ND/C, Merit overall. SQAH, Individual consideration. SQAV, Individual consideration. IB, 24 points.

Mix Information Technology and Statistics & Operational Research
UCAS Code: GG45 • Mode: 3 Years Equal Combination
Qualifications/Requirements: GCE, A level grades: DD. Maths at A level. AGNVQ, Individual consideration. ND/C, Individual consideration. SQAH, Individual consideration. SQAV, Individual consideration. IB, 24 points.

Mix Information Technology and Tourism & Leisure Studies
UCAS Code: GP57 • Mode: 3 Years Equal Combination
Qualifications/Requirements: GCE, A level grades: CC. AGNVQ, Merit. ND/C, Merit overall. SQAH, Individual consideration. SQAV, Individual consideration. IB, 24 points.

HND Computing
UCAS Code: 005G • Mode: 2 Years Single Subject
Qualifications/Requirements: GCE, A level grades: D. AGNVQ, Pass. ND/C, Merit overall. SQAH, Individual consideration. SQAV, Individual consideration. IB, 24 points.

DipHE Business Information Management

UCAS Code: GN51 • Mode: 2 Years Equal Combination

Qualifications/Requirements: GCE, A level grades: C. AGNVQ, Merit. ND/C, Merit overall. SQAH, Individual consideration. SQAV, Individual consideration. IB, 24 points.

CARDIFF UNIVERSITY C15

BSc Computer Science

UCAS Code: G500 • Mode: 3 Years Single Subject

Qualifications/Requirements: GCE, A/AS: 22-24 points. AGNVQ, Merit with A/AS. ND/C, Merit overall and 3 Distinctions. SQAH, Grades AAABB. SQAV, Individual consideration. IB, Individual consideration.

BSc Computing and Physics

UCAS Code: FG35 • Mode: 3 Years Equal Combination

Qualifications/Requirements: GCE, A/AS: 22 points. Physics and Maths. AGNVQ, Individual consideration. ND/C, Individual consideration. SQAH, Individual consideration. SQAV, Individual consideration. IB, Individual consideration.

UNIVERSITY OF WALES INSTITUTE, CARDIFF C20

BSc Business Information Systems

UCAS Code: G5N1 • Mode: 3 Years Major/Minor

Qualifications/Requirements: GCE, A/AS: 10 points. AGNVQ, Merit in Information Technology. ND/C, Merit overall. SQAH, Grades CCCC. SQAV, Individual consideration. IB, Individual consideration.

HND Computing

UCAS Code: 105G • Mode: 2 Years Single Subject

Qualifications/Requirements: GCE, A/AS: 2 points. AGNVQ, Merit (in specific programmes). ND/C, 3 Merits. SQAH, Grades CC. SQAV, Individual consideration. IB, Individual consideration.

CARMARTHENSHIRE COLLEGE C22

HND Computing (Information Systems)

UCAS Code: 025G • Mode: 2 Years Single Subject

Qualifications/Requirements: AGNVQ, Individual consideration. ND/C, Individual consideration. SQAH, Individual consideration. SQAV, Individual consideration. IB, Individual consideration.

UNIVERSITY OF CENTRAL ENGLAND IN BIRMINGHAM C25

BA Business Administration with Business Information Systems

UCAS Code: N150 • Mode: 3 Years Single Subject

Qualifications/Requirements: GCE, A/AS: 14 points. AGNVQ, Distinction. ND/C, Merits and 3 Distinctions. SQAH, Grades CCCC. SQAV, Individual consideration. IB, 24 points.

BA Business Information and Systems

UCAS Code: GP52 • Mode: 3 Years Equal Combination

Qualifications/Requirements: GCE, A/AS: 12 points. AGNVQ, Merit. ND/C, Merit overall. SQAH, Grades CCCC. SQAV, Individual consideration. IB, 24 points.

BA Information Studies

UCAS Code: G523 • Mode: 3 Years Single Subject

Qualifications/Requirements: GCE, A/AS: 12 points. AGNVQ, Merit. ND/C, Merit overall. SQAH, Individual consideration. IB, 24 points.

BSc Computing

UCAS Code: G500 • Mode: 3 Years Single Subject

Qualifications/Requirements: GCE, A/AS: 12 points. AGNVQ, Merit. ND/C, Merit overall. SQAH, Individual consideration. SQAV, Individual consideration. IB, Individual consideration.

BSc Computing & Electronics

UCAS Code: GH56 • Mode: 3/4 Years Equal Combination

Qualifications/Requirements: GCE, A/AS: 16 points. AGNVQ, Merit. ND/C, 5 Merits. SQAH, Grades BBCCC. SQAV, Individual consideration. IB, 28 points.

BSc Computing and Electronics (Foundation Year)

UCAS Code: GH5P • Mode: 4/5 Years Equal Combination

Qualifications/Requirements: GCE, A/AS: 4 points. AGNVQ, Pass. ND/C, 1 Merit. SQAH, Grades CC. SQAV, Individual consideration. IB, 24 points.

BSc Industrial Information Technology

UCAS Code: G560 • Mode: 3/4 Years Single Subject

Qualifications/Requirements: GCE, A/AS: 16 points. AGNVQ, Merit. ND/C, 5 Merits. SQAH, Grades BBCCC. SQAV, Individual consideration. IB, 28 points.

BSc Information Systems

UCAS Code: G520 • Mode: 3 Years Single Subject

Qualifications/Requirements: GCE, A/AS: 12 points. AGNVQ, Merit. ND/C, Merit overall. SQAH, Individual consideration. SQAV, Individual consideration. IB, Individual consideration.

BEng Industrial Information Technology Foundation Year

UCAS Code: G568 • Mode: 4/5 Years Single Subject

Qualifications/Requirements: GCE, A/AS: 4 points. AGNVQ, Pass. ND/C, 1 Merit. SQAH, Grades CC. SQAV, Individual consideration. IB, 24 points.

HND Computer Studies
UCAS Code: 205G • Mode: 2 Years Single Subject

Qualifications/Requirements: GCE, A/AS: 4 points. AGNVQ, Pass. ND/C, 2 Merits. SQAH, Individual consideration. SQAV, Individual consideration. IB, Individual consideration.

HND Computing
UCAS Code: 005G • Mode: 2 Years Single Subject

Qualifications/Requirements: GCE, A/AS: 4 points. AGNVQ, Pass. ND/C, 2 Merits. SQAH, Individual consideration. SQAV, Individual consideration. IB, Individual consideration.

HND Computing and Electronics
UCAS Code: 65HG • Mode: 2/3 Years Equal Combination

Qualifications/Requirements: GCE, A/AS: 4 points. AGNVQ, Pass. ND/C, 1 Merit. SQAH, Grades CC. SQAV, Individual consideration. IB, 24 points.

UNIVERSITY OF CENTRAL LANCASHIRE C30

BA Accounting and Business Information Systems
UCAS Code: GN54 • Mode: 3 Years Equal Combination

Qualifications/Requirements: GCE, A/AS: 12 points. AGNVQ, Distinction. ND/C, Merit overall. SQAH, Grades BBC. SQAV, Individual consideration. IB, 26 points.

BA Business and Business Information Systems
UCAS Code: NG15 • Mode: 3 Years Equal Combination

Qualifications/Requirements: GCE, A/AS: 14 points. AGNVQ, Distinction in Business. ND/C, Merit overall. SQAH, Grades BBBC. SQAV, Individual consideration. IB, 28 points.

BA Business Information Systems and Accounting
UCAS Code: GN5K • Mode: 3 Years Equal Combination

Qualifications/Requirements: GCE, A/AS: 12 points. AGNVQ, Merit in Business with 6 additional units or with A/AS. ND/C, Merit overall. SQAH, Grades BBB. SQAV, Individual consideration. IB, 26 points.

BA Business Information Systems and Business
UCAS Code: GN5C • Mode: 3 Years Equal Combination

Qualifications/Requirements: GCE, A/AS: 14 points. AGNVQ, Distinction in Business. ND/C, Merit overall. SQAH, Grades BBBC. SQAV, Individual consideration. IB, 28 points.

BA Business Information Systems and Management
UCAS Code: GN5D • Mode: 3 Years Equal Combination

Qualifications/Requirements: GCE, A/AS: 14 points. AGNVQ, Distinction in Business. ND/C, Merit overall. SQAH, Grades BBBC. SQAV, Individual consideration. IB, 28 points.

BA Business Information Systems and Statistics
UCAS Code: GG4M • Mode: 3 Years Equal Combination

Qualifications/Requirements: GCE, A/AS: 14 points. AGNVQ, Distinction in Business. ND/C, Merit overall. SQAH, Grades BBBC. SQAV, Individual consideration. IB, 28 points.

BA Computing and Accounting
UCAS Code: NG45 • Mode: 3 Years Equal Combination

Qualifications/Requirements: GCE, A/AS: 14 points. AGNVQ, Distinction (in specific programmes) with 6 additional units or with A/AS. ND/C, Merit overall. SQAH, Grades BBB. SQAV, Individual consideration. IB, 26 points.

BA Economics and Business Information Systems
UCAS Code: LG15 • Mode: 3 Years Equal Combination

Qualifications/Requirements: GCE, A/AS: 14 points. AGNVQ, Distinction (in specific programmes). ND/C, Merit overall. SQAH, Grades BBBC. SQAV, Individual consideration. IB, 28 points.

BA Management and Business Information Systems
UCAS Code: NG1M • Mode: 3 Years Equal Combination

Qualifications/Requirements: GCE, A/AS: 14 points. AGNVQ, Distinction in Business with 6 additional units or with A/AS. ND/C, Merit overall. SQAH, Grades BBBC. SQAV, Individual consideration. IB, 28 points.

BSc Applied Physics and Computing
UCAS Code: FG35 • Mode: 3 Years Equal Combination

Qualifications/Requirements: GCE, A/AS: 12 points. Maths and Physics at A level. AGNVQ, Merit in Science with 6 additional units or with A level. ND/C, Merit overall. SQAH, Grades BBB including specific subject(s). SQAV, Individual consideration. IB, 26 points.

BSc Astronomy and Computing
UCAS Code: FG55 • Mode: 3 Years Equal Combination

Qualifications/Requirements: GCE, A/AS: 14 points. Physics and Maths. AGNVQ, Distinction in Science with 6 additional units or with A level. ND/C, Merit overall. SQAH, Grades BBBC including specific subject(s). SQAV, Individual consideration. IB, 26 points.

BSc Business Computing
UCAS Code: GN51 • Mode: 3 Years Equal Combination

Qualifications/Requirements: GCE, A/AS: 14 points. AGNVQ, Distinction with 6 additional units or with A/AS. ND/C, Merit overall. SQAH, Grades BBCC. SQAV, Individual consideration. IB, 26 points.

BSc Business Information Technology (4 year sandwich)
UCAS Code: G5N1 • Mode: 4 Years Major/Minor

Qualifications/Requirements: GCE, A/AS: 14 points. AGNVQ, Distinction (in specific programmes) with 6 additional units or with A/AS. ND/C, Merit overall and 2 Distinctions. SQAH, Grades BBCC. SQAV, Individual consideration. IB, 26 points.

BSc Computing (3/4 years)
UCAS Code: G500 • Mode: 3/4 Years Single Subject

Qualifications/Requirements: GCE, A/AS: 14 points. AGNVQ, Distinction with 6 additional units or with A/AS. ND/C, Merit overall. SQAH, Grades BBCC. SQAV, Individual consideration. IB, 26 points.

BSc Computing and Business
UCAS Code: GNM1 • Mode: 3 Years Equal Combination

Qualifications/Requirements: GCE, A/AS: 16 points. AGNVQ, Distinction (in specific programmes) with 6 additional units or with A level. ND/C, Merit overall and 3 Distinctions. SQAH, Grades BBBC. SQAV, Individual consideration. IB, 28 points.

BSc Computing and Mathematics
UCAS Code: GG51 • Mode: 3 Years Equal Combination

Qualifications/Requirements: GCE, A/AS: 14 points. Maths. AGNVQ, Distinction (in specific programmes) with 6 additional units with A level. ND/C, Merit overall. SQAH, Grades BBBC including specific subject(s). SQAV, Individual consideration. IB, 26 points.

BSc Computing and Media Technology
UCAS Code: GP5K • Mode: 3 Years Equal Combination

Qualifications/Requirements: GCE, A/AS: 14 points. AGNVQ, Distinction in Science with 6 additional units or with A/AS. ND/C, Merit overall. SQAH, Grades BBBC. SQAV, Individual consideration. IB, 26 points.

BSc Computing and Psychology
UCAS Code: CG85 • Mode: 3 Years Equal Combination

Qualifications/Requirements: GCE, A/AS: 16 points. AGNVQ, Distinction (in specific programmes) with 6 additional units or with A level. ND/C, Merit overall and 3 Distinctions. SQAH, Grades BBBC. SQAV, Individual consideration. IB, 28 points.

BSc Information Systems Design
UCAS Code: G520 • Mode: 3/4 Years Single Subject

Qualifications/Requirements: GCE, A/AS: 14 points. AGNVQ, Distinction with 6 additional units or with A/AS. ND/C, Merit overall. SQAH, Grades BBCC. IB, 26 points.

BSc Management Information Systems
UCAS Code: G521 • Mode: 1 Year Single Subject

Qualifications/Requirements: ND/C, Higher National.

BSc Mathematics and Business Information Systems
UCAS Code: GG1M • Mode: 3 Years Equal Combination

Qualifications/Requirements: GCE, A/AS: 12 points. Maths. AGNVQ, Merit (in specific programmes) with 6 additional units or with A level. ND/C, Merit overall. SQAH, Grades BBB including specific subject(s). SQAV, Individual consideration. IB, 26 points.

BSc Mathematics and Computing
UCAS Code: GG15 • Mode: 3 Years Equal Combination

Qualifications/Requirements: GCE, A/AS: 12 points. Maths. AGNVQ, Merit (in specific programmes) with 6 additional units or with A level. ND/C, Merit overall. SQAH, Grades BBB including specific subject(s). SQAV, Individual consideration. IB, 26 points.

BSc Statistics and Business Information Systems
UCAS Code: GG45 • Mode: 3 Years Equal Combination

Qualifications/Requirements: GCE, A/AS: 12 points. AGNVQ, Merit (in specific programmes) with 6 additional units or with A level. ND/C, Merit overall. SQAH, Grades BBB. SQAV, Individual consideration. IB, 24 points.

HND Business Information Technology
UCAS Code: 265G • Mode: 2 Years Single Subject

Qualifications/Requirements: GCE, A/AS: 6 points. AGNVQ, Merit. ND/C, 3 Merits. SQAH, Grades CC. SQAV, Individual consideration. IB, 24 points.

HND Computing
UCAS Code: 205G • Mode: 2 Years Single Subject

Qualifications/Requirements: GCE, A/AS: 4 points. AGNVQ, Merit. ND/C, 2 Merits. SQAH, Grades CC. SQAV, Individual consideration. IB, 24 points.

HND Computing (Information System Applications)
UCAS Code: 015G • Mode: 2 Years Single Subject

Qualifications/Requirements: GCE, A level grades: E. AGNVQ, Pass (in specific programmes). ND/C, National. SQAH, Grades CC. SQAV, Individual consideration. IB, 24 points.

HND Computing (Information Systems Design)
UCAS Code: 025G • Mode: 2 Years Single Subject

Qualifications/Requirements: GCE, A/AS: 4 points. AGNVQ, Merit. ND/C, 2 Merits. SQAH, Grades CC. SQAV, Individual consideration. IB, 24 points.

CHELTENHAM & GLOUCESTER COLLEGE OF HIGHER EDUCATION C50

BSc Business Computer Systems and Business Information Technology
UCAS Code: GNNC • Mode: 3 Years Equal Combination

Qualifications/Requirements: GCE, A/AS: 8-12 points. AGNVQ, Merit. ND/C, Merit overall. SQAH, Grades CCCC. SQAV, Individual consideration. IB, 24 points.

BSc Business Computer Systems and Computing
UCAS Code: GN5C • Mode: 3 Years Equal Combination

Qualifications/Requirements: GCE, A/AS: 8 points. AGNVQ, Merit. ND/C, Merit overall. SQAH, Grades CCCC. SQAV, Individual consideration. IB, 24 points.

BSc Business Computer Systems and Geography
UCAS Code: GFMW • Mode: 3 Years Equal Combination
Qualifications/Requirements: GCE, A/AS: 8-12 points. AGNVQ, Merit. ND/C, Merit overall. SQAH, Grades CCCC. SQAV, Individual consideration. IB, 24 points.

BSc Business Computer Systems and Geology
UCAS Code: GF5P • Mode: 3 Years Equal Combination
Qualifications/Requirements: GCE, A/AS: 8 points. AGNVQ, Merit. ND/C, Merit overall. SQAH, Grades CCCC. SQAV, Individual consideration. IB, 26 points.

BSc Business Computer Systems and Multimedia
UCAS Code: G520 • Mode: 3 Years Single Subject
Qualifications/Requirements: GCE, A/AS: 8-12 points. AGNVQ, Merit. ND/C, Merit overall. SQAH, Grades CCCC. SQAV, Individual consideration. IB, 24 points.

BSc Business Computer Systems and Physical Geography
UCAS Code: FG8M • Mode: 3 Years Equal Combination
Qualifications/Requirements: GCE, A/AS: 8-12 points. AGNVQ, Merit. ND/C, Merit overall. SQAH, Grades CCCC. SQAV, Individual consideration. IB, 26 points.

BSc Business Computer Systems and Psychology
UCAS Code: GLN7 • Mode: 3 Years Equal Combination
Qualifications/Requirements: GCE, A/AS: 8-12 points. AGNVQ, Merit. ND/C, Merit overall. SQAH, Grades CCCC. SQAV, Individual consideration. IB, 24 points.

BSc Business Computer Systems with Accounting & Financial Management
UCAS Code: NGH5 • Mode: 3 Years Equal Combination
Qualifications/Requirements: GCE, A/AS: 8 points. AGNVQ, Merit. ND/C, Merit overall. SQAH, Grades CCCC. SQAV, Individual consideration. IB, 26 points.

BSc Business Computer Systems with Business Information Technology
UCAS Code: G525 • Mode: 3 Years Single Subject
Qualifications/Requirements: GCE, A/AS: 8 points. AGNVQ, Merit. ND/C, Merit overall. SQAH, Grades CCCC. SQAV, Individual consideration. IB, 24 points.

BSc Business Computer Systems with Business Management
UCAS Code: G5ND • Mode: 3 Years Major/Minor
Qualifications/Requirements: GCE, A/AS: 8-12 points. AGNVQ, Merit in Business with 3 additional units. ND/C, Merit overall. SQAH, Grades CCCC. SQAV, Individual consideration. IB, 24 points.

BSc Business Computer Systems with Computing
UCAS Code: N1GM • Mode: 3 Years Major/Minor
Qualifications/Requirements: GCE, A/AS: 8 points. AGNVQ, Merit. ND/C, Merit overall. SQAH, Grades CCCC. SQAV, Individual consideration. IB, 24 points.

BSc Business Computer Systems with Environmental Policy
UCAS Code: G5F9 • Mode: 3 Years Major/Minor
Qualifications/Requirements: GCE, A/AS: 8 points. AGNVQ, Merit. ND/C, Merit overall. SQAH, Grades CCCC. SQAV, Individual consideration. IB, 24 points.

BSc Business Computer Systems with Financial Services Management
UCAS Code: G5NH • Mode: 3 Years Major/Minor
Qualifications/Requirements: GCE, A/AS: 8 points. AGNVQ, Merit. ND/C, Merit overall. SQAH, Grades CCCC. SQAV, Individual consideration. IB, 24 points.

BSc Business Computer Systems with Geography
UCAS Code: GFMV • Mode: 3 Years Equal Combination
Qualifications/Requirements: GCE, A/AS: 8 points. AGNVQ, Merit. ND/C, Merit overall. SQAH, Grades CCCC. SQAV, Individual consideration. IB, 24 points.

BSc Business Computer Systems with Geology
UCAS Code: G5FQ • Mode: 3 Years Major/Minor
Qualifications/Requirements: GCE, A/AS: 8 points. AGNVQ, Merit. ND/C, Merit overall. SQAH, Grades CCCC. SQAV, Individual consideration. IB, 26 points.

BSc Business Computer Systems with Hospitality Management (Hotel)
UCAS Code: G5NR • Mode: 3 Years Major/Minor
Qualifications/Requirements: GCE, A/AS: 8-12 points. AGNVQ, Merit. ND/C, Merit overall. SQAH, Grades CCCC. SQAV, Individual consideration. IB, 24 points.

BSc Business Computer Systems with Human Resource Management
UCAS Code: GNMC • Mode: 3 Years Equal Combination
Qualifications/Requirements: GCE, A/AS: 8-12 points. AGNVQ, Merit. ND/C, Merit overall. SQAH, Grades CCCC. SQAV, Individual consideration. IB, 26 points.

BSc Business Computer Systems with Leisure Management
UCAS Code: G5N7 • Mode: 3 Years Major/Minor
Qualifications/Requirements: GCE, A/AS: 8-12 points. AGNVQ, Merit. ND/C, Merit overall. SQAH, Grades CCCC. SQAV, Individual consideration. IB, 24 points.

BSc Business Computer Systems with Marketing Management
UCAS Code: NG55 • Mode: 3 Years Equal Combination
Qualifications/Requirements: GCE, A/AS: 8-12 points. AGNVQ, Merit. ND/C, Merit overall. SQAH, Grades CCCC. SQAV, Individual consideration. IB, 26 points.

BSc Business Computer Systems with Media Communications
UCAS Code: G5P4 • Mode: 3 Years Major/Minor
Qualifications/Requirements: GCE, A/AS: 8-12 points. AGNVQ, Merit. ND/C, Merit overall. SQAH, Grades CCCC. SQAV, Individual consideration. IB, 24 points.

BSc Business Computer Systems with Modern Languages
UCAS Code: G5T9 • Mode: 3 Years Major/Minor
Qualifications/Requirements: GCE, A/AS: 8-12 points. AGNVQ, Merit. ND/C, Merit overall. SQAH, Grades CCCC. SQAV, Individual consideration. IB, 24 points.

BSc Business Computer Systems with Multimedia
UCAS Code: G521 • Mode: 3 Years Single Subject
Qualifications/Requirements: GCE, A/AS: 8 points. AGNVQ, Merit. ND/C, Merit overall. SQAH, Grades CCCC. SQAV, Individual consideration. IB, 26 points.

BSc Business Computer Systems with Physical Geography
UCAS Code: G5FW • Mode: 3 Years Major/Minor
Qualifications/Requirements: GCE, A/AS: 8 points. AGNVQ, Merit. ND/C, Merit overall. SQAH, Grades CCCC. SQAV, Individual consideration. IB, 26 points.

BSc Business Computer Systems with Psychology
UCAS Code: G5LT • Mode: 3 Years Major/Minor
Qualifications/Requirements: GCE, A/AS: 8-12 points. AGNVQ, Merit. ND/C, Merit overall. SQAH, Grades CCCC. SQAV, Individual consideration. IB, 24 points.

BSc Business Computer Systems with Religious Studies
UCAS Code: G5V8 • Mode: 3 Years Major/Minor
Qualifications/Requirements: GCE, A/AS: 8 points. AGNVQ, Merit. ND/C, Merit overall. SQAH, Grades CCCC. SQAV, Individual consideration. IB, 24 points.

BSc Business Computer Systems with Women's Studies
UCAS Code: G5MX • Mode: 3 Years Major/Minor
Qualifications/Requirements: GCE, A/AS: 8-12 points. AGNVQ, Merit. ND/C, Merit overall. SQAH, Grades CCCC. SQAV, Individual consideration. IB, 24 points.

BSc Business Information Technoloogy with Financial Services Management
UCAS Code: G5N3 • Mode: 3 Years Major/Minor
Qualifications/Requirements: GCE, A/AS: 8-12 points. AGNVQ, Merit. ND/C, Merit overall. SQAH, Grades CCCC. SQAV, Individual consideration. IB, 24 points.

BSc Business Information Technoloogy with Hospitality Management (Catering)
UCAS Code: G5NT • Mode: 3 Years Major/Minor
Qualifications/Requirements: GCE, A/AS: 8-12 points. AGNVQ, Merit. ND/C, Merit overall. SQAH, Grades CCCC. SQAV, Individual consideration. IB, 24 points.

BSc Business Information Technoloogy with Hospitality Management (Hotel)
UCAS Code: G501 • Mode: 3 Years Single Subject
Qualifications/Requirements: GCE, A/AS: 8-12 points. AGNVQ, Merit. ND/C, Merit overall. SQAH, Grades CCCC. SQAV, Individual consideration. IB, 24 points.

BSc Business Information Technology with Accounting & Financial Management
UCAS Code: NGJ5 • Mode: 3 Years Equal Combination
Qualifications/Requirements: GCE, A/AS: 8-12 points. AGNVQ, Merit. ND/C, Merit overall. SQAH, Grades CCCC. SQAV, Individual consideration. IB, 26 points.

BSc Business Information Technology with Business Computer Systems
UCAS Code: G523 • Mode: 3 Years Single Subject
Qualifications/Requirements: GCE, A/AS: 8 points. AGNVQ, Merit. ND/C, Merit overall. SQAH, Grades CCCC. SQAV, Individual consideration. IB, 24 points.

BSc Business Information Technology with Human Resource Management
UCAS Code: GNMD • Mode: 3 Years Equal Combination
Qualifications/Requirements: GCE, A/AS: 8-12 points. AGNVQ, Merit. ND/C, Merit overall. SQAH, Grades CCCC. SQAV, Individual consideration. IB, 26 points.

BSc Business Information Technology with Leisure Management
UCAS Code: GNNR • Mode: 3/4 Years Equal Combination
Qualifications/Requirements: GCE, A/AS: 8-12 points. AGNVQ, Merit. ND/C, Merit overall. SQAH, Grades CCCC. SQAV, Individual consideration. IB, 24 points.

BSc Business Information Technology with Marketing Management
UCAS Code: GN55 • Mode: 3 Years Equal Combination
Qualifications/Requirements: GCE, A/AS: 8-12 points. AGNVQ, Merit. ND/C, Merit overall. SQAH, Grades CCCC. SQAV, Individual consideration. IB, 26 points.

BSc Business Information Technology with Tourism Management
UCAS Code: G5P7 • Mode: 3/4 Years Major/Minor
Qualifications/Requirements: GCE, A/AS: 8-12 points. AGNVQ, Merit. ND/C, Merit overall. SQAH, Grades CCCC. SQAV, Individual consideration. IB, 24 points.

BSc Computing and Geography
UCAS Code: GFN8 • Mode: 3 Years Equal Combination
Qualifications/Requirements: GCE, A/AS: 8-12 points. AGNVQ, Merit. ND/C, Merit overall. SQAH, Grades CCCC. SQAV, Individual consideration. IB, 24 points.

BSc Computing and Geology
UCAS Code: GF56 • Mode: 3 Years Equal Combination
Qualifications/Requirements: GCE, A/AS: 8 points. AGNVQ, Pass with 3 additional units. ND/C, Merit overall. SQAH, Grades CCCC. SQAV, Individual consideration. IB, 24 points.

BSc Computing and Hospitality Management (Catering)
UCAS Code: GNNT • Mode: 3 Years Equal Combination
Qualifications/Requirements: Please refer to prospectus.

BSc Computing and Information Technology
UCAS Code: G527 • Mode: 3 Years Single Subject
Qualifications/Requirements: GCE, A/AS: 8 points. AGNVQ, Merit. ND/C, Merit overall. SQAH, Grades CCCC. SQAV, Individual consideration. IB, 24 points.

BSc Computing and Multimedia
UCAS Code: G561 • Mode: 3 Years Single Subject
Qualifications/Requirements: GCE, A/AS: 8-12 points. AGNVQ, Merit. ND/C, Merit overall. SQAH, Grades CCCC. SQAV, Individual consideration. IB, 26 points.

BSc Computing and Physical Geography
UCAS Code: GF58 • Mode: 3 Years Equal Combination
Qualifications/Requirements: GCE, A/AS: 8-12 points. AGNVQ, Merit. ND/C, Merit overall. SQAH, Grades CCCC. SQAV, Individual consideration. IB, 26 points.

BSc Computing and Psychology
UCAS Code: GLMT • Mode: 3 Years Equal Combination
Qualifications/Requirements: GCE, A/AS: 8-12 points. AGNVQ, Merit. ND/C, Merit overall. SQAH, Grades CCCC. SQAV, Individual consideration. IB, 24 points.

BSc Computing and Water Resource Management
UCAS Code: GN59 • Mode: 3 Years Equal Combination
Qualifications/Requirements: GCE, A/AS: 8-12 points. AGNVQ, Merit. ND/C, Merit overall. SQAH, Grades CCCC. SQAV, Individual consideration. IB, 24 points.

BSc Computing with Accounting & Financial Management
UCAS Code: NG3M • Mode: 3 Years Equal Combination
Qualifications/Requirements: GCE, A/AS: 8 points. AGNVQ, Merit. ND/C, Merit overall. SQAH, Grades CCCC. SQAV, Individual consideration. IB, 26 points.

BSc Computing with Business Computer Systems
UCAS Code: G5NC • Mode: 3 Years Major/Minor
Qualifications/Requirements: GCE, A/AS: 8 points. AGNVQ, Pass with 3 additional units. ND/C, Merit overall. SQAH, Grades CCCC. SQAV, Individual consideration. IB, 24 points.

BSc Computing with Business Management
UCAS Code: G502 • Mode: 3 Years Single Subject
Qualifications/Requirements: GCE, A/AS: 8 points. AGNVQ, Merit in Business with 3 additional units. ND/C, Merit overall. SQAH, Grades CCCC. SQAV, Individual consideration. IB, 24 points.

BSc Computing with Environmental Policy
UCAS Code: G503 • Mode: 3 Years Single Subject
Qualifications/Requirements: GCE, A/AS: 8-10 points. AGNVQ, Merit. ND/C, Merit overall. SQAH, Grades CCCC. SQAV, Individual consideration. IB, 24 points.

BSc Computing with Financial Services Management
UCAS Code: G5NJ • Mode: 3 Years Major/Minor
Qualifications/Requirements: GCE, A/AS: 8-10 points. AGNVQ, Merit. ND/C, Merit overall. SQAH, Grades CCCC. SQAV, Individual consideration. IB, 24 points.

BSc Computing with Geography
UCAS Code: GFNW • Mode: 3 Years Equal Combination
Qualifications/Requirements: GCE, A/AS: 8 points. AGNVQ, Merit with 3 additional units. ND/C, Merit overall. SQAH, Grades CCCC. SQAV, Individual consideration. IB, 24 points.

BSc Computing with Geology
UCAS Code: G5FP • Mode: 3 Years Major/Minor
Qualifications/Requirements: GCE, A/AS: 8 points. AGNVQ, Pass with 3 additional units. ND/C, Merit overall. SQAH, Grades CCCC. SQAV, Individual consideration. IB, 24 points.

BSc Computing with Hospitality Management (Catering)
UCAS Code: NGTM • Mode: 3 Years Equal Combination
Qualifications/Requirements: GCE, A/AS: 8 points. AGNVQ, Merit. ND/C, Merit overall. SQAH, Grades CCCC. SQAV, Individual consideration. IB, 24 points.

BSc Computing with Hospitality Management (Hotel)
UCAS Code: G504 • Mode: 3 Years Single Subject
Qualifications/Requirements: GCE, A/AS: 8-12 points. AGNVQ, Merit. ND/C, Merit overall. SQAH, Grades CCCC. SQAV, Individual consideration. IB, 24 points.

BSc Computing with Human Geography
UCAS Code: G5LV • Mode: 3 Years Major/Minor
Qualifications/Requirements: GCE, A/AS: 8 points. AGNVQ, Merit. ND/C, Merit overall. SQAH, Grades CCCC. SQAV, Individual consideration. IB, 26 points.

BSc Computing with Human Resource Management
UCAS Code: GNN1 • Mode: 3 Years Equal Combination
Qualifications/Requirements: GCE, A/AS: 8 points. AGNVQ, Merit. ND/C, Merit overall. SQAH, Grades CCCC. SQAV, Individual consideration. IB, 26 points.

BSc Computing with Information Technology
UCAS Code: G526 • Mode: 3 Years Single Subject
Qualifications/Requirements: GCE, A/AS: 8 points. AGNVQ, Merit. ND/C, Merit overall. SQAH, Grades CCCC. SQAV, Individual consideration. IB, 24 points.

BSc Computing with International Business Management

UCAS Code: G7ND • Mode: 3 Years Major/Minor

Qualifications/Requirements: GCE, A/AS: 10 points. AGNVQ, Merit. ND/C, Merit overall and 3 Distinctions. SQAH, Grades CCCC. SQAV, Individual consideration. IB, 26 points.

BSc Computing with International Marketing Management

UCAS Code: G7NN • Mode: 3 Years Major/Minor

Qualifications/Requirements: GCE, A/AS: 10 points. AGNVQ, Merit. ND/C, Merit overall and 3 Distinctions. SQAH, Grades CCCC. SQAV, Individual consideration. IB, 26 points.

BSc Computing with Marketing Management

UCAS Code: GNM5 • Mode: 3 Years Equal Combination

Qualifications/Requirements: GCE, A/AS: 8 points. AGNVQ, Merit. ND/C, Merit overall. SQAH, Grades CCCC. SQAV, Individual consideration. IB, 26 points.

BSc Computing with Media Communications

UCAS Code: G5PL • Mode: 3 Years Major/Minor

Qualifications/Requirements: GCE, A/AS: 8-12 points. AGNVQ, Merit. ND/C, Merit overall. SQAH, Grades CCCC. SQAV, Individual consideration. IB, 24 points.

BSc Computing with Modern Languages (French)

UCAS Code: G5RD • Mode: 3 Years Major/Minor

Qualifications/Requirements: GCE, A/AS: 8 points. AGNVQ, Merit. ND/C, Merit overall. SQAH, Grades CCCC. SQAV, Individual consideration. IB, 24 points.

BSc Computing with Multimedia

UCAS Code: G505 • Mode: 3 Years Single Subject

Qualifications/Requirements: GCE, A/AS: 8 points. AGNVQ, Merit. ND/C, Merit overall. SQAH, Grades CCCC. SQAV, Individual consideration. IB, 26 points.

BSc Computing with Physical Geography

UCAS Code: G5FV • Mode: 3 Years Major/Minor

Qualifications/Requirements: GCE, A/AS: 8 points. AGNVQ, Merit. ND/C, Merit overall. SQAH, Grades CCCC. SQAV, Individual consideration. IB, 26 points.

BSc Computing with Psychology

UCAS Code: GL5T • Mode: 3 Years Equal Combination

Qualifications/Requirements: GCE, A/AS: 8-12 points. AGNVQ, Merit. ND/C, Merit overall. SQAH, Grades CCCC. SQAV, Individual consideration. IB, 24 points.

BSc Computing with Tourism Management

UCAS Code: G5PR • Mode: 3 Years Major/Minor

Qualifications/Requirements: GCE, A/AS: 8-12 points. AGNVQ, Merit. ND/C, Merit overall. SQAH, Grades CCCC. SQAV, Individual consideration. IB, 24 points.

BSc Computing with Water Resource Management

UCAS Code: G5N9 • Mode: 3 Years Major/Minor

Qualifications/Requirements: GCE, A/AS: 8-12 points. AGNVQ, Merit. ND/C, Merit overall. SQAH, Grades CCCC. SQAV, Individual consideration. IB, 24 points.

BSc Computing with Women's Studies

UCAS Code: G5MY • Mode: 3 Years Major/Minor

Qualifications/Requirements: GCE, A/AS: 8-12 points. AGNVQ, Merit. ND/C, Merit overall. SQAH, Grades CCCC. SQAV, Individual consideration. IB, 24 points.

BSc Environmental Science and Information Technology

UCAS Code: FGX7 • Mode: 3 Years Equal Combination

Qualifications/Requirements: GCE, A/AS: 8-12 points. AGNVQ, Merit. ND/C, Merit overall. SQAH, Grades CCCC. SQAV, Individual consideration. IB, 24 points.

BSc Geography and Information Technology

UCAS Code: GFNV • Mode: 3 Years Equal Combination

Qualifications/Requirements: GCE, A/AS: 12 points. AGNVQ, Merit. ND/C, Merit overall. SQAH, Grades CCCC. SQAV, Individual consideration. IB, 24 points.

BSc Geology and Information Technology

UCAS Code: FG65 • Mode: 3 Years Equal Combination

Qualifications/Requirements: GCE, A/AS: 8 points. AGNVQ, Merit with 3 additional units. ND/C, Merit overall. SQAH, Grades CCCC. SQAV, Individual consideration. IB, 26 points.

BSc Information Technology and Multimedia

UCAS Code: G560 • Mode: 3 Years Single Subject

Qualifications/Requirements: GCE, A/AS: 8-12 points. AGNVQ, Merit with 3 additional units. ND/C, Merit overall. SQAH, Grades CCCC. SQAV, Individual consideration. IB, 26 points.

BSc Information Technology and Natural Resource Management

UCAS Code: GF5Y • Mode: 3 Years Equal Combination

Qualifications/Requirements: GCE, A/AS: 8 points. AGNVQ, Merit with 3 additional units. ND/C, Merit overall. SQAH, Grades CCCC. SQAV, Individual consideration. IB, 26 points.

BSc Information Technology and Physical Geography

UCAS Code: FG85 • Mode: 3/4 Years Equal Combination

Qualifications/Requirements: GCE, A/AS: 8-12 points. AGNVQ, Merit with 3 additional units. ND/C, Merit overall. SQAH, Grades CCCC. SQAV, Individual consideration. IB, 24 points.

BSc Information Technology and Water Resource Management
UCAS Code: GN79 • Mode: 3 Years Equal Combination
Qualifications/Requirements: GCE, A/AS: 8-12 points. AGNVQ, Merit. ND/C, Merit overall. SQAH, Grades CCCC. SQAV, Individual consideration. IB, 24 points.

BSc Information Technology with Computing
UCAS Code: G524 • Mode: 3/4 Years Single Subject
Qualifications/Requirements: GCE, A/AS: 8 points. AGNVQ, Merit with 3 additional units. ND/C, Merit overall. SQAH, Grades CCCC. SQAV, Individual consideration. IB, 24 points.

BSc Information Technology with Environmental Policy
UCAS Code: G5FY • Mode: 3 Years Major/Minor
Qualifications/Requirements: GCE, A/AS: 8-12 points. AGNVQ, Merit with 3 additional units. ND/C, Merit overall. SQAH, Grades CCCC. SQAV, Individual consideration. IB, 24 points.

BSc Information Technology with Environmental Science
UCAS Code: G7FX • Mode: 3 Years Major/Minor
Qualifications/Requirements: GCE, A/AS: 8-12 points. AGNVQ, Merit. ND/C, Merit overall. SQAH, Grades CCCC. SQAV, Individual consideration. IB, 24 points.

BSc Information Technology with Geography
UCAS Code: FGVN • Mode: 3 Years Equal Combination
Qualifications/Requirements: GCE, A/AS: 8-12 points. AGNVQ, Merit with 3 additional units. ND/C, Merit overall. SQAH, Grades CCCC. SQAV, Individual consideration. IB, 24 points.

BSc Information Technology with Geology
UCAS Code: G5F6 • Mode: 3/4 Years Major/Minor
Qualifications/Requirements: GCE, A/AS: 8 points. AGNVQ, Merit with 3 additional units. ND/C, Merit overall. SQAH, Grades CCCC. SQAV, Individual consideration. IB, 24 points.

BSc Information Technology with Human Geography
UCAS Code: G5L8 • Mode: 3 Years Major/Minor
Qualifications/Requirements: GCE, A/AS: 8-12 points. AGNVQ, Merit with 3 additional units. ND/C, Merit overall. SQAH, Grades CCCC. SQAV, Individual consideration. IB, 24 points.

BSc Information Technology with Media Communications
UCAS Code: G5PK • Mode: 3 Years Major/Minor
Qualifications/Requirements: GCE, A/AS: 8-12 points. AGNVQ, Merit with 3 additional units. ND/C, Merit overall. SQAH, Grades CCCC. SQAV, Individual consideration. IB, 24 points.

BSc Information Technology with Modern Languages (French)
UCAS Code: G5RC • Mode: 3 Years Major/Minor
Qualifications/Requirements: GCE, A/AS: 8 points. French. AGNVQ, Merit with 3 additional units. ND/C, Merit overall. SQAH, Grades CCCC. SQAV, Individual consideration. IB, 26 points.

BSc Information Technology with Multimedia
UCAS Code: G562 • Mode: 3 Years Single Subject
Qualifications/Requirements: GCE, A/AS: 8 points. AGNVQ, Merit with 3 additional units. ND/C, Merit overall. SQAH, Grades CCCC. SQAV, Individual consideration. IB, 26 points.

BSc Information Technology with Natural Resource Management
UCAS Code: G5FX • Mode: 3 Years Major/Minor
Qualifications/Requirements: GCE, A/AS: 8 points. AGNVQ, Merit with 3 additional units. ND/C, Merit overall. SQAH, Grades CCCC. SQAV, Individual consideration. IB, 26 points.

BSc Information Technology with Physical Geography
UCAS Code: G5F8 • Mode: 3 Years Major/Minor
Qualifications/Requirements: GCE, A/AS: 8-10 points. AGNVQ, Merit with 3 additional units. ND/C, Merit overall. SQAH, Grades CCCC. SQAV, Individual consideration. IB, 24 points.

BSc Information Technology with Religious Studies
UCAS Code: G5VV • Mode: 3 Years Major/Minor
Qualifications/Requirements: GCE, A/AS: 8 points. AGNVQ, Merit with 3 additional units. ND/C, Merit overall. SQAH, Grades CCCC. SQAV, Individual consideration. IB, 24 points.

BSc Information Technology with Theology
UCAS Code: G5VW • Mode: 3 Years Major/Minor
Qualifications/Requirements: GCE, A/AS: 12 points. AGNVQ, Merit. ND/C, Merit overall and 3 Distinctions. SQAH, Grades CCCC. SQAV, Individual consideration. IB, 26 points.

BSc Information Technology with Water Resource Management
UCAS Code: G7N9 • Mode: 3 Years Major/Minor
Qualifications/Requirements: GCE, A/AS: 8-12 points. AGNVQ, Merit. ND/C, Merit overall. SQAH, Grades CCCC. SQAV, Individual consideration. IB, 24 points.

BSc Information Technology with Women's Studies
UCAS Code: G5M9 • Mode: 3 Years Major/Minor
Qualifications/Requirements: GCE, A/AS: 8-10 points. AGNVQ, Merit with 3 additional units. ND/C, Merit overall. SQAH, Grades CCCC. SQAV, Individual consideration. IB, 24 points.

Mod Business Computer Systems and Business Management
UCAS Code: NG15 • Mode: 3 Years Equal Combination

Qualifications/Requirements: GCE, A/AS: 12 points. AGNVQ, Merit in Business with 3 additional units. ND/C, Merit overall and 2 Distinctions. SQAH, Grades CCCC. SQAV, Individual consideration. IB, 24 points.

Mod Business Computer Systems and Environmental Policy
UCAS Code: GF59 • Mode: 3 Years Equal Combination

Qualifications/Requirements: GCE, A/AS: 8-12 points. AGNVQ, Merit. ND/C, Merit overall. SQAH, Grades CCCC. SQAV, Individual consideration. IB, 24 points.

Mod Business Computer Systems and Financial Services Management
UCAS Code: NG35 • Mode: 3 Years Equal Combination

Qualifications/Requirements: GCE, A/AS: 8-12 points. AGNVQ, Merit. ND/C, Merit overall. SQAH, Grades CCCC. SQAV, Individual consideration. IB, 24 points.

Mod Business Computer Systems and Hospitality Management (Hotel)
UCAS Code: GN5R • Mode: 3 Years Equal Combination

Qualifications/Requirements: GCE, A/AS: 8-12 points. AGNVQ, Merit. ND/C, Merit overall. SQAH, Grades CCCC. SQAV, Individual consideration. IB, 24 points.

Mod Business Computer Systems and Human Resource Management
UCAS Code: GNND • Mode: 3 Years Equal Combination

Qualifications/Requirements: GCE, A/AS: 8-12 points. AGNVQ, Merit. ND/C, Merit overall. SQAH, Grades CCCC. SQAV, Individual consideration. IB, 26 points.

Mod Business Computer Systems and Leisure Management
UCAS Code: GN57 • Mode: 3 Years Equal Combination

Qualifications/Requirements: GCE, A/AS: 8-12 points. AGNVQ, Merit in Leisure and Tourism with 3 additional units. ND/C, Merit overall and 2 Distinctions. SQAH, Grades CCCC. SQAV, Individual consideration. IB, 24 points.

Mod Business Computer Systems and Marketing Management
UCAS Code: GNN5 • Mode: 3 Years Equal Combination

Qualifications/Requirements: GCE, A/AS: 8-12 points. AGNVQ, Merit in Business with 3 additional units. ND/C, Merit overall. SQAH, Grades CCCC. SQAV, Individual consideration. IB, 26 points.

Mod Business Computer Systems and Media Communications
UCAS Code: GP54 • Mode: 3 Years Equal Combination

Qualifications/Requirements: GCE, A/AS: 8-12 points. AGNVQ, Merit. ND/C, Merit overall. SQAH, Grades CCCC. SQAV, Individual consideration. IB, 24 points.

Mod Business Computer Systems and Religious Studies
UCAS Code: VG85 • Mode: 3 Years Equal Combination

Qualifications/Requirements: GCE, A/AS: 8 points. AGNVQ, Merit. ND/C, Merit overall. SQAH, Grades CCCC. SQAV, Individual consideration. IB, 24 points.

Mod Business Computer Systems and Women's Studies
UCAS Code: GM5X • Mode: 3 Years Equal Combination

Qualifications/Requirements: GCE, A/AS: 8-12 points. AGNVQ, Merit. ND/C, Merit overall. SQAH, Grades CCCC. SQAV, Individual consideration. IB, 24 points.

Mod Business Information Technology and Business Management
UCAS Code: GN51 • Mode: 3 Years Equal Combination

Qualifications/Requirements: GCE, A/AS: 12 points. AGNVQ, Merit in Business with 3 additional units. ND/C, Merit overall. SQAH, Grades CCCC. SQAV, Individual consideration. IB, 24 points.

Mod Business Information Technology and Financial Services Management
UCAS Code: GN53 • Mode: 3 Years Equal Combination

Qualifications/Requirements: GCE, A/AS: 8-12 points. AGNVQ, Merit. ND/C, Merit overall. SQAH, Grades CCCC. SQAV, Individual consideration. IB, 24 points.

Mod Business Information Technology and Human Resource Management
UCAS Code: NGC5 • Mode: 3 Years Equal Combination

Qualifications/Requirements: GCE, A/AS: 8-12 points. AGNVQ, Merit. ND/C, Merit overall. SQAH, Grades CCCC. SQAV, Individual consideration. IB, 26 points.

Mod Business Information Technology and Leisure Management
UCAS Code: NGR5 • Mode: 3/4 Years Equal Combination

Qualifications/Requirements: GCE, A/AS: 10-14 points. AGNVQ, Merit in Leisure and Tourism with 3 additional units. ND/C, Merit overall and 2 Distinctions. SQAH, Grades CCCC. SQAV, Individual consideration. IB, 24 points.

Mod Business Information Technology and Marketing Management
UCAS Code: GN5M • Mode: 3 Years Equal Combination

Qualifications/Requirements: GCE, A/AS: 8-12 points. AGNVQ, Merit. ND/C, Merit overall. SQAH, Grades CCCC. SQAV, Individual consideration. IB, 26 points.

Mod Business Information Technology and Tourism Management
UCAS Code: GP57 • Mode: 3/4 Years Equal Combination

Qualifications/Requirements: GCE, A/AS: 8-12 points. AGNVQ, Merit in Leisure and Tourism with 3 additional units. ND/C, Merit overall and 2 Distinctions. SQAH, Grades CCCC. SQAV, Individual consideration. IB, 24 points.

Mod Business Management and Computing
UCAS Code: NG1M • Mode: 3 Years Equal Combination
Qualifications/Requirements: GCE, A/AS: 12 points. AGNVQ, Merit in Business with 3 additional units. ND/C, Merit overall and 2 Distinctions. SQAH, Grades CCCC. SQAV, Individual consideration. IB, 26 points.

Mod Computing and Environmental Policy
UCAS Code: GF5X • Mode: 3 Years Equal Combination
Qualifications/Requirements: GCE, A/AS: 8-12 points. AGNVQ, Pass with 3 additional units. ND/C, Merit overall. SQAH, Grades CCCC. SQAV, Individual consideration. IB, 24 points.

Mod Computing and Financial Services Management
UCAS Code: GN5H • Mode: 3 Years Equal Combination
Qualifications/Requirements: GCE, A/AS: 8-12 points. AGNVQ, Pass with 3 additional units. ND/C, Merit overall. SQAH, Grades CCCC. SQAV, Individual consideration. IB, 24 points.

Mod Computing and Hospitality Management (Hotel)
UCAS Code: GN5T • Mode: 3 Years Equal Combination
Qualifications/Requirements: GCE, A/AS: 8-12 points. AGNVQ, Merit. ND/C, Merit overall. SQAH, Grades CCCC. SQAV, Individual consideration. IB, 24 points.

Mod Computing and Human Geography
UCAS Code: LG85 • Mode: 3 Years Equal Combination
Qualifications/Requirements: GCE, A/AS: 8-12 points. AGNVQ, Merit. ND/C, Merit overall. SQAH, Grades CCCC. SQAV, Individual consideration. IB, 26 points.

Mod Computing and Human Resource Management
UCAS Code: NGD5 • Mode: 3 Years Equal Combination
Qualifications/Requirements: GCE, A/AS: 8-12 points. AGNVQ, Merit. ND/C, Merit overall. SQAH, Grades CCCC. SQAV, Individual consideration. IB, 26 points.

Mod Computing and Marketing Management
UCAS Code: GN5N • Mode: 3 Years Equal Combination
Qualifications/Requirements: GCE, A/AS: 8-12 points. AGNVQ, Merit. ND/C, Merit overall. SQAH, Grades CCCC. SQAV, Individual consideration. IB, 26 points.

Mod Computing and Media Communications
UCAS Code: GP5L • Mode: 3 Years Equal Combination
Qualifications/Requirements: GCE, A/AS: 8-12 points. AGNVQ, Merit. ND/C, Merit overall and 2 Distinctions. SQAH, Grades CCCC. SQAV, Individual consideration. IB, 24 points.

Mod Computing and Tourism Management
UCAS Code: GPM7 • Mode: 3 Years Equal Combination
Qualifications/Requirements: GCE, A/AS: 8-12 points. AGNVQ, Merit in Leisure and Tourism with 3 additional units. ND/C, Merit overall and 2 Distinctions. SQAH, Grades CCCC. SQAV, Individual consideration. IB, 24 points.

Mod Computing and Women's Studies
UCAS Code: GM5Y • Mode: 3 Years Equal Combination
Qualifications/Requirements: GCE, A/AS: 8-12 points. AGNVQ, Merit. ND/C, Merit overall. SQAH, Grades CCCC. SQAV, Individual consideration. IB, 24 points.

Mod Environmental Policy and Information Technology
UCAS Code: FG95 • Mode: 3 Years Equal Combination
Qualifications/Requirements: GCE, A/AS: 8-12 points. AGNVQ, Merit with 3 additional units. ND/C, Merit overall. SQAH, Grades CCCC. SQAV, Individual consideration. IB, 24 points.

Mod Human Geography and Information Technology
UCAS Code: GL58 • Mode: 3/4 Years Equal Combination
Qualifications/Requirements: GCE, A/AS: 8-12 points. AGNVQ, Merit with 3 additional units. ND/C, Merit overall. SQAH, Grades CCCC. SQAV, Individual consideration. IB, 24 points.

Mod Information Technology and Media Communications
UCAS Code: GP5K • Mode: 3 Years Equal Combination
Qualifications/Requirements: GCE, A/AS: 8-12 points. AGNVQ, Merit in Media with 3 additional units. ND/C, Merit overall and 2 Distinctions. SQAH, Grades CCCC. SQAV, Individual consideration. IB, 24 points.

Mod Information Technology and Religious Studies
UCAS Code: GV58 • Mode: 3/4 Years Equal Combination
Qualifications/Requirements: GCE, A/AS: 8 points. AGNVQ, Merit with 3 additional units. ND/C, Merit overall. SQAH, Grades CCCC. SQAV, Individual consideration. IB, 24 points.

Mod Information Technology and Theology
UCAS Code: GV5V • Mode: 3 Years Equal Combination
Qualifications/Requirements: GCE, A/AS: 8 points. AGNVQ, Merit. ND/C, Merit overall and 3 Distinctions. SQAH, Grades CCCC. SQAV, Individual consideration. IB, 26 points.

Mod Information Technology and Women's Studies
UCAS Code: GM59 • Mode: 3/4 Years Equal Combination
Qualifications/Requirements: GCE, A/AS: 8 points. AGNVQ, Merit. ND/C, Merit overall. SQAH, Grades CCCC. SQAV, Individual consideration. IB, 24 points.

HND Business Information Technology
UCAS Code: 165G • Mode: 2 Years Single Subject

Qualifications/Requirements: GCE, A/AS: 4 points. AGNVQ, Pass. ND/C, 4 Merits. SQAH, Individual consideration. SQAV, Individual consideration. IB, Individual consideration.

HND Business Information Technology
UCAS Code: 265G • Mode: 3 Years Single Subject

Qualifications/Requirements: GCE, A/AS: 4 points. AGNVQ, Pass. ND/C, 4 Merits. SQAH, Individual consideration. SQAV, Individual consideration. IB, Individual consideration.

HND Computing
UCAS Code: 005G • Mode: 2 Years Single Subject

Qualifications/Requirements: GCE, A/AS: 2 points. AGNVQ, Pass. ND/C, 2 Merits. SQAH, Individual consideration. SQAV, Individual consideration. IB, Individual consideration.

HND Computing
UCAS Code: 105G • Mode: 3 Years Single Subject

Qualifications/Requirements: GCE, A/AS: 2 points. AGNVQ, Pass. ND/C, 2 Merits. SQAH, Individual consideration. SQAV, Individual consideration. IB, Individual consideration.

HND Information Systems
UCAS Code: 025G • Mode: 2 Years Single Subject

Qualifications/Requirements: GCE, A/AS: 2 points. AGNVQ, Pass. ND/C, National. SQAH, Individual consideration. SQAV, Individual consideration. IB, Individual consideration.

HND Information Systems
UCAS Code: 325G • Mode: 3 Years Single Subject

Qualifications/Requirements: GCE, A/AS: 2 points. AGNVQ, Pass. ND/C, National. SQAH, Individual consideration. SQAV, Individual consideration. IB, Individual consideration.

CHESTER: A COLLEGE OF THE UNIVERSITY OF LIVERPOOL C55

BA Art and Computer Science/IS
UCAS Code: WG95 • Mode: 3 Years Equal Combination

Qualifications/Requirements: GCE, A/AS: 12 points. AGNVQ, Merit. ND/C, Merit overall. SQAH, Grades CCCC. SQAV, National. IB, 24 points.

BA Art with Computer Science/IS
UCAS Code: W9G5 • Mode: 3 Years Major/Minor

Qualifications/Requirements: GCE, A/AS: 12 points. AGNVQ, Merit. ND/C, Merit overall. SQAH, Grades CCCC. SQAV, National. IB, 24 points.

BA Computer Science/IS and Counselling Skills
UCAS Code: GL55 • Mode: 3 Years Equal Combination

Qualifications/Requirements: GCE, A/AS: 14 points. AGNVQ, Distinction (in specific programmes). ND/C, Distinction Overall. SQAH, Grades CCCC. SQAV, National. IB, 24 points.

BA Computer Science/IS and Dance
UCAS Code: GW5K • Mode: 3 Years Equal Combination

Qualifications/Requirements: GCE, A/AS: 12 points. AGNVQ, Merit (in specific programmes). ND/C, Merit overall. SQAH, Grades CCCC. SQAV, National. IB, 24 points.

BA Computer Science/IS and Drama and Theatre Studies
UCAS Code: GW54 • Mode: 3 Years Equal Combination

Qualifications/Requirements: GCE, A/AS: 12 points. AGNVQ, Merit. ND/C, Merit overall. SQAH, Grades CCCC. SQAV, National. IB, 24 points.

BA Computer Science/IS and English Literature
UCAS Code: GQ53 • Mode: 3 Years Equal Combination

Qualifications/Requirements: GCE, A/AS: 12 points. English at A level. AGNVQ, Not normally sufficient. ND/C, Not normally sufficient. SQAH, Grades CCCC. SQAV, Not normally sufficient. IB, 24 points.

BA Computer Science/IS and French
UCAS Code: GR51 • Mode: 3 Years Equal Combination

Qualifications/Requirements: GCE, A/AS: 12 points. AGNVQ, Merit (in specific programmes). ND/C, Merit overall. SQAH, Grades CCCC. IB, 24 points.

BA Computer Science/IS and German
UCAS Code: GR52 • Mode: 3 Years Equal Combination

Qualifications/Requirements: GCE, A/AS: 12 points. AGNVQ, Merit (in specific programmes). ND/C, Merit overall. SQAH, Grades CCCC. IB, 24 points.

BA Computer Science/IS and Health Studies
UCAS Code: GL5L • Mode: 3 Years Equal Combination

Qualifications/Requirements: GCE, A/AS: 12 points. AGNVQ, Merit (in specific programmes). ND/C, Merit overall. SQAH, Grades CCCC. SQAV, National. IB, 24 points.

BA Computer Science/IS and History
UCAS Code: GV51 • Mode: 3 Years Equal Combination

Qualifications/Requirements: GCE, A/AS: 12 points. History, Economics or Sociology at A level. AGNVQ, Merit (in specific programmes). ND/C, Merit overall. SQAH, Grades CCCC. SQAV, National. IB, 24 points.

BA Computer Science/IS and Spanish
UCAS Code: GR54 • Mode: 3 Years Equal Combination

Qualifications/Requirements: GCE, A/AS: 12 points. AGNVQ, Merit (in specific programmes). ND/C, Merit overall. SQAH, Grades CCCC. IB, 24 points.

BA Computer Science/IS and Theology and Religious Studies
UCAS Code: GV58 • Mode: 3 Years Equal Combination
Qualifications/Requirements: GCE, A/AS: 12 points. AGNVQ, Merit. ND/C, Merit overall. SQAH, Grades CCCC. SQAV, National. IB, 24 points.

BA Computer Science/IS with Art
UCAS Code: G5W9 • Mode: 3 Years Major/Minor
Qualifications/Requirements: GCE, A/AS: 12 points. AGNVQ, Merit (in specific programmes). ND/C, Merit overall. SQAH, Grades CCCC. SQAV, National. IB, 24 points.

BA Computer Science/IS with Counselling Skills
UCAS Code: G5L5 • Mode: 3 Years Major/Minor
Qualifications/Requirements: GCE, A/AS: 14 points. AGNVQ, Distinction (in specific programmes). ND/C, Distinction Overall. SQAH, Grades CCCC. SQAV, National. IB, 24 points.

BA Computer Science/IS with Dance
UCAS Code: G5WK • Mode: 3 Years Major/Minor
Qualifications/Requirements: GCE, A/AS: 12 points. AGNVQ, Merit (in specific programmes). ND/C, Merit overall. SQAH, Grades CCCC. SQAV, National. IB, 24 points.

BA Computer Science/IS with Drama and Theatre Studies
UCAS Code: G5W4 • Mode: 3 Years Major/Minor
Qualifications/Requirements: GCE, A/AS: 12 points. AGNVQ, Merit. ND/C, Merit overall. SQAH, Grades CCCC. SQAV, National. IB, 24 points.

BA Computer Science/IS with English Literature
UCAS Code: G5Q3 • Mode: 3 Years Major/Minor
Qualifications/Requirements: GCE, A/AS: 12 points. English at A level. AGNVQ, Not normally sufficient. ND/C, Not normally sufficient. SQAH, Grades CCCC. SQAV, National. IB, 24 points.

BA Computer Science/IS with French
UCAS Code: G5R1 • Mode: 3 Years Major/Minor
Qualifications/Requirements: GCE, A/AS: 12 points. AGNVQ, Merit. ND/C, Merit overall. SQAH, Grades CCCC. SQAV, National. IB, 24 points.

BA Computer Science/IS with German
UCAS Code: G5R2 • Mode: 3 Years Major/Minor
Qualifications/Requirements: GCE, A/AS: 12 points. AGNVQ, Merit (in specific programmes). ND/C, Merit overall. SQAH, Grades CCCC. SQAV, National. IB, 24 points.

BA Computer Science/IS with Health Studies
UCAS Code: G5LK • Mode: 3 Years Major/Minor
Qualifications/Requirements: GCE, A/AS: 12 points. AGNVQ, Merit (in specific programmes). ND/C, Merit overall. SQAH, Grades CCCC. SQAV, National. IB, 24 points.

BA Computer Science/IS with History
UCAS Code: G5V1 • Mode: 3 Years Major/Minor
Qualifications/Requirements: GCE, A/AS: 12 points. History, Economics or Sociology at A level. AGNVQ, Merit (in specific programmes). ND/C, Merit overall. SQAH, Grades CCCC. SQAV, National. IB, 24 points.

BA Computer Science/IS with Spanish
UCAS Code: G5R4 • Mode: 3 Years Major/Minor
Qualifications/Requirements: GCE, A/AS: 12 points. AGNVQ, Merit (in specific programmes). ND/C, Merit overall. SQAH, Grades CCCC. IB, 24 points.

BA Computer Science/IS with Theology and Religious Studies
UCAS Code: G5V8 • Mode: 3 Years Major/Minor
Qualifications/Requirements: GCE, A/AS: 12 points. AGNVQ, Merit. ND/C, Merit overall. SQAH, Grades CCCC. SQAV, National. IB, 24 points.

BSc Biology and Computer Science/IS
UCAS Code: CG15 • Mode: 3 Years Equal Combination
Qualifications/Requirements: GCE, A/AS: 12 points. Biology at A level. AGNVQ, Merit (in specific programmes). ND/C, Merit overall. SQAH, Grades CCC. SQAV, National. IB, 24 points.

BSc Business Information Systems
UCAS Code: GN51 • Mode: 3 Years Equal Combination
Qualifications/Requirements: GCE, A/AS: 14 points. AGNVQ, Merit (in specific programmes). ND/C, Merit overall. SQAH, Grades CCCC. SQAV, National. IB, 24 points.

BSc Computer Science/IS
UCAS Code: G500 • Mode: 3 Years Single Subject
Qualifications/Requirements: GCE, A/AS: 12 points. AGNVQ, Merit (in specific programmes). ND/C, Merit overall. SQAH, Grades CCCC. SQAV, National. IB, 24 points.

BSc Computer Science/IS and Geography
UCAS Code: GF58 • Mode: 3 Years Equal Combination
Qualifications/Requirements: GCE, A/AS: 12 points. Geography or Geology at A level. AGNVQ, Merit (in specific programmes). ND/C, Merit overall. SQAH, Grades CCC. SQAV, National. IB, 24 points.

BSc Computer Science/IS and Health Sciences

UCAS Code: BG95 • Mode: 3 Years Equal Combination

Qualifications/Requirements: GCE, A/AS: 12 points. AGNVQ, Merit (in specific programmes). ND/C, Merit overall. SQAH, Grades CCC. SQAV, National. IB, 24 points.

BSc Computer Science/IS and Mathematics

UCAS Code: GG51 • Mode: 3 Years Equal Combination

Qualifications/Requirements: GCE, A/AS: 12 points. Maths at A level. AGNVQ, Not normally sufficient. ND/C, Not normally sufficient. SQAH, Grades CCC. SQAV, National. IB, 24 points.

BSc Computer Science/IS and Physical Education/Sports Science

UCAS Code: GB56 • Mode: 3 Years Equal Combination

Qualifications/Requirements: GCE, A/AS: 12 points. AGNVQ, Merit (in specific programmes). ND/C, Merit overall. SQAH, Grades CCC. SQAV, National. IB, 24 points.

BSc Computer Science/IS and Psychology

UCAS Code: GL57 • Mode: 3 Years Equal Combination

Qualifications/Requirements: GCE, A/AS: 12 points. AGNVQ, Merit (in specific programmes). ND/C, Merit overall. SQAH, Grades CCC. SQAV, National. IB, 24 points.

BSc Computer Science/IS with Biology

UCAS Code: G5C1 • Mode: 3 Years Major/Minor

Qualifications/Requirements: GCE, A/AS: 12 points. Biology at A level. AGNVQ, Merit (in specific programmes). ND/C, Merit overall. SQAH, Grades CCC. SQAV, National. IB, 24 points.

BSc Computer Science/IS with Geography

UCAS Code: G5F8 • Mode: 3 Years Major/Minor

Qualifications/Requirements: GCE, A/AS: 12 points. Geography or Geology at A level. AGNVQ, Merit (in specific programmes). ND/C, Merit overall. SQAH, Grades CCCC. SQAV, National. IB, 24 points.

BSc Computer Science/IS with Health Sciences

UCAS Code: G5B9 • Mode: 3 Years Major/Minor

Qualifications/Requirements: GCE, A/AS: 12 points. AGNVQ, Merit (in specific programmes). ND/C, Merit overall. SQAH, Grades CCCC. SQAV, National. IB, 24 points.

BSc Computer Science/IS with Mathematics

UCAS Code: G5G1 • Mode: 3 Years Major/Minor

Qualifications/Requirements: GCE, A/AS: 12 points. Maths at A level. AGNVQ, Not normally sufficient. ND/C, Not normally sufficient. SQAH, Grades CCCC. SQAV, Not normally sufficient. IB, 24 points.

BSc Computer Science/IS with Physical Education/Sports Science

UCAS Code: G5B6 • Mode: 3 Years Major/Minor

Qualifications/Requirements: GCE, A/AS: 12 points. AGNVQ, Merit (in specific programmes). ND/C, Merit overall. SQAH, Grades CCC. SQAV, National. IB, 24 points.

BSc Computer Science/IS with Psychology

UCAS Code: G5L7 • Mode: 3 Years Major/Minor

Qualifications/Requirements: GCE, A/AS: 12 points. AGNVQ, Merit. ND/C, Merit overall. SQAH, Grades CCCC. SQAV, National. IB, 24 points.

BSc Computer Science/IS with Social Science

UCAS Code: G5L3 • Mode: 3 Years Major/Minor

Qualifications/Requirements: GCE, A/AS: 12 points. Sociology, Psychology or Geography. AGNVQ, Merit (in specific programmes). ND/C, Merit overall. SQAH, Grades CCCC. SQAV, National. IB, 24 points.

BSc Computer Science/IS with Statistics

UCAS Code: G5G4 • Mode: 3 Years Major/Minor

Qualifications/Requirements: GCE, A/AS: 12 points. AGNVQ, Merit (in specific programmes). ND/C, Merit overall. SQAH, Grades CCCC. IB, 24 points.

BSc Mathematics, Statistics and Computing

UCAS Code: G900 • Mode: 3 Years Single Subject

Qualifications/Requirements: GCE, A/AS: 12 points. Maths at A level. AGNVQ, Not normally sufficient. ND/C, Not normally sufficient. SQAH, Grades CCCC. SQAV, Not normally sufficient. IB, 24 points.

HND Computing

UCAS Code: 105G • Mode: 2 Years Single Subject

Qualifications/Requirements: GCE, A/AS: 6 points. AGNVQ, Pass (in specific programmes). ND/C, National. SQAH, Grades DDD. SQAV, National. IB, 24 points.

UNIVERSITY COLLEGE CHICHESTER C58

BA Information and Communications Technology with Education (QTS)

UCAS Code: G5X5 • Mode: 4 Years Major/Minor

Qualifications/Requirements: GCE, A/AS: 12 points. AGNVQ, Merit (in specific programmes). ND/C, Merits and Distinction. SQAH, Individual consideration. SQAV, Individual consideration. IB, Individual consideration.

CITY UNIVERSITY C60

BSc Business Computing Systems (4 years SW)

UCAS Code: G522 • Mode: 4 Years Single Subject

Qualifications/Requirements: GCE, A level grades: BBC-BCC. AGNVQ, Distinction (in specific programmes). ND/C, 2 Merits and 3 Distinctions. SQAH, Individual consideration. SQAV, Individual consideration. IB, Individual consideration.

BEng Computing (3 or 4 years)
UCAS Code: G500 • Mode: 3/4 Years Single Subject

Qualifications/Requirements: GCE, A/AS: 22 points. Maths, Physics, Computing or Science. AGNVQ, Distinction (in specific programmes) with A level. ND/C, Merit overall and 3 Distinctions. SQAH, Grades ABBBB. SQAV, Individual consideration. IB, 28 points.

BEng Computing (Distributed Information Systems & Communications) (3 or 4 years)
UCAS Code: G520 • Mode: 3/4 Years Single Subject

Qualifications/Requirements: GCE, A/AS: 22 points. Maths, Physics, Computing or Science. AGNVQ, Distinction (in specific programmes) with A level. ND/C, Merit overall and 3 Distinctions. SQAH, Grades ABBBB. SQAV, Individual consideration. IB, 28 points.

CITY COLLEGE, BIRMINGHAM C62

HND Business and Information Technology
UCAS Code: 15NG • Mode: 2 Years Equal Combination

Qualifications/Requirements: Please refer to prospectus.

HND Computing
UCAS Code: 005G • Mode: 2 Years Single Subject

Qualifications/Requirements: Please refer to prospectus.

CITY OF BRISTOL COLLEGE C63

HND Computing (Computing, Software Engineering or Business IT)
UCAS Code: 65GG • Mode: 2 Years Equal Combination

Qualifications/Requirements: An approved subject from restricted list. AGNVQ, Individual consideration. ND/C, Individual consideration. SQAH, Individual consideration. SQAV, Individual consideration. IB, Individual consideration.

COLCHESTER INSTITUTE C75

HND Computing and Information Technology
UCAS Code: 005G • Mode: 2 Years Single Subject

Qualifications/Requirements: GCE, A/AS: 8 points. Italian. AGNVQ, Merit. ND/C, Merit overall. SQAH, Individual consideration. SQAV, Individual consideration. IB, Individual consideration.

CORNWALL COLLEGE WITH DUCHY COLLEGE C78

HND Information Technology
UCAS Code: 065G • Mode: 2 Years Single Subject

Qualifications/Requirements: GCE, A/AS: 4 points. AGNVQ, Individual consideration. ND/C, Merits and Distinction. SQAH, Individual consideration. SQAV, Individual consideration. IB, Individual consideration.

COVENTRY UNIVERSITY C85

BA Business Computing
UCAS Code: G560 • Mode: 3/4 Years Single Subject

Qualifications/Requirements: GCE, A/AS: 12 points. AGNVQ, Merit. ND/C, 4 Merits. SQAH, Grades CCCC. SQAV, Individual consideration. IB, Individual consideration.

BA Business Information Technology
UCAS Code: G561 • Mode: 4 Years Single Subject

Qualifications/Requirements: GCE, A/AS: 4 points. AGNVQ, Pass. ND/C, 4 Merits. SQAH, Individual consideration. SQAV, Individual consideration. IB, Individual consideration.

BA Business Information Technology
UCAS Code: GN51 • Mode: 3/4 Years Equal Combination

Qualifications/Requirements: GCE, A/AS: 12 points. AGNVQ, Merit. ND/C, 4 Merits. SQAH, Grades CCCC. SQAV, Individual consideration. IB, Individual consideration.

BA Business Information Technology (Foundation)
UCAS Code: GNM1 • Mode: 4/5 Years Equal Combination

Qualifications/Requirements: GCE, A/AS: 2 points. AGNVQ, Pass. ND/C, National. SQAH, Individual consideration. SQAV, Individual consideration. IB, Individual consideration.

BA Business Information Technology with Accounting
UCAS Code: G5N4 • Mode: 3/4 Years Major/Minor

Qualifications/Requirements: GCE, A/AS: 12 points. AGNVQ, Merit. ND/C, 4 Merits. SQAH, Grades CCCC. SQAV, Individual consideration. IB, Individual consideration.

BA Business Information Technology with Human Resource Management
UCAS Code: G5N1 • Mode: 3/4 Years Major/Minor

Qualifications/Requirements: GCE, A/AS: 12 points. AGNVQ, Merit. ND/C, 4 Merits. SQAH, Grades CCCC. SQAV, Individual consideration. IB, Individual consideration.

BA Business Information Technology with International Business
UCAS Code: G5NC • Mode: 3/4 Years Major/Minor

Qualifications/Requirements: GCE, A/AS: 12 points. AGNVQ, Merit. ND/C, 4 Merits. SQAH, Grades CCCC. SQAV, Individual consideration. IB, Individual consideration.

BA Business Information Technology with Marketing
UCAS Code: G5N5 • Mode: 3/4 Years Major/Minor

Qualifications/Requirements: GCE, A/AS: 12 points. AGNVQ, Merit. ND/C, 4 Merits. SQAH, Grades CCCC. SQAV, Individual consideration. IB, Individual consideration.

BA Business Information Technology with Supply Chain Management
UCAS Code: G5NM • Mode: 3/4 Years Major/Minor

Qualifications/Requirements: GCE, A/AS: 12 points. AGNVQ, Merit. ND/C, 4 Merits. SQAH, Grades CCCC. SQAV, Individual consideration. IB, Individual consideration.

BA Business Information Technology with Tourism
UCAS Code: G5P7 • Mode: 3/4 Years Major/Minor

Qualifications/Requirements: GCE, A/AS: 12 points. AGNVQ, Merit. ND/C, 4 Merits. SQAH, Grades CCCC. SQAV, Individual consideration. IB, Individual consideration.

BA Extended Business Information Technology (Overseas)
UCAS Code: GN5C • Mode: 4 Years Equal Combination

Qualifications/Requirements: GCE, A/AS: 2 points. AGNVQ, Pass. ND/C, National. SQAH, Individual consideration. SQAV, Individual consideration. IB, Pass in Diploma.

BA Information Technology and English (for non-native English speakers)
UCAS Code: GQ53 • Mode: 3 Years Equal Combination

Qualifications/Requirements: AGNVQ, Merit. ND/C, 4 Merits.

BA Information Technology and Law
UCAS Code: GM53 • Mode: 3/4 Years Equal Combination

Qualifications/Requirements: GCE, A/AS: 12 points. AGNVQ, Merit. ND/C, 4 Merits. SQAH, Grades CCCC. SQAV, Individual consideration. IB, Individual consideration.

BSc Biological Sciences and Computing
UCAS Code: GC51 • Mode: 3/4 Years Equal Combination

Qualifications/Requirements: GCE, A/AS: 12 points. Biology at A level. AGNVQ, Merit in Science. ND/C, 3 Merits (in specific programmes). SQAH, Grades CCC including specific subject(s). SQAV, Individual consideration. IB, Individual consideration.

BSc Chemistry and Computing
UCAS Code: FG15 • Mode: 3/4 Years Equal Combination

Qualifications/Requirements: GCE, A/AS: 10 points. Chemistry at A level. AGNVQ, Merit in Science. ND/C, 5 Merits. SQAH, Individual consideration. SQAV, Individual consideration. IB, Individual consideration.

BSc Chemistry and Computing
UCAS Code: FG1N • Mode: 4/5 Years Equal Combination

Qualifications/Requirements: AGNVQ, Individual consideration. ND/C, Individual consideration. SQAH, Individual consideration. SQAV, Individual consideration. IB, Individual consideration.

BSc Computer Science
UCAS Code: G500 • Mode: 3/4 Years Single Subject

Qualifications/Requirements: GCE, A/AS: 12 points. AGNVQ, Merit. ND/C, 4 Merits. SQAH, Grades CCCC. SQAV, Individual consideration. IB, Individual consideration.

BSc Computing
UCAS Code: G503 • Mode: 4 Years Single Subject

Qualifications/Requirements: GCE, A/AS: 4 points. AGNVQ, Pass. ND/C, 2 Merits. SQAH, Individual consideration. SQAV, Individual consideration. IB, Individual consideration.

BSc Computing
UCAS Code: G504 • Mode: 3/4 Years Single Subject

Qualifications/Requirements: GCE, A/AS: 12 points. AGNVQ, Merit. ND/C, 4 Merits. SQAH, Grades CCCC. SQAV, Individual consideration. IB, Individual consideration.

BSc Computing and Biological Sciences
UCAS Code: GC5C • Mode: 4/5 Years Equal Combination

Qualifications/Requirements: AGNVQ, Individual consideration. ND/C, Individual consideration. SQAH, Individual consideration. SQAV, Individual consideration. IB, Individual consideration.

BSc Computing and Economics
UCAS Code: GL51 • Mode: 3/4 Years Equal Combination

Qualifications/Requirements: GCE, A/AS: 12 points. AGNVQ, Merit. ND/C, 4 Merits. SQAH, Individual consideration. SQAV, Individual consideration. IB, Individual consideration.

BSc Computing with European Studies
UCAS Code: G5T2 • Mode: 4 Years Major/Minor

Qualifications/Requirements: GCE, A/AS: 12 points. AGNVQ, Merit. ND/C, 4 Merits. SQAH, Grades CCCC. SQAV, Individual consideration. IB, Individual consideration.

BSc Geographical Information Systems
UCAS Code: G562 • Mode: 3/4 Years Single Subject

Qualifications/Requirements: GCE, A/AS: 12 points. AGNVQ, Merit. ND/C, 4 Merits. SQAH, Individual consideration. SQAV, Individual consideration. IB, Individual consideration.

BSc Geography and Computing
UCAS Code: LG85 • Mode: 3/4 Years Equal Combination

Qualifications/Requirements: GCE, A/AS: 12 points. AGNVQ, Merit. ND/C, 4 Merits. SQAH, Individual consideration. SQAV, Individual consideration. IB, Individual consideration.

BSc Mathematics and Computing
UCAS Code: GG15 • Mode: 3/4 Years Equal Combination

Qualifications/Requirements: GCE, A/AS: 12-16 points. Maths or Computing at A level. AGNVQ, Merit. ND/C, 3 Merits (in specific programmes). SQAH, Individual consideration. SQAV, Individual consideration. IB, Individual consideration.

BSc Network Computing

UCAS Code: G530 • Mode: 3/4 Years Single Subject

Qualifications/Requirements: GCE, A/AS: 12 points. AGNVQ, Merit. ND/C, 4 Merits. SQAH, Grades CCCC. SQAV, Individual consideration. IB, Individual consideration.

BEng Computers, Networking and Communication Technology

UCAS Code: H621 • Mode: 3/4 Years Single Subject

Qualifications/Requirements: GCE, A/AS: 10 points. Maths or Science. AGNVQ, Merit. ND/C, 4 Merits. SQAH, Individual consideration. SQAV, Individual consideration. IB, Individual consideration.

HND Business Information Technology

UCAS Code: 15NG • Mode: 2 Years Equal Combination

Qualifications/Requirements: GCE, A/AS: 4 points. AGNVQ, Pass. ND/C, 2 Merits. SQAH, Individual consideration. SQAV, Individual consideration. IB, Individual consideration.

HND Computing

UCAS Code: 105G • Mode: 2 Years Single Subject

Qualifications/Requirements: GCE, A/AS: 4 points. AGNVQ, Pass. ND/C, 2 Merits. SQAH, Individual consideration. SQAV, Individual consideration. IB, Individual consideration.

CRANFIELD UNIVERSITY C90

BSc Business Information Systems

UCAS Code: N1G5 • Mode: 3 Years Major/Minor

Qualifications/Requirements: GCE, A level grades: CC-CCC. AGNVQ, Merit (in specific programmes). ND/C, 3 Merits. SQAH, CSYS required. SQAV, Individual consideration. IB, Individual consideration.

BSc Information Systems Management

UCAS Code: GN5C • Mode: 3 Years Equal Combination

Qualifications/Requirements: GCE, A level grades: CC-CCC. AGNVQ, Merit (in specific programmes). ND/C, 3 Merits. SQAH, CSYS required. SQAV, Individual consideration. IB, Individual consideration.

DE MONTFORT UNIVERSITY D26

BA Business Information Systems

UCAS Code: G521 • Mode: 4 Years Single Subject

Qualifications/Requirements: GCE, A/AS: 16 points. AGNVQ, Merit. ND/C, 2 Merits and 2 Distinctions. SQAH, Grades BBBC including specific subject(s). SQAV, Individual consideration. IB, 30 points.

BA Computing and Finance

UCAS Code: GN53 • Mode: 3 Years Equal Combination

Qualifications/Requirements: Please refer to prospectus.

BSc Computer and Information Systems

UCAS Code: G520 • Mode: 3/4 Years Single Subject

Qualifications/Requirements: GCE, A/AS: 12-18 points. AGNVQ, Distinction. ND/C, 2 Merits and 2 Distinctions. SQAH, Grades BBBC including specific subject(s). SQAV, Individual consideration. IB, 30 points.

BSc Computer Science (4 years SW)

UCAS Code: G500 • Mode: 4 Years Single Subject

Qualifications/Requirements: GCE, A/AS: 12-18 points. AGNVQ, Merit. ND/C, 2 Merits and 2 Distinctions. SQAH, Grades BBBC including specific subject(s). SQAV, Individual consideration. IB, 30 points.

BSc Computing

UCAS Code: G501 • Mode: 3/4 Years Single Subject

Qualifications/Requirements: GCE, A/AS: 8-14 points. AGNVQ, Pass. ND/C, 4 Merits (in specific programmes). SQAH, Grades BBCC including specific subject(s). SQAV, Individual consideration. IB, 28 points.

BSc Computing and Electronics

UCAS Code: HG65 • Mode: 3 Years Equal Combination

Qualifications/Requirements: GCE, A/AS: 8-14 points. AGNVQ, Merit. ND/C, 4 Merits (in specific programmes). SQAH, Grades BBCC including specific subject(s). SQAV, Individual consideration. IB, 28 points.

BSc Computing and Music Technology

UCAS Code: GW53 • Mode: 3 Years Equal Combination

Qualifications/Requirements: GCE, A/AS: 8-14 points. AGNVQ, Merit. ND/C, 4 Merits (in specific programmes). SQAH, Grades BBCC including specific subject(s). SQAV, Individual consideration. IB, 28 points.

BSc Computing and Psychology

UCAS Code: CG85 • Mode: 3 Years Equal Combination

Qualifications/Requirements: GCE, A/AS: 8-14 points. AGNVQ, Merit. ND/C, 4 Merits (in specific programmes). SQAH, Grades BBBC including specific subject(s). SQAV, Individual consideration. IB, 28 points.

BSc Computing and Technology

UCAS Code: GJ55 • Mode: 3 Years Equal Combination

Qualifications/Requirements: GCE, A/AS: 8-14 points. AGNVQ, Merit. ND/C, 4 Merits (in specific programmes). SQAH, Grades BBCC including specific subject(s). SQAV, Individual consideration. IB, 28 points.

BSc Economics and Information Systems

UCAS Code: GL51 • Mode: 3 Years Equal Combination

Qualifications/Requirements: GCE, A/AS: 12 points. AGNVQ, Merit (in specific programmes). ND/C, 3 Merits and 3 Distinctions. SQAH, Individual consideration. SQAV, Individual consideration. IB, Individual consideration.

BSc Information Systems and Politics

UCAS Code: GM5D • Mode: 3/4 Years Equal Combination

Qualifications/Requirements: GCE, A/AS: 12-18 points. AGNVQ, Distinction. ND/C, 2 Merits and 2 Distinctions. SQAH, Grades BBBC including specific subject(s). SQAV, Individual consideration. IB, 30 points.

BSc Information Systems and Psychology
UCAS Code: GL57 • Mode: 3 Years Equal Combination

Qualifications/Requirements: Please refer to prospectus.

BSc Information Systems and Social Psychology
UCAS Code: GL5R • Mode: 3/4 Years Equal Combination

Qualifications/Requirements: GCE, A/AS: 12-18 points. AGNVQ, Distinction. ND/C, 2 Merits and 2 Distinctions. SQAH, Grades BBBC including specific subject(s). SQAV, Individual consideration. IB, 30 points.

BSc Information Systems and Sociology
UCAS Code: GL53 • Mode: 3/4 Years Equal Combination

Qualifications/Requirements: GCE, A/AS: 12-18 points. AGNVQ, Distinction. ND/C, 2 Merits and 2 Distinctions. SQAH, Grades BBBC including specific subject(s). SQAV, Individual consideration. IB, 30 points.

BSc Information Systems with Management
UCAS Code: G5N1 • Mode: 3/4 Years Major/Minor

Qualifications/Requirements: GCE, A/AS: 12-18 points. AGNVQ, Distinction. ND/C, 2 Merits and 2 Distinctions. SQAH, Grades BBBC including specific subject(s). SQAV, Individual consideration. IB, 30 points.

BSc Information Technology
UCAS Code: G560 • Mode: 4 Years Single Subject

Qualifications/Requirements: GCE, A/AS: 8-16 points. AGNVQ, Individual consideration. ND/C, 4 Merits. SQAH, Grades BBBB. SQAV, Individual consideration. IB, 30 points.

BSc Mathematics and Technology
UCAS Code: GJ15 • Mode: 4 Years Equal Combination

Qualifications/Requirements: GCE, A/AS: 12 points. Maths. AGNVQ, Merit (in specific programmes). ND/C, 3 Merits and 1 Distinction. SQAH, Individual consideration. SQAV, Individual consideration. IB, Individual consideration.

Mod Accounting (N-V) and Computing
UCAS Code: NG45 • Mode: 3 Years Equal Combination

Qualifications/Requirements: Please refer to prospectus.

Mod Computing and French
UCAS Code: GR51 • Mode: 3 Years Equal Combination

Qualifications/Requirements: GCE, A/AS: 8-14 points. AGNVQ, Merit. ND/C, 4 Merits (in specific programmes). SQAH, Grades BBCC including specific subject(s). SQAV, Individual consideration. IB, 28 points.

Mod Computing and German
UCAS Code: GR52 • Mode: 3 Years Equal Combination

Qualifications/Requirements: GCE, A/AS: 8-14 points. AGNVQ, Merit. ND/C, 4 Merits (in specific programmes). SQAH, Grades BBCC including specific subject(s). SQAV, Individual consideration. IB, 28 points.

Mod Computing and Human Resource Management
UCAS Code: GN51 • Mode: 3 Years Equal Combination

Qualifications/Requirements: GCE, A/AS: 8-14 points. AGNVQ, Merit. ND/C, 4 Merits (in specific programmes). SQAH, Grades BBCC including specific subject(s). SQAV, Individual consideration. IB, 28 points.

Mod Computing and Management Science
UCAS Code: GG59 • Mode: 3 Years Equal Combination

Qualifications/Requirements: GCE, A/AS: 8-14 points. AGNVQ, Merit. ND/C, 4 Merits (in specific programmes). SQAH, Grades BBBC including specific subject(s). SQAV, Individual consideration. IB, 28 points.

Mod Computing and Marketing
UCAS Code: GN55 • Mode: 3 Years Equal Combination

Qualifications/Requirements: GCE, A/AS: 8-14 points. AGNVQ, Merit. ND/C, 4 Merits (in specific programmes). SQAH, Grades BBBC including specific subject(s). SQAV, Individual consideration. IB, 28 points.

Mod Computing and Media Studies
UCAS Code: GP5L • Mode: 3 Years Equal Combination

Qualifications/Requirements: GCE, A/AS: 12-14 points. AGNVQ, Merit. ND/C, 4 Merits (in specific programmes). SQAH, Grades BBBB including specific subject(s). SQAV, Individual consideration. IB, 30 points.

Mod Computing and Spanish
UCAS Code: GR54 • Mode: 3 Years Equal Combination

Qualifications/Requirements: GCE, A/AS: 8-14 points. AGNVQ, Merit. ND/C, 4 Merits (in specific programmes). SQAH, Grades BBBC including specific subject(s). SQAV, Individual consideration. IB, 28 points.

HND Computing
UCAS Code: 105G • Mode: 2 Years Single Subject

Qualifications/Requirements: GCE, A/AS: 4-6 points. AGNVQ, Pass. ND/C, 3 Merits (in specific programmes). SQAH, Grades CCCC including specific subject(s). SQAV, Individual consideration. IB, 26 points.

UNIVERSITY OF DERBY D39

BSc Computer Studies
UCAS Code: G501 • Mode: 4 Years Single Subject

Qualifications/Requirements: GCE, A/AS: 12-14 points. AGNVQ, Merit (in specific programmes). ND/C, Merit overall (in specific programmes). SQAH, Grades CCCC. SQAV, Individual consideration. IB, 26 points.

BSc Computer Studies (Artificial Intelligence)
UCAS Code: G800 • Mode: 4 Years Single Subject

Qualifications/Requirements: GCE, A/AS: 12-14 points. AGNVQ, Merit (in specific programmes). ND/C, Merit overall (in specific programmes). SQAH, Grades BCCC. SQAV, Individual consideration. IB, 26 points.

BSc Computer Studies (Networks)
UCAS Code: G506 • Mode: 4 Years Single Subject

Qualifications/Requirements: GCE, A/AS: 12-14 points. AGNVQ, Merit (in specific programmes). ND/C, Merit overall (in specific programmes). SQAH, Grades BCCC. SQAV, Individual consideration. IB, 26 points.

BSc Computer Studies (Software Engineering)
UCAS Code: G5G7 • Mode: 4 Years Major/Minor

Qualifications/Requirements: GCE, A/AS: 12-14 points. AGNVQ, Merit (in specific programmes). ND/C, Merit overall (in specific programmes). SQAH, Grades BCCC. SQAV, Individual consideration. IB, 26 points.

BSc Computer Studies (Visualisation)
UCAS Code: G700 • Mode: 4 Years Single Subject

Qualifications/Requirements: GCE, A/AS: 12-14 points. AGNVQ, Merit (in specific programmes). ND/C, Merit overall (in specific programmes). SQAH, Grades BCCC. SQAV, Individual consideration. IB, 26 points.

BSc Information Systems
UCAS Code: G520 • Mode: 4 Years Single Subject

Qualifications/Requirements: GCE, A/AS: 12-14 points. AGNVQ, Merit (in specific programmes). ND/C, Merit overall (in specific programmes). SQAH, Grades BCCC. SQAV, Individual consideration. IB, 26 points.

BSc Information Systems (Artificial Intelligence)
UCAS Code: G5G8 • Mode: 4 Years Major/Minor

Qualifications/Requirements: GCE, A/AS: 12-14 points. AGNVQ, Merit (in specific programmes). ND/C, Merit overall (in specific programmes). SQAH, Grades BCCC. SQAV, Individual consideration. IB, 26 points.

BSc Information Systems (Networks)
UCAS Code: G5G6 • Mode: 4 Years Major/Minor

Qualifications/Requirements: GCE, A/AS: 12-14 points. AGNVQ, Merit (in specific programmes). ND/C, Merit overall (in specific programmes). SQAH, Grades BCCC. SQAV, Individual consideration. IB, 26 points.

BSc Mathematical & Computer Studies
UCAS Code: GG51 • Mode: 4 Years Equal Combination

Qualifications/Requirements: GCE, A/AS: 10 points. Maths at A level. AGNVQ, Merit with A level. ND/C, 3 Merits (in specific programmes). SQAH, Grades CCCD including specific subject(s). SQAV, Individual consideration. IB, Pass in Diploma including specific subjects.

BSc Mathematics, Statistics and Computing
UCAS Code: G900 • Mode: 4 Years Single Subject

Qualifications/Requirements: GCE, A/AS: 10 points. Maths at A level. AGNVQ, Merit with A level. ND/C, 3 Merits (in specific programmes). SQAH, Grades CCCD including specific subject(s). SQAV, Individual consideration. IB, Pass in Diploma including specific subjects.

BEng Computer Science
UCAS Code: G502 • Mode: 3/4 Years Single Subject

Qualifications/Requirements: GCE, A/AS: 18 points. Maths. AGNVQ, Distinction (in specific programmes). ND/C, Merit overall and 3 Distinctions (in specific programmes). SQAH, Grades BBCC including specific subject(s). SQAV, Individual consideration. IB, 30 points.

HND Computer Studies
UCAS Code: 105G • Mode: 2 Years Single Subject

Qualifications/Requirements: GCE, A/AS: 4 points. AGNVQ, Pass. ND/C, National. SQAH, Grades CCC. SQAV, Individual consideration. IB, Pass in Diploma.

DONCASTER COLLEGE D52

BSc Business Computing
UCAS Code: G520 • Mode: 3 Years Single Subject

Qualifications/Requirements: Please refer to prospectus.

BSc Integrated Business Technology
UCAS Code: GN51 • Mode: 1 Year Equal Combination

Qualifications/Requirements: ND/C, Higher National. SQAH, Individual consideration. SQAV, Higher National. IB, Individual consideration.

HND Business & Management Information Systems
UCAS Code: 51GN • Mode: 2 Years Equal Combination

Qualifications/Requirements: GCE, A/AS: 6 points. AGNVQ, Pass. ND/C, 4 Merits.

HND Business Information Technology
UCAS Code: 025G • Mode: 2 Years Single Subject

Qualifications/Requirements: GCE, A level grades: EE. AGNVQ, Pass. ND/C, National. SQAH, Individual consideration. SQAV, Individual consideration. IB, Individual consideration.

HND Computer Studies/Software Engineering
UCAS Code: 005G • Mode: 2 Years Single Subject

Qualifications/Requirements: GCE, A level grades: EE. AGNVQ, Pass. ND/C, National. SQAH, Individual consideration. SQAV, Individual consideration. IB, Individual consideration.

DUDLEY COLLEGE OF TECHNOLOGY D58

HND Computing
UCAS Code: 005G • Mode: 2 Years Single Subject

Qualifications/Requirements: GCE, A/AS: 2 points. AGNVQ, Individual consideration. ND/C, Individual consideration.

UNIVERSITY OF DUNDEE D65

BSc Applied Computing

UCAS Code: G510 • Mode: 4 Years Single Subject

Qualifications/Requirements: GCE, A/AS: 18 points. Science at A level. AGNVQ, Merit in Science. ND/C, 5 Merits (in specific programmes). SQAH, Grades BBBB including specific subject(s). SQAV, National including specific subject(s). IB, 25 points.

BSc Applied Computing

UCAS Code: G511 • Mode: 3 Years Single Subject

Qualifications/Requirements: GCE, A/AS: 10 points. Science at A level. AGNVQ, Merit in Science. ND/C, 5 Merits (in specific programmes). SQAH, Grades BBCC including specific subject(s). SQAV, National including specific subject(s). IB, 25 points.

BSc Applied Computing and Economics

UCAS Code: GL51 • Mode: 4 Years Equal Combination

Qualifications/Requirements: GCE, A/AS: 14 points. Science at A level. AGNVQ, Merit in Science. ND/C, 5 Merits (in specific programmes). SQAH, Grades BBBC including specific subject(s). SQAV, National including specific subject(s). IB, 25 points.

BSc Applied Computing and Financial Economics

UCAS Code: GL5C • Mode: 4 Years Equal Combination

Qualifications/Requirements: GCE, A/AS: 14 points. Science at A level. AGNVQ, Merit in Science. ND/C, 5 Merits (in specific programmes). SQAH, Grades BBBC including specific subject(s). SQAV, National including specific subject(s). IB, 25 points.

BSc Chemistry and Applied Computing

UCAS Code: FG15 • Mode: 4 Years Equal Combination

Qualifications/Requirements: GCE, A/AS: 14 points. Two Sciences at A level. AGNVQ, Merit in Science. ND/C, 5 Merits (in specific programmes). SQAH, Grades BBBC including specific subject(s). SQAV, National including specific subject(s). IB, 25 points.

BSc Computing and Cognitive Science

UCAS Code: CG85 • Mode: 4 Years Equal Combination

Qualifications/Requirements: GCE, A/AS: 14 points. Science at A level. AGNVQ, Merit in Science. ND/C, 5 Merits (in specific programmes). SQAH, Grades BBBC including specific subject(s). SQAV, National including specific subject(s). IB, 25 points.

BSc Mathematics and Applied Computing

UCAS Code: GG51 • Mode: 3 Years Equal Combination

Qualifications/Requirements: GCE, A/AS: 10 points. Maths. AGNVQ, Merit in Science with A level. ND/C, 5 Merits (in specific programmes). SQAH, Grades BBCC including specific subject(s). SQAV, National including specific subject(s). IB, 25 points.

BSc Physics and Applied Computing

UCAS Code: FG35 • Mode: 4 Years Equal Combination

Qualifications/Requirements: GCE, A/AS: 14 points. Maths and Science at A level. AGNVQ, Merit in Science with A level. ND/C, 5 Merits (in specific programmes). SQAH, Grades BBBC including specific subject(s). SQAV, National including specific subject(s). IB, 25 points.

BSc Psychology and Applied Computing

UCAS Code: LG75 • Mode: 4 Years Equal Combination

Qualifications/Requirements: GCE, A/AS: 14 points. Science at A level. AGNVQ, Merit in Science. ND/C, 5 Merits (in specific programmes). SQAH, Grades BBBC including specific subject(s). SQAV, National including specific subject(s). IB, 25 points.

THE UNIVERSITY OF DURHAM D86

BSc Computer Science

UCAS Code: G500 • Mode: 3 Years Single Subject

Qualifications/Requirements: GCE, A/AS: 22 points. Two Sciences. AGNVQ, Individual consideration. ND/C, Individual consideration. SQAH, Grades AABBB including specific subject(s). SQAV, Individual consideration. IB, 30 points.

BSc Computer Science (European Studies)

UCAS Code: G501 • Mode: 3 Years Single Subject

Qualifications/Requirements: GCE, A/AS: 22 points. Two Sciences. AGNVQ, Individual consideration. ND/C, Individual consideration. SQAH, Grades AABBB including specific subject(s). SQAV, Individual consideration. IB, 30 points.

BSc Computer Science and Mathematics

UCAS Code: GG51 • Mode: 3 Years Equal Combination

Qualifications/Requirements: GCE, A level grades: ABC-ABB. Maths at A level and Science. AGNVQ, Not normally sufficient. ND/C, Not normally sufficient. SQAH, CSYS required. SQAV, Not normally sufficient. IB, 32 points.

BSc Information Systems Management

UCAS Code: G520 • Mode: 3 Years Single Subject

Qualifications/Requirements: GCE, A/AS: 22 points. Two Sciences at A level. AGNVQ, Individual consideration. ND/C, Individual consideration. SQAH, Grades AABBB including specific subject(s). SQAV, Individual consideration. IB, 30 points.

UNIVERSITY OF EAST ANGLIA E14

BSc Accounting Information Systems

UCAS Code: NG45 • Mode: 3 Years Equal Combination

Qualifications/Requirements: GCE, A level grades: BBB-BBC. AGNVQ, Distinction with 6 additional units. ND/C, 3 Merits and 3 Distinctions. SQAH, Grades BBBBB. SQAV, Individual consideration. IB, 30 points.

BSc Business Information Systems
UCAS Code: GN54 • Mode: 3 Years Equal Combination

Qualifications/Requirements: GCE, A level grades: BBB-BBC. AGNVQ, Not normally sufficient. ND/C, Individual consideration. SQAH, Grades BBBBB. SQAV, Individual consideration. IB, 30 points.

BSc Computing and Mathematics
UCAS Code: GG51 • Mode: 3 Years Equal Combination

Qualifications/Requirements: GCE, A level grades: BC-CCC. Maths at A level. AGNVQ, Pass. ND/C, Individual consideration. SQAH, Grades BBBC including specific subject(s). SQAV, Not normally sufficient. IB, 32 points.

BSc Computing for Computer Graphics
UCAS Code: G505 • Mode: 3 Years Single Subject

Qualifications/Requirements: GCE, A level grades: BBC-BCC. AGNVQ, Distinction with A level. ND/C, Merit overall and 3 Distinctions. SQAH, Grades BBBB. IB, 28 points.

BSc Computing Science
UCAS Code: G500 • Mode: 3 Years Single Subject

Qualifications/Requirements: GCE, A level grades: BBC-BCC. AGNVQ, Distinction with A level. ND/C, Merit overall and 3 Distinctions. SQAH, Grades BBBB. SQAV, Higher National. IB, 28 points.

BSc Computing Science and Systems
UCAS Code: G507 • Mode: 3 Years Single Subject

Qualifications/Requirements: GCE, A level grades: BBC-BCC. AGNVQ, Distinction with A level. ND/C, Merit overall and 3 Distinctions. SQAH, Grades BBBB. IB, 28 points.

BSc Computing Science with a year in North America
UCAS Code: G502 • Mode: 3 Years Single Subject

Qualifications/Requirements: GCE, A level grades: BBB-BBC. Maths at A level. AGNVQ, Not normally sufficient. ND/C, Individual consideration. SQAH, Grades BBBBB. SQAV, Higher National. IB, 30 points.

BSc Computing, Applied
UCAS Code: G510 • Mode: 3 Years Single Subject

Qualifications/Requirements: GCE, A level grades: BBC-BCC. AGNVQ, Distinction with A level. ND/C, Merit overall and 3 Distinctions. SQAH, Grades BBBB. SQAV, Individual consideration. IB, 28 points.

BSc Computing, Applied (2nd year in North America)
UCAS Code: G511 • Mode: 3 Years Single Subject

Qualifications/Requirements: GCE, A level grades: BBB. AGNVQ, Not normally sufficient. ND/C, Individual consideration. SQAH, Grades BBBBB. SQAV, Higher National. IB, 30 points.

BSc Information Systems and Software Engineering
UCAS Code: G515 • Mode: 3 Years Single Subject

Qualifications/Requirements: GCE, A level grades: BBC-BCC. AGNVQ, Distinction with A level. ND/C, Merit overall and 3 Distinctions. SQAH, Grades BBBB. IB, 28 points.

UNIVERSITY OF EAST LONDON E28

BA Accounting & Finance/Information Technology
UCAS Code: GN54 • Mode: 3 Years Equal Combination

Qualifications/Requirements: Please refer to prospectus.

BA Business Studies/Computing & Business Information Systems
UCAS Code: N1G5 • Mode: 3 Years Major/Minor

Qualifications/Requirements: GCE, A/AS: 10 points. AGNVQ, Merit in Business. ND/C, 6. SQAH, Individual consideration. SQAV, Individual consideration. IB, 24 points.

BA Business Studies/Information Technology
UCAS Code: NG15 • Mode: 3 Years Equal Combination

Qualifications/Requirements: GCE, A/AS: 12 points. AGNVQ, Merit. ND/C, Merit overall. SQAH, Individual consideration. SQAV, Individual consideration. IB, Individual consideration.

BA Communication Studies/Information Technology
UCAS Code: GP53 • Mode: 3 Years Equal Combination

Qualifications/Requirements: GCE, A/AS: 12 points. AGNVQ, Merit (in specific programmes). ND/C, Merit overall. SQAH, Individual consideration. SQAV, Individual consideration. IB, Individual consideration.

BA Computing & Business Information Systems/Business Studies
UCAS Code: G5N1 • Mode: 3 Years Major/Minor

Qualifications/Requirements: GCE, A/AS: 10 points. AGNVQ, Merit (in specific programmes). ND/C, 6 Merits. SQAH, Individual consideration. SQAV, Individual consideration. IB, 24 points.

BA Computing & Business Information Systems/French
UCAS Code: GR5C • Mode: 3 Years Equal Combination

Qualifications/Requirements: GCE, A/AS: 12 points. AGNVQ, Merit. ND/C, Merit overall. SQAH, Individual consideration. SQAV, Individual consideration. IB, Individual consideration.

BA Computing & Business Information Systems/German
UCAS Code: GR5F • Mode: 3 Years Equal Combination

Qualifications/Requirements: GCE, A/AS: 12 points. AGNVQ, Merit. ND/C, Merit overall. SQAH, Individual consideration. SQAV, Individual consideration. IB, Individual consideration.

BA Computing & Business Information Systems/Italian

UCAS Code: G5RH • Mode: 3 Years Major/Minor

Qualifications/Requirements: GCE, A/AS: 12 points. AGNVQ, Merit. ND/C, Merit overall. SQAH, Individual consideration. SQAV, Individual consideration. IB, Individual consideration.

BA Computing & Business Information Systems/Spanish

UCAS Code: GR5K • Mode: 3 Years Equal Combination

Qualifications/Requirements: GCE, A/AS: 12 points. AGNVQ, Merit. ND/C, Merit overall. SQAH, Individual consideration. SQAV, Individual consideration. IB, Individual consideration.

BA Cultural Studies/Information Technology

UCAS Code: GL5P • Mode: 3 Years Equal Combination

Qualifications/Requirements: GCE, A/AS: 14 points. AGNVQ, Merit. ND/C, Merit overall. SQAH, Individual consideration. SQAV, Individual consideration. IB, Individual consideration.

BA Education & Community Studies/Information Technology

UCAS Code: GX59 • Mode: 3 Years Equal Combination

Qualifications/Requirements: GCE, A/AS: 12 points. AGNVQ, Merit (in specific programmes). ND/C, Merit overall. SQAH, Individual consideration. SQAV, Individual consideration. IB, Individual consideration.

BA European Studies/Information Technology

UCAS Code: T2G5 • Mode: 3 Years Major/Minor

Qualifications/Requirements: GCE, A/AS: 12 points. AGNVQ, Merit. ND/C, Merit overall. SQAH, Individual consideration. SQAV, Individual consideration. IB, Individual consideration.

BA Information Technology/Gender & Women's Studies

UCAS Code: GM59 • Mode: 3 Years Equal Combination

Qualifications/Requirements: GCE, A/AS: 12 points. AGNVQ, Merit. ND/C, Merit overall. SQAH, Individual consideration. SQAV, Individual consideration. IB, Individual consideration.

BA Information Technology/Linguistics

UCAS Code: GQ51 • Mode: 3 Years Equal Combination

Qualifications/Requirements: GCE, A/AS: 12 points. AGNVQ, Not normally sufficient. ND/C, Merit overall. SQAH, Individual consideration. SQAV, Individual consideration. IB, Individual consideration.

BA Information Technology/New Technology

UCAS Code: G525 • Mode: 3 Years Single Subject

Qualifications/Requirements: Please refer to prospectus.

BA Information Technology/Politics

UCAS Code: MG15 • Mode: 3 Years Equal Combination

Qualifications/Requirements: GCE, A/AS: 12 points. AGNVQ, Merit. ND/C, Merit overall. SQAH, Individual consideration. SQAV, Individual consideration. IB, Individual consideration.

BA Information Technology/Social Research

UCAS Code: GL54 • Mode: 3 Years Equal Combination

Qualifications/Requirements: GCE, A/AS: 12 points. AGNVQ, Merit. ND/C, Merit overall. SQAH, Individual consideration. SQAV, Individual consideration. IB, Individual consideration.

BA Information Technology/Spanish

UCAS Code: GR54 • Mode: 3 Years Equal Combination

Qualifications/Requirements: GCE, A/AS: 12 points. AGNVQ, Merit with A/AS. ND/C, Merit overall. SQAH, Individual consideration. SQAV, Individual consideration. IB, Individual consideration.

BA Law/Information Technology

UCAS Code: GM53 • Mode: 3 Years Equal Combination

Qualifications/Requirements: GCE, A/AS: 12 points. AGNVQ, Distinction. ND/C, Merit overall. SQAH, Individual consideration. SQAV, Individual consideration. IB, Individual consideration.

BA Social Sciences/Information Technology

UCAS Code: GL5H • Mode: 3 Years Equal Combination

Qualifications/Requirements: GCE, A/AS: 12 points. AGNVQ, Merit in Business. ND/C, Merit overall. SQAH, Individual consideration. SQAV, Individual consideration. IB, Individual consideration.

BA Third World Development/ Information Technology

UCAS Code: GM5Y • Mode: 3 Years Equal Combination

Qualifications/Requirements: GCE, A/AS: 12 points. AGNVQ, Merit. ND/C, Merit overall. SQAH, Individual consideration. SQAV, Individual consideration. IB, Individual consideration.

BSc Anthropology/Information Technology

UCAS Code: LG6M • Mode: 3 Years Equal Combination

Qualifications/Requirements: GCE, A/AS: 12 points. AGNVQ, Merit. ND/C, Merit overall. SQAH, Individual consideration. SQAV, Individual consideration. IB, Individual consideration.

BSc Archaeological Sciences/Information Technology

UCAS Code: FG45 • Mode: 3 Years Equal Combination

Qualifications/Requirements: GCE, A/AS: 12 points. AGNVQ, Merit (in specific programmes). ND/C, Merit overall. SQAH, Individual consideration. SQAV, Individual consideration. IB, Individual consideration.

BSc Biology/Information Technology
UCAS Code: CG15 • Mode: 3 Years Equal Combination
Qualifications/Requirements: GCE, A/AS: 12 points. AGNVQ, Merit (in specific programmes). ND/C, Merit overall. SQAH, Individual consideration. SQAV, Individual consideration. IB, Individual consideration.

BSc Business Information Systems
UCAS Code: G520 • Mode: 3/4 Years Single Subject
Qualifications/Requirements: GCE, A/AS: 10 points. AGNVQ, Merit in Information Technology. ND/C, 6 Merits. SQAH, Individual consideration. SQAV, Individual consideration. IB, 24 points.

BSc Computing
UCAS Code: G500 • Mode: 3 Years Single Subject
Qualifications/Requirements: GCE, A/AS: 10 points. AGNVQ, Merit in Information Technology. ND/C, 6 Merits. SQAH, Individual consideration. SQAV, Individual consideration. IB, 24 points.

BSc Computing & Business Information Systems with English Language Studies
UCAS Code: G5Q3 • Mode: 3 Years Major/Minor
Qualifications/Requirements: Please refer to prospectus.

BSc Computing and Electronics
UCAS Code: GH56 • Mode: 3/4 Years Equal Combination
Qualifications/Requirements: GCE, A/AS: 10 points. Maths or Physics. AGNVQ, Merit in Engineering. ND/C, 3 Merits. SQAH, Individual consideration. SQAV, Individual consideration. IB, Individual consideration.

BSc Distributed Information Systems
UCAS Code: G521 • Mode: 4 Years Single Subject
Qualifications/Requirements: GCE, A/AS: 10 points. AGNVQ, Merit in Information Technology. ND/C, 6 Merits. SQAH, Individual consideration. SQAV, Individual consideration. IB, 24 points.

BSc Environmental Sciences/Information Technology
UCAS Code: GF59 • Mode: 3 Years Equal Combination
Qualifications/Requirements: GCE, A/AS: 12 points. AGNVQ, Merit (in specific programmes). ND/C, Merit overall. SQAH, Individual consideration. SQAV, Individual consideration. IB, Individual consideration.

BSc Geographical Information Systems
UCAS Code: G5F8 • Mode: 3 Years Major/Minor
Qualifications/Requirements: GCE, A/AS: 12 points. AGNVQ, Merit. ND/C, 3 Merits. SQAH, Individual consideration. SQAV, Individual consideration. IB, Individual consideration.

BSc Information Systems
UCAS Code: G522 • Mode: 4 Years Single Subject
Qualifications/Requirements: GCE, A/AS: 10 points. AGNVQ, Merit (in specific programmes). ND/C, 6 Merits. SQAH, Individual consideration. SQAV, Individual consideration. IB, Individual consideration.

BSc Information Technology and Training
UCAS Code: G560 • Mode: 3 Years Single Subject
Qualifications/Requirements: Please refer to prospectus.

BSc Information Technology with English Language Studies
UCAS Code: G5QH • Mode: 3 Years Major/Minor
Qualifications/Requirements: Please refer to prospectus.

BSc Information Technology/Economics
UCAS Code: G5L1 • Mode: 3 Years Major/Minor
Qualifications/Requirements: GCE, A/AS: 12 points. AGNVQ, Merit. ND/C, Merit overall. SQAH, Individual consideration. SQAV, Individual consideration. IB, Individual consideration.

BSc Information Technology/European Studies
UCAS Code: G5T2 • Mode: 3 Years Major/Minor
Qualifications/Requirements: GCE, A/AS: 12 points. AGNVQ, Merit. ND/C, Merit overall. SQAH, Individual consideration. SQAV, Individual consideration. IB, Individual consideration.

BSc Information Technology/Health Studies
UCAS Code: G5B9 • Mode: 3 Years Major/Minor
Qualifications/Requirements: GCE, A/AS: 12 points. AGNVQ, Merit. ND/C, Merit overall. SQAH, Individual consideration. SQAV, Individual consideration. IB, Individual consideration.

BSc Information Technology/Italian
UCAS Code: G5R3 • Mode: 3 Years Major/Minor
Qualifications/Requirements: GCE, A/AS: 12 points. AGNVQ, Merit. ND/C, Merit overall. SQAH, Individual consideration. SQAV, Individual consideration. IB, Individual consideration.

BSc Information Technology/Printed Textiles and Surface Decoration
UCAS Code: G5J4 • Mode: 3 Years Major/Minor
Qualifications/Requirements: GCE, A/AS: 12 points. AGNVQ, Merit. ND/C, Merit overall. SQAH, Individual consideration. SQAV, Individual consideration. IB, Individual consideration.

BSc Information Technology/Psychosocial Studies
UCAS Code: GL5T • Mode: 3 Years Equal Combination
Qualifications/Requirements: GCE, A/AS: 12 points. SQAH, Individual consideration. SQAV, Individual consideration. IB, Individual consideration.

BSc Spatial Business Informatics
UCAS Code: G523 • Mode: 3 Years Single Subject
Qualifications/Requirements: GCE, A/AS: 12 points. AGNVQ, Merit. ND/C, 3 Merits. SQAH, Individual consideration. SQAV, Individual consideration. IB, Individual consideration.

HND Business and Computing
UCAS Code: 51GN • Mode: 2 Years Equal Combination

Qualifications/Requirements: GCE, A/AS: 4 points. AGNVQ, Pass (in specific programmes). ND/C, 1 Merit. SQAH, Individual consideration. SQAV, Individual consideration. IB, Individual consideration.

HND Computing
UCAS Code: 105G • Mode: 2 Years Single Subject

Qualifications/Requirements: GCE, A/AS: 4 points. AGNVQ, Pass in Information Technology. ND/C, 1 Merit. SQAH, Individual consideration. SQAV, Individual consideration. IB, Individual consideration.

EDGE HILL COLLEGE OF HIGHER EDUCATION E42

BA Information Systems and Communication & Media
UCAS Code: GP53 • Mode: 3 Years Equal Combination

Qualifications/Requirements: GCE, A level grades: CC. AGNVQ, Merit or Pass with A/AS. ND/C, 3 Merits and 3 Distinctions. SQAH, Grades BBCC. SQAV, Individual consideration. IB, Pass in Diploma.

BSc Business Information Systems
UCAS Code: G522 • Mode: 3 Years Single Subject

Qualifications/Requirements: GCE, A level grades: CC. AGNVQ, Merit or Pass with A/AS. ND/C, 3 Merits and 3 Distinctions. SQAH, Grades BBCC. SQAV, Individual consideration. IB, Pass in Diploma.

BSc Geography and Information Systems
UCAS Code: GL58 • Mode: 3 Years Equal Combination

Qualifications/Requirements: GCE, A level grades: CD. Geography and Environmental Science at A level. AGNVQ, Pass with A level. ND/C, Individual consideration. SQAH, Grades BBCC including specific subject(s). SQAV, Individual consideration. IB, Pass in Diploma.

BSc Information Systems
UCAS Code: G520 • Mode: 3 Years Single Subject

Qualifications/Requirements: GCE, A level grades: CD. AGNVQ, Merit or Pass with A/AS. ND/C, 3 Merits and 3 Distinctions. SQAH, Grades BBCC. SQAV, Individual consideration. IB, Pass in Diploma.

BSc Information Systems and Mathematics
UCAS Code: GG15 • Mode: 3 Years Equal Combination

Qualifications/Requirements: GCE, A level grades: CD. Maths at A level. AGNVQ, Merit or Pass with A/AS. ND/C, 3 Merits and 3 Distinctions. SQAH, Grades BBCC. SQAV, Individual consideration. IB, Pass in Diploma.

BSc Information Systems with Human Geography
UCAS Code: G5L8 • Mode: 3 Years Major/Minor

Qualifications/Requirements: GCE, A level grades: CC. AGNVQ, Merit or Pass with A/AS. ND/C, 3 Merits and 3 Distinctions. SQAH, Grades BBCC. SQAV, Individual consideration. IB, Pass in Diploma.

BSc Information Systems with Mathematics
UCAS Code: G5G1 • Mode: 3 Years Major/Minor

Qualifications/Requirements: GCE, A level grades: CC. AGNVQ, Merit or Pass with A/AS. ND/C, 3 Merits and 3 Distinctions. SQAH, Grades BBCC. SQAV, Individual consideration. IB, Pass in Diploma.

BSc Information Systems with Women's Studies
UCAS Code: G5M9 • Mode: 3 Years Major/Minor

Qualifications/Requirements: GCE, A level grades: CC. AGNVQ, Merit or Pass with A/AS. ND/C, 3 Merits and 3 Distinctions. SQAH, Grades BBCC. SQAV, Individual consideration. IB, Pass in Diploma.

BSc Information Systems with Writing
UCAS Code: G5W9 • Mode: 3 Years Major/Minor

Qualifications/Requirements: GCE, A level grades: CC. AGNVQ, Merit or Pass with A/AS. ND/C, 3 Merits and 3 Distinctions. SQAH, Grades BBCC. SQAV, Individual consideration. IB, Pass in Diploma.

BSc Information Technology
UCAS Code: X6G5 • Mode: 3 Years Major/Minor

Qualifications/Requirements: GCE, A level grades: DD. AGNVQ, Merit in Manufacturing or Pass (in specific programmes) with A/AS. ND/C, 4 Merits and 2 Distinctions. SQAH, Grades BBCC. SQAV, Individual consideration. IB, Pass in Diploma.

BSc Marketing and Information Systems
UCAS Code: NG15 • Mode: 3 Years Equal Combination

Qualifications/Requirements: GCE, A level grades: CC. AGNVQ, Merit or Pass with A/AS. ND/C, Individual consideration. SQAH, Grades BBCC. SQAV, Individual consideration. IB, Pass in Diploma.

BSc Sports Studies and Information Systems
UCAS Code: BG65 • Mode: 3 Years Equal Combination

Qualifications/Requirements: GCE, A level grades: BC. AGNVQ, Distinction or Merit with A/AS. ND/C, 2 Merits and 4 Distinctions. SQAH, Grades BBCC. SQAV, Individual consideration. IB, Pass in Diploma.

THE UNIVERSITY OF EDINBURGH E56

BSc Computer Science

UCAS Code: G500 • Mode: 4 Years Single Subject

Qualifications/Requirements: GCE, A level grades: CCC. Maths at A level. AGNVQ, Pass. ND/C, Merit overall (in specific programmes). SQAH, Grades BBBB including specific subject(s). SQAV, National including specific subject(s). IB, Pass in Diploma including specific subjects.

BSc Computer Science and Management Science

UCAS Code: GN51 • Mode: 4 Years Equal Combination

Qualifications/Requirements: GCE, A level grades: CCC. Maths at A level. AGNVQ, Pass. ND/C, Merit overall (in specific programmes). SQAH, Grades BBBB including specific subject(s). SQAV, National including specific subject(s). IB, Pass in Diploma including specific subjects.

BSc Computer Science and Physics

UCAS Code: GF53 • Mode: 4 Years Equal Combination

Qualifications/Requirements: GCE, A level grades: CCC. Maths and Physics at A level. AGNVQ, Pass. ND/C, Merit overall (in specific programmes). SQAH, Grades BBBB including specific subject(s). SQAV, National including specific subject(s). IB, Pass in Diploma including specific subjects.

BEng Computer Science

UCAS Code: G501 • Mode: 4 Years Single Subject

Qualifications/Requirements: GCE, A level grades: CCC. Maths at A level. AGNVQ, Pass. ND/C, Merit overall (in specific programmes). SQAH, Grades BBBB including specific subject(s). SQAV, National including specific subject(s). IB, Pass in Diploma including specific subjects.

BEng Computer Science and Electronics

UCAS Code: GH56 • Mode: 4 Years Equal Combination

Qualifications/Requirements: GCE, A level grades: CCC. Maths and Physics at A level. AGNVQ, Pass. ND/C, Merit overall (in specific programmes). SQAH, Grades BBBB including specific subject(s). SQAV, National including specific subject(s). IB, Pass in Diploma including specific subjects.

BEng Electronics and Computer Science

UCAS Code: HG65 • Mode: 4 Years Equal Combination

Qualifications/Requirements: GCE, A level grades: CCC. Maths and Physics at A level. AGNVQ, Pass. ND/C, Merit overall (in specific programmes). SQAH, Grades BBBB including specific subject(s). SQAV, National including specific subject(s). IB, Pass in Diploma including specific subjects.

THE UNIVERSITY OF ESSEX E70

BSc Computer Science

UCAS Code: G500 • Mode: 3 Years Single Subject

Qualifications/Requirements: GCE, A/AS: 20 points. AGNVQ, Distinction. ND/C, Merit overall and 2 Distinctions. SQAH, Grades BBBB. SQAV, Individual consideration. IB, 28 points.

BSc Computer Science (Artificial Intelligence)

UCAS Code: G800 • Mode: 3 Years Single Subject

Qualifications/Requirements: GCE, A/AS: 20 points. AGNVQ, Distinction. ND/C, Merit overall and 2 Distinctions. SQAH, Grades BBBB. SQAV, Individual consideration. IB, 28 points.

BSc Computer Science (Robotics and Intelligent Machines)

UCAS Code: G5G8 • Mode: 3 Years Major/Minor

Qualifications/Requirements: GCE, A/AS: 20 points. AGNVQ, Distinction. ND/C, Merit overall and 2 Distinctions. SQAH, Grades BBBB. SQAV, Individual consideration. IB, 28 points.

BSc Computer Science (Software Engineering)

UCAS Code: G700 • Mode: 3 Years Single Subject

Qualifications/Requirements: GCE, A/AS: 20 points. AGNVQ, Distinction. ND/C, Merit overall and 2 Distinctions. SQAH, Grades BBBB. SQAV, Individual consideration. IB, 28 points.

BSc Computing and Economics

UCAS Code: GL51 • Mode: 3 Years Equal Combination

Qualifications/Requirements: GCE, A/AS: 20 points. AGNVQ, Distinction. ND/C, Merit overall and 2 Distinctions. SQAH, Grades BBBB. SQAV, Individual consideration. IB, 28 points.

BSc Computing and Management

UCAS Code: G5N3 • Mode: 3 Years Major/Minor

Qualifications/Requirements: GCE, A level grades: BCC. AGNVQ, Distinction. ND/C, Merit overall and 2 Distinctions. SQAH, Grades BBBB. SQAV, Individual consideration. IB, 28 points.

BSc Mathematics and Computing

UCAS Code: GG15 • Mode: 3 Years Equal Combination

Qualifications/Requirements: GCE, A/AS: 20 points. Maths at A level. AGNVQ, Distinction with A level. ND/C, Merit overall and 2 Distinctions (in specific programmes). SQAH, CSYS required. SQAV, Individual consideration. IB, 28 points.

UNIVERSITY OF EXETER E84

BSc Computer Science

UCAS Code: G500 • Mode: 3 Years Single Subject

Qualifications/Requirements: GCE, A/AS: 20 points. AGNVQ, Merit or Distinction (in specific programmes). ND/C, Merit overall and 2 Distinctions. SQAH, Grades BBBBB. SQAV, Individual consideration. IB, 29 points.

BSc Computer Science with European Study

UCAS Code: G5T2 • Mode: 4 Years Major/Minor

Qualifications/Requirements: GCE, A/AS: 20 points. AGNVQ, Merit or Distinction (in specific programmes). ND/C, Merit overall and 2 Distinctions. SQAH, Grades BBBBB. SQAV, Individual consideration. IB, 29 points.

BSc Computer Science with Mathematical Studies
UCAS Code: G5GD • Mode: 3 Years Major/Minor
Qualifications/Requirements: GCE, A/AS: 20 points. Maths at A level. AGNVQ, Merit (in specific programmes) with A level. ND/C, Merit overall and 2 Distinctions. SQAH, Grades BBBBB. SQAV, Individual consideration. IB, 29 points.

HND Computing (Truro College)
UCAS Code: 005G • Mode: 2 Years Single Subject
Qualifications/Requirements: GCE, A/AS: 4 points. AGNVQ, Pass. ND/C, 1 Merit. SQAH, Grades CCCC. SQAV, Individual consideration. IB, 24 points.

FARNBOROUGH COLLEGE OF TECHNOLOGY F66

BSc Computing
UCAS Code: G500 • Mode: 3 Years Single Subject
Qualifications/Requirements: GCE, A level grades: DE. AGNVQ, Pass. ND/C, Individual consideration. SQAH, Individual consideration. SQAV, Individual consideration. IB, Individual consideration.

HND Business Information Technology
UCAS Code: 265G • Mode: 2 Years Single Subject
Qualifications/Requirements: GCE, A level grades: E. AGNVQ, Pass. ND/C, National. SQAH, Individual consideration. SQAV, Individual consideration. IB, Individual consideration.

HND Computing
UCAS Code: 005G • Mode: 2 Years Single Subject
Qualifications/Requirements: GCE, A level grades: E. AGNVQ, Pass. ND/C, National. SQAH, Individual consideration. SQAV, Individual consideration. IB, Individual consideration.

UNIVERSITY OF GLAMORGAN G14

BSc Accounting and Information Systems
UCAS Code: G5N4 • Mode: 3/4 Years Major/Minor
Qualifications/Requirements: GCE, A/AS: 12 points. AGNVQ, Merit (in specific programmes). ND/C, Merit overall. SQAH, Individual consideration. SQAV, Individual consideration. IB, Individual consideration.

BSc Computer Studies
UCAS Code: G501 • Mode: 3/4 Years Single Subject
Qualifications/Requirements: GCE, A/AS: 12 points. AGNVQ, Merit (in specific programmes). ND/C, Individual consideration. SQAH, Individual consideration. SQAV, Individual consideration. IB, Individual consideration.

BSc Computer Studies (Foundation Year)
UCAS Code: G508 • Mode: 4 Years Single Subject
Qualifications/Requirements: AGNVQ, Individual consideration. ND/C, Individual consideration. SQAH, Individual consideration. SQAV, Individual consideration. IB, Individual consideration.

BSc Computing and Business
UCAS Code: GN51 • Mode: 3/4 Years Equal Combination
Qualifications/Requirements: GCE, A/AS: 12 points. AGNVQ, Merit (in specific programmes). ND/C, Individual consideration. SQAH, Individual consideration. SQAV, Individual consideration. IB, Individual consideration.

BSc Computing and Geographical Information Systems
UCAS Code: GG58 • Mode: 3/4 Years Equal Combination
Qualifications/Requirements: GCE, A/AS: 12 points. AGNVQ, Merit (in specific programmes). ND/C, Individual consideration. SQAH, Individual consideration. SQAV, Individual consideration. IB, Individual consideration.

BSc Computing with Business
UCAS Code: G5N1 • Mode: 3/4 Years Major/Minor
Qualifications/Requirements: GCE, A/AS: 12 points. AGNVQ, Merit (in specific programmes). ND/C, Individual consideration. SQAH, Individual consideration. SQAV, Individual consideration. IB, Individual consideration.

BSc Information Systems
UCAS Code: G521 • Mode: 3/4 Years Single Subject
Qualifications/Requirements: GCE, A/AS: 12 points. AGNVQ, Merit (in specific programmes). ND/C, Individual consideration. SQAH, Individual consideration. SQAV, Individual consideration. IB, Individual consideration.

BSc Information Systems and Business Studies
UCAS Code: GN5C • Mode: 3/4 Years Equal Combination
Qualifications/Requirements: Please refer to prospectus.

BSc Information Systems with Business Studies
UCAS Code: G5NC • Mode: 3/4 Years Major/Minor
Qualifications/Requirements: Please refer to prospectus.

BSc Information Technology
UCAS Code: G560 • Mode: 3/4 Years Single Subject
Qualifications/Requirements: GCE, A level grades: CD. Maths, Science or Computing at A level. AGNVQ, Merit. ND/C, 3 Merits. SQAH, Individual consideration. SQAV, Individual consideration. IB, Individual consideration.

BSc Information Technology
UCAS Code: G562 • Mode: 3 Years Single Subject
Qualifications/Requirements: GCE, A level grades: D. AGNVQ, Pass. ND/C, National. SQAH, Individual consideration. SQAV, Individual consideration. IB, Individual consideration.

BSc Information Technology with European Business Studies
UCAS Code: G564 • Mode: 3/4 Years Single Subject
Qualifications/Requirements: GCE, A level grades: CD. Maths, Science or Computing at A level. AGNVQ, Merit. ND/C, 3 Merits. SQAH, Individual consideration. SQAV, Individual consideration. IB, Individual consideration.

HND Business Information Technology
UCAS Code: 1N5G • Mode: 2 Years Major/Minor

Qualifications/Requirements: GCE, A/AS: 4 points. AGNVQ, Pass (in specific programmes). ND/C, 3 Merits. SQAH, Individual consideration. SQAV, Individual consideration. IB, Individual consideration.

HND Computing
UCAS Code: 105G • Mode: 2 Years Single Subject

Qualifications/Requirements: GCE, A/AS: 4 points. AGNVQ, Pass (in specific programmes). ND/C, 3 Merits. SQAH, Individual consideration. SQAV, Individual consideration. IB, Individual consideration.

HND Computing (Information Systems)
UCAS Code: 025G • Mode: 2 Years Single Subject

Qualifications/Requirements: GCE, A/AS: 4 points. AGNVQ, Pass (in specific programmes). ND/C, 3 Merits. SQAH, Individual consideration. SQAV, Individual consideration. IB, Individual consideration.

HND Information Technology
UCAS Code: 065G • Mode: 2 Years Single Subject

Qualifications/Requirements: GCE, A level grades: D. AGNVQ, Pass. ND/C, National. SQAH, Individual consideration. SQAV, Individual consideration. IB, Individual consideration.

UNIVERSITY OF GLASGOW G28

BSc Applied Mathematics and Computer Science
UCAS Code: GG51 • Mode: 4 Years Equal Combination

Qualifications/Requirements: Please refer to prospectus.

BSc Archaeology/Computing Science
UCAS Code: GV56 • Mode: 4 Years Equal Combination

Qualifications/Requirements: GCE, A level grades: BBC-CCC. Two Sciences. AGNVQ, Merit. ND/C, National. SQAH, Grades BBBB including specific subject(s). SQAV, National. IB, 24 points.

BSc Computing Science
UCAS Code: G500 • Mode: 4 Years Single Subject

Qualifications/Requirements: GCE, A level grades: BBC-CCC. Two Sciences. AGNVQ, Merit. ND/C, National. SQAH, Grades BBBB including specific subject(s). SQAV, National. IB, 24 points.

BSc Computing Science and Economics
UCAS Code: GL51 • Mode: 4 Years Equal Combination

Qualifications/Requirements: GCE, A level grades: BBC-CCC. Two Sciences. AGNVQ, Merit. ND/C, National. SQAH, Grades BBB including specific subject(s). SQAV, National. IB, 24 points.

BSc Computing Science and Mathematics
UCAS Code: GG5C • Mode: 4 Years Equal Combination

Qualifications/Requirements: Please refer to prospectus.

BSc Computing Science/Earth Science
UCAS Code: FG65 • Mode: 4 Years Equal Combination

Qualifications/Requirements: GCE, A level grades: BBC-CCC. Two Sciences. AGNVQ, Merit. ND/C, National. SQAH, Grades BBBB including specific subject(s). SQAV, National. IB, 24 points.

BSc Computing Science/Geography
UCAS Code: FG85 • Mode: 4 Years Equal Combination

Qualifications/Requirements: GCE, A level grades: BBC-CCC. Two Sciences. AGNVQ, Merit. ND/C, National. SQAH, Grades BBBB including specific subject(s). SQAV, National. IB, 24 points.

BSc Computing Science/Management Studies
UCAS Code: NG15 • Mode: 4 Years Equal Combination

Qualifications/Requirements: GCE, A level grades: BBC-CCC. Two Sciences. AGNVQ, Merit. ND/C, National. SQAH, Grades BBBB including specific subject(s). SQAV, National. IB, 24 points.

BSc Computing Science/Physiology (Neuroinformatics)
UCAS Code: GB51 • Mode: 4 Years Equal Combination

Qualifications/Requirements: GCE, A level grades: BBC-CCC. Two Sciences. AGNVQ, Merit. ND/C, National. SQAH, Grades BBBB including specific subject(s). SQAV, National. IB, 24 points.

BSc Computing Science/Psychology
UCAS Code: CG85 • Mode: 4 Years Equal Combination

Qualifications/Requirements: GCE, A level grades: BBC-CCC. Two Sciences. AGNVQ, Merit. ND/C, National. SQAH, Grades BBBB including specific subject(s). SQAV, National. IB, 24 points.

MA Archaeology/Computing
UCAS Code: VG65 • Mode: 4 Years Equal Combination

Qualifications/Requirements: GCE, A level grades: BBC. AGNVQ, Merit. ND/C, Higher National. SQAH, Grades BBBB. SQAV, Higher National. IB, 30 points.

MA Business Economics/Computing Science
UCAS Code: LGC5 • Mode: 4 Years Equal Combination

Qualifications/Requirements: GCE, A level grades: BBC. AGNVQ, Merit. ND/C, Merit. SQAH, Grades ABBB. SQAV, Higher National. IB, 30 points.

MA Celtic Civilisation/Computing
UCAS Code: GQ5M • Mode: 4 Years Equal Combination

Qualifications/Requirements: GCE, A level grades: BBC. AGNVQ, Merit. ND/C, Higher National. SQAH, Grades BBBB. SQAV, Higher National. IB, 30 points.

MA Celtic/Computing
UCAS Code: GQ55 • Mode: 4 Years Equal Combination

Qualifications/Requirements: GCE, A level grades: BBC. AGNVQ, Merit. ND/C, Higher National. SQAH, Grades BBBB. SQAV, Higher National. IB, 30 points.

MA Classical Civilisation/Computing
UCAS Code: GQ58 • Mode: 4 Years Equal Combination

Qualifications/Requirements: GCE, A level grades: BBC. AGNVQ, Merit. ND/C, Higher National. SQAH, Grades BBBB. SQAV, Higher National. IB, 30 points.

MA Computing Science/Economic & Social History
UCAS Code: VG35 • Mode: 4 Years Equal Combination

Qualifications/Requirements: GCE, A level grades: BBC. AGNVQ, Merit. ND/C, Merit. SQAH, Grades ABBB. SQAV, Higher National. IB, 30 points.

MA Computing Science/Economics
UCAS Code: GLN1 • Mode: 4 Years Equal Combination

Qualifications/Requirements: GCE, A level grades: BBC. AGNVQ, Merit. ND/C, Merit. SQAH, Grades ABBB. SQAV, Higher National. IB, 30 points.

MA Computing Science/Geography
UCAS Code: LG85 • Mode: 4 Years Equal Combination

Qualifications/Requirements: GCE, A level grades: BBC. AGNVQ, Merit. ND/C, Merit. SQAH, Grades ABBB. SQAV, Higher National. IB, 30 points.

MA Computing Science/Management Studies
UCAS Code: GN51 • Mode: 4 Years Equal Combination

Qualifications/Requirements: GCE, A level grades: BBC. AGNVQ, Merit. ND/C, Merit. SQAH, Grades ABBB. SQAV, Higher National. IB, 30 points.

MA Computing Science/Philosophy
UCAS Code: GVM7 • Mode: 4 Years Equal Combination

Qualifications/Requirements: GCE, A level grades: BBC. AGNVQ, Merit. ND/C, Merit. SQAH, Grades ABBB. SQAV, Higher National. IB, 30 points.

MA Computing Science/Politics
UCAS Code: MG15 • Mode: 4 Years Equal Combination

Qualifications/Requirements: GCE, A level grades: BBC. AGNVQ, Merit. ND/C, Merit. SQAH, Grades ABBB. SQAV, Higher National. IB, 30 points.

MA Computing Science/Psychology
UCAS Code: CG8M • Mode: 4 Years Equal Combination

Qualifications/Requirements: GCE, A level grades: BBC. AGNVQ, Merit. ND/C, Merit. SQAH, Grades ABBB. SQAV, Higher National. IB, 30 points.

MA Computing Science/Sociology
UCAS Code: LG35 • Mode: 4 Years Equal Combination

Qualifications/Requirements: GCE, A level grades: BBC. AGNVQ, Merit. ND/C, Merit. SQAH, Grades ABBB. SQAV, Higher National. IB, 30 points.

MA Computing/Czech
UCAS Code: GT51 • Mode: 5 Years Equal Combination

Qualifications/Requirements: GCE, A level grades: BBC. AGNVQ, Merit. ND/C, Higher National. SQAH, Grades BBBB. SQAV, Higher National. IB, 30 points.

MA Computing/Economic & Social History
UCAS Code: GV53 • Mode: 4 Years Equal Combination

Qualifications/Requirements: GCE, A level grades: BBC. AGNVQ, Merit. ND/C, Higher National. SQAH, Grades BBBB. SQAV, Higher National. IB, 30 points.

MA Computing/Economics
UCAS Code: GL5C • Mode: 4 Years Equal Combination

Qualifications/Requirements: GCE, A level grades: BBC. AGNVQ, Merit. ND/C, Higher National. SQAH, Grades BBBB. SQAV, Higher National. IB, 30 points.

MA Computing/English
UCAS Code: GQ53 • Mode: 4 Years Equal Combination

Qualifications/Requirements: GCE, A level grades: BBC. AGNVQ, Merit. ND/C, Higher National. SQAH, Grades BBBB. SQAV, Higher National. IB, 30 points.

MA Computing/French
UCAS Code: GR51 • Mode: 5 Years Equal Combination

Qualifications/Requirements: GCE, A level grades: BBC. AGNVQ, Merit. ND/C, Higher National. SQAH, Grades BBBB. SQAV, Higher National. IB, 30 points.

MA Computing/Geography
UCAS Code: GL58 • Mode: 4 Years Equal Combination

Qualifications/Requirements: GCE, A level grades: BBC. AGNVQ, Merit. ND/C, Higher National. SQAH, Grades BBBB. SQAV, Higher National. IB, 30 points.

MA Computing/Greek
UCAS Code: GQ57 • Mode: 4 Years Equal Combination

Qualifications/Requirements: GCE, A level grades: BBC. AGNVQ, Merit. ND/C, Higher National. SQAH, Grades BBBB. SQAV, Higher National. IB, 30 points.

MA Computing/History
UCAS Code: GV51 • Mode: 4 Years Equal Combination

Qualifications/Requirements: GCE, A level grades: BBC. AGNVQ, Merit. ND/C, Higher National. SQAH, Grades BBBB. SQAV, Higher National. IB, 30 points.

MA Computing/History of Art
UCAS Code: GV54 • Mode: 4 Years Equal Combination

Qualifications/Requirements: GCE, A level grades: BBC. AGNVQ, Merit. ND/C, Higher National. SQAH, Grades BBBB. SQAV, Higher National. IB, 30 points.

MA Computing/Latin
UCAS Code: GQ56 • Mode: 4 Years Equal Combination

Qualifications/Requirements: GCE, A level grades: BBC. AGNVQ, Merit. ND/C, Higher National. SQAH, Grades BBBB. SQAV, Higher National. IB, 30 points.

MA Computing/Management Studies
UCAS Code: NGC5 • Mode: 4 Years Equal Combination

Qualifications/Requirements: GCE, A level grades: BBC. AGNVQ, Merit. ND/C, Higher National. SQAH, Grades BBBB. SQAV, Higher National. IB, 30 points.

MA Computing/Music
UCAS Code: GW53 • Mode: 4 Years Equal Combination

Qualifications/Requirements: GCE, A level grades: BBB. Music. AGNVQ, Merit. ND/C, Higher National. SQAH, Grades ABBB. SQAV, Higher National. IB, 30 points.

MA Computing/Philosophy
UCAS Code: GV57 • Mode: 4 Years Equal Combination

Qualifications/Requirements: GCE, A level grades: BBC. AGNVQ, Merit. ND/C, Higher National. SQAH, Grades BBBB. SQAV, Higher National. IB, 30 points.

MA Computing/Politics
UCAS Code: GM51 • Mode: 4 Years Equal Combination

Qualifications/Requirements: GCE, A level grades: BBC. AGNVQ, Merit. ND/C, Higher National. SQAH, Grades BBBB. SQAV, Higher National. IB, 30 points.

MA Computing/Psychology
UCAS Code: GC58 • Mode: 4 Years Equal Combination

Qualifications/Requirements: GCE, A level grades: BBC. AGNVQ, Merit. ND/C, Higher National. SQAH, Grades BBBB. SQAV, Higher National. IB, 30 points.

MA Computing/Russian
UCAS Code: GR58 • Mode: 5 Years Equal Combination

Qualifications/Requirements: GCE, A level grades: BBC. AGNVQ, Merit. ND/C, Higher National. SQAH, Grades BBBB. SQAV, Higher National. IB, 30 points.

MA Computing/Scottish History
UCAS Code: GV5C • Mode: 4 Years Equal Combination

Qualifications/Requirements: GCE, A level grades: BBC. AGNVQ, Merit. ND/C, Higher National. SQAH, Grades BBBB. SQAV, Higher National. IB, 30 points.

MA Computing/Scottish Literature
UCAS Code: GQ52 • Mode: 4 Years Equal Combination

Qualifications/Requirements: GCE, A level grades: BBC. AGNVQ, Merit. ND/C, Higher National. SQAH, Grades BBBB. SQAV, Higher National. IB, 30 points.

MA Computing/Social and Urban Policy
UCAS Code: GL5K • Mode: 4 Years Equal Combination

Qualifications/Requirements: GCE, A level grades: BBC. AGNVQ, Merit. ND/C, 8 Merits. SQAH, Grades BBBB. SQAV, Higher National. IB, 30 points.

MA Computing/Sociology
UCAS Code: GL53 • Mode: 4 Years Equal Combination

Qualifications/Requirements: GCE, A level grades: BBC. AGNVQ, Merit. ND/C, Higher National. SQAH, Grades BBBB. SQAV, Higher National. IB, 30 points.

MA Computing/Theatre Studies
UCAS Code: GW54 • Mode: 4 Years Equal Combination

Qualifications/Requirements: GCE, A level grades: BBC. AGNVQ, Merit. ND/C, Higher National. SQAH, Grades BBBB. SQAV, Higher National. IB, 30 points.

GLASGOW CALEDONIAN UNIVERSITY G42

BA Business and Information Management
UCAS Code: N1G5 • Mode: 3/4 Years Major/Minor

Qualifications/Requirements: GCE, A level grades: CC. English at A level. AGNVQ, Individual consideration. ND/C, Individual consideration. SQAH, Grades BBB including specific subject(s). SQAV, Individual consideration. IB, Individual consideration.

BSc Computer Studies
UCAS Code: G500 • Mode: 3/5 Years Single Subject

Qualifications/Requirements: GCE, A level grades: CC-CDD. Maths, English, a modern foreign language and Computing. AGNVQ, Individual consideration. ND/C, Individual consideration. SQAH, Grades BBCC including specific subject(s). SQAV, Individual consideration. IB, Individual consideration.

BSc Information Technology Studies
UCAS Code: GG57 • Mode: 1/2 Years Equal Combination

Qualifications/Requirements: AGNVQ, Not normally sufficient. ND/C, Individual consideration. SQAH, Not normally sufficient. SQAV, Higher National. IB, Individual consideration.

GLOUCESTERSHIRE COLLEGE OF ARTS AND TECHNOLOGY G45

HND Computing
UCAS Code: 005G • Mode: 2 Years Single Subject

Qualifications/Requirements: Please refer to prospectus.

GOLDSMITHS COLLEGE (UNIVERSITY OF LONDON) G56

BSc Computer Science and Statistics
UCAS Code: GG54 • Mode: 3/4 Years Equal Combination

Qualifications/Requirements: GCE, A level grades: DD. Maths at A level. AGNVQ, Merit. ND/C, Merit overall. SQAH, Grades BBBBC. SQAV, National. IB, Pass in Diploma including specific subjects.

BSc Computing and Information Systems
UCAS Code: G520 • Mode: 3/4 Years Single Subject

Qualifications/Requirements: GCE, A level grades: CCC. AGNVQ, Merit. ND/C, Merit overall. SQAH, Grades BBBBC. SQAV, National. IB, Pass in Diploma including specific subjects.

BSc Computing, Operational Research and Statistics for Business
UCAS Code: G5N2 • Mode: 3/4 Years Major/Minor

Qualifications/Requirements: GCE, A level grades: DD. AGNVQ, Merit. ND/C, Merit overall. IB, Pass in Diploma including specific subjects.

BSc Mathematics and Computer Science

UCAS Code: GG15 • Mode: 3/4 Years Equal Combination

Qualifications/Requirements: GCE, A level grades: DD. Maths at A level. AGNVQ, Merit. ND/C, Merit overall. SQAH, Grades BCCCC. SQAV, National. IB, Pass in Diploma.

UNIVERSITY OF GREENWICH G70

BA Law & Information Systems

UCAS Code: GM53 • Mode: 3 Years Equal Combination

Qualifications/Requirements: Please refer to prospectus.

BSc Computing

UCAS Code: G504 • Mode: 3 Years Single Subject

Qualifications/Requirements: GCE, A/AS: 8-10 points. AGNVQ, Merit (in specific programmes). ND/C, 8 Merits. SQAH, Grades CCC. SQAV, Individual consideration. IB, Individual consideration.

BSc Computing & Mathematics

UCAS Code: GG51 • Mode: 3/4 Years Equal Combination

Qualifications/Requirements: GCE, A level grades: CE. Maths at A level. AGNVQ, Individual consideration. ND/C, 3 Merits (in specific programmes). SQAH, Grades CCC including specific subject(s). SQAV, Individual consideration. IB, Individual consideration.

BSc Computing & Statistics

UCAS Code: GG54 • Mode: 3/4 Years Equal Combination

Qualifications/Requirements: GCE, A level grades: CE. Maths at A level. AGNVQ, Individual consideration. ND/C, 3 Merits (in specific programmes). SQAH, Grades CCC. SQAV, Individual consideration. IB, Individual consideration.

BSc Computing Science

UCAS Code: G500 • Mode: 3/4 Years Single Subject

Qualifications/Requirements: GCE, A/AS: 8-10 points. AGNVQ, Merit (in specific programmes). ND/C, 8 Merits. SQAH, Grades CCC. SQAV, Individual consideration. IB, Individual consideration.

BSc Computing with Business Management

UCAS Code: G5NC • Mode: 3/4 Years Major/Minor

Qualifications/Requirements: GCE, A/AS: 8-10 points. AGNVQ, Merit (in specific programmes). ND/C, 8 Merits. SQAH, Grades CCC. SQAV, Individual consideration. IB, Individual consideration.

BSc Computing with French

UCAS Code: G5R1 • Mode: 3/4 Years Major/Minor

Qualifications/Requirements: GCE, A/AS: 8-10 points. AGNVQ, Merit. ND/C, 8 Merits. SQAH, Grades CCC. SQAV, Individual consideration. IB, Individual consideration.

BSc Computing with German

UCAS Code: G5R2 • Mode: 3/4 Years Major/Minor

Qualifications/Requirements: GCE, A/AS: 8-10 points. AGNVQ, Merit. ND/C, 8 Merits. SQAH, Grades CCC. SQAV, Individual consideration. IB, Individual consideration.

BSc Computing with Italian

UCAS Code: G5R3 • Mode: 3 Years Major/Minor

Qualifications/Requirements: Please refer to prospectus.

BSc Computing with Law

UCAS Code: G5M3 • Mode: 3 Years Major/Minor

Qualifications/Requirements: GCE, A/AS: 8-10 points. AGNVQ, Merit (in specific programmes). ND/C, 8 Merits. SQAH, Grades CCC. SQAV, Individual consideration. IB, Individual consideration.

BSc Computing with Mathematics

UCAS Code: G5G1 • Mode: 3/4 Years Major/Minor

Qualifications/Requirements: GCE, A/AS: 8-10 points. AGNVQ, Merit (in specific programmes). ND/C, 8 Merits. SQAH, Grades CCC. SQAV, Individual consideration. IB, Individual consideration.

BSc Computing with Spanish

UCAS Code: G5R4 • Mode: 3/4 Years Major/Minor

Qualifications/Requirements: GCE, A/AS: 8-10 points. AGNVQ, Merit. ND/C, 8 Merits. SQAH, Grades CCC. SQAV, Individual consideration. IB, Individual consideration.

BSc Computing with Statistics

UCAS Code: G5G4 • Mode: 3/4 Years Major/Minor

Qualifications/Requirements: GCE, A/AS: 8-10 points. AGNVQ, Merit (in specific programmes). ND/C, 8 Merits. SQAH, Grades CCC. SQAV, Individual consideration. IB, Individual consideration.

BSc Geographical Information Systems

UCAS Code: FG8M • Mode: 4/5 Years Equal Combination

Qualifications/Requirements: GCE, A/AS: 4 points. AGNVQ, Individual consideration. ND/C, Individual consideration. SQAH, Individual consideration. SQAV, Individual consideration. IB, Individual consideration.

BSc Geographical Information Systems and Remote Sensing

UCAS Code: FGV5 • Mode: 3 Years Equal Combination

Qualifications/Requirements: Please refer to prospectus.

BSc Geographical Information Systems with Remote Sensing

UCAS Code: FG8N • Mode: 3 Years Equal Combination

Qualifications/Requirements: Please refer to prospectus.

BSc Information Systems

UCAS Code: G561 • Mode: 3/4 Years Single Subject

Qualifications/Requirements: GCE, A/AS: 8-10 points. AGNVQ, Pass (in specific programmes). ND/C, 4 Merits. SQAH, Grades CCC. SQAV, Individual consideration. IB, Individual consideration.

BSc Information Systems with Business Management

UCAS Code: G5N1 • Mode: 3/4 Years Major/Minor

Qualifications/Requirements: GCE, A/AS: 8-10 points. AGNVQ, Merit (in specific programmes). ND/C, 8 Merits. SQAH, Grades CCC. SQAV, Individual consideration. IB, Individual consideration.

BSc Information Systems with French

UCAS Code: G5RC • Mode: 3/4 Years Major/Minor

Qualifications/Requirements: GCE, A/AS: 8-10 points. AGNVQ, Pass (in specific programmes). ND/C, 4 Merits. SQAH, Grades CCC. SQAV, Individual consideration. IB, Individual consideration.

BSc Information Systems with German

UCAS Code: G5RF • Mode: 3/4 Years Major/Minor

Qualifications/Requirements: GCE, A/AS: 8-10 points. AGNVQ, Pass (in specific programmes). ND/C, 4 Merits. SQAH, Grades CCC. SQAV, Individual consideration. IB, Individual consideration.

BSc Information Systems with Law

UCAS Code: G5MH • Mode: 3/4 Years Major/Minor

Qualifications/Requirements: GCE, A/AS: 8-10 points. AGNVQ, Merit (in specific programmes). ND/C, 8 Merits. SQAH, Grades CCC. SQAV, Individual consideration. IB, Individual consideration.

BSc Information Systems with Spanish

UCAS Code: G5RK • Mode: 3/4 Years Major/Minor

Qualifications/Requirements: GCE, A/AS: 8-10 points. AGNVQ, Pass (in specific programmes). ND/C, 4 Merits. SQAH, Grades CCC. SQAV, Individual consideration. IB, Individual consideration.

BSc Mathematics, Statistics and Computing (FT/SW)

UCAS Code: G900 • Mode: 3/4 Years Single Subject

Qualifications/Requirements: GCE, A level grades: CE. Maths at A level. AGNVQ, Individual consideration. ND/C, 3 Merits (in specific programmes). SQAH, Grades CCC including specific subject(s). SQAV, Individual consideration. IB, Individual consideration.

HND Computing

UCAS Code: 005G • Mode: 2 Years Single Subject

Qualifications/Requirements: GCE, A/AS: 2 points. AGNVQ, Pass (in specific programmes). ND/C, 2 Merits. SQAH, Grades C. SQAV, Individual consideration. IB, Individual consideration.

HND Computing and Information Systems

UCAS Code: 025G • Mode: 2 Years Single Subject

Qualifications/Requirements: GCE, A/AS: 2 points. AGNVQ, Pass (in specific programmes). ND/C, 2 Merits. SQAH, Grades C. SQAV, Individual consideration. IB, Individual consideration.

HND Mathematics, Statistics and Computing

UCAS Code: 009G • Mode: 2 Years Single Subject

Qualifications/Requirements: GCE, A/AS: 2 points. Maths at A level. AGNVQ, Individual consideration. ND/C, 1 Merit (in specific programmes). SQAH, Grades C including specific subject(s). SQAV, Individual consideration. IB, Individual consideration.

GREENWICH SCHOOL OF MANAGEMENT G74

BA Business Management and Information Technology

UCAS Code: NG15 • Mode: 2 Years Equal Combination

Qualifications/Requirements: Please refer to prospectus.

HEREFORDSHIRE COLLEGE OF TECHNOLOGY H16

HND Computing

UCAS Code: 005G • Mode: 2 Years Single Subject

Qualifications/Requirements: GCE, A/AS: 2 points. AGNVQ, Pass. ND/C, Individual consideration. SQAH, Grades CC. SQAV, Individual consideration.

HERIOT-WATT UNIVERSITY, EDINBURGH H24

BSc Computer Science

UCAS Code: G500 • Mode: 4 Years Single Subject

Qualifications/Requirements: GCE, A level grades: CCD. Maths. AGNVQ, Merit (in specific programmes). ND/C, Merit overall (in specific programmes). SQAH, Grades BBBC including specific subject(s). SQAV, National including specific subject(s). IB, 28 points.

BSc Computer Science (Artificial Intelligence)

UCAS Code: G800 • Mode: 4 Years Single Subject

Qualifications/Requirements: GCE, A level grades: CCD. Maths. AGNVQ, Merit (in specific programmes). ND/C, Merit overall (in specific programmes). SQAH, Grades BBBC including specific subject(s). SQAV, National including specific subject(s). IB, 28 points.

BSc Computer Science (Human Computer Interaction)

UCAS Code: G570 • Mode: 4 Years Single Subject

Qualifications/Requirements: GCE, A level grades: CCD. Maths. AGNVQ, Merit (in specific programmes). ND/C, Merit overall (in specific programmes). SQAH, Grades BBBC including specific subject(s). SQAV, National including specific subject(s). IB, 28 points.

BSc Computer Science (Information Systems)

UCAS Code: G520 • Mode: 4 Years Single Subject

Qualifications/Requirements: GCE, A level grades: CCD. Maths. AGNVQ, Merit (in specific programmes). ND/C, Merit overall (in specific programmes). SQAH, Grades BBBC including specific subject(s). SQAV, National including specific subject(s). IB, 28 points.

BSc Computer Science (Multimedia Systems)

UCAS Code: G540 • Mode: 4 Years Single Subject

Qualifications/Requirements: GCE, A level grades: CCD. Maths. AGNVQ, Merit (in specific programmes). ND/C, Merit overall (in specific programmes). SQAH, Grades BBBC including specific subject(s). SQAV, National including specific subject(s). IB, 28 points.

BSc Computer Science (Software Engineering)

UCAS Code: G700 • Mode: 4 Years Single Subject

Qualifications/Requirements: GCE, A level grades: CCD. Maths. AGNVQ, Merit (in specific programmes). ND/C, Merit overall (in specific programmes). SQAH, Grades BBBC including specific subject(s). SQAV, National including specific subject(s). IB, 28 points.

BSc Computing for Industry

UCAS Code: GH57 • Mode: 4 Years Equal Combination

Qualifications/Requirements: GCE, A level grades: CC. AGNVQ, Merit (in specific programmes). ND/C, Individual consideration. SQAH, Grades BBC including specific subject(s). SQAV, Higher National. IB, Pass in Diploma.

BEng Computing and Electronics

UCAS Code: GH56 • Mode: 4 Years Equal Combination

Qualifications/Requirements: GCE, A level grades: CDD. Maths and Physics. AGNVQ, Merit (in specific programmes). ND/C, Merit overall. SQAH, Grades BBBC including specific subject(s). SQAV, Individual consideration. IB, 28 points.

BEng Information Systems Engineering

UCAS Code: HG66 • Mode: 4 Years Equal Combination

Qualifications/Requirements: GCE, A level grades: CDD. Maths and Physics. AGNVQ, Merit (in specific programmes) with A/AS. ND/C, Merit overall. SQAH, Grades BBBC including specific subject(s). SQAV, Individual consideration. IB, 28 points.

MEng Computing and Electronics

UCAS Code: GH5P • Mode: 5 Years Equal Combination

Qualifications/Requirements: GCE, A level grades: CDD. Maths and Physics. AGNVQ, Merit (in specific programmes). ND/C, (in specific programmes). SQAH, Grades BBBC including specific subject(s). SQAV, National including specific subject(s). IB, 28 points.

MEng Computing and Electronics

UCAS Code: GH5Q • Mode: 4 Years Equal Combination

Qualifications/Requirements: GCE, A level grades: BCC. Maths and Physics. AGNVQ, Not normally sufficient. ND/C, Higher National. SQAH, CSYS required. SQAV, Higher National. IB, Individual consideration.

HND Computing

UCAS Code: 005G • Mode: 2 Years Single Subject

Qualifications/Requirements: GCE, A level grades: D. AGNVQ, Individual consideration. ND/C, National. SQAH, Grades CC. SQAV, Individual consideration. IB, Individual consideration.

UNIVERSITY OF HERTFORDSHIRE H36

BA Marketing/Information Systems

UCAS Code: NG55 • Mode: 3/4 Years Equal Combination

Qualifications/Requirements: GCE, A/AS: 18 points. AGNVQ, Merit or Distinction in Business with A/AS. ND/C, Distinction Overall. SQAH, Grades BBBB. SQAV, Individual consideration. IB, 28 points.

BSc Accounting and Management Information Systems

UCAS Code: GN54 • Mode: 3/4 Years Equal Combination

Qualifications/Requirements: GCE, A/AS: 18 points. AGNVQ, Merit or Distinction in Business with A/AS. ND/C, Distinction Overall. SQAH, Grades BBBB. IB, 28 points.

BSc Astronomy/Computing

UCAS Code: F5G5 • Mode: 3/4 Years Major/Minor

Qualifications/Requirements: GCE, A/AS: 14 points. Maths. AGNVQ, Merit (in specific programmes). ND/C, Merit overall (in specific programmes). SQAH, Grades BCCC including specific subject(s). SQAV, Individual consideration. IB, 24 points.

BSc Business Statistics/Information Systems

UCAS Code: GGK5 • Mode: 3/4 Years Equal Combination

Qualifications/Requirements: GCE, A/AS: 18 points. AGNVQ, Merit or Distinction in Business with A/AS. ND/C, Distinction Overall. SQAH, Grades BBBB. SQAV, Individual consideration. IB, 28 points.

BSc Business/Computing

UCAS Code: N1G5 • Mode: 3/4 Years Major/Minor

Qualifications/Requirements: GCE, A/AS: 18 points. AGNVQ, Merit (in specific programmes) with 6 additional units. ND/C, 6 Merits and 2 Distinctions. SQAH, Grades BBCC. SQAV, Individual consideration. IB, 26 points.

BSc Chemistry/Computing

UCAS Code: F1G5 • Mode: 3/4 Years Major/Minor

Qualifications/Requirements: GCE, A/AS: 14 points. Chemistry. AGNVQ, Merit in Science. ND/C, Merit overall (in specific programmes). SQAH, Grades BCCC including specific subject(s). SQAV, Individual consideration. IB, 24 points.

BSc Computer Science (HND top-up)
UCAS Code: G503 • Mode: 1 Year Single Subject

Qualifications/Requirements: AGNVQ, Not normally sufficient. ND/C, Higher National (in specific programmes). SQAH, Not normally sufficient. SQAV, Higher National including specific subject(s). IB, Not normally sufficient.

BSc Computer Science (SW)
UCAS Code: G500 • Mode: 4 Years Single Subject

Qualifications/Requirements: GCE, A/AS: 16 points. AGNVQ, Merit. ND/C, Merit overall. SQAH, Individual consideration. SQAV, Individual consideration. IB, 28 points.

BSc Computing
UCAS Code: G505 • Mode: 3/4 Years Single Subject

Qualifications/Requirements: GCE, A/AS: 14 points. AGNVQ, Merit (in specific programmes). ND/C, Merit overall. SQAH, Grades BCCC. SQAV, Individual consideration. IB, 24 points.

BSc Computing and Networks
UCAS Code: G521 • Mode: 4 Years Single Subject

Qualifications/Requirements: GCE, A/AS: 16 points. AGNVQ, Merit. ND/C, Merit overall. SQAH, Individual consideration. SQAV, Individual consideration. IB, 28 points.

BSc Computing/Astronomy
UCAS Code: G5F5 • Mode: 3/4 Years Major/Minor

Qualifications/Requirements: GCE, A/AS: 14 points. Maths. AGNVQ, Merit (in specific programmes). ND/C, 3 Merits (in specific programmes). SQAH, Grades BCCC including specific subject(s). SQAV, Individual consideration. IB, 24 points.

BSc Computing/Business
UCAS Code: G5N1 • Mode: 3/4 Years Major/Minor

Qualifications/Requirements: GCE, A/AS: 18 points. AGNVQ, Merit (in specific programmes) with 6 additional units. ND/C, 6 Merits and 2 Distinctions. SQAH, Grades BBCC. SQAV, Individual consideration. IB, 26 points.

BSc Computing/Chemistry
UCAS Code: G5F1 • Mode: 3/4 Years Major/Minor

Qualifications/Requirements: GCE, A/AS: 14 points. Chemistry. AGNVQ, Merit in Science. ND/C, Merit overall (in specific programmes). SQAH, Grades BCCC including specific subject(s). SQAV, Individual consideration. IB, 24 points.

BSc Computing/Electronic Music
UCAS Code: G5W3 • Mode: 3/4 Years Major/Minor

Qualifications/Requirements: GCE, A/AS: 14 points. Music. AGNVQ, Merit (in specific programmes) with A level. ND/C, Merit overall (in specific programmes). SQAH, Grades BCCC including specific subject(s). SQAV, Individual consideration. IB, 26 points.

BSc Computing/Electronics
UCAS Code: G5H6 • Mode: 3/4 Years Major/Minor

Qualifications/Requirements: GCE, A/AS: 14 points. AGNVQ, Merit (in specific programmes). ND/C, Merit overall (in specific programmes). SQAH, Grades BCCC including specific subject(s). SQAV, Individual consideration. IB, 24 points.

BSc Computing/European Studies
UCAS Code: G5TF • Mode: 3/4 Years Major/Minor

Qualifications/Requirements: GCE, A/AS: 14 points. AGNVQ, Merit (in specific programmes). ND/C, Merit overall. SQAH, Grades BCCC. SQAV, Individual consideration. IB, 26 points.

BSc Computing/Geology
UCAS Code: G5F6 • Mode: 3/4 Years Major/Minor

Qualifications/Requirements: GCE, A/AS: 14 points. AGNVQ, Merit (in specific programmes). ND/C, Merit overall (in specific programmes). SQAH, Grades BCCC including specific subject(s). SQAV, Individual consideration. IB, 24 points.

BSc Computing/Human Biology
UCAS Code: G5B1 • Mode: 3/4 Years Major/Minor

Qualifications/Requirements: GCE, A/AS: 14 points. Science. AGNVQ, Merit in Science. ND/C, Merit overall (in specific programmes). SQAH, Grades BCCC including specific subject(s). SQAV, Individual consideration. IB, 24 points.

BSc Computing/Law
UCAS Code: G5M3 • Mode: 3/4 Years Major/Minor

Qualifications/Requirements: GCE, A/AS: 20 points. AGNVQ, Distinction (in specific programmes). ND/C, 4 Merits and 4 Distinctions. SQAH, Grades BBBC. SQAV, Individual consideration. IB, 26 points.

BSc Computing/Linguistic Sciences
UCAS Code: G5Q1 • Mode: 3/4 Years Major/Minor

Qualifications/Requirements: GCE, A/AS: 14 points. AGNVQ, Merit (in specific programmes). ND/C, Merit overall. SQAH, Grades BCCC. SQAV, Individual consideration. IB, 24 points.

BSc Computing/Manufacturing Systems
UCAS Code: G5H7 • Mode: 3/4 Years Major/Minor

Qualifications/Requirements: GCE, A/AS: 14 points. AGNVQ, Merit (in specific programmes). ND/C, Merit overall. SQAH, Grades BCCC. SQAV, Individual consideration. IB, 24 points.

BSc Computing/Mathematics
UCAS Code: G5G1 • Mode: 3/4 Years Major/Minor

Qualifications/Requirements: GCE, A/AS: 14 points. Maths. AGNVQ, Merit (in specific programmes) with A level. ND/C, Merit overall (in specific programmes). SQAH, Grades BCCC including specific subject(s). SQAV, Individual consideration. IB, 24 points.

BSc Computing/Philosophy
UCAS Code: G5V7 • Mode: 3/4 Years Major/Minor

Qualifications/Requirements: GCE, A/AS: 14 points. AGNVQ, Merit (in specific programmes). ND/C, Merit overall. SQAH, Grades BCCC. SQAV, Individual consideration. IB, 24 points.

BSc Computing/Physics
UCAS Code: G5F3 • Mode: 3/4 Years Major/Minor

Qualifications/Requirements: GCE, A/AS: 14 points. Maths and Physics. AGNVQ, Merit (in specific programmes). ND/C, Merit overall (in specific programmes). SQAH, Grades BCCC including specific subject(s). SQAV, Individual consideration. IB, 24 points.

BSc Computing/Psychology
UCAS Code: G5C8 • Mode: 3/4 Years Major/Minor

Qualifications/Requirements: GCE, A/AS: 20 points. AGNVQ, Distinction (in specific programmes). ND/C, 4 Merits and 4 Distinctions. SQAH, Grades BBBC. SQAV, Individual consideration. IB, 26 points.

BSc Computing/Statistics
UCAS Code: G5G4 • Mode: 3/4 Years Major/Minor

Qualifications/Requirements: GCE, A/AS: 14 points. AGNVQ, Merit (in specific programmes). ND/C, Merit overall (in specific programmes). SQAH, Grades BCCC. SQAV, Individual consideration. IB, 24 points.

BSc Electronic Music/Computing
UCAS Code: W3G5 • Mode: 3/4 Years Major/Minor

Qualifications/Requirements: GCE, A/AS: 14 points. Music. AGNVQ, Merit (in specific programmes) with A level. ND/C, Merit overall (in specific programmes). SQAH, Grades BCCC including specific subject(s). SQAV, Individual consideration. IB, 26 points.

BSc European Studies/Computing
UCAS Code: T2G5 • Mode: 3/4 Years Major/Minor

Qualifications/Requirements: GCE, A/AS: 14 points. AGNVQ, Merit (in specific programmes). ND/C, Merit overall. SQAH, Grades BCCC. SQAV, Individual consideration. IB, 24 points.

BSc Geology/Computing
UCAS Code: F6G5 • Mode: 3/4 Years Major/Minor

Qualifications/Requirements: GCE, A/AS: 14 points. AGNVQ, Merit (in specific programmes). ND/C, Merit overall (in specific programmes). SQAH, Grades BCCC including specific subject(s). SQAV, Individual consideration. IB, 24 points.

BSc Human Biology/Computing
UCAS Code: B1G5 • Mode: 3/4 Years Major/Minor

Qualifications/Requirements: GCE, A/AS: 14 points. Science. AGNVQ, Merit in Science. ND/C, Merit overall (in specific programmes). SQAH, Grades BCCC including specific subject(s). SQAV, Individual consideration. IB, 24 points.

BSc Information Systems
UCAS Code: G520 • Mode: 4 Years Single Subject

Qualifications/Requirements: GCE, A/AS: 16 points. AGNVQ, Merit. ND/C, Merit overall. SQAH, Individual consideration. SQAV, Individual consideration. IB, 28 points.

BSc Information Technology for Health Care
UCAS Code: G560 • Mode: 3 Years Single Subject

Qualifications/Requirements: Please refer to prospectus.

BSc Law/Computing
UCAS Code: M3G5 • Mode: 3/4 Years Major/Minor

Qualifications/Requirements: GCE, A/AS: 20 points. AGNVQ, Distinction (in specific programmes). ND/C, 4 Merits and 4 Distinctions. SQAH, Grades BBBC. SQAV, Individual consideration. IB, 26 points.

BSc Manufacturing Systems/Computing
UCAS Code: H7G5 • Mode: 3/4 Years Major/Minor

Qualifications/Requirements: GCE, A/AS: 14 points. AGNVQ, Merit (in specific programmes). ND/C, Merit overall. SQAH, Grades BCCC. SQAV, Individual consideration. IB, 24 points.

BSc Mathematics/Computing
UCAS Code: G1G5 • Mode: 3/4 Years Major/Minor

Qualifications/Requirements: GCE, A/AS: 14 points. Maths. AGNVQ, Merit (in specific programmes) with A level. ND/C, Merit overall (in specific programmes). SQAH, Grades BCCC including specific subject(s). SQAV, Individual consideration. IB, 24 points.

BSc Philosophy/Computing
UCAS Code: V7G5 • Mode: 3/4 Years Major/Minor

Qualifications/Requirements: GCE, A/AS: 14 points. AGNVQ, Merit (in specific programmes). ND/C, Merit overall. SQAH, Grades BCCC. SQAV, Individual consideration. IB, 24 points.

BSc Physics/Computing
UCAS Code: F3G5 • Mode: 3/4 Years Major/Minor

Qualifications/Requirements: GCE, A/AS: 14 points. Maths and Physics. AGNVQ, Merit (in specific programmes). ND/C, Merit overall (in specific programmes). SQAH, Grades BCCC including specific subject(s). SQAV, Individual consideration. IB, 24 points.

BSc Psychology/Computing
UCAS Code: C8G5 • Mode: 3/4 Years Major/Minor

Qualifications/Requirements: GCE, A/AS: 20 points. AGNVQ, Distinction (in specific programmes). ND/C, 4 Merits and 4 Distinctions. SQAH, Grades BBBC. SQAV, Individual consideration. IB, 26 points.

HND Business Information Technology
UCAS Code: 265G • Mode: 2 Years Single Subject

Qualifications/Requirements: GCE, A/AS: 4 points. AGNVQ, Pass. ND/C, 2 Merits. SQAH, Individual consideration. SQAV, Individual consideration. IB, 24 points.

HND Computing
UCAS Code: 105G • Mode: 2 Years Single Subject
Qualifications/Requirements: GCE, A/AS: 4 points. AGNVQ, Pass. ND/C, 2 Merits. SQAH, Individual consideration. SQAV, Individual consideration. IB, Individual consideration.

HND Computing (Business Systems)
UCAS Code: 205G • Mode: 2 Years Single Subject
Qualifications/Requirements: GCE, A/AS: 4 points. AGNVQ, Pass. ND/C, 2 Merits. SQAH, Individual consideration. SQAV, Individual consideration. IB, Individual consideration.

HND Computing (Multimedia)
UCAS Code: 065G • Mode: 2 Years Single Subject
Qualifications/Requirements: GCE, A/AS: 4 points. AGNVQ, Pass. ND/C, 2 Merits. SQAH, Individual consideration. SQAV, Individual consideration. IB, Individual consideration.

UHIp H49

BSc Computing
UCAS Code: G500 • Mode: 3 Years Single Subject
Qualifications/Requirements: Please refer to prospectus.

HNC Business Administration with Information Technology
UCAS Code: 5G1N • Mode: 1 Year Major/Minor
Qualifications/Requirements: Please refer to prospectus.

HNC Cursa Comais
UCAS Code: 105G • Mode: 1 Year Single Subject
Qualifications/Requirements: Please refer to prospectus.

THE UNIVERSITY OF HUDDERSFIELD H60

BA Business Computing with Artificial Intelligence
UCAS Code: G5G8 • Mode: 4 Years Major/Minor
Qualifications/Requirements: GCE, A/AS: 16 points. AGNVQ, Merit. ND/C, Merit overall. SQAH, Grades BBBB. SQAV, Individual consideration. IB, Individual consideration.

BA Business Computing with Computing Science
UCAS Code: G502 • Mode: 4 Years Single Subject
Qualifications/Requirements: GCE, A/AS: 16 points. AGNVQ, Merit. ND/C, Merit overall. SQAH, Grades BBBB. SQAV, Individual consideration. IB, Individual consideration.

BA Business Computing with French
UCAS Code: G5R1 • Mode: 4 Years Major/Minor
Qualifications/Requirements: GCE, A/AS: 16 points. AGNVQ, Merit. ND/C, Merit overall. SQAH, Grades BBBB. SQAV, Individual consideration. IB, Individual consideration.

BA Business Computing with German
UCAS Code: G5R2 • Mode: 4 Years Major/Minor
Qualifications/Requirements: GCE, A/AS: 16 points. AGNVQ, Merit. ND/C, Merit overall. SQAH, Grades BBBB. SQAV, Individual consideration. IB, Individual consideration.

BA Business Computing with Human/ Computer Interaction
UCAS Code: G503 • Mode: 4 Years Single Subject
Qualifications/Requirements: GCE, A/AS: 16 points. AGNVQ, Merit. ND/C, Merit overall. SQAH, Grades BBBB. SQAV, Individual consideration. IB, Individual consideration.

BA Business Computing with Multimedia
UCAS Code: G504 • Mode: 4 Years Single Subject
Qualifications/Requirements: GCE, A/AS: 16 points. AGNVQ, Merit. ND/C, Merit overall. SQAH, Grades BBBB. SQAV, Individual consideration. IB, Individual consideration.

BA Business Computing with Operational Research
UCAS Code: G5N2 • Mode: 4 Years Major/Minor
Qualifications/Requirements: GCE, A/AS: 16 points. AGNVQ, Merit. ND/C, Merit overall. SQAH, Grades BBBB. SQAV, Individual consideration. IB, Individual consideration.

BA Business Computing with Psychology
UCAS Code: G5L7 • Mode: 4 Years Major/Minor
Qualifications/Requirements: GCE, A/AS: 16 points. AGNVQ, Merit. ND/C, Merit overall. SQAH, Grades BBBB. SQAV, Individual consideration. IB, Individual consideration.

BA Business Computing with Software Development
UCAS Code: G5G7 • Mode: 4 Years Major/Minor
Qualifications/Requirements: GCE, A/AS: 16 points. AGNVQ, Merit. ND/C, Merit overall. SQAH, Grades BBBB. SQAV, Individual consideration. IB, Individual consideration.

BA Business Computing with Statistics
UCAS Code: G5G4 • Mode: 4 Years Major/Minor
Qualifications/Requirements: GCE, A/AS: 16 points. AGNVQ, Merit. ND/C, Merit overall. SQAH, Grades BBBB. SQAV, Individual consideration. IB, Individual consideration.

BA Computing and Business Analysis
UCAS Code: GN51 • Mode: 4 Years Equal Combination
Qualifications/Requirements: GCE, A/AS: 16 points. AGNVQ, Merit. ND/C, Merit overall. SQAH, Grades BBBB. SQAV, Individual consideration. IB, Individual consideration.

BA Computing in Business
UCAS Code: G523 • Mode: 4 Years Single Subject

Qualifications/Requirements: GCE, A/AS: 16 points. AGNVQ, Merit. ND/C, Merit overall. SQAH, Grades BBBB. SQAV, Individual consideration. IB, Individual consideration.

BA Interactive Media
UCAS Code: G5P4 • Mode: 4 Years Major/Minor

Qualifications/Requirements: GCE, A/AS: 16 points. AGNVQ, Merit. ND/C, Merit overall. SQAH, Grades BBBB. SQAV, Individual consideration. IB, Individual consideration.

BEd Information Technology
UCAS Code: XG75 • Mode: 2 Years Equal Combination

Qualifications/Requirements: AGNVQ, Not normally sufficient. ND/C, Higher National. SQAH, Not normally sufficient. SQAV, Higher National. IB, Not normally sufficient.

BSc Architectural Computer-Aided Technology
UCAS Code: GK51 • Mode: 3 Years Equal Combination

Qualifications/Requirements: GCE, A/AS: 12 points. AGNVQ, Merit (in specific programmes) with 4 additional units or with A/AS. ND/C, 5 Merits. SQAH, Grades BBB. SQAV, Individual consideration. IB, Individual consideration.

BSc Computing and Management Sciences
UCAS Code: GN5C • Mode: 4 Years Equal Combination

Qualifications/Requirements: GCE, A/AS: 16 points. AGNVQ, Merit. ND/C, Merit overall. SQAH, Grades BBBB. SQAV, Individual consideration. IB, Individual consideration.

BSc Computing and Mathematics
UCAS Code: GG51 • Mode: 3/4 Years Equal Combination

Qualifications/Requirements: GCE, A/AS: 16 points. Maths at A level. AGNVQ, Merit with A level. ND/C, Merit overall. SQAH, Grades BBBB. SQAV, Individual consideration. IB, Individual consideration.

BSc Computing and Statistics
UCAS Code: GG54 • Mode: 3/4 Years Equal Combination

Qualifications/Requirements: GCE, A/AS: 16 points. AGNVQ, Merit. ND/C, Merit overall. SQAH, Grades BBBB. SQAV, Individual consideration. IB, Individual consideration.

BSc Computing Science
UCAS Code: G500 • Mode: 3/4 Years Single Subject

Qualifications/Requirements: GCE, A/AS: 16 points. AGNVQ, Merit. ND/C, Merit overall. SQAH, Grades BBBB. SQAV, Individual consideration. IB, Individual consideration.

HND Business Information Technology
UCAS Code: 265G • Mode: 3 Years Single Subject

Qualifications/Requirements: GCE, A/AS: 6 points. AGNVQ, Pass. ND/C, National. SQAH, Grades CCC. SQAV, Individual consideration. IB, Individual consideration.

HND Computing
UCAS Code: 105G • Mode: 3 Years Single Subject

Qualifications/Requirements: GCE, A/AS: 6 points. AGNVQ, Pass. ND/C, National. SQAH, Grades CCC. SQAV, Individual consideration. IB, Individual consideration.

THE UNIVERSITY OF HULL H72

BSc Computer and Management Sciences
UCAS Code: GN51 • Mode: 3 Years Equal Combination

Qualifications/Requirements: GCE, A level grades: BCC-CCC. Maths at A level. AGNVQ, Merit (in specific programmes) with A level. ND/C, Merit overall and Distinction (in specific programmes). SQAH, Grades BBBCC. SQAV, Individual consideration. IB, 26 points.

BSc Computer Graphics and Mathematical Modelling
UCAS Code: GG5C • Mode: 3 Years Equal Combination

Qualifications/Requirements: GCE, A/AS: 18 points. Science and Maths at A level. AGNVQ, Merit (in specific programmes). ND/C, Merit overall. SQAH, Grades BBBCC. SQAV, Individual consideration. IB, 26 points.

BSc Computer Graphics and Mathematical Modelling
UCAS Code: GG5D • Mode: 4 Years Equal Combination

Qualifications/Requirements: GCE, A/AS: 10 points. AGNVQ, Pass. ND/C, National. SQAH, Grades CCCCD. SQAV, Individual consideration. IB, 24 points.

BSc Computer Science
UCAS Code: G500 • Mode: 3 Years Single Subject

Qualifications/Requirements: GCE, A/AS: 18 points. Maths at A level. AGNVQ, Merit (in specific programmes) with A level. ND/C, Merit overall (in specific programmes). SQAH, Grades BBBCC. SQAV, Individual consideration. IB, 26 points.

BSc Computer Science
UCAS Code: G501 • Mode: 4 Years Single Subject

Qualifications/Requirements: GCE, A level grades: CD. AGNVQ, Pass (in specific programmes). ND/C, National. SQAH, Grades CCCCD. SQAV, Individual consideration. IB, 24 points.

BSc Computer Science with Information Engineering
UCAS Code: G560 • Mode: 3 Years Single Subject

Qualifications/Requirements: GCE, A/AS: 18 points. AGNVQ, Merit (in specific programmes) with A level. ND/C, Merit overall. SQAH, Grades BBBCC. SQAV, Individual consideration. IB, 26 points.

BSc Computer Science with Information Engineering
UCAS Code: G568 • Mode: 4 Years Single Subject
Qualifications/Requirements: GCE, A level grades: CD. AGNVQ, Merit (in specific programmes) with A level. ND/C, National. SQAH, Grades CCCCD. SQAV, Individual consideration. IB, 24 points.

BSc Information Management
UCAS Code: N1G5 • Mode: 3 Years Major/Minor
Qualifications/Requirements: GCE, A level grades: BCC. AGNVQ, Distinction with 6 additional units or with A/AS. ND/C, Merit overall and 3 Distinctions. SQAH, Grades AABBB. SQAV, Individual consideration. IB, 28 points.

BSc Information Management (International)
UCAS Code: N1GM • Mode: 4 Years Major/Minor
Qualifications/Requirements: GCE, A level grades: BCC. AGNVQ, Distinction with 6 additional units or with A/AS. ND/C, Merit overall and 3 Distinctions. SQAH, Grades AABBB. SQAV, Individual consideration. IB, 28 points.

IMPERIAL COLLEGE OF SCIENCE, TECHNOLOGY AND MEDICINE (UNIVERSITY OF LONDON) I50

BSc Mathematics and Computer Science (f) (h)
UCAS Code: GG15 • Mode: 3 Years Equal Combination
Qualifications/Requirements: GCE, A level grades: ABB. Maths at A level. AGNVQ, Not normally sufficient. ND/C, Individual consideration. SQAH, Individual consideration. SQAV, Individual consideration. IB, Individual consideration.

BEng Computing (h)
UCAS Code: G500 • Mode: 3 Years Single Subject
Qualifications/Requirements: GCE, A level grades: AAB. Maths at A level. AGNVQ, Pass. ND/C, Individual consideration. SQAH, Individual consideration. SQAV, Individual consideration. IB, Individual consideration.

BEng Information Systems Engineering (h) (g)
UCAS Code: HG65 • Mode: 3 Years Equal Combination
Qualifications/Requirements: GCE, A level grades: AAB-ABB. Maths and Physics at A level. AGNVQ, Not normally sufficient. ND/C, Higher National or Distinction Overall (in specific programmes). SQAH, CSYS required. SQAV, Higher National including specific subject(s). IB, Individual consideration.

MEng Computing (h)
UCAS Code: G501 • Mode: 4 Years Single Subject
Qualifications/Requirements: GCE, A level grades: AAB. Maths at A level. AGNVQ, Pass. ND/C, Individual consideration. SQAH, Individual consideration. SQAV, Individual consideration. IB, Individual consideration.

MEng Computing (Computational Management) (h)
UCAS Code: G520 • Mode: 4 Years Single Subject
Qualifications/Requirements: GCE, A level grades: AAB. Maths at A level. AGNVQ, Pass. ND/C, Individual consideration. SQAH, Individual consideration. SQAV, Individual consideration. IB, Individual consideration.

MEng Computing (European Programme of Study) (h)
UCAS Code: G502 • Mode: 4 Years Single Subject
Qualifications/Requirements: GCE, A level grades: AAB. Maths at A level. AGNVQ, Pass. ND/C, Individual consideration. SQAH, Individual consideration. SQAV, Individual consideration. IB, Individual consideration.

MEng Computing (Mathematical Foundations) (h)
UCAS Code: G550 • Mode: 4 Years Single Subject
Qualifications/Requirements: GCE, A level grades: AAB. Maths at A level. AGNVQ, Pass. ND/C, Individual consideration. SQAH, Individual consideration. SQAV, Individual consideration. IB, Individual consideration.

MEng Information Systems Engineering (g) (h)
UCAS Code: GH56 • Mode: 4 Years Equal Combination
Qualifications/Requirements: GCE, A level grades: AAB-ABB. Maths and Physics at A level. AGNVQ, Not normally sufficient. ND/C, Higher National or Distinction Overall (in specific programmes). SQAH, CSYS required. SQAV, Higher National including specific subject(s). IB, Individual consideration.

MSci Mathematics and Computer Science (f) (h)
UCAS Code: GG51 • Mode: 4 Years Equal Combination
Qualifications/Requirements: GCE, A level grades: ABB. Maths at A level. AGNVQ, Not normally sufficient. ND/C, Individual consideration. SQAH, Individual consideration. SQAV, Individual consideration. IB, Individual consideration.

KEELE UNIVERSITY K12

BA/BSc Computer Science and International Politics
UCAS Code: GM5C • Mode: 3 Years Equal Combination
Qualifications/Requirements: GCE, A level grades: BCC-CCC. Maths or Science. AGNVQ, Merit (in specific programmes) with A level. ND/C, Individual consideration. SQAH, CSYS required. SQAV, Individual consideration. IB, 26 points.

BSc Astrophysics and Computer Science
UCAS Code: FG55 • Mode: 3 Years Equal Combination
Qualifications/Requirements: GCE, A level grades: BCC-CCD. Physics. AGNVQ, Merit (in specific programmes) with A level. ND/C, Individual consideration. SQAH, CSYS required. SQAV, Individual consideration. IB, 26 points.

BSc Biochemistry and Computer Science
UCAS Code: CG75 • Mode: 3 Years Equal Combination

Qualifications/Requirements: GCE, A level grades: BCC-CCD. Chemistry. AGNVQ, Merit (in specific programmes) with A level. ND/C, Individual consideration. SQAH, CSYS required. SQAV, Individual consideration. IB, 26 points.

BSc Biology and Computer Science
UCAS Code: CG15 • Mode: 3 Years Equal Combination

Qualifications/Requirements: GCE, A level grades: BCC-CCD. Science. AGNVQ, Merit (in specific programmes) with A level. ND/C, Individual consideration. SQAH, CSYS required. SQAV, Individual consideration. IB, 26 points.

BSc Computer Science
UCAS Code: G500 • Mode: 3 Years Single Subject

Qualifications/Requirements: GCE, A level grades: BCC-CCD. Maths or Science. AGNVQ, Merit (in specific programmes) with A level. ND/C, Individual consideration. SQAH, CSYS required. SQAV, Individual consideration. IB, 26 points.

BSc Computer Science and Geology
UCAS Code: FG65 • Mode: 3 Years Equal Combination

Qualifications/Requirements: GCE, A level grades: BCC-CCD. Science. AGNVQ, Merit (in specific programmes) with A level. ND/C, Individual consideration. SQAH, CSYS required. SQAV, Individual consideration. IB, 26 points.

BSc Computer Science and Neuroscience
UCAS Code: BG15 • Mode: 3 Years Equal Combination

Qualifications/Requirements: GCE, A level grades: BCC-CCD. Science at A level. AGNVQ, Merit in Science with A level. ND/C, Individual consideration. SQAH, CSYS required. SQAV, Individual consideration. IB, 26 points.

BSc Computer Science and Physical Geography
UCAS Code: FG85 • Mode: 3 Years Equal Combination

Qualifications/Requirements: GCE, A level grades: BCC-CCC. Science. AGNVQ, Merit (in specific programmes) with A level. ND/C, Individual consideration. SQAH, CSYS required. SQAV, Individual consideration. IB, 26 points.

BSc Computer Science and Physics
UCAS Code: FG35 • Mode: 3 Years Equal Combination

Qualifications/Requirements: GCE, A level grades: BCC-CCD. Physics. AGNVQ, Merit (in specific programmes) with A level. ND/C, Individual consideration. SQAH, CSYS required. SQAV, Individual consideration. IB, 26 points.

BSc Computer Science and Psychology
UCAS Code: CG85 • Mode: 3 Years Equal Combination

Qualifications/Requirements: GCE, A level grades: BCC. Maths or Science. AGNVQ, Distinction (in specific programmes). ND/C, Individual consideration. SQAH, CSYS required. SQAV, Individual consideration. IB, 28 points.

MSci Computer Science and Astrophysics
UCAS Code: GFM5 • Mode: 4 Years Equal Combination

Qualifications/Requirements: GCE, A level grades: BCC-CCD. Physics. AGNVQ, Merit (in specific programmes) with A level. ND/C, Individual consideration. SQAH, CSYS required. SQAV, Individual consideration. IB, 26 points.

MSci Computer Science and Mathematics
UCAS Code: GG5C • Mode: 4 Years Equal Combination

Qualifications/Requirements: GCE, A level grades: BCC-CCD. Maths at A level. AGNVQ, Merit (in specific programmes) with A level. ND/C, Individual consideration. SQAH, CSYS required. SQAV, Individual consideration. IB, 26 points.

MSci Geology and Computer Science
UCAS Code: GF5P • Mode: 4 Years Equal Combination

Qualifications/Requirements: GCE, A level grades: BCC-CCD. Science. AGNVQ, Merit (in specific programmes) with A level. ND/C, Individual consideration. SQAH, CSYS required. SQAV, Individual consideration. IB, 26 points.

MSci Physics and Computer Science
UCAS Code: GF5H • Mode: 4 Years Equal Combination

Qualifications/Requirements: GCE, A level grades: BCC-CCD. Physics. AGNVQ, Merit (in specific programmes) with A level. ND/C, Individual consideration. SQAH, CSYS required. SQAV, Individual consideration. IB, 26 points.

Mod Ancient History and Computer Science
UCAS Code: GV5D • Mode: 3 Years Equal Combination

Qualifications/Requirements: GCE, A level grades: BCC-CCC. Maths or Science. AGNVQ, Distinction (in specific programmes) with A level. ND/C, Individual consideration. SQAH, CSYS required. SQAV, Individual consideration. IB, 28 points.

Mod Business Administration and Computer Science
UCAS Code: GN59 • Mode: 3 Years Equal Combination

Qualifications/Requirements: GCE, A level grades: BCC-CCC. Maths or Science. AGNVQ, Distinction (in specific programmes). ND/C, Individual consideration. SQAH, CSYS required. SQAV, Individual consideration. IB, 28 points.

Mod Classical Studies and Computer Science
UCAS Code: GQ58 • Mode: 3 Years Equal Combination

Qualifications/Requirements: GCE, A level grades: BCC-CCC. Maths or Science. AGNVQ, Distinction (in specific programmes) with A level. ND/C, Individual consideration. SQAH, CSYS required. SQAV, Individual consideration. IB, 28 points.

Mod Computer Science and Criminology
UCAS Code: GM5H • Mode: 3 Years Equal Combination

Qualifications/Requirements: GCE, A level grades: BCC. Maths or Science. AGNVQ, Distinction (in specific programmes) with A level. ND/C, Individual consideration. SQAH, CSYS required. SQAV, Individual consideration. IB, 28 points.

Mod Computer Science and English
UCAS Code: GQ53 • Mode: 3 Years Equal Combination

Qualifications/Requirements: GCE, A level grades: BCC. English, English Language, English Literature at A level and Maths or Science. AGNVQ, Distinction (in specific programmes) with A level. ND/C, Individual consideration. SQAH, CSYS required. SQAV, Individual consideration. IB, 28 points.

Mod Computer Science and Finance
UCAS Code: GN53 • Mode: 3 Years Equal Combination

Qualifications/Requirements: GCE, A level grades: BCC-CCC. Maths or Science. AGNVQ, Distinction (in specific programmes). ND/C, Individual consideration. SQAH, CSYS required. SQAV, Individual consideration. IB, 28 points.

Mod Computer Science and Geography
UCAS Code: LG8M • Mode: 3 Years Equal Combination

Qualifications/Requirements: GCE, A level grades: BCC-CCC. Geography. AGNVQ, Merit (in specific programmes) with A level. ND/C, Individual consideration. SQAH, CSYS required. SQAV, Individual consideration. IB, 26 points.

Mod Computer Science and German
UCAS Code: GR52 • Mode: 3 Years Equal Combination

Qualifications/Requirements: GCE, A level grades: BCC-CCC. German at A level and Maths or Science. AGNVQ, Distinction (in specific programmes) with A level. ND/C, Individual consideration. SQAH, CSYS required. SQAV, Individual consideration. IB, 28 points.

Mod Computer Science and History
UCAS Code: GV51 • Mode: 3 Years Equal Combination

Qualifications/Requirements: GCE, A level grades: BCC-CCC. Maths or Science. AGNVQ, Distinction (in specific programmes) with A level. ND/C, Individual consideration. SQAH, CSYS required. SQAV, Individual consideration. IB, 28 points.

Mod Computer Science and Human Geography
UCAS Code: GL5V • Mode: 3 Years Equal Combination

Qualifications/Requirements: GCE, A level grades: BCC-CCC. Geography. AGNVQ, Merit (in specific programmes) with A level. ND/C, Individual consideration. SQAH, CSYS required. SQAV, Individual consideration. IB, 26 points.

Mod Computer Science and Human Resource Management
UCAS Code: GN56 • Mode: 3 Years Equal Combination

Qualifications/Requirements: GCE, A level grades: BCC-CCC. Maths or Science. AGNVQ, Distinction (in specific programmes). ND/C, Individual consideration. SQAH, CSYS required. SQAV, Individual consideration. IB, 28 points.

Mod Computer Science and International History
UCAS Code: GV5C • Mode: 3 Years Equal Combination

Qualifications/Requirements: GCE, A level grades: BCC-CCC. Maths or Science. AGNVQ, Distinction (in specific programmes) with A level. ND/C, Individual consideration. SQAH, CSYS required. SQAV, Individual consideration. IB, 28 points.

Mod Computer Science and Latin
UCAS Code: GQ56 • Mode: 3 Years Equal Combination

Qualifications/Requirements: GCE, A level grades: BCC-CCC. Latin at A level and Maths or Science. AGNVQ, Distinction (in specific programmes) with A level. ND/C, Individual consideration. SQAH, CSYS required. SQAV, Individual consideration. IB, 28 points.

Mod Computer Science and Law
UCAS Code: GM53 • Mode: 3 Years Equal Combination

Qualifications/Requirements: GCE, A level grades: BCC. Maths or Science. AGNVQ, Distinction (in specific programmes). ND/C, Individual consideration. SQAH, CSYS required. SQAV, Individual consideration. IB, 28 points.

Mod Computer Science and Management Science
UCAS Code: GN51 • Mode: 3 Years Equal Combination

Qualifications/Requirements: GCE, A level grades: BCC-CCC. Maths or Science. AGNVQ, Distinction (in specific programmes). ND/C, Individual consideration. SQAH, CSYS required. SQAV, Individual consideration. IB, 28 points.

Mod Computer Science and Music
UCAS Code: GW53 • Mode: 3 Years Equal Combination

Qualifications/Requirements: GCE, A level grades: BCC-CCC. Music at A level and Maths or Science. AGNVQ, Distinction (in specific programmes) with A level. ND/C, Individual consideration. SQAH, CSYS required. SQAV, Individual consideration. IB, 28 points.

Mod Computer Science and Music (Electronic Music)
UCAS Code: GW5J • Mode: 3 Years Equal Combination

Qualifications/Requirements: GCE, A level grades: BCC-CCC. Music at A level and Maths or Science. AGNVQ, Distinction (in specific programmes) with A level. ND/C, Individual consideration. SQAH, CSYS required. SQAV, Individual consideration. IB, 28 points.

Mod Computer Science and Philosophy

UCAS Code: GV57 • Mode: 3 Years Equal Combination

Qualifications/Requirements: GCE, A level grades: BCC-CCC. Maths or Science. AGNVQ, Distinction (in specific programmes) with A level. ND/C, Individual consideration. SQAH, CSYS required. SQAV, Individual consideration. IB, 28 points.

Mod Computer Science and Politics

UCAS Code: GM51 • Mode: 3 Years Equal Combination

Qualifications/Requirements: GCE, A level grades: BCC-CCC. Maths or Science. AGNVQ, Distinction (in specific programmes) with A level. ND/C, Individual consideration. SQAH, CSYS required. SQAV, Individual consideration. IB, 28 points.

Mod Computer Science and Russian

UCAS Code: GR58 • Mode: 3 Years Equal Combination

Qualifications/Requirements: GCE, A level grades: BCC-CCC. Russian at A level and Maths or Science. AGNVQ, Distinction (in specific programmes) with A level. ND/C, Individual consideration. SQAH, CSYS required. SQAV, Individual consideration. IB, 28 points.

Mod Computer Science and Russian Studies

UCAS Code: GRM8 • Mode: 3 Years Equal Combination

Qualifications/Requirements: GCE, A level grades: BCC-CCC. Maths or Science. AGNVQ, Distinction (in specific programmes) with A level. ND/C, Individual consideration. SQAH, CSYS required. SQAV, Individual consideration. IB, 28 points.

Mod Computer Science and Sociology & Social Anthropology

UCAS Code: LG3C • Mode: 3 Years Equal Combination

Qualifications/Requirements: GCE, A level grades: BCC-CCC. Maths or Science. AGNVQ, Distinction (in specific programmes). ND/C, Individual consideration. SQAH, CSYS required. SQAV, Individual consideration. IB, 28 points.

Mod Computer Science and Visual Arts

UCAS Code: GW51 • Mode: 3 Years Equal Combination

Qualifications/Requirements: GCE, A level grades: BCC-CCC. Maths or Science. AGNVQ, Distinction (in specific programmes) with A level. ND/C, Individual consideration. SQAH, CSYS required. SQAV, Individual consideration. IB, 28 points.

THE UNIVERSITY OF KENT AT CANTERBURY K24

BA Computing and Accounting & Finance

UCAS Code: GN54 • Mode: 3 Years Equal Combination

Qualifications/Requirements: GCE, A/AS: 22 points. AGNVQ, Distinction (in specific programmes). ND/C, 2 Merits and 4 Distinctions. SQAH, Grades AAABB. SQAV, Individual consideration. IB, 31 points.

BA Computing and Economics

UCAS Code: GLM1 • Mode: 3 Years Equal Combination

Qualifications/Requirements: GCE, A/AS: 22 points. AGNVQ, Distinction (in specific programmes). ND/C, 2 Merits and 4 Distinctions. SQAH, Grades AAABB. SQAV, Individual consideration. IB, 31 points.

BA Computing and Social Psychology

UCAS Code: GL57 • Mode: 3 Years Equal Combination

Qualifications/Requirements: GCE, A level grades: BBB. AGNVQ, Distinction (in specific programmes). ND/C, 1 Merit and 5 Distinctions. SQAH, Grades AAAAB. SQAV, Individual consideration. IB, 33 points.

BA Computing/Classical Studies

UCAS Code: QG85 • Mode: 3 Years Equal Combination

Qualifications/Requirements: GCE, A/AS: 20 points. AGNVQ, Individual consideration. ND/C, 3 Merits and 3 Distinctions. SQAH, Individual consideration. SQAV, Individual consideration. IB, 28 points.

BA Computing/Comparative Literary Studies

UCAS Code: QG25 • Mode: 3 Years Equal Combination

Qualifications/Requirements: GCE, A/AS: 20 points. AGNVQ, Individual consideration. ND/C, 3 Merits and 3 Distinctions. SQAH, Individual consideration. SQAV, Individual consideration. IB, 28 points.

BA Drama/Computing

UCAS Code: WG45 • Mode: 3 Years Equal Combination

Qualifications/Requirements: GCE, A/AS: 22 points. AGNVQ, Individual consideration. ND/C, 2 Merits and 4 Distinctions. SQAH, Individual consideration. SQAV, Individual consideration. IB, 30 points.

BA European Studies/Computing

UCAS Code: TG25 • Mode: 4 Years Equal Combination

Qualifications/Requirements: GCE, A/AS: 20 points. AGNVQ, Individual consideration. ND/C, 3 Merits and 3 Distinctions. SQAH, Individual consideration. SQAV, Individual consideration. IB, 28 points.

BA History and Theory of Art/Computing

UCAS Code: VG45 • Mode: 3 Years Equal Combination

Qualifications/Requirements: GCE, A/AS: 20 points. AGNVQ, Individual consideration. ND/C, 3 Merits and 3 Distinctions. SQAH, Individual consideration. SQAV, Individual consideration. IB, 28 points.

BA History/Computing

UCAS Code: VG15 • Mode: 3 Years Equal Combination

Qualifications/Requirements: GCE, A/AS: 22 points. AGNVQ, Individual consideration. ND/C, 2 Merits and 4 Distinctions. SQAH, Individual consideration. SQAV, Individual consideration. IB, 30 points.

BA Philosophy/Computing

UCAS Code: VG75 • Mode: 3 Years Equal Combination

Qualifications/Requirements: GCE, A/AS: 20 points. AGNVQ, Individual consideration. ND/C, 3 Merits and 3 Distinctions. SQAH, Individual consideration. SQAV, Individual consideration. IB, 28 points.

BA Theology/Computing

UCAS Code: VG85 • Mode: 3 Years Equal Combination

Qualifications/Requirements: GCE, A/AS: 20 points. AGNVQ, Individual consideration. ND/C, 3 Merits and 3 Distinctions. SQAH, Individual consideration. SQAV, Individual consideration. IB, 28 points.

BSc Computer Science

UCAS Code: G500 • Mode: 3 Years Single Subject

Qualifications/Requirements: GCE, A level grades: BBC. AGNVQ, Distinction in Information Technology. ND/C, 3 Merits and 3 Distinctions. SQAH, Grades ABBB. SQAV, Individual consideration. IB, 30 points.

BSc Computer Science and Business Administration

UCAS Code: GN51 • Mode: 3 Years Equal Combination

Qualifications/Requirements: GCE, A level grades: BBC. AGNVQ, Distinction in Information Technology. ND/C, 3 Merits and 3 Distinctions. SQAH, Grades ABBB. SQAV, Individual consideration. IB, 30 points.

BSc Computer Science and Business Administration with a year in industry

UCAS Code: GN5C • Mode: 4 Years Equal Combination

Qualifications/Requirements: GCE, A level grades: BBC. AGNVQ, Distinction in Information Technology. ND/C, 3 Merits and 3 Distinctions. SQAH, Grades ABBB. SQAV, Individual consideration. IB, 30 points.

BSc Computer Science with a year in industry

UCAS Code: G504 • Mode: 4 Years Single Subject

Qualifications/Requirements: GCE, A level grades: BBC. AGNVQ, Distinction in Information Technology. ND/C, 3 Merits and 3 Distinctions. SQAH, Grades ABBB. SQAV, Individual consideration. IB, 30 points.

BSc Computer Science with Management Science

UCAS Code: G5N1 • Mode: 3 Years Major/Minor

Qualifications/Requirements: GCE, A level grades: BBC. Maths at A level. AGNVQ, Distinction in Information Technology with A level. ND/C, Individual consideration. SQAH, Grades ABBB including specific subject(s). SQAV, Individual consideration. IB, 30 points.

BSc Computer Science with Management Science and a year in industry

UCAS Code: G5NC • Mode: 4 Years Major/Minor

Qualifications/Requirements: GCE, A level grades: BBC. Maths at A level. AGNVQ, Distinction in Information Technology with A level. ND/C, Individual consideration. SQAH, Grades ABBB including specific subject(s). SQAV, Individual consideration. IB, 30 points.

BSc Computing and Business Administration

UCAS Code: GN5D • Mode: 3 Years Equal Combination

Qualifications/Requirements: GCE, A level grades: BBC. AGNVQ, Individual consideration. ND/C, Individual consideration. SQAH, Grades ABBB. SQAV, Individual consideration. IB, 30 points.

BSc Computing and Business Administration with a year in industry

UCAS Code: GNM1 • Mode: 4 Years Equal Combination

Qualifications/Requirements: GCE, A level grades: BBC. AGNVQ, Individual consideration. ND/C, Individual consideration. SQAH, Grades ABBB. SQAV, Individual consideration. IB, 30 points.

BSc Computing and Social Psychology

UCAS Code: CG85 • Mode: 3 Years Equal Combination

Qualifications/Requirements: GCE, A level grades: BBB. AGNVQ, Distinction (in specific programmes). ND/C, 1 Merit and 5 Distinctions. SQAH, Grades AAAAB. SQAV, Individual consideration. IB, 33 points.

BSc Computing and Social Statistics

UCAS Code: GG54 • Mode: 3 Years Equal Combination

Qualifications/Requirements: GCE, A/AS: 20 points. Maths. AGNVQ, Individual consideration. ND/C, Individual consideration. SQAH, Grades AABBB including specific subject(s). SQAV, Individual consideration. IB, 29 points.

BSc Management Science and Computing (3 or 4 years)

UCAS Code: NG15 • Mode: 3/4 Years Equal Combination

Qualifications/Requirements: GCE, A level grades: BCC. Maths. AGNVQ, Merit. ND/C, 3 Merits and 3 Distinctions. SQAH, Individual consideration. SQAV, Individual consideration. IB, 29 points.

BSc Mathematics and Computer Science

UCAS Code: GG15 • Mode: 3 Years Equal Combination

Qualifications/Requirements: GCE, A/AS: 20 points. Maths at A level. AGNVQ, Merit with A level. ND/C, Individual consideration. SQAH, Grades BBBB including specific subject(s). SQAV, Individual consideration. IB, 28 points.

BSc Mathematics and Computer Science with a year in industry

UCAS Code: GG1M • Mode: 4 Years Equal Combination

Qualifications/Requirements: GCE, A/AS: 20 points. Maths at A level. AGNVQ, Merit with A level. ND/C, Individual consideration. SQAH, Grades BBBB including specific subject(s). SQAV, Individual consideration. IB, 28 points.

MEng Computer Science

UCAS Code: G502 • Mode: 4 Years Single Subject

Qualifications/Requirements: GCE, A level grades: BBB. AGNVQ, Distinction in Information Technology. ND/C, 2 Merits and 4 Distinctions. SQAH, Grades AABB. SQAV, Individual consideration. IB, 32 points.

HND Computing and Management Science

UCAS Code: 51GN • Mode: 2 Years Equal Combination

Qualifications/Requirements: GCE, A/AS: 2 points. Science. AGNVQ, Pass in Science. ND/C, National.

KING'S COLLEGE LONDON (UNIVERSITY OF LONDON) K60

BSc Computer Science

UCAS Code: G500 • Mode: 3/4 Years Single Subject

Qualifications/Requirements: GCE, A/AS: 24 points. Maths. AGNVQ, Not normally sufficient. ND/C, Not normally sufficient. SQAH, Grades AAABB. SQAV, Not normally sufficient. IB, 32 points.

BSc Computer Science with Management

UCAS Code: G5N1 • Mode: 3/4 Years Major/Minor

Qualifications/Requirements: GCE, A/AS: 24 points. Maths. AGNVQ, Not normally sufficient. ND/C, Not normally sufficient. SQAH, Grades AAABB. SQAV, Not normally sufficient. IB, 32 points.

BSc Mathematics and Computer Science

UCAS Code: GG15 • Mode: 3 Years Equal Combination

Qualifications/Requirements: GCE, A/AS: 24 points. Maths at A level. AGNVQ, Individual consideration. ND/C, Not normally sufficient. SQAH, Grades ABBBB including specific subject(s). SQAV, Not normally sufficient. IB, 28 points.

BSc Mathematics and Computer Science (Management)

UCAS Code: GG1N • Mode: 3 Years Equal Combination

Qualifications/Requirements: GCE, A/AS: 24 points. Maths at A level. AGNVQ, Individual consideration. ND/C, Not normally sufficient. SQAH, Grades ABBBB including specific subject(s). SQAV, Not normally sufficient. IB, 28 points.

MSci Computer Science

UCAS Code: G504 • Mode: 4 Years Single Subject

Qualifications/Requirements: Please refer to prospectus.

KINGSTON UNIVERSITY K84

BA Mathematics & Information Technology (7-11 years)

UCAS Code: XG69 • Mode: 3 Years Equal Combination

Qualifications/Requirements: GCE, A/AS: 12 points. AGNVQ, Individual consideration. ND/C, Individual consideration. SQAH, Individual consideration. SQAV, Individual consideration. IB, Individual consideration.

BA Mathematics and Information Technology (3-8 years)

UCAS Code: XG29 • Mode: 3 Years Equal Combination

Qualifications/Requirements: GCE, A/AS: 12 points. AGNVQ, Individual consideration. ND/C, Individual consideration. SQAH, Individual consideration. SQAV, Individual consideration. IB, Individual consideration.

BSc Business Information Technology

UCAS Code: G562 • Mode: 4 Years Single Subject

Qualifications/Requirements: GCE, A level grades: CCD. AGNVQ, Individual consideration. ND/C, 4 Merits and 4 Distinctions. SQAH, Individual consideration. SQAV, Individual consideration. IB, 30 points.

BSc Chemistry & Computing

UCAS Code: FG15 • Mode: 3 Years Equal Combination

Qualifications/Requirements: GCE, A/AS: 12-14 points. Chemistry. AGNVQ, Individual consideration. ND/C, 3 Merits (in specific programmes). SQAH, Grades CCC. SQAV, Individual consideration. IB, Individual consideration.

BSc Computer Information Systems Design

UCAS Code: G561 • Mode: 4 Years Single Subject

Qualifications/Requirements: GCE, A/AS: 18 points. AGNVQ, Merit with 3 additional units. ND/C, 5 Distinctions (in specific programmes). SQAH, Individual consideration. SQAV, Individual consideration. IB, Individual consideration.

BSc Computer Science

UCAS Code: G500 • Mode: 4 Years Single Subject

Qualifications/Requirements: GCE, A/AS: 18 points. AGNVQ, Merit (in specific programmes) with 3 additional units or with A/AS. ND/C, Individual consideration. SQAH, Individual consideration. SQAV, Individual consideration. IB, 26 points.

BSc Computer Science (Digital Imaging)

UCAS Code: GH5Q • Mode: 4 Years Equal Combination

Qualifications/Requirements: GCE, A/AS: 18 points. AGNVQ, Merit in Engineering with 3 additional units or with A/AS. ND/C, Individual consideration. SQAH, Individual consideration. SQAV, Individual consideration. IB, 26 points.

BSc Computer Science (Network Communications)

UCAS Code: GH5P • Mode: 4 Years Equal Combination

Qualifications/Requirements: GCE, A/AS: 18 points. AGNVQ, Merit in Engineering with 3 additional units or with A/AS. ND/C, Individual consideration. SQAH, Individual consideration. SQAV, Individual consideration. IB, 26 points.

BSc Geographical Information Systems

UCAS Code: GL58 • Mode: 3 Years Equal Combination

Qualifications/Requirements: GCE, A/AS: 10-12 points. Geography or Italian. AGNVQ, Individual consideration. ND/C, (in specific programmes). SQAH, Grades BCCC. SQAV, Individual consideration. IB, Individual consideration.

BSc Geographical Information Systems (Foundation)

UCAS Code: LG85 • Mode: 4 Years Equal Combination

Qualifications/Requirements: GCE, A/AS: 12 points. AGNVQ, Individual consideration. ND/C, Individual consideration. SQAH, Individual consideration. SQAV, Individual consideration. IB, Individual consideration.

BSc Geography & Computing
UCAS Code: FG85 • Mode: 3 Years Equal Combination

Qualifications/Requirements: GCE, A/AS: 14-16 points. Geography. AGNVQ, Individual consideration. ND/C, (in specific programmes). SQAH, Grades BCCC. SQAV, Individual consideration. IB, Individual consideration.

BSc Geology & Computing
UCAS Code: FG65 • Mode: 3 Years Equal Combination

Qualifications/Requirements: GCE, A/AS: 12-14 points. Science. AGNVQ, Individual consideration. ND/C, 3 Merits (in specific programmes). SQAH, Grades CCC. SQAV, Individual consideration. IB, Individual consideration.

BSc Statistics & Computing
UCAS Code: GG54 • Mode: 3 Years Equal Combination

Qualifications/Requirements: GCE, A/AS: 12-14 points. AGNVQ, Individual consideration. ND/C, 3 Merits (in specific programmes). SQAH, Grades CCC. SQAV, Individual consideration. IB, Individual consideration.

HND Geographical Information Systems
UCAS Code: 85LG • Mode: 2 Years Equal Combination

Qualifications/Requirements: GCE, A/AS: 6 points. Geography or Italian. AGNVQ, Individual consideration. ND/C, (in specific programmes). SQAH, Grades CC. SQAV, Individual consideration. IB, Individual consideration.

THE UNIVERSITY OF WALES, LAMPETER L07

BA Business Management and Information Technology
UCAS Code: GN51 • Mode: 1 Year Equal Combination

Qualifications/Requirements: GCE, A/AS: 16 points. AGNVQ, Individual consideration. ND/C, Individual consideration. SQAH, Individual consideration. SQAV, Individual consideration. IB, Individual consideration.

BA Information Technology and Ancient History
UCAS Code: GV5C • Mode: 3 Years Equal Combination

Qualifications/Requirements: GCE, A/AS: 14-16 points. AGNVQ, Individual consideration. ND/C, Individual consideration. SQAH, Individual consideration. SQAV, Individual consideration. IB, Individual consideration.

BA Information Technology and Anthropology
UCAS Code: GL56 • Mode: 3 Years Equal Combination

Qualifications/Requirements: GCE, A/AS: 14-16 points. AGNVQ, Individual consideration. ND/C, Individual consideration. SQAH, Individual consideration. SQAV, Individual consideration. IB, Individual consideration.

BA Information Technology and Archaeology
UCAS Code: GV56 • Mode: 3 Years Equal Combination

Qualifications/Requirements: GCE, A/AS: 14-16 points. AGNVQ, Individual consideration. ND/C, Individual consideration. SQAH, Individual consideration. SQAV, Individual consideration. IB, Individual consideration.

BA Information Technology and Australian Studies
UCAS Code: GL5P • Mode: 3 Years Equal Combination

Qualifications/Requirements: AGNVQ, Individual consideration. ND/C, Individual consideration. SQAH, Individual consideration. SQAV, Individual consideration. IB, Individual consideration.

BA Information Technology and Church History
UCAS Code: GV51 • Mode: 3 Years Equal Combination

Qualifications/Requirements: GCE, A/AS: 14 points. AGNVQ, Individual consideration. ND/C, Individual consideration. SQAH, Individual consideration. SQAV, Individual consideration. IB, Individual consideration.

BA Information Technology and Classical Studies
UCAS Code: GQ58 • Mode: 3 Years Equal Combination

Qualifications/Requirements: GCE, A/AS: 16 points. AGNVQ, Individual consideration. ND/C, Individual consideration. SQAH, Individual consideration. SQAV, Individual consideration. IB, Individual consideration.

BA Information Technology and Cultural Studies in Geography
UCAS Code: GL5V • Mode: 3 Years Equal Combination

Qualifications/Requirements: GCE, A/AS: 16 points. AGNVQ, Individual consideration. ND/C, Individual consideration. SQAH, Individual consideration. SQAV, Individual consideration. IB, Individual consideration.

BA Information Technology and English Literature
UCAS Code: GQ53 • Mode: 3 Years Equal Combination

Qualifications/Requirements: GCE, A/AS: 18 points. English at A level. AGNVQ, Individual consideration. ND/C, Individual consideration. SQAH, Individual consideration. SQAV, Individual consideration. IB, Individual consideration.

BA Information Technology and French
UCAS Code: GR51 • Mode: 4 Years Equal Combination

Qualifications/Requirements: GCE, A/AS: 14-16 points. French at A level. AGNVQ, Individual consideration. ND/C, Individual consideration. SQAH, Individual consideration. SQAV, Individual consideration. IB, Individual consideration.

BA Information Technology and Geography
UCAS Code: GL58 • Mode: 3 Years Equal Combination

Qualifications/Requirements: GCE, A/AS: 16 points. Geography at A level. AGNVQ, Individual consideration. ND/C, Individual consideration. SQAH, Individual consideration. SQAV, Individual consideration. IB, Individual consideration.

BA Information Technology and Greek

UCAS Code: GQ57 • Mode: 3 Years Equal Combination

Qualifications/Requirements: GCE, A/AS: 14-16 points. AGNVQ, Individual consideration. ND/C, Individual consideration. SQAH, Individual consideration. SQAV, Individual consideration. IB, Individual consideration.

BA Information Technology and History

UCAS Code: GV5D • Mode: 3 Years Equal Combination

Qualifications/Requirements: GCE, A/AS: 14-16 points. History at A level. AGNVQ, Individual consideration. ND/C, Individual consideration. SQAH, Individual consideration. SQAV, Individual consideration. IB, Individual consideration.

BA Information Technology and North American Studies

UCAS Code: GQ54 • Mode: 3 Years Equal Combination

Qualifications/Requirements: GCE, A/AS: 16 points. AGNVQ, Individual consideration. ND/C, Individual consideration. SQAH, Individual consideration. SQAV, Individual consideration. IB, Individual consideration.

BA Medieval Studies and Information Technology

UCAS Code: VG1M • Mode: 3 Years Equal Combination

Qualifications/Requirements: GCE, A/AS: 16 points. AGNVQ, Individual consideration. ND/C, Individual consideration. SQAH, Individual consideration. SQAV, Individual consideration. IB, Individual consideration.

BA Modern Historical Studies and Information Technology

UCAS Code: VG1N • Mode: 3 Years Equal Combination

Qualifications/Requirements: GCE, A/AS: 16 points. History. AGNVQ, Individual consideration. ND/C, Individual consideration. SQAH, Individual consideration. SQAV, Individual consideration. IB, Individual consideration.

BA Philosophical Studies and Information Technology

UCAS Code: GV57 • Mode: 3 Years Equal Combination

Qualifications/Requirements: GCE, A/AS: 16 points. AGNVQ, Individual consideration. ND/C, Individual consideration. SQAH, Individual consideration. SQAV, Individual consideration. IB, Individual consideration.

BA Theology and Information Technology

UCAS Code: GV5V • Mode: 3 Years Equal Combination

Qualifications/Requirements: GCE, A/AS: 14 points. AGNVQ, Individual consideration. ND/C, Individual consideration. SQAH, Individual consideration. SQAV, Individual consideration. IB, Individual consideration.

Mod Diploma in Information Technology/Management

UCAS Code: GP52 • Mode: 2/3 Years Equal Combination

Qualifications/Requirements: GCE, A/AS: 14 points. AGNVQ, Individual consideration. ND/C, Individual consideration. SQAH, Individual consideration. SQAV, Individual consideration. IB, Individual consideration.

LANCASTER UNIVERSITY L14

BSc Accounting, Finance and Computer Science

UCAS Code: NG45 • Mode: 3 Years Equal Combination

Qualifications/Requirements: GCE, A level grades: BBC. Maths at A level. AGNVQ, Distinction with 6 additional units or with A/AS. ND/C, Merits and Distinction (in specific programmes). SQAH, Grades ABBBB including specific subject(s). SQAV, Individual consideration. IB, 32 points.

BSc Business Computing and Information Systems

UCAS Code: NG25 • Mode: 3 Years Equal Combination

Qualifications/Requirements: GCE, A level grades: BCC. Maths or Science at A level. AGNVQ, Distinction with 6 additional units or with A/AS. ND/C, Merits and Distinction (in specific programmes). SQAH, Grades BBBBB including specific subject(s). SQAV, Individual consideration. IB, 30 points.

BSc Computer Science

UCAS Code: G500 • Mode: 3 Years Single Subject

Qualifications/Requirements: GCE, A level grades: BCC. AGNVQ, Distinction with 6 additional units or with A/AS. ND/C, Merits and Distinction (in specific programmes). SQAH, Grades BBBBB. SQAV, Individual consideration. IB, 30 points.

BSc Computer Science and Music

UCAS Code: GW53 • Mode: 3 Years Equal Combination

Qualifications/Requirements: GCE, A level grades: BCC. Music at A level. AGNVQ, Distinction with 6 additional units or with A/AS. ND/C, Individual consideration. SQAH, Grades BBBBB including specific subject(s). SQAV, Individual consideration. IB, 30 points.

BSc Computer Science with Multimedia Systems

UCAS Code: G5P4 • Mode: 3 Years Major/Minor

Qualifications/Requirements: GCE, A level grades: BCC. AGNVQ, Distinction with 6 additional units or with A/AS. ND/C, Merits and Distinction (in specific programmes). SQAH, Grades BBBBB. SQAV, Individual consideration. IB, 30 points.

BSc Computer Science with Software Engineering

UCAS Code: G5G7 • Mode: 3 Years Major/Minor

Qualifications/Requirements: GCE, A level grades: BCC. AGNVQ, Distinction with 6 additional units or with A/AS. ND/C, Individual consideration. SQAH, Grades BBBBB. SQAV, Individual consideration. IB, 30 points.

BSc Computing and European Languages

UCAS Code: GT52 • Mode: 4 Years Equal Combination

Qualifications/Requirements: GCE, A level grades: BCC. Two modern foreign languages. AGNVQ, Distinction with 6 additional units or with A/AS. ND/C, Individual consideration. SQAH, Grades BBBBB including specific subject(s). SQAV, Individual consideration. IB, 30 points.

UNIVERSITY OF LEEDS L23

BA Computing/Linguistics

UCAS Code: QG15 • Mode: 3 Years Equal Combination

Qualifications/Requirements: GCE, A level grades: BBC. AGNVQ, Individual consideration. ND/C, Individual consideration. SQAH, Grades ABBBB. SQAV, Individual consideration. IB, 30 points.

BA Computing/Philosophy

UCAS Code: VG75 • Mode: 3 Years Equal Combination

Qualifications/Requirements: GCE, A level grades: BBC. AGNVQ, Individual consideration. ND/C, Individual consideration. SQAH, Grades ABBBB. SQAV, Individual consideration. IB, 30 points.

BA Computing/Russian Civilisation

UCAS Code: RG85 • Mode: 3 Years Equal Combination

Qualifications/Requirements: GCE, A level grades: BBC. AGNVQ, Individual consideration. ND/C, Individual consideration. SQAH, Individual consideration. SQAV, Individual consideration. IB, Individual consideration.

BSc Accounting/Computing

UCAS Code: GN54 • Mode: 3/4 Years Equal Combination

Qualifications/Requirements: GCE, A level grades: BBC. AGNVQ, Individual consideration. ND/C, Individual consideration. SQAH, Individual consideration. SQAV, Individual consideration. IB, 30 points.

BSc Accounting/Information Systems

UCAS Code: G5NK • Mode: 3/4 Years Major/Minor

Qualifications/Requirements: GCE, A level grades: BBC. AGNVQ, Individual consideration. ND/C, Individual consideration. SQAH, Individual consideration. SQAV, Individual consideration. IB, 30 points.

BSc Chemistry/Computer Science

UCAS Code: FG15 • Mode: 3/4 Years Equal Combination

Qualifications/Requirements: GCE, A level grades: BCC. Chemistry and Maths at A level. AGNVQ, Individual consideration. ND/C, 1 Merit and 5 Distinctions (in specific programmes). SQAH, Individual consideration. SQAV, Individual consideration. IB, 28 points.

BSc Computer Science

UCAS Code: G500 • Mode: 3/4 Years Single Subject

Qualifications/Requirements: GCE, A level grades: BBC. Maths at A level. AGNVQ, Individual consideration. ND/C, Individual consideration. SQAH, Grades ABBBB. SQAV, Individual consideration. IB, 30 points.

BSc Computer Science/Economics

UCAS Code: GL51 • Mode: 3 Years Equal Combination

Qualifications/Requirements: GCE, A level grades: BBC. Maths at A level. AGNVQ, Individual consideration. ND/C, Individual consideration. SQAH, Individual consideration. SQAV, Individual consideration. IB, 30 points.

BSc Computer Science-Music

UCAS Code: GW53 • Mode: 3/4 Years Equal Combination

Qualifications/Requirements: GCE, A level grades: BBC. Maths and Music at A level. AGNVQ, Individual consideration. ND/C, 1 Merit and 5 Distinctions (in specific programmes). SQAH, Individual consideration. SQAV, Individual consideration. IB, 30 points.

BSc Computer Science/Philosophy

UCAS Code: GV57 • Mode: 3 Years Equal Combination

Qualifications/Requirements: GCE, A level grades: BBC. Maths at A level. AGNVQ, Individual consideration. ND/C, Individual consideration. SQAH, Individual consideration. SQAV, Individual consideration. IB, Individual consideration.

BSc Computer Science/Physics

UCAS Code: FG35 • Mode: 3/4 Years Equal Combination

Qualifications/Requirements: GCE, A level grades: BBC. Maths and Physics at A level. AGNVQ, Individual consideration. ND/C, 1 Merit and 5 Distinctions (in specific programmes). SQAH, Individual consideration. SQAV, Individual consideration. IB, 30 points.

BSc Computing

UCAS Code: G501 • Mode: 3 Years Single Subject

Qualifications/Requirements: GCE, A level grades: BBC. AGNVQ, Individual consideration. ND/C, Individual consideration. SQAH, Individual consideration. SQAV, Individual consideration. IB, Individual consideration.

BSc Computing/French

UCAS Code: GR51 • Mode: 4 Years Equal Combination

Qualifications/Requirements: GCE, A level grades: BBC. French at A level. AGNVQ, Individual consideration. SQAH, Individual consideration. SQAV, Individual consideration. IB, 30 points.

BSc Computing/German

UCAS Code: GR52 • Mode: 4 Years Equal Combination

Qualifications/Requirements: GCE, A level grades: BBC. German at A level. AGNVQ, Individual consideration. SQAH, Individual consideration. SQAV, Individual consideration. IB, 30 points.

BSc Computing/Management Studies

UCAS Code: GN51 • Mode: 3/4 Years Equal Combination

Qualifications/Requirements: GCE, A level grades: BBC. AGNVQ, Individual consideration. ND/C, 1 Merit and 5 Distinctions (in specific programmes). SQAH, Individual consideration. SQAV, Individual consideration. IB, 30 points.

BSc Information Systems

UCAS Code: G520 • Mode: 3/4 Years Single Subject

Qualifications/Requirements: GCE, A level grades: BBC. AGNVQ, Individual consideration. ND/C, 1 Merit and 5 Distinctions (in specific programmes). SQAH, Grades ABBBB. SQAV, Individual consideration. IB, 30 points.

BSc Information Systems-Management Studies
UCAS Code: G5NC • Mode: 3/4 Years Major/Minor

Qualifications/Requirements: GCE, A level grades: BBC. AGNVQ, Individual consideration. ND/C, 1 Merit and 5 Distinctions (in specific programmes). SQAH, Individual consideration. SQAV, Individual consideration. IB, 30 points.

LEEDS METROPOLITAN UNIVERSITY L27

BA Business Information Management
UCAS Code: GP52 • Mode: 3/4 Years Equal Combination

Qualifications/Requirements: GCE, A/AS: 12-14 points. A modern foreign language at A level. AGNVQ, Distinction (in specific programmes). ND/C, 3 Merits and 1 Distinction. SQAH, Grades BBCC. SQAV, Individual consideration. IB, 26 points.

BSc Accounting and Information Systems
UCAS Code: NG45 • Mode: 3 Years Equal Combination

Qualifications/Requirements: GCE, A/AS: 16 points. AGNVQ, Distinction in Business or Merit in Business with 6 additional units or with A/AS. ND/C, 4 Merits and 3 Distinctions. SQAH, Grades BBBCC. SQAV, Individual consideration. IB, 28 points.

BSc Business Information Systems
UCAS Code: G520 • Mode: 3/4 Years Single Subject

Qualifications/Requirements: GCE, A/AS: 12-14 points. AGNVQ, Distinction (in specific programmes). ND/C, 3 Merits and 2 Distinctions. SQAH, Grades BBCC. SQAV, Individual consideration. IB, 26 points.

BSc Business Information Technology
UCAS Code: G562 • Mode: 1 Year Single Subject

Qualifications/Requirements: AGNVQ, Not normally sufficient. ND/C, Higher National. SQAH, Not normally sufficient. SQAV, Higher National. IB, Not normally sufficient.

BSc Computing
UCAS Code: G501 • Mode: 3/4 Years Single Subject

Qualifications/Requirements: GCE, A/AS: 12-14 points. AGNVQ, Distinction in Information Technology or Merit in Information Technology. ND/C, 3 Merits and 1 Distinction. SQAH, Grades BBBC. SQAV, Individual consideration. IB, 26 points.

BSc Information Systems
UCAS Code: G521 • Mode: 3/4 Years Single Subject

Qualifications/Requirements: GCE, A/AS: 12-14 points. AGNVQ, Distinction (in specific programmes). ND/C, 3 Merits and 1 Distinction. SQAH, Grades BBBC. SQAV, Individual consideration. IB, 26 points.

HND Business Information Technology
UCAS Code: 265G • Mode: 2/3 Years Single Subject

Qualifications/Requirements: GCE, A/AS: 6-8 points. AGNVQ, Merit (in specific programmes). ND/C, 4 Merits. SQAH, Grades BCC. SQAV, Individual consideration. IB, Individual consideration.

HND Computing
UCAS Code: 105G • Mode: 2/3 Years Single Subject

Qualifications/Requirements: GCE, A/AS: 6-8 points. AGNVQ, Merit in Information Technology or Pass in Information Technology. ND/C, 2 to 3 Merits (in specific programmes). SQAH, Grades BCC. SQAV, Individual consideration. IB, Individual consideration.

UNIVERSITY OF LEICESTER L34

BSc Computer Science
UCAS Code: G500 • Mode: 3 Years Single Subject

Qualifications/Requirements: GCE, A/AS: 20 points. AGNVQ, Distinction (in specific programmes) with A/AS. ND/C, Individual consideration. SQAH, Grades BBBBC including specific subject(s). SQAV, Individual consideration. IB, 28 points.

BSc Mathematics and Computer Science
UCAS Code: GG15 • Mode: 4 Years Equal Combination

Qualifications/Requirements: GCE, A/AS: 20 points. Maths at A level. AGNVQ, Distinction (in specific programmes) with A level. ND/C, Individual consideration. SQAH, Grades BBBBC including specific subject(s). SQAV, Individual consideration. IB, 28 points.

BSc Mathematics and Computer Science (European Communications)
UCAS Code: GG1M • Mode: 4 Years Equal Combination

Qualifications/Requirements: GCE, A/AS: 20 points. Maths at A level. AGNVQ, Distinction (in specific programmes) with A level. ND/C, Individual consideration. SQAH, Grades BBBBC including specific subject(s). SQAV, Individual consideration. IB, 28 points.

UNIVERSITY OF LINCOLNSHIRE AND HUMBERSIDE L39

BA/BSc Accountancy and Information Technology
UCAS Code: GN54 • Mode: 3 Years Equal Combination

Qualifications/Requirements: GCE, A/AS: 12 points. AGNVQ, Merit. ND/C, 3 Merits and 1 Distinction. SQAH, Grades CCCC. SQAV, Individual consideration. IB, 24 points.

BA/BSc Communications and Information Systems
UCAS Code: GP5H • Mode: 3 Years Equal Combination

Qualifications/Requirements: GCE, A/AS: 16 points. AGNVQ, Distinction. ND/C, 1 Merit and 3 Distinctions. SQAH, Grades BBCCC. SQAV, Individual consideration. IB, 24 points.

BA/BSc Computing and Accountancy
UCAS Code: GN5K • Mode: 3 Years Equal Combination

Qualifications/Requirements: GCE, A/AS: 12 points. AGNVQ, Merit. ND/C, 3 Merits and 1 Distinction. SQAH, Grades CCCC. SQAV, Individual consideration. IB, 24 points.

BA/BSc Computing and Business
UCAS Code: NG1M • Mode: 3 Years Equal Combination
Qualifications/Requirements: GCE, A/AS: 12 points. AGNVQ, Merit. ND/C, 3 Merits and 1 Distinction. SQAH, Grades CCCC. SQAV, Individual consideration. IB, 24 points.

BA/BSc Computing and European Studies
UCAS Code: GT52 • Mode: 3 Years Equal Combination
Qualifications/Requirements: GCE, A/AS: 12 points. AGNVQ, Merit. ND/C, 3 Merits and 1 Distinction. SQAH, Grades CCCC. SQAV, Individual consideration. IB, 24 points.

BA/BSc Computing and Human Resource Management
UCAS Code: GN56 • Mode: 3 Years Equal Combination
Qualifications/Requirements: GCE, A/AS: 12 points. AGNVQ, Merit. ND/C, 3 Merits and 1 Distinction. SQAH, Grades CCCC. SQAV, Individual consideration. IB, 24 points.

BA/BSc Computing and Information Technology
UCAS Code: G523 • Mode: 3 Years Single Subject
Qualifications/Requirements: GCE, A/AS: 12 points. AGNVQ, Merit. ND/C, 3 Merits and 1 Distinction. SQAH, Grades CCCC. SQAV, Individual consideration. IB, 24 points.

BA/BSc Computing and Marketing
UCAS Code: GN5M • Mode: 3 Years Equal Combination
Qualifications/Requirements: GCE, A/AS: 12 points. AGNVQ, Merit. ND/C, 3 Merits and 1 Distinction. SQAH, Grades CCCC. SQAV, Individual consideration. IB, 24 points.

BA/BSc Computing and Media Technology
UCAS Code: GP54 • Mode: 3 Years Equal Combination
Qualifications/Requirements: GCE, A/AS: 12 points. AGNVQ, Merit. ND/C, 3 Merits and 1 Distinction. SQAH, Grades CCCC. SQAV, Individual consideration. IB, 24 points.

BA/BSc Computing and Modern Languages
UCAS Code: GT5X • Mode: 3 Years Equal Combination
Qualifications/Requirements: GCE, A/AS: 12 points. A modern foreign language. AGNVQ, Merit. ND/C, 3 Merits and 1 Distinction. SQAH, Grades CCCC. SQAV, Individual consideration. IB, 24 points.

BA/BSc Criminology and Information Systems
UCAS Code: GM5H • Mode: 3 Years Equal Combination
Qualifications/Requirements: GCE, A/AS: 16 points. AGNVQ, Distinction. ND/C, 1 Merit and 3 Distinctions. SQAH, Grades BBCCC. SQAV, Individual consideration. IB, 24 points.

BA/BSc Economics and Information Systems
UCAS Code: GL51 • Mode: 3 Years Equal Combination
Qualifications/Requirements: GCE, A/AS: 16 points. AGNVQ, Distinction. ND/C, 1 Merit and 3 Distinctions. SQAH, Grades BBCCC. SQAV, Individual consideration. IB, 24 points.

BA/BSc European Studies and Information Technology
UCAS Code: GT5F • Mode: 3 Years Equal Combination
Qualifications/Requirements: GCE, A/AS: 12 points. AGNVQ, Merit. ND/C, 3 Merits and 1 Distinction. SQAH, Grades CCCC. SQAV, Individual consideration. IB, 24 points.

BA/BSc Human Resource Management and Information Technology
UCAS Code: GN5P • Mode: 3 Years Equal Combination
Qualifications/Requirements: GCE, A/AS: 12 points. AGNVQ, Merit. ND/C, 3 Merits and 1 Distinction. SQAH, Grades CCCC. SQAV, Individual consideration. IB, 24 points.

BA/BSc Information Systems and International Relations
UCAS Code: GM51 • Mode: 3 Years Equal Combination
Qualifications/Requirements: GCE, A/AS: 16 points. AGNVQ, Distinction. ND/C, 1 Merit and 3 Distinctions. SQAH, Grades BBCCC. SQAV, Individual consideration. IB, 24 points.

BA/BSc Information Systems and Journalism
UCAS Code: GP56 • Mode: 3 Years Equal Combination
Qualifications/Requirements: GCE, A/AS: 18 points. AGNVQ, Distinction. ND/C, 1 Merit and 4 Distinctions. SQAH, Grades BBBCC. SQAV, Individual consideration. IB, 26 points.

BA/BSc Information Systems and Law
UCAS Code: GM53 • Mode: 3 Years Equal Combination
Qualifications/Requirements: GCE, A/AS: 16 points. AGNVQ, Distinction. ND/C, 1 Merit and 3 Distinctions. SQAH, Grades BBCCC. SQAV, Individual consideration. IB, 24 points.

BA/BSc Information Systems and Marketing
UCAS Code: GN5N • Mode: 3 Years Equal Combination
Qualifications/Requirements: GCE, A/AS: 14 points. AGNVQ, Merit. ND/C, 2 Merits and 2 Distinctions. SQAH, Grades BCCC. SQAV, Individual consideration. IB, 24 points.

BA/BSc Information Systems and Media Production
UCAS Code: GP5L • Mode: 3 Years Equal Combination
Qualifications/Requirements: GCE, A/AS: 18 points. AGNVQ, Distinction. ND/C, 1 Merit and 4 Distinctions. SQAH, Grades BBBCC. SQAV, Individual consideration. IB, 26 points.

BA/BSc Information Systems and Politics

UCAS Code: GM5C • Mode: 3 Years Equal Combination

Qualifications/Requirements: GCE, A/AS: 14 points. AGNVQ, Merit. ND/C, 2 Merits and 2 Distinctions. SQAH, Grades BCCC. SQAV, Individual consideration. IB, 24 points.

BA/BSc Information Systems and Psychology

UCAS Code: CG8N • Mode: 3 Years Equal Combination

Qualifications/Requirements: GCE, A/AS: 18 points. AGNVQ, Distinction. ND/C, 1 Merit and 4 Distinctions. SQAH, Grades BBBCC. SQAV, Individual consideration. IB, 26 points.

BA/BSc Information Systems and Social Policy

UCAS Code: GL5K • Mode: 3 Years Equal Combination

Qualifications/Requirements: GCE, A/AS: 14 points. AGNVQ, Merit. ND/C, 2 Merits and 2 Distinctions. SQAH, Grades BCCC. SQAV, Individual consideration. IB, 24 points.

BA/BSc Information Systems and Tourism

UCAS Code: GP57 • Mode: 3 Years Equal Combination

Qualifications/Requirements: GCE, A/AS: 14 points. AGNVQ, Merit. ND/C, 2 Merits and 2 Distinctions. SQAH, Grades BCCC. SQAV, Individual consideration. IB, 24 points.

BA/BSc Information Technology and Marketing

UCAS Code: GN55 • Mode: 3 Years Equal Combination

Qualifications/Requirements: GCE, A/AS: 12 points. AGNVQ, Merit. ND/C, 3 Merits and 1 Distinction. SQAH, Grades CCCC. SQAV, Individual consideration. IB, 24 points.

BA/BSc Information Technology and Media Technology

UCAS Code: GP5K • Mode: 3 Years Equal Combination

Qualifications/Requirements: GCE, A/AS: 12 points. AGNVQ, Merit. ND/C, 3 Merits and 1 Distinction. SQAH, Grades CCCC. SQAV, Individual consideration. IB, 24 points.

BA/BSc Information Technology and Modern Languages

UCAS Code: GT5Y • Mode: 3 Years Equal Combination

Qualifications/Requirements: GCE, A/AS: 12 points. A modern foreign language. AGNVQ, Merit. ND/C, 3 Merits and 1 Distinction. SQAH, Grades CCCC. SQAV, Individual consideration. IB, 24 points.

BA/BSc Management and Information Systems

UCAS Code: NG1N • Mode: 3 Years Equal Combination

Qualifications/Requirements: GCE, A/AS: 14 points. AGNVQ, Merit. ND/C, 2 Merits and 2 Distinctions. SQAH, Grades BCCC. SQAV, Individual consideration. IB, 24 points.

BSc Computing

UCAS Code: G500 • Mode: 3/4 Years Single Subject

Qualifications/Requirements: GCE, A/AS: 12 points. AGNVQ, Merit. ND/C, 3 Merits and 1 Distinction. SQAH, Grades CCCC. SQAV, Individual consideration. IB, 24 points.

BSc Computing (Foundation)

UCAS Code: G508 • Mode: 1 Year Single Subject

Qualifications/Requirements: AGNVQ, Individual consideration. ND/C, Individual consideration. SQAH, Individual consideration. SQAV, Individual consideration. IB, Individual consideration.

BSc Computing (Games, Simulation and Virtual Reality)

UCAS Code: G501 • Mode: 3 Years Single Subject

Qualifications/Requirements: GCE, A/AS: 12 points. AGNVQ, Merit. ND/C, 3 Merits and 1 Distinction. SQAH, Grades CCCC. SQAV, Individual consideration. IB, 24 points.

BSc Computing (Internet Technologies)

UCAS Code: G502 • Mode: 3 Years Single Subject

Qualifications/Requirements: GCE, A/AS: 12 points. AGNVQ, Merit. ND/C, 3 Merits and 1 Distinction. SQAH, Grades CCCC. SQAV, Individual consideration. IB, 24 points.

BSc Computing (Multimedia and Systems Development)

UCAS Code: G503 • Mode: 3 Years Single Subject

Qualifications/Requirements: GCE, A/AS: 12 points. AGNVQ, Merit. ND/C, 3 Merits and 1 Distinction. SQAH, Grades CCCC. SQAV, Individual consideration. IB, 24 points.

BSc Computing (Network Systems Support and Management)

UCAS Code: G504 • Mode: 3 Years Single Subject

Qualifications/Requirements: GCE, A/AS: 12 points. AGNVQ, Merit. ND/C, 3 Merits and 1 Distinction. SQAH, Grades CCCC. SQAV, Individual consideration. IB, 24 points.

BSc Information Technology

UCAS Code: G560 • Mode: 3 Years Single Subject

Qualifications/Requirements: GCE, A/AS: 12 points. AGNVQ, Merit. ND/C, 3 Merits and 1 Distinction. SQAH, Grades CCCC. SQAV, Individual consideration. IB, 24 points.

Mod Advertising and Computing

UCAS Code: PG35 • Mode: 3 Years Equal Combination

Qualifications/Requirements: AGNVQ, Individual consideration. ND/C, Individual consideration. SQAH, Individual consideration. SQAV, Individual consideration. IB, Individual consideration.

Mod Advertising and Information Technology
UCAS Code: PG3M • Mode: 3 Years Equal Combination
Qualifications/Requirements: AGNVQ, Individual consideration. ND/C, Individual consideration. SQAH, Individual consideration. SQAV, Individual consideration. IB, Individual consideration.

Mod Animation and Computing
UCAS Code: WGF5 • Mode: 3 Years Equal Combination
Qualifications/Requirements: AGNVQ, Individual consideration. ND/C, Individual consideration. SQAH, Individual consideration. SQAV, Individual consideration. IB, Individual consideration.

Mod Animation and Information Technology
UCAS Code: WGFM • Mode: 3 Years Equal Combination
Qualifications/Requirements: AGNVQ, Individual consideration. ND/C, Individual consideration. SQAH, Individual consideration. SQAV, Individual consideration. IB, Individual consideration.

Mod Applied Social Science and Information Technology
UCAS Code: LGHM • Mode: 3 Years Equal Combination
Qualifications/Requirements: AGNVQ, Individual consideration. ND/C, Individual consideration. SQAH, Individual consideration. SQAV, Individual consideration. IB, Individual consideration.

MOd Architectural Technology and Computing
UCAS Code: KG25 • Mode: 3 Years Equal Combination
Qualifications/Requirements: AGNVQ, Individual consideration. ND/C, Individual consideration. SQAH, Individual consideration. SQAV, Individual consideration. IB, Individual consideration.

Mod Architectural Technology and Information Technology
UCAS Code: KG2M • Mode: 3 Years Equal Combination
Qualifications/Requirements: AGNVQ, Individual consideration. ND/C, Individual consideration. SQAH, Individual consideration. SQAV, Individual consideration. IB, Individual consideration.

Mod Computing and Animation
UCAS Code: GW5F • Mode: 3 Years Equal Combination
Qualifications/Requirements: AGNVQ, Individual consideration. ND/C, Individual consideration. SQAH, Individual consideration. SQAV, Individual consideration. IB, Individual consideration.

Mod Computing and Electronic Commerce
UCAS Code: GNMN • Mode: 3 Years Equal Combination
Qualifications/Requirements: AGNVQ, Individual consideration. ND/C, Individual consideration. SQAH, Individual consideration. SQAV, Individual consideration. IB, Individual consideration.

Mod Computing and Fine Art
UCAS Code: GW51 • Mode: 3 Years Equal Combination
Qualifications/Requirements: AGNVQ, Individual consideration. ND/C, Individual consideration. SQAH, Individual consideration. SQAV, Individual consideration. IB, Individual consideration.

Mod Computing and Interactive Design
UCAS Code: GG5N • Mode: 3 Years Equal Combination
Qualifications/Requirements: AGNVQ, Individual consideration. ND/C, Individual consideration. SQAH, Individual consideration. SQAV, Individual consideration. IB, Individual consideration.

Mod Computing and Interior Design
UCAS Code: GW59 • Mode: 3 Years Equal Combination
Qualifications/Requirements: AGNVQ, Individual consideration. ND/C, Individual consideration. SQAH, Individual consideration. SQAV, Individual consideration. IB, Individual consideration.

Mod Computing and Museum and Exhibition Design
UCAS Code: GP51 • Mode: 3 Years Equal Combination
Qualifications/Requirements: AGNVQ, Individual consideration. ND/C, Individual consideration. SQAH, Individual consideration. SQAV, Individual consideration. IB, Individual consideration.

Mod English and Information Systems
UCAS Code: GQM3 • Mode: 3 Years Equal Combination
Qualifications/Requirements: GCE, A/AS: 16 points. AGNVQ, Distinction. ND/C, 1 Merit and 3 Distinctions. SQAH, Grades BBCCC. SQAV, Individual consideration. IB, 24 points.

Mod Environmental Biology and Information Systems
UCAS Code: CG95 • Mode: 3 Years Equal Combination
Qualifications/Requirements: AGNVQ, Individual consideration. ND/C, Individual consideration. SQAH, Individual consideration. SQAV, Individual consideration. IB, Individual consideration.

Mod Environmental Studies and Information Systems
UCAS Code: FGYM • Mode: 3 Years Equal Combination
Qualifications/Requirements: AGNVQ, Individual consideration. ND/C, Individual consideration. SQAH, Individual consideration. SQAV, Individual consideration. IB, Individual consideration.

Mod Fine Art and Information Technology
UCAS Code: WG1M • Mode: 3 Years Equal Combination
Qualifications/Requirements: AGNVQ, Individual consideration. ND/C, Individual consideration. SQAH, Individual consideration. SQAV, Individual consideration. IB, Individual consideration.

Mod Food Studies and Information Systems
UCAS Code: DGKM • Mode: 3 Years Equal Combination
Qualifications/Requirements: AGNVQ, Individual consideration. ND/C, Individual consideration. SQAH, Individual consideration. SQAV, Individual consideration. IB, Individual consideration.

Mod Forensic Science and Information Systems
UCAS Code: BG1M • Mode: 3 Years Equal Combination
Qualifications/Requirements: AGNVQ, Individual consideration. ND/C, Individual consideration. SQAH, Individual consideration. SQAV, Individual consideration. IB, Individual consideration.

Mod Graphic Design and Information Technology
UCAS Code: WG2M • Mode: 3 Years Equal Combination
Qualifications/Requirements: AGNVQ, Individual consideration. ND/C, Individual consideration. SQAH, Individual consideration. SQAV, Individual consideration. IB, Individual consideration.

Mod Graphic Design and Interactive Design
UCAS Code: WG2N • Mode: 3 Years Equal Combination
Qualifications/Requirements: AGNVQ, Individual consideration. ND/C, Individual consideration. SQAH, Individual consideration. SQAV, Individual consideration. IB, Individual consideration.

Mod Illustration and Information Technology
UCAS Code: WGGM • Mode: 3 Years Equal Combination
Qualifications/Requirements: AGNVQ, Individual consideration. ND/C, Individual consideration. SQAH, Individual consideration. SQAV, Individual consideration. IB, Individual consideration.

Mod Information Systems and Public Relations
UCAS Code: GP5J • Mode: 3 Years Equal Combination
Qualifications/Requirements: AGNVQ, Individual consideration. ND/C, Individual consideration. SQAH, Individual consideration. SQAV, Individual consideration. IB, Individual consideration.

Mod Information Systems and Social Anthropology
UCAS Code: GLM6 • Mode: 3 Years Equal Combination
Qualifications/Requirements: AGNVQ, Individual consideration. ND/C, Individual consideration. SQAH, Individual consideration. SQAV, Individual consideration. IB, Individual consideration.

Mod Information Technology and Interactive Design
UCAS Code: GGMN • Mode: 3 Years Equal Combination
Qualifications/Requirements: AGNVQ, Individual consideration. ND/C, Individual consideration. SQAH, Individual consideration. SQAV, Individual consideration. IB, Individual consideration.

Mod Information Technology and Interior Design
UCAS Code: GWM9 • Mode: 3 Years Equal Combination
Qualifications/Requirements: AGNVQ, Individual consideration. ND/C, Individual consideration. SQAH, Individual consideration. SQAV, Individual consideration. IB, Individual consideration.

Mod Information Technology and Museum and Exhibition Design
UCAS Code: GPM1 • Mode: 3 Years Equal Combination
Qualifications/Requirements: AGNVQ, Individual consideration. ND/C, Individual consideration. SQAH, Individual consideration. SQAV, Individual consideration. IB, Individual consideration.

HND Business and Information Systems
UCAS Code: 51GN • Mode: 2 Years Equal Combination
Qualifications/Requirements: Please refer to prospectus.

HND Business and Information Technology
UCAS Code: 1MNG • Mode: 2 Years Equal Combination
Qualifications/Requirements: GCE, A/AS: 4 points. AGNVQ, Pass. ND/C, 2 Merits. SQAH, Grades C. SQAV, Individual consideration. IB, 24 points.

HND Computing
UCAS Code: 105G • Mode: 2 Years Single Subject
Qualifications/Requirements: GCE, A/AS: 4 points. AGNVQ, Pass. ND/C, 2 Merits. SQAH, Grades C. SQAV, Individual consideration. IB, 24 points.

THE UNIVERSITY OF LIVERPOOL L41

BA Accounting and Computer Science
UCAS Code: GN54 • Mode: 3 Years Equal Combination
Qualifications/Requirements: GCE, A level grades: CCC. Maths, Economics or Business Studies. AGNVQ, Individual consideration. ND/C, Individual consideration. SQAH, Individual consideration. SQAV, Individual consideration. IB, Individual consideration.

BA Business Economics and Computer Science
UCAS Code: LG15 • Mode: 3 Years Equal Combination
Qualifications/Requirements: GCE, A level grades: CCC. Maths, Economics or Business Studies. AGNVQ, Individual consideration. ND/C, Individual consideration. SQAH, Individual consideration. SQAV, Individual consideration. IB, Individual consideration.

BA Economics and Computer Science
UCAS Code: GL51 • Mode: 3 Years Equal Combination
Qualifications/Requirements: GCE, A level grades: CCC. Maths, Economics or Business Studies. AGNVQ, Individual consideration. ND/C, Individual consideration. SQAH, Individual consideration. SQAV, Individual consideration. IB, Individual consideration.

BSc Computer Information Systems
UCAS Code: G520 • Mode: 3 Years Single Subject
Qualifications/Requirements: GCE, A/AS: 20 points. AGNVQ, Distinction in Science with A level. ND/C, Merit overall and 4 Distinctions (in specific programmes). SQAH, Grades BBBBC including specific subject(s). SQAV, Individual consideration. IB, 30 points.

BSc Computer Information Systems (1+2)
UCAS Code: G523 • Mode: 3 Years Single Subject
Qualifications/Requirements: GCE, A/AS: 20 points. AGNVQ, Individual consideration. ND/C, Merit overall and 4 Distinctions (in specific programmes). SQAH, Grades BBBBC including specific subject(s). SQAV, Individual consideration. IB, 30 points.

BSc Computer Information Systems (1+3)
UCAS Code: G521 • Mode: 4 Years Single Subject
Qualifications/Requirements: GCE, A/AS: 8-10 points. AGNVQ, Individual consideration. ND/C, Merit overall (in specific programmes). SQAH, Grades CCCDD including specific subject(s). SQAV, Individual consideration. IB, 24 points.

BSc Computer Information Systems (2+2)
UCAS Code: G522 • Mode: 4 Years Single Subject
Qualifications/Requirements: GCE, A/AS: 8-10 points. AGNVQ, Individual consideration. ND/C, Merit overall (in specific programmes). SQAH, Grades CCCDD including specific subject(s). SQAV, Individual consideration. IB, 24 points.

BSc Computer Science
UCAS Code: G500 • Mode: 3 Years Single Subject
Qualifications/Requirements: GCE, A/AS: 20 points. Maths at A level. AGNVQ, Distinction in Science with A level. ND/C, Merit overall and 4 Distinctions (in specific programmes). SQAH, Grades BBBBC including specific subject(s). SQAV, Individual consideration. IB, 30 points.

BSc Computer Science (Foundation) (1+3)
UCAS Code: G508 • Mode: 4 Years Single Subject
Qualifications/Requirements: GCE, A/AS: 10 points. Maths and Science at A level. AGNVQ, Individual consideration. ND/C, Merit overall (in specific programmes). SQAH, Grades CCCDD including specific subject(s). SQAV, Individual consideration. IB, 24 points.

BSc Computer Science with a European Language
UCAS Code: G5T2 • Mode: 4 Years Major/Minor
Qualifications/Requirements: GCE, A/AS: 20 points. Maths at A level. AGNVQ, Individual consideration. ND/C, Merit overall and 4 Distinctions (in specific programmes). SQAH, Grades BBBBC including specific subject(s). SQAV, Individual consideration. IB, 30 points.

BSc Mathematics and Computer Science
UCAS Code: GG15 • Mode: 3 Years Equal Combination
Qualifications/Requirements: GCE, A/AS: 22 points. Maths at A level. AGNVQ, Distinction (in specific programmes) with A level. ND/C, Merit overall (in specific programmes). SQAH, Grades BBBCC including specific subject(s). SQAV, Individual consideration. IB, 31 points.

BSc Physics and Computer Science
UCAS Code: FG35 • Mode: 3 Years Equal Combination
Qualifications/Requirements: GCE, A/AS: 20 points. Maths and Physics at A level. AGNVQ, Distinction in Science with A level. ND/C, Merit overall (in specific programmes). SQAH, Grades BBBCC including specific subject(s). SQAV, Individual consideration. IB, 31 points.

LIVERPOOL COMMUNITY COLLEGE L43

HND Business Information Technology
UCAS Code: 15NG • Mode: 2 Years Equal Combination
Qualifications/Requirements: Please refer to prospectus.

LIVERPOOL HOPE L46

BA Business and Community Enterprise/Information Technology
UCAS Code: NG15 • Mode: 3 Years Equal Combination
Qualifications/Requirements: GCE, A/AS: 12 points. AGNVQ, Merit. ND/C, 8 Merits. SQAH, Individual consideration. SQAV, Individual consideration. IB, Individual consideration.

BA Cities, Communities & Regeneration and Information Technology
UCAS Code: GL55 • Mode: 3 Years Equal Combination
Qualifications/Requirements: Please refer to prospectus.

BA Identity Studies and Information Technology
UCAS Code: LG35 • Mode: 3 Years Equal Combination
Qualifications/Requirements: Please refer to prospectus.

BA Information Technology/Drama & Theatre Studies
UCAS Code: GW54 • Mode: 3 Years Equal Combination
Qualifications/Requirements: GCE, A/AS: 12 points. AGNVQ, Merit. ND/C, 8 Merits. SQAH, Individual consideration. SQAV, Individual consideration. IB, Individual consideration.

BA Information Technology/English
UCAS Code: GQ53 • Mode: 3 Years Equal Combination
Qualifications/Requirements: GCE, A/AS: 12 points. English Literature at A level. AGNVQ, Pass with A level. ND/C, 8 Merits. SQAH, Individual consideration. SQAV, Individual consideration. IB, Individual consideration.

BA Information Technology/European Studies
UCAS Code: TG25 • Mode: 3 Years Equal Combination
Qualifications/Requirements: GCE, A/AS: 12 points. AGNVQ, Merit. ND/C, 8 Merits. SQAH, Individual consideration. SQAV, Individual consideration. IB, Individual consideration.

BA Information Technology/French

UCAS Code: GR51 • Mode: 3 Years Equal Combination

Qualifications/Requirements: GCE, A/AS: 12 points. French at A level. AGNVQ, Pass with A level. ND/C, 8 Merits. SQAH, Individual consideration. SQAV, Individual consideration. IB, Individual consideration.

BA Information Technology/History

UCAS Code: GV51 • Mode: 3 Years Equal Combination

Qualifications/Requirements: GCE, A/AS: 12 points. History at A level. AGNVQ, Pass with A level. ND/C, 8 Merits. SQAH, Individual consideration. SQAV, Individual consideration. IB, Individual consideration.

BA Sociology/Information Technology

UCAS Code: GL53 • Mode: 3 Years Equal Combination

Qualifications/Requirements: GCE, A/AS: 12 points. AGNVQ, Merit. ND/C, 8 Merits. SQAH, Individual consideration. SQAV, Individual consideration. IB, Individual consideration.

BEd Information Technology

UCAS Code: X2G5 • Mode: 4 Years Major/Minor

Qualifications/Requirements: GCE, A/AS: 10 points. AGNVQ, Merit. ND/C, 6 Merits. SQAH, Individual consideration. SQAV, Individual consideration. IB, Individual consideration.

BEd Information Technology

UCAS Code: X4G5 • Mode: 4 Years Major/Minor

Qualifications/Requirements: GCE, A/AS: 10 points. AGNVQ, Merit. ND/C, 6 Merits. SQAH, Individual consideration. SQAV, Individual consideration. IB, Individual consideration.

BSc Information Technology/ Environmental Studies

UCAS Code: GF59 • Mode: 3 Years Equal Combination

Qualifications/Requirements: GCE, A/AS: 10 points. A level: Biology, Geography, Geology, Environmental Science or History. Specific GCSEs required. AGNVQ, Merit (in specific programmes). ND/C, 6 Merits. SQAH, Individual consideration. SQAV, Individual consideration. IB, Individual consideration.

BSc Information Technology/Geography

UCAS Code: GF58 • Mode: 3 Years Equal Combination

Qualifications/Requirements: GCE, A/AS: 10 points. Geography, Geology or Environmental Science at A level. AGNVQ, Merit (in specific programmes). ND/C, 6 Merits. SQAH, Individual consideration. SQAV, Individual consideration. IB, Individual consideration.

BSc Information Technology/Human & Applied Biology

UCAS Code: CG15 • Mode: 3 Years Equal Combination

Qualifications/Requirements: GCE, A/AS: 10 points. Biology at A level. AGNVQ, Merit (in specific programmes) or Pass with A level. GCSEs or equivalent needed. ND/C, 6 Merits. SQAH, Individual consideration. SQAV, Individual consideration. IB, Individual consideration.

BSc Mathematics/Information Technology

UCAS Code: GG51 • Mode: 3 Years Equal Combination

Qualifications/Requirements: GCE, A/AS: 10 points. Maths. AGNVQ, Pass with A level. ND/C, 6 Merits. SQAH, Individual consideration. SQAV, Individual consideration. IB, Individual consideration.

BSc Psychology/Information Technology

UCAS Code: GC58 • Mode: 3 Years Equal Combination

Qualifications/Requirements: GCE, A/AS: 10 points. AGNVQ, Merit. ND/C, 6 Merits. SQAH, Individual consideration. SQAV, Individual consideration. IB, Individual consideration.

LIVERPOOL JOHN MOORES UNIVERSITY L51

BEd Information Technology

UCAS Code: X7G5 • Mode: 2 Years Major/Minor

Qualifications/Requirements: Please refer to prospectus.

BSc Business Information Systems

UCAS Code: G522 • Mode: 4 Years Single Subject

Qualifications/Requirements: GCE, A/AS: 12-14 points. AGNVQ, Distinction in Business. ND/C, Distinction Overall. SQAH, Grades CCC. SQAV, Individual consideration. IB, 25 points.

BSc Computer Studies

UCAS Code: G501 • Mode: 4 Years Single Subject

Qualifications/Requirements: GCE, A/AS: 12 points. AGNVQ, Distinction (in specific programmes). ND/C, Merit overall and 1 Distinction. SQAH, Individual consideration. SQAV, Individual consideration. IB, 26 points.

BSc Information Systems Management

UCAS Code: GN51 • Mode: 4 Years Equal Combination

Qualifications/Requirements: GCE, A/AS: 12 points. AGNVQ, Distinction (in specific programmes). ND/C, Merit overall and Distinction. SQAH, Individual consideration. SQAV, Individual consideration. IB, 26 points.

BSc Mathematics, Statistics and Computing

UCAS Code: G920 • Mode: 4 Years Single Subject

Qualifications/Requirements: GCE, A/AS: 10 points. Maths. AGNVQ, Distinction (in specific programmes) with A level. ND/C, 3 Merits (in specific programmes). SQAH, Individual consideration. SQAV, Individual consideration. IB, 26 points.

HND Computing

UCAS Code: 105G • Mode: 2 Years Single Subject

Qualifications/Requirements: GCE, A/AS: 4 points. AGNVQ, Merit (in specific programmes). ND/C, 1 Merit (in specific programmes). SQAH, Individual consideration. SQAV, Individual consideration. IB, Individual consideration.

LLANDRILLO COLLEGE, NORTH WALES L53

HND Computing
UCAS Code: 005G • Mode: 2 Years Single Subject

Qualifications/Requirements: GCE, A/AS: 6-8 points. AGNVQ, Pass. ND/C, National. SQAH, Grades CCCC. SQAV, Individual consideration. IB, 26 points.

HND Information Systems
UCAS Code: 025G • Mode: 2 Years Single Subject

Qualifications/Requirements: Please refer to prospectus.

HND Personal Computer Technology
UCAS Code: 065G • Mode: 2 Years Single Subject

Qualifications/Requirements: Please refer to prospectus.

LONDON GUILDHALL UNIVERSITY L55

BSc Computing and Information Systems
UCAS Code: G520 • Mode: 3 Years Single Subject

Qualifications/Requirements: GCE, A level grades: CD. AGNVQ, Merit (in specific programmes). ND/C, 4 Merits. SQAH, Individual consideration. SQAV, Individual consideration. IB, 24 points.

Mod Accounting & Business Information Technology
UCAS Code: GN74 • Mode: 3 Years Equal Combination

Qualifications/Requirements: GCE, A level grades: DD. AGNVQ, Merit (in specific programmes). ND/C, Merit overall. SQAH, Individual consideration. SQAV, Individual consideration. IB, 24 points.

Mod Accounting & Computing
UCAS Code: GN54 • Mode: 3 Years Equal Combination

Qualifications/Requirements: GCE, A level grades: DD. AGNVQ, Merit (in specific programmes). ND/C, Merit overall. SQAH, Individual consideration. SQAV, Individual consideration. IB, 24 points.

Mod American Studies & Business Information Technology
UCAS Code: GQ74 • Mode: 3 Years Equal Combination

Qualifications/Requirements: GCE, A level grades: CD. AGNVQ, Merit. ND/C, Merit overall and 2 Distinctions. SQAH, Individual consideration. SQAV, Individual consideration. IB, 26 points.

Mod Asia Studies & Business Information Technology
UCAS Code: GT75 • Mode: 3 Years Equal Combination

Qualifications/Requirements: GCE, A level grades: CD. AGNVQ, Merit. ND/C, Merit overall and 2 Distinctions. SQAH, Individual consideration. SQAV, Individual consideration. IB, 26 points.

Mod Asia Studies & Computing
UCAS Code: GT55 • Mode: 3 Years Equal Combination

Qualifications/Requirements: GCE, A level grades: CD. AGNVQ, Merit. ND/C, Merit overall and 2 Distinctions. SQAH, Individual consideration. SQAV, Individual consideration. IB, 26 points.

Mod Banking & Computing
UCAS Code: NG35 • Mode: 3 Years Equal Combination

Qualifications/Requirements: GCE, A level grades: DD. AGNVQ, Merit (in specific programmes). ND/C, Merit overall. SQAH, Individual consideration. SQAV, Individual consideration. IB, 24 points.

Mod Business & Computing
UCAS Code: GN51 • Mode: 3 Years Equal Combination

Qualifications/Requirements: GCE, A level grades: CD-DDD. AGNVQ, Merit (in specific programmes). ND/C, Merit overall and 2 Distinctions. SQAH, Individual consideration. SQAV, Individual consideration. IB, 24 points.

Mod Business Economics & Computing
UCAS Code: GL5C • Mode: 3 Years Equal Combination

Qualifications/Requirements: GCE, A level grades: DD. AGNVQ, Merit (in specific programmes). ND/C, Merit overall. SQAH, Individual consideration. SQAV, Individual consideration. IB, 24 points.

Mod Business Information Technology & Computing
UCAS Code: GG57 • Mode: 3 Years Equal Combination

Qualifications/Requirements: GCE, A level grades: DD. AGNVQ, Merit (in specific programmes). ND/C, Merit overall. SQAH, Individual consideration. SQAV, Individual consideration. IB, 24 points.

Mod Business Information Technology & Mathematics
UCAS Code: GG17 • Mode: 3 Years Equal Combination

Qualifications/Requirements: GCE, A level grades: DD. AGNVQ, Merit (in specific programmes). ND/C, Merit overall. SQAH, Individual consideration. SQAV, Individual consideration. IB, 24 points.

Mod Communications & Computing
UCAS Code: GP54 • Mode: 3 Years Equal Combination

Qualifications/Requirements: GCE, A level grades: CC-CDD. AGNVQ, Distinction (in specific programmes). ND/C, Merit overall and 4 Distinctions. SQAH, Individual consideration. SQAV, Individual consideration. IB, 26 points.

Mod Computing & 3D/Spatial Design
UCAS Code: GW5F • Mode: 3 Years Equal Combination

Qualifications/Requirements: GCE, A level grades: DD. OTHER. AGNVQ, Merit (in specific programmes). ND/C, Merit overall. SQAH, Individual consideration. SQAV, Individual consideration. IB, 24 points.

Mod Computing & Criminology
UCAS Code: GM5H • Mode: 3 Years Equal Combination

Qualifications/Requirements: GCE, A level grades: CD. AGNVQ, Merit. ND/C, Merit overall and 2 Distinctions. SQAH, Individual consideration. SQAV, Individual consideration. IB, 26 points.

Mod Computing & Design Studies
UCAS Code: GW52 • Mode: 3 Years Equal Combination

Qualifications/Requirements: GCE, A level grades: CD-DDD. AGNVQ, Merit (in specific programmes). ND/C, Merit overall and 2 Distinctions. SQAH, Individual consideration. SQAV, Individual consideration. IB, 24 points.

Mod Computing & Development Studies
UCAS Code: GM59 • Mode: 3 Years Equal Combination

Qualifications/Requirements: GCE, A level grades: DD. AGNVQ, Merit (in specific programmes). ND/C, Merit overall. SQAH, Individual consideration. SQAV, Individual consideration. IB, 24 points.

Mod Computing & Economics
UCAS Code: GL51 • Mode: 3 Years Equal Combination

Qualifications/Requirements: GCE, A level grades: DD. AGNVQ, Merit (in specific programmes). ND/C, Merit overall. SQAH, Individual consideration. SQAV, Individual consideration. IB, 24 points.

Mod Computing & English Studies
UCAS Code: GQ53 • Mode: 3 Years Equal Combination

Qualifications/Requirements: GCE, A level grades: CD-DDD. AGNVQ, Merit (in specific programmes). ND/C, Merit overall and 2 Distinctions. SQAH, Individual consideration. SQAV, Individual consideration. IB, 24 points.

Mod Computing & European Studies
UCAS Code: GT52 • Mode: 3 Years Equal Combination

Qualifications/Requirements: GCE, A level grades: DD. AGNVQ, Merit (in specific programmes). ND/C, Merit overall. SQAH, Individual consideration. SQAV, Individual consideration. IB, 24 points.

Mod Computing & Financial Services
UCAS Code: GN53 • Mode: 3 Years Equal Combination

Qualifications/Requirements: GCE, A level grades: DD. AGNVQ, Merit (in specific programmes). ND/C, Merit overall. SQAH, Individual consideration. SQAV, Individual consideration. IB, 24 points.

Mod Computing & Fine Art
UCAS Code: GW51 • Mode: 3 Years Equal Combination

Qualifications/Requirements: GCE, A level grades: CC-CDD. OTHER. AGNVQ, Merit (in specific programmes). ND/C, Merit overall and 2 Distinctions. SQAH, Individual consideration. SQAV, Individual consideration. IB, 26 points.

Mod Computing & French
UCAS Code: GR51 • Mode: 4 Years Equal Combination

Qualifications/Requirements: GCE, A level grades: DD. AGNVQ, Merit (in specific programmes). ND/C, Merit overall. SQAH, Individual consideration. SQAV, Individual consideration. IB, 24 points.

Mod Computing & German
UCAS Code: GR52 • Mode: 4 Years Equal Combination

Qualifications/Requirements: GCE, A level grades: DD. AGNVQ, Merit (in specific programmes). ND/C, Merit overall. SQAH, Individual consideration. SQAV, Individual consideration. IB, 24 points.

Mod Computing & Insurance
UCAS Code: GN5J • Mode: 3 Years Equal Combination

Qualifications/Requirements: GCE, A level grades: DD. AGNVQ, Merit (in specific programmes). ND/C, Merit overall. SQAH, Individual consideration. SQAV, Individual consideration. IB, 24 points.

Mod Computing & International Relations
UCAS Code: GM5C • Mode: 3 Years Equal Combination

Qualifications/Requirements: GCE, A level grades: DD. AGNVQ, Merit (in specific programmes). ND/C, Merit overall. SQAH, Individual consideration. SQAV, Individual consideration. IB, 24 points.

Mod Computing & Investment
UCAS Code: NGJ5 • Mode: 3 Years Equal Combination

Qualifications/Requirements: GCE, A level grades: DD. AGNVQ, Merit (in specific programmes). ND/C, Merit overall. SQAH, Individual consideration. SQAV, Individual consideration. IB, 24 points.

Mod Computing & Law
UCAS Code: GM53 • Mode: 3 Years Equal Combination

Qualifications/Requirements: GCE, A level grades: CC-CDD. AGNVQ, Merit (in specific programmes). ND/C, Merit overall and 2 Distinctions. SQAH, Individual consideration. SQAV, Individual consideration. IB, 26 points.

Mod Computing & Marketing
UCAS Code: GN55 • Mode: 3 Years Equal Combination

Qualifications/Requirements: GCE, A level grades: CD-DDD. AGNVQ, Merit (in specific programmes). ND/C, Merit overall and 2 Distinctions. SQAH, Individual consideration. SQAV, Individual consideration. IB, 26 points.

Mod Computing & Modern History
UCAS Code: GV51 • Mode: 3 Years Equal Combination

Qualifications/Requirements: GCE, A level grades: DD. AGNVQ, Merit (in specific programmes). ND/C, Merit overall. SQAH, Individual consideration. SQAV, Individual consideration. IB, 24 points.

Mod Computing & Multimedia Systems
UCAS Code: GG5M • Mode: 3 Years Equal Combination

Qualifications/Requirements: GCE, A level grades: DD. AGNVQ, Merit (in specific programmes). ND/C, Merit overall. SQAH, Individual consideration. SQAV, Individual consideration. IB, 24 points.

Mod Computing & Philosophy
UCAS Code: GV57 • Mode: 3 Years Equal Combination

Qualifications/Requirements: GCE, A level grades: CD. AGNVQ, Merit. ND/C, Merit overall and 2 Distinctions. SQAH, Individual consideration. SQAV, Individual consideration. IB, 26 points.

Mod Computing & Politics
UCAS Code: GM51 • Mode: 3 Years Equal Combination

Qualifications/Requirements: GCE, A level grades: DD. AGNVQ, Merit (in specific programmes). ND/C, Merit overall. SQAH, Individual consideration. SQAV, Individual consideration. IB, 24 points.

Mod Computing & Product Development & Manufacture
UCAS Code: GJ54 • Mode: 3 Years Equal Combination

Qualifications/Requirements: GCE, A level grades: DD. AGNVQ, Merit (in specific programmes). ND/C, Merit overall. SQAH, Individual consideration. SQAV, Individual consideration. IB, 24 points.

Mod Computing & Psychology
UCAS Code: CG85 • Mode: 3 Years Equal Combination

Qualifications/Requirements: GCE, A level grades: CD-DDD. AGNVQ, Merit (in specific programmes). ND/C, Merit overall and 2 Distinctions. SQAH, Individual consideration. SQAV, Individual consideration. IB, 26 points.

Mod Computing & Public Policy
UCAS Code: GM5D • Mode: 3 Years Equal Combination

Qualifications/Requirements: GCE, A level grades: CD. AGNVQ, Merit. ND/C, Merit overall and 2 Distinctions. SQAH, Individual consideration. SQAV, Individual consideration. IB, 26 points.

Mod Computing & Social Policy & Management
UCAS Code: GL54 • Mode: 3 Years Equal Combination

Qualifications/Requirements: GCE, A level grades: CD-DDD. AGNVQ, Merit (in specific programmes). ND/C, Merit overall. SQAH, Individual consideration. SQAV, Individual consideration. IB, 24 points.

Mod Computing & Sociology
UCAS Code: GL53 • Mode: 3 Years Equal Combination

Qualifications/Requirements: GCE, A level grades: CD-DDD. AGNVQ, Merit (in specific programmes). ND/C, Merit overall. SQAH, Individual consideration. SQAV, Individual consideration. IB, 24 points.

Mod Computing & Spanish
UCAS Code: GR54 • Mode: 4 Years Equal Combination

Qualifications/Requirements: GCE, A level grades: DD. AGNVQ, Merit (in specific programmes). ND/C, Merit overall. SQAH, Individual consideration. SQAV, Individual consideration. IB, 24 points.

Mod Computing & Taxation
UCAS Code: GN5H • Mode: 3 Years Equal Combination

Qualifications/Requirements: GCE, A level grades: DD. AGNVQ, Merit (in specific programmes). ND/C, Merit overall. SQAH, Individual consideration. SQAV, Individual consideration. IB, 24 points.

Mod Computing & Textile Furnishing Design
UCAS Code: GJ5K • Mode: 3 Years Equal Combination

Qualifications/Requirements: GCE, A level grades: DD. OTHER. AGNVQ, Merit (in specific programmes). ND/C, Merit overall. SQAH, Individual consideration. SQAV, Individual consideration. IB, 24 points.

Mod Computing & Transport
UCAS Code: GN59 • Mode: 3 Years Equal Combination

Qualifications/Requirements: GCE, A level grades: CD. AGNVQ, Merit. ND/C, Merit overall and 2 Distinctions. SQAH, Individual consideration. SQAV, Individual consideration. IB, 26 points.

Mod Insurance & Multimedia Systems
UCAS Code: GNMJ • Mode: 3 Years Equal Combination

Qualifications/Requirements: GCE, A level grades: CD. AGNVQ, Merit (in specific programmes). ND/C, Merit overall and 2 Distinctions. SQAH, Individual consideration. SQAV, Individual consideration. IB, 24 points.

HND Computing
UCAS Code: 105G • Mode: 2 Years Single Subject

Qualifications/Requirements: GCE, A level grades: C-DD. AGNVQ, Merit (in specific programmes). ND/C, 2 Merits. SQAH, Individual consideration. SQAV, Individual consideration. IB, 24 points.

LOUGHBOROUGH UNIVERSITY L79

BSc Computer Science
UCAS Code: G500 • Mode: 3 Years Single Subject

Qualifications/Requirements: GCE, A/AS: 22 points. Maths or Physics at A level. AGNVQ, Distinction in Information Technology with 6 additional units or with A level. ND/C, 3 Distinctions. SQAH, Individual consideration. SQAV, Individual consideration. IB, 28 points.

BSc Computer Science (4 years SW)
UCAS Code: G501 • Mode: 4 Years Single Subject

Qualifications/Requirements: GCE, A/AS: 22 points. Maths or Physics at A level. AGNVQ, Distinction in Information Technology with 6 additional units or with A level. ND/C, 3 Distinctions. SQAH, Individual consideration. SQAV, Individual consideration. IB, 28 points.

BSc Computing and Management
UCAS Code: GN5D • Mode: 3 Years Equal Combination
Qualifications/Requirements: GCE, A/AS: 22 points. AGNVQ, Distinction in Information Technology with 6 additional units or with A level. ND/C, 3 Distinctions. SQAH, Individual consideration. SQAV, Individual consideration. IB, 28 points.

BSc Computing and Management (4 years SW)
UCAS Code: GN51 • Mode: 4 Years Equal Combination
Qualifications/Requirements: GCE, A/AS: 22 points. AGNVQ, Distinction in Information Technology with 6 additional units or with A level. ND/C, 3 Distinctions. SQAH, Individual consideration. SQAV, Individual consideration. IB, 28 points.

BSc Information Management and Computing
UCAS Code: G562 • Mode: 3 Years Single Subject
Qualifications/Requirements: GCE, A/AS: 20 points. AGNVQ, Distinction with 6 additional units or with A level. ND/C, 2 Merits and 3 Distinctions. SQAH, Individual consideration. SQAV, Individual consideration. IB, 28 points.

BSc Information Management and Computing (4 years SW)
UCAS Code: G563 • Mode: 4 Years Single Subject
Qualifications/Requirements: GCE, A/AS: 20 points. AGNVQ, Distinction with 6 additional units or with A level. ND/C, 2 Merits and 3 Distinctions. SQAH, Individual consideration. SQAV, Individual consideration. IB, 28 points.

BSc Mathematics and Computing
UCAS Code: GG15 • Mode: 3 Years Equal Combination
Qualifications/Requirements: GCE, A level grades: BCC. Maths at A level. AGNVQ, Pass. SQAH, Individual consideration. IB, 28 points.

BSc Mathematics and Computing (4 years SW)
UCAS Code: GG51 • Mode: 4 Years Equal Combination
Qualifications/Requirements: GCE, A level grades: BCC. Maths at A level. AGNVQ, Pass. SQAH, Individual consideration. IB, 28 points.

UNIVERSITY OF LUTON L93

BSc Business Decision Management
UCAS Code: GN51 • Mode: 3 Years Equal Combination
Qualifications/Requirements: GCE, A/AS: 12 points. AGNVQ, Distinction or Merit with 6 additional units or with A/AS. ND/C, Merits and Distinction. SQAH, Grades BBCC. SQAV, Individual consideration. IB, 32 points.

BSc Business Information Technology
UCAS Code: G561 • Mode: 3 Years Single Subject
Qualifications/Requirements: Please refer to prospectus.

BSc Computer Science
UCAS Code: G500 • Mode: 3 Years Single Subject
Qualifications/Requirements: GCE, A/AS: 12 points. AGNVQ, Distinction or Merit with 6 additional units or with A/AS. ND/C, 5 Merits. SQAH, Grades BBCC. SQAV, Individual consideration. IB, 32 points.

BSc Computer Visualisation and Animation
UCAS Code: G540 • Mode: 3 Years Single Subject
Qualifications/Requirements: GCE, A/AS: 12 points. AGNVQ, Distinction or Merit with 6 additional units or with A/AS. ND/C, 5 Merits. SQAH, Grades BBCC. SQAV, Individual consideration. IB, 32 points.

BSc Computing
UCAS Code: G509 • Mode: 4 Years Single Subject
Qualifications/Requirements: AGNVQ, Individual consideration. ND/C, Individual consideration. SQAH, Individual consideration. SQAV, Individual consideration. IB, Individual consideration.

BSc Human Centred Computing
UCAS Code: G530 • Mode: 3 Years Single Subject
Qualifications/Requirements: GCE, A/AS: 12 points. AGNVQ, Distinction or Merit with 6 additional units or with A/AS. ND/C, 5 Merits. SQAH, Grades BBCC. SQAV, Individual consideration. IB, 32 points.

BSc Information Systems Development
UCAS Code: G520 • Mode: 3 Years Single Subject
Qualifications/Requirements: GCE, A/AS: 12 points. AGNVQ, Distinction or Merit with 6 additional units or with A/AS. ND/C, 5 Merits. SQAH, Grades BBCC. SQAV, Individual consideration. IB, 32 points.

BSc Information Systems Networking
UCAS Code: G521 • Mode: 3 Years Single Subject
Qualifications/Requirements: GCE, A/AS: 12 points. AGNVQ, Distinction or Merit with 6 additional units or with A/AS. ND/C, 5 Merits. SQAH, Grades BBCC. SQAV, Individual consideration. IB, 32 points.

HND Business Decision Management
UCAS Code: 15NG • Mode: 2 Years Equal Combination
Qualifications/Requirements: GCE, A/AS: 6 points. AGNVQ, Merit or Pass with 6 additional units or with A/AS. ND/C, Merit. SQAH, Grades CCCC. SQAV, Individual consideration. IB, 26 points.

HND Business Information Technology
UCAS Code: 265G • Mode: 2 Years Single Subject
Qualifications/Requirements: GCE, A/AS: 6 points. AGNVQ, Merit or Pass with 6 additional units or with A/AS. ND/C, Merit. SQAH, Grades CCCC. SQAV, Individual consideration. IB, 26 points.

HND Computing
UCAS Code: 105G • Mode: 2 Years Single Subject
Qualifications/Requirements: GCE, A/AS: 6 points. AGNVQ, Merit or Pass with 6 additional units or with A/AS. ND/C, Merit. SQAH, Grades CCCC. SQAV, Individual consideration. IB, 26 points.

HND Geographical Information Systems

UCAS Code: 58GF • Mode: 2 Years Equal Combination

Qualifications/Requirements: GCE, A/AS: 6 points. AGNVQ, Merit or Pass with 6 additional units or with A/AS. ND/C, Merit. SQAH, Grades CCCC. SQAV, Individual consideration. IB, 26 points.

THE UNIVERSITY OF MANCHESTER M20

BSc Computer Science

UCAS Code: G500 • Mode: 3 Years Single Subject

Qualifications/Requirements: GCE, A level grades: BBB. Maths at A level. AGNVQ, Distinction in Science with A level. ND/C, Individual consideration. SQAH, Grades BBBBB including specific subject(s). SQAV, Individual consideration. IB, 30 points.

BSc Computer Science with Business and Management

UCAS Code: G5N1 • Mode: 3 Years Major/Minor

Qualifications/Requirements: GCE, A level grades: BBB. Maths at A level. AGNVQ, Distinction in Science with A level. ND/C, Individual consideration. SQAH, Grades BBBBB including specific subject(s). SQAV, Not normally sufficient. IB, 30 points.

BSc Computer Science with Industrial Experience

UCAS Code: G505 • Mode: 4 Years Single Subject

Qualifications/Requirements: GCE, A level grades: BBB. Maths at A level. AGNVQ, Distinction in Science with A level. ND/C, Individual consideration. SQAH, Grades BBBBB including specific subject(s). SQAV, Not normally sufficient. IB, 30 points.

BSc Computing and Information Systems with Industrial Experience

UCAS Code: G507 • Mode: 4 Years Single Subject

Qualifications/Requirements: GCE, A level grades: BBB. Maths at A level. AGNVQ, Distinction in Science with A level. ND/C, Individual consideration. SQAH, Grades BBBBB including specific subject(s). SQAV, Not normally sufficient. IB, 30 points.

BSc Computing and Information Systems

UCAS Code: G506 • Mode: 3 Years Single Subject

Qualifications/Requirements: GCE, A level grades: BBB. Maths at A level. AGNVQ, Distinction in Science with A level. ND/C, Individual consideration. SQAH, Grades BBBBB including specific subject(s). SQAV, Not normally sufficient. IB, 30 points.

MEng Computer Science

UCAS Code: G501 • Mode: 4 Years Single Subject

Qualifications/Requirements: GCE, A level grades: AAB. Maths at A level. AGNVQ, Not normally sufficient. ND/C, Not normally sufficient. SQAH, Grades AAABB including specific subject(s). SQAV, Not normally sufficient. IB, 33 points.

THE UNIVERSITY OF MANCHESTER INSTITUTE OF SCIENCE AND TECHNOLOGY (UMIST) M25

BSc Biological and Computational Science (Bioinformatics)

UCAS Code: CG15 • Mode: 3 Years Equal Combination

Qualifications/Requirements: GCE, A/AS: 24 points. Biology and two Sciences. AGNVQ, Individual consideration. ND/C, 4 Merits (in specific programmes). SQAH, CSYS required. SQAV, National including specific subject(s). IB, 30 points.

BSc Biological and Computational Science (Bioinformatics) with Industrial Experience

UCAS Code: CG1M • Mode: 4 Years Equal Combination

Qualifications/Requirements: Please refer to prospectus.

BSc Computation

UCAS Code: G510 • Mode: 3 Years Single Subject

Qualifications/Requirements: GCE, A/AS: 18 points. AGNVQ, Distinction (in specific programmes). ND/C, 3 Merits and 2 Distinctions. SQAH, Grades ABBCC. SQAV, Individual consideration. IB, 30 points.

BSc Computation with Industrial Experience

UCAS Code: G511 • Mode: 4 Years Single Subject

Qualifications/Requirements: GCE, A/AS: 18 points. AGNVQ, Distinction (in specific programmes). ND/C, 3 Merits and 2 Distinctions. SQAH, Grades ABBCC. SQAV, Individual consideration. IB, 30 points.

BSc Computing and Geography

UCAS Code: GF58 • Mode: 3 Years Equal Combination

Qualifications/Requirements: GCE, A level grades: BBC. AGNVQ, Distinction in Information Technology. ND/C, Merits and Distinction (in specific programmes). SQAH, CSYS required. SQAV, Individual consideration. IB, 30 points.

BSc Computing Science

UCAS Code: G500 • Mode: 3 Years Single Subject

Qualifications/Requirements: GCE, A/AS: 18 points. AGNVQ, Distinction (in specific programmes). ND/C, 3 Merits and 2 Distinctions. SQAH, Grades ABBCC. SQAV, Individual consideration. IB, 30 points.

BSc Computing Science with Industrial Experience

UCAS Code: G501 • Mode: 4 Years Single Subject

Qualifications/Requirements: GCE, A/AS: 18 points. AGNVQ, Distinction (in specific programmes). ND/C, 3 Merits and 2 Distinctions. SQAH, Grades ABBCC. SQAV, Individual consideration. IB, 30 points.

BSc Information Systems Engineering

UCAS Code: G560 • Mode: 3 Years Single Subject

Qualifications/Requirements: GCE, A/AS: 18 points. Maths. AGNVQ, Distinction (in specific programmes). ND/C, 3 Merits and 2 Distinctions. SQAH, Grades ABBCC. SQAV, Individual consideration. IB, 30 points.

BSc Information Systems Engineering with Industrial Experience

UCAS Code: G561 • Mode: 4 Years Single Subject

Qualifications/Requirements: GCE, A/AS: 18 points. Maths. AGNVQ, Distinction (in specific programmes). ND/C, 3 Merits and 2 Distinctions. SQAH, Grades ABBCC. SQAV, Individual consideration. IB, 30 points.

BSc Management and Information Technology

UCAS Code: GN51 • Mode: 3 Years Equal Combination

Qualifications/Requirements: GCE, A level grades: BBB. AGNVQ, Distinction (in specific programmes) with A/AS. ND/C, 3 Merits and 4 Distinctions. SQAH, CSYS required. SQAV, Individual consideration. IB, 32 points.

MNeuro Neuroscience and Computation

UCAS Code: BG15 • Mode: 4 Years Equal Combination

Qualifications/Requirements: GCE, A level grades: ABB.

THE MANCHESTER METROPOLITAN UNIVERSITY M40

BA/BSc Information and Communication

UCAS Code: G563 • Mode: 3 Years Single Subject

Qualifications/Requirements: GCE, A/AS: 10-12 points. AGNVQ, Individual consideration. ND/C, Individual consideration. SQAH, Individual consideration. SQAV, Individual consideration. IB, Individual consideration.

BSc Applicable Mathematics/Information Systems

UCAS Code: GG51 • Mode: 3 Years Equal Combination

Qualifications/Requirements: GCE, A/AS: 16 points. Maths. AGNVQ, Merit (in specific programmes). ND/C, 1 Merit and 3 Distinctions (in specific programmes). SQAH, Grades BBBCC including specific subject(s). SQAV, Individual consideration. IB, 28 points.

BSc Business Economics/Computing Science

UCAS Code: GL5D • Mode: 3 Years Equal Combination

Qualifications/Requirements: GCE, A/AS: 16 points. AGNVQ, Merit (in specific programmes). ND/C, 1 Merit and 3 Distinctions. SQAH, Grades BBBCC. SQAV, Individual consideration. IB, 28 points.

BSc Business Economics/Information Systems

UCAS Code: LG15 • Mode: 3 Years Equal Combination

Qualifications/Requirements: GCE, A/AS: 16 points. AGNVQ, Merit (in specific programmes). ND/C, 1 Merit and 3 Distinctions. SQAH, Grades BBBCC including specific subject(s). SQAV, Individual consideration. IB, 28 points.

BSc Business Information Technology

UCAS Code: G562 • Mode: 4 Years Single Subject

Qualifications/Requirements: GCE, A/AS: 16 points. AGNVQ, Distinction. ND/C, Distinction Overall. SQAH, Grades CCCCC. SQAV, Individual consideration. IB, Individual consideration.

BSc Business Mathematics/Information Systems

UCAS Code: GG5C • Mode: 3 Years Equal Combination

Qualifications/Requirements: GCE, A/AS: 16 points. Maths at A level, Physics or Economics. AGNVQ, Merit (in specific programmes). ND/C, 1 Merit and 3 Distinctions (in specific programmes). SQAH, Grades BBBCC including specific subject(s). SQAV, Individual consideration. IB, 28 points.

BSc Chemistry/Computing Science

UCAS Code: FG15 • Mode: 3 Years Equal Combination

Qualifications/Requirements: GCE, A/AS: 12 points. Chemistry. AGNVQ, Merit (in specific programmes). ND/C, 5 Merits (in specific programmes). SQAH, Grades BCCCC including specific subject(s). SQAV, Individual consideration. IB, 27 points.

BSc Chemistry/Information Systems

UCAS Code: GF51 • Mode: 3 Years Equal Combination

Qualifications/Requirements: GCE, A/AS: 12 points. Chemistry. AGNVQ, Merit (in specific programmes). ND/C, 1 Merit and 3 Distinctions (in specific programmes). SQAH, Grades BBBCC including specific subject(s). SQAV, Individual consideration. IB, 28 points.

BSc Computer and Network Technology

UCAS Code: G530 • Mode: 3/4 Years Single Subject

Qualifications/Requirements: Please refer to prospectus.

BSc Computing

UCAS Code: G500 • Mode: 4 Years Single Subject

Qualifications/Requirements: GCE, A/AS: 12-18 points. AGNVQ, Distinction (in specific programmes) or Merit (in specific programmes) with 4 additional units or with A/AS. ND/C, Merits and Distinction. SQAH, Grades BBCC. SQAV, Individual consideration. IB, 24 points.

BSc Computing (Foundation)

UCAS Code: G508 • Mode: 5 Years Single Subject

Qualifications/Requirements: GCE, A level grades: E. Maths or Physics. AGNVQ, Pass (in specific programmes). ND/C, 2 Merits (in specific programmes). SQAH, Individual consideration. SQAV, Individual consideration. IB, Individual consideration.

BSc Computing Science and Management Systems

UCAS Code: GN51 • Mode: 3 Years Equal Combination

Qualifications/Requirements: Please refer to prospectus.

BSc Computing Science and Sociological Studies

UCAS Code: GL5J • Mode: 3 Years Equal Combination

Qualifications/Requirements: Please refer to prospectus.

BSc Computing Science/Economics

UCAS Code: GL51 • Mode: 3 Years Equal Combination

Qualifications/Requirements: GCE, A/AS: 16 points. AGNVQ, Merit (in specific programmes). ND/C, 1 Merit and 3 Distinctions. SQAH, Grades BBBCC. SQAV, Individual consideration. IB, 28 points.

BSc Computing Science/Environmental Studies

UCAS Code: FG95 • Mode: 3 Years Equal Combination

Qualifications/Requirements: GCE, A/AS: 16 points. AGNVQ, Merit (in specific programmes). ND/C, 1 Merit and 3 Distinctions. SQAH, Grades BBBCC. SQAV, Individual consideration. IB, 28 points.

BSc Computing Science/European Studies

UCAS Code: GT52 • Mode: 3 Years Equal Combination

Qualifications/Requirements: GCE, A/AS: 16 points. AGNVQ, Merit (in specific programmes). ND/C, 1 Merit and 3 Distinctions. SQAH, Grades BBBCC. SQAV, Individual consideration. IB, 28 points.

BSc Computing Science/Geography

UCAS Code: GL58 • Mode: 3 Years Equal Combination

Qualifications/Requirements: GCE, A/AS: 18 points. AGNVQ, Merit (in specific programmes). ND/C, 2 Merits and 4 Distinctions. SQAH, Grades BBBBC. SQAV, Individual consideration. IB, 29 points.

BSc Computing Science/Materials Science

UCAS Code: FG25 • Mode: 3 Years Equal Combination

Qualifications/Requirements: GCE, A/AS: 12 points. Maths, Physics or Chemistry. AGNVQ, Merit (in specific programmes). ND/C, 5 Merits (in specific programmes). SQAH, Grades BCCCC including specific subject(s). SQAV, Individual consideration. IB, 27 points.

BSc Computing Science/Psychology

UCAS Code: GL57 • Mode: 3 Years Equal Combination

Qualifications/Requirements: GCE, A/AS: 18 points. AGNVQ, Distinction (in specific programmes). ND/C, 2 Merits and 4 Distinctions. SQAH, Grades BBBBC. SQAV, Individual consideration. IB, 29 points.

BSc Economics/Information Systems

UCAS Code: GL5C • Mode: 3 Years Equal Combination

Qualifications/Requirements: GCE, A/AS: 16 points. AGNVQ, Merit (in specific programmes). ND/C, 1 Merit and 3 Distinctions. SQAH, Grades BBBCC. SQAV, Individual consideration. IB, 28 points.

BSc Environmental Studies/Information Systems

UCAS Code: GF59 • Mode: 3 Years Equal Combination

Qualifications/Requirements: GCE, A/AS: 16 points. AGNVQ, Merit (in specific programmes). ND/C, 1 Merit and 3 Distinctions. SQAH, Grades BBBCC. SQAV, Individual consideration. IB, 28 points.

BSc European Studies/Information Systems

UCAS Code: GT5F • Mode: 3 Years Equal Combination

Qualifications/Requirements: GCE, A/AS: 16 points. AGNVQ, Merit (in specific programmes). ND/C, 1 Merit and 3 Distinctions. SQAH, Grades BBBCC. SQAV, Individual consideration. IB, 28 points.

BSc Geography/Information Systems

UCAS Code: GL5V • Mode: 3 Years Equal Combination

Qualifications/Requirements: GCE, A/AS: 16 points. AGNVQ, Merit (in specific programmes). ND/C, 1 Merit and 3 Distinctions. SQAH, Grades BBBCC. SQAV, Individual consideration. IB, 28 points.

BSc Information Systems

UCAS Code: G521 • Mode: 4 Years Single Subject

Qualifications/Requirements: GCE, A/AS: 12-18 points. AGNVQ, Distinction (in specific programmes) or Merit (in specific programmes) with 4 additional units or with A/AS. ND/C, Merits and Distinction. SQAH, Grades BBCC. SQAV, Individual consideration. IB, 24 points.

BSc Information Systems (Foundation)

UCAS Code: G528 • Mode: 5 Years Single Subject

Qualifications/Requirements: GCE, A level grades: E. Maths or Physics. AGNVQ, Pass (in specific programmes). ND/C, 2 Merits (in specific programmes). SQAH, Individual consideration. SQAV, Individual consideration. IB, Individual consideration.

BSc Information Systems and Management Systems

UCAS Code: GN5C • Mode: 3 Years Equal Combination

Qualifications/Requirements: Please refer to prospectus.

BSc Information Systems and Sociological Studies

UCAS Code: GLMH • Mode: 3 Years Equal Combination

Qualifications/Requirements: Please refer to prospectus.

BSc Information Systems/Materials Science

UCAS Code: GF52 • Mode: 3 Years Equal Combination

Qualifications/Requirements: GCE, A/AS: 12 points. Maths, Physics or Chemistry. AGNVQ, Merit (in specific programmes). ND/C, 5 Merits (in specific programmes). SQAH, Grades BCCCC including specific subject(s). SQAV, Individual consideration. IB, 27 points.

BSc Information Systems/Psychology

UCAS Code: GL5R • Mode: 3 Years Equal Combination

Qualifications/Requirements: GCE, A/AS: 18 points. AGNVQ, Distinction (in specific programmes). ND/C, 2 Merits and 4 Distinctions. SQAH, Grades BBBCC including specific subject(s). SQAV, Individual consideration. IB, 29 points.

BSc Information Technology

UCAS Code: G560 • Mode: 4 Years Single Subject

Qualifications/Requirements: GCE, A/AS: 16 points. AGNVQ, Distinction in Information Technology. ND/C, Merit overall (in specific programmes). SQAH, Grades BBBB including specific subject(s). SQAV, Individual consideration. IB, 26 points.

BSc Information Technology (Foundation)

UCAS Code: G568 • Mode: 5 Years Single Subject

Qualifications/Requirements: GCE, A level grades: E. Maths or Physics. AGNVQ, Pass (in specific programmes). ND/C, 2 Merits (in specific programmes). SQAH, Individual consideration. SQAV, Individual consideration. IB, Individual consideration.

HND Business Information Technology

UCAS Code: 265G • Mode: 2 Years Single Subject

Qualifications/Requirements: GCE, A/AS: 6 points. AGNVQ, Merit. ND/C, Merit overall. SQAH, Individual consideration. SQAV, Individual consideration. IB, Individual consideration.

HND Computing

UCAS Code: 105G • Mode: 2 Years Single Subject

Qualifications/Requirements: GCE, A/AS: 6-10 points. AGNVQ, Merit (in specific programmes). ND/C, Merit overall. SQAH, Grades CCC. SQAV, Individual consideration. IB, Pass in Diploma.

BEd Business and Information Technology Education

UCAS Code: XN71 • Mode: 2 Years Equal Combination

Qualifications/Requirements: AGNVQ, Not normally sufficient. ND/C, Higher National (in specific programmes). SQAH, Not normally sufficient. SQAV, Not normally sufficient. IB, Not normally sufficient.

MID-CHESHIRE COLLEGE M77

HND Computing

UCAS Code: 005G • Mode: 2 Years Single Subject

Qualifications/Requirements: GCE, A level grades: E. AGNVQ, Individual consideration. ND/C, National. SQAH, Individual consideration. SQAV, Individual consideration. IB, Individual consideration.

MIDDLESEX UNIVERSITY M80

BSc Applied Computing

UCAS Code: G523 • Mode: 3 Years Single Subject

Qualifications/Requirements: GCE, A/AS: 12-16 points. AGNVQ, Merit (in specific programmes). ND/C, 5 Merits. SQAH, Grades CCCC. SQAV, Individual consideration. IB, 26 points.

BSC Applied Computing plus minor subject

UCAS Code: G5Y4 • Mode: 3 Years Major/Minor

Qualifications/Requirements: GCE, A/AS: 12-16 points. AGNVQ, Merit (in specific programmes). ND/C, 5 Merits. SQAH, Grades CCCC. SQAV, Individual consideration. IB, 26 points.

BSc Business Information Systems

UCAS Code: G522 • Mode: 3 Years Single Subject

Qualifications/Requirements: GCE, A/AS: 12-16 points. AGNVQ, Merit (in specific programmes). ND/C, 5 Merits. SQAH, Individual consideration. SQAV, Individual consideration. IB, 26 points.

BSc Business Information Systems plus minor subject

UCAS Code: G5YK • Mode: 3 Years

Qualifications/Requirements: GCE, A/AS: 12-16 points. AGNVQ, Merit (in specific programmes). ND/C, 5 Merits. SQAH, Grades CCCC. SQAV, Individual consideration. IB, 24 points.

BSc Computing Science

UCAS Code: G501 • Mode: 3 Years Single Subject

Qualifications/Requirements: GCE, A/AS: 12-16 points. AGNVQ, Merit (in specific programmes). ND/C, 5 Merits. SQAH, Individual consideration. SQAV, Individual consideration. IB, 28 points.

BSc Computing with International Foundation

UCAS Code: GQ53 • Mode: 4 Years Equal Combination

Qualifications/Requirements: Please refer to prospectus.

BSc Information Technology

UCAS Code: G561 • Mode: 3 Years Single Subject

Qualifications/Requirements: GCE, A/AS: 12-16 points. AGNVQ, Merit (in specific programmes). ND/C, 5 Merits. SQAH, Individual consideration. SQAV, Individual consideration. IB, 28 points.

BSc Information Technology plus minor subject

UCAS Code: G5YL • Mode: 3 Years Major/Minor

Qualifications/Requirements: GCE, A/AS: 12-16 points. AGNVQ, Merit (in specific programmes). ND/C, 5 Merits. SQAH, Grades CCCC. SQAV, Individual consideration. IB, 24 points.

Mix Computing Foundation

UCAS Code: G508 • Mode: 4 Years Single Subject

Qualifications/Requirements: GCE, A level grades: E. AGNVQ, Pass. ND/C, National. SQAH, Grades CC. SQAV, Individual consideration. IB, Individual consideration.

HND Computing

UCAS Code: 105G • Mode: 2 Years Single Subject

Qualifications/Requirements: GCE, A/AS: 4 points. AGNVQ, Merit. ND/C, National. SQAH, Grades CC. SQAV, Individual consideration. IB, Individual consideration.

HND Networking and Computer Systems

UCAS Code: 65GG • Mode: 2 Years Equal Combination

Qualifications/Requirements: GCE, A level grades: E. AGNVQ, Pass (in specific programmes). ND/C, 3 Merits. SQAH, Individual consideration. SQAV, Individual consideration. IB, 24 points.

NAPIER UNIVERSITY N07

BA Librarianship and Information Studies

UCAS Code: P100 • Mode: 3/4 Years Single Subject

Qualifications/Requirements: GCE, A level grades: CD. English. AGNVQ, Individual consideration. ND/C, Individual consideration. SQAH, Grades BCC. SQAV, Individual consideration. IB, Individual consideration.

BSc Computing

UCAS Code: G500 • Mode: 3/4/5 Years Single Subject

Qualifications/Requirements: GCE, A level grades: DD. AGNVQ, Individual consideration. ND/C, Individual consideration. SQAH, Grades BBC. SQAV, Individual consideration. IB, Individual consideration.

BSc Information Systems

UCAS Code: G520 • Mode: 3/4/5 Years Single Subject

Qualifications/Requirements: GCE, A level grades: DD. AGNVQ, Individual consideration. ND/C, Individual consideration. SQAH, Grades BBC. SQAV, Individual consideration. IB, Individual consideration.

BSc Management and Information Technology

UCAS Code: NG15 • Mode: 3/4 Years Equal Combination

Qualifications/Requirements: GCE, A level grades: DD. ND/C, Individual consideration. SQAH, Grades BBC. SQAV, Individual consideration.

BSc Network Computing (HND top-up)

UCAS Code: G530 • Mode: 1 Year Single Subject

Qualifications/Requirements: Please refer to prospectus.

BEng Computing and Electronic Systems

UCAS Code: GH5P • Mode: 3/4 Years Equal Combination

Qualifications/Requirements: GCE, A level grades: DD. AGNVQ, Individual consideration. ND/C, Individual consideration. SQAH, Grades BBC. SQAV, Individual consideration. IB, Individual consideration.

HND Computing

UCAS Code: 005G • Mode: 2 Years Single Subject

Qualifications/Requirements: GCE, A level grades: D. AGNVQ, Individual consideration. ND/C, Individual consideration. SQAH, Grades CC. SQAV, Individual consideration. IB, Individual consideration.

UNIVERSITY COLLEGE NORTHAMPTON N14

BA Accounting/Information Systems

UCAS Code: N4G5 • Mode: 3 Years Major/Minor

Qualifications/Requirements: GCE, A/AS: 10 points. AGNVQ, Merit. ND/C, Merits and 1 Distinction. SQAH, Grades BCC. SQAV, Individual consideration. IB, 24 points.

BA Business Information Systems

UCAS Code: G521 • Mode: 3 Years Single Subject

Qualifications/Requirements: GCE, A/AS: 8 points. AGNVQ, Merit. ND/C, 5 Merits. SQAH, Grades CCC. SQAV, Individual consideration. IB, 24 points.

BA Education/Information Systems

UCAS Code: X9G5 • Mode: 3 Years Major/Minor

Qualifications/Requirements: GCE, A/AS: 10-12 points. AGNVQ, Merit. ND/C, 5 Merits. SQAH, Grades BCC. SQAV, Individual consideration. IB, 24 points.

BA Law/Information Systems

UCAS Code: M3G5 • Mode: 3 Years Major/Minor

Qualifications/Requirements: GCE, A/AS: 10 points. AGNVQ, Merit. ND/C, 3 Merits and 2 Distinctions. SQAH, Grades BCC. SQAV, Individual consideration. IB, 24 points.

BSc Computer Science (Computer Communications)

UCAS Code: G560 • Mode: 3 Years Single Subject

Qualifications/Requirements: GCE, A level grades: DD. AGNVQ, Merit. ND/C, 4 Merits. SQAH, Grades CCC. SQAV, Individual consideration. IB, 24 points.

BSc Computing

UCAS Code: G500 • Mode: 3 Years Single Subject

Qualifications/Requirements: GCE, A/AS: 8 points. AGNVQ, Merit. ND/C, 5 Merits. SQAH, Grades CCC. SQAV, Individual consideration. IB, 24 points.

Mod Business Administration/Information Systems

UCAS Code: N1G5 • Mode: 3 Years Major/Minor

Qualifications/Requirements: GCE, A/AS: 10 points. AGNVQ, Merit. ND/C, Merits and 1 Distinction. SQAH, Grades BCC. SQAV, Individual consideration. IB, 24 points.

Mod Drama/Information Systems

UCAS Code: W4G5 • Mode: 3 Years Major/Minor

Qualifications/Requirements: GCE, A/AS: 12 points. AGNVQ, Merit. ND/C, 5 Merits and 1 Distinction. SQAH, Grades BCC. SQAV, Individual consideration. IB, 24 points.

Mod Ecology/Information Systems

UCAS Code: C9G5 • Mode: 3 Years Major/Minor

Qualifications/Requirements: GCE, A/AS: 8 points. AGNVQ, Merit. ND/C, 5 Merits. SQAH, Grades CCC. SQAV, Individual consideration. IB, 24 points.

Mod Energy Management/Information Systems

UCAS Code: J9G5 • Mode: 3 Years Major/Minor

Qualifications/Requirements: GCE, A/AS: 8 points. AGNVQ, Merit. ND/C, 5 Merits. SQAH, Grades CCC. SQAV, Individual consideration. IB, 24 points.

Mod Environmental Chemistry/Information Systems
UCAS Code: F1G5 • Mode: 3 Years Major/Minor
Qualifications/Requirements: GCE, A/AS: 8 points. AGNVQ, Merit. ND/C, 5 Merits. SQAH, Grades CCC. SQAV, Individual consideration. IB, 24 points.

Mod Geography/Information Systems
UCAS Code: F8G5 • Mode: 3 Years Major/Minor
Qualifications/Requirements: GCE, A/AS: 10-12 points. Geography. AGNVQ, Merit. ND/C, 5 Merits. SQAH, Grades CCC. SQAV, Individual consideration. IB, 24 points.

Mod Human Biological Studies/Information Systems
UCAS Code: B1G5 • Mode: 3 Years Major/Minor
Qualifications/Requirements: GCE, A/AS: 8-10 points. Science. AGNVQ, Merit. ND/C, 5 Merits. SQAH, Grades CCC. SQAV, Individual consideration. IB, 24 points.

Mod Industrial Archaeology/Information Systems
UCAS Code: V6G5 • Mode: 3 Years Major/Minor
Qualifications/Requirements: GCE, A/AS: 10 points. AGNVQ, Merit. ND/C, 5 Merits. SQAH, Grades CCC. SQAV, Individual consideration. IB, 24 points.

Mod Industrial Enterprise/Information Systems
UCAS Code: H1G5 • Mode: 3 Years Major/Minor
Qualifications/Requirements: GCE, A/AS: 8 points. AGNVQ, Merit. ND/C, 5 Merits. SQAH, Grades CCC. SQAV, Individual consideration. IB, 24 points.

Mod Information Systems/Accounting
UCAS Code: G5N4 • Mode: 3 Years Major/Minor
Qualifications/Requirements: GCE, A/AS: 10-12 points. AGNVQ, Merit. ND/C, 5 Merits. SQAH, Grades CCC. SQAV, Individual consideration. IB, 24 points.

Mod Information Systems/Architectural Studies
UCAS Code: G5V4 • Mode: 3 Years Major/Minor
Qualifications/Requirements: GCE, A/AS: 10-12 points. AGNVQ, Merit. ND/C, 5 Merits. SQAH, Grades CCC. SQAV, Individual consideration. IB, 24 points.

Mod Information Systems/Business Administration
UCAS Code: G5N1 • Mode: 3 Years Major/Minor
Qualifications/Requirements: GCE, A/AS: 10-12 points. AGNVQ, Merit. ND/C, 5 Merits. SQAH, Grades CCC. SQAV, Individual consideration. IB, 24 points.

Mod Information Systems/Drama
UCAS Code: G5W4 • Mode: 3 Years Major/Minor
Qualifications/Requirements: GCE, A/AS: 10-12 points. AGNVQ, Merit. ND/C, 5 Merits. SQAH, Grades CCC. SQAV, Individual consideration. IB, Individual consideration.

Mod Information Systems/Ecology
UCAS Code: G5C9 • Mode: 3 Years Major/Minor
Qualifications/Requirements: GCE, A/AS: 10-12 points. AGNVQ, Merit. ND/C, 5 Merits. SQAH, Grades CCC. SQAV, Individual consideration. IB, 24 points.

Mod Information Systems/Education
UCAS Code: G5X9 • Mode: 3 Years Major/Minor
Qualifications/Requirements: GCE, A/AS: 10-12 points. AGNVQ, Merit. ND/C, 5 Merits. SQAH, Grades CCC. SQAV, Individual consideration. IB, 24 points.

Mod Information Systems/Energy Management
UCAS Code: G5J9 • Mode: 3 Years Major/Minor
Qualifications/Requirements: GCE, A/AS: 10-12 points. AGNVQ, Merit. ND/C, 5 Merits. SQAH, Grades CCC. SQAV, Individual consideration. IB, 24 points.

Mod Information Systems/English
UCAS Code: G5Q3 • Mode: 3 Years Major/Minor
Qualifications/Requirements: GCE, A/AS: 10-12 points. AGNVQ, Merit. ND/C, 5 Merits. SQAH, Grades CCC. SQAV, Individual consideration. IB, 24 points.

Mod Information Systems/Environmental Chemistry
UCAS Code: G5F1 • Mode: 3 Years Major/Minor
Qualifications/Requirements: GCE, A/AS: 10-12 points. AGNVQ, Merit. ND/C, 5 Merits. SQAH, Grades CCC. SQAV, Individual consideration. IB, 24 points.

Mod Information Systems/Equine Studies
UCAS Code: G5D2 • Mode: 3 Years
Qualifications/Requirements: GCE, A/AS: 10-12 points. AGNVQ, Merit. ND/C, 5 Merits. SQAH, Grades CCC. SQAV, Individual consideration. IB, 24 points.

Mod Information Systems/European Union Studies
UCAS Code: G5T2 • Mode: 3 Years Major/Minor
Qualifications/Requirements: GCE, A/AS: 10-12 points. AGNVQ, Merit. ND/C, 5 Merits. SQAH, Grades CCC. SQAV, Individual consideration. IB, 24 points.

Mod Information Systems/Fine Art
UCAS Code: G5W1 • Mode: 3 Years Major/Minor
Qualifications/Requirements: GCE, A/AS: 10-12 points. AGNVQ, Merit. ND/C, 5 Merits. SQAH, Grades CCC. SQAV, Individual consideration. IB, 24 points.

Mod Information Systems/French
UCAS Code: G5R1 • Mode: 3 Years Major/Minor
Qualifications/Requirements: GCE, A/AS: 10-12 points. French. AGNVQ, Merit. ND/C, 5 Merits. SQAH, Grades CCC. SQAV, Individual consideration. IB, 24 points.

Mod Information Systems/Geography
UCAS Code: G5F8 • Mode: 3 Years Major/Minor
Qualifications/Requirements: GCE, A/AS: 10-12 points. AGNVQ, Merit. ND/C, 5 Merits. SQAH, Grades CCC. SQAV, Individual consideration. IB, 24 points.

Mod Information Systems/History
UCAS Code: G5V1 • Mode: 3 Years Major/Minor

Qualifications/Requirements: GCE, A/AS: 10-12 points. AGNVQ, Merit. ND/C, 5 Merits. SQAH, Grades CCC. SQAV, Individual consideration. IB, 24 points.

Mod Information Systems/Industrial Enterprise
UCAS Code: G5H1 • Mode: 3 Years Major/Minor

Qualifications/Requirements: GCE, A/AS: 10-12 points. AGNVQ, Merit. ND/C, 5 Merits. SQAH, Grades CCC. SQAV, Individual consideration. IB, 24 points.

Mod Information Systems/Italian
UCAS Code: G5R3 • Mode: 3 Years Major/Minor

Qualifications/Requirements: GCE, A/AS: 10-12 points. AGNVQ, Merit. ND/C, 5 Merits. SQAH, Grades CCC. SQAV, Individual consideration. IB, 24 points.

Mod Information Systems/Law
UCAS Code: G5M3 • Mode: 3 Years Major/Minor

Qualifications/Requirements: GCE, A/AS: 10-12 points. AGNVQ, Merit. ND/C, 5 Merits. SQAH, Grades CCC. SQAV, Individual consideration. IB, 24 points.

Mod Information Systems/Management Science
UCAS Code: G5G4 • Mode: 3 Years Major/Minor

Qualifications/Requirements: GCE, A/AS: 10-12 points. AGNVQ, Merit. ND/C, 5 Merits. SQAH, Grades CCC. SQAV, Individual consideration. IB, 24 points.

Mod Information Systems/Mathematics
UCAS Code: G5G1 • Mode: 3 Years Major/Minor

Qualifications/Requirements: GCE, A/AS: 10-12 points. Maths. AGNVQ, Merit. ND/C, 5 Merits. SQAH, Grades CCC. SQAV, Individual consideration. IB, 24 points.

Mod Information Systems/Media and Popular Culture
UCAS Code: G5P4 • Mode: 3 Years Major/Minor

Qualifications/Requirements: GCE, A/AS: 10-12 points. AGNVQ, Merit. ND/C, 5 Merits. SQAH, Grades CCC. SQAV, Individual consideration. IB, 24 points.

Mod Information Systems/People in Organisations
UCAS Code: G5N6 • Mode: 3 Years Major/Minor

Qualifications/Requirements: GCE, A/AS: 10-12 points. AGNVQ, Merit. ND/C, 5 Merits. SQAH, Grades CCC. SQAV, Individual consideration. IB, 24 points.

Mod Information Systems/Philosophy
UCAS Code: G5V7 • Mode: 3 Years Major/Minor

Qualifications/Requirements: GCE, A/AS: 8 points. AGNVQ, Merit. ND/C, 5 Merits. SQAH, Grades CCC. SQAV, Individual consideration. IB, 24 points.

Mod Information Systems/Sociology
UCAS Code: G5L3 • Mode: 3 Years Major/Minor

Qualifications/Requirements: GCE, A/AS: 10-12 points. AGNVQ, Merit. ND/C, 5 Merits. SQAH, Grades CCC. SQAV, Individual consideration. IB, 24 points.

Mod Information Systems/Spanish
UCAS Code: G5R4 • Mode: 3 Years Major/Minor

Qualifications/Requirements: GCE, A/AS: 10-12 points. AGNVQ, Merit. ND/C, 5 Merits. SQAH, Grades CCC. SQAV, Individual consideration. IB, 24 points.

Mod Information Systems/Sport Studies
UCAS Code: G5B6 • Mode: 3 Years Major/Minor

Qualifications/Requirements: GCE, A/AS: 10-12 points. AGNVQ, Merit. ND/C, 5 Merits. SQAH, Grades CCC. SQAV, Individual consideration. IB, 24 points.

Mod Information Systems/Third World Development
UCAS Code: G5M9 • Mode: 3 Years Major/Minor

Qualifications/Requirements: GCE, A/AS: 10-12 points. AGNVQ, Merit. ND/C, 5 Merits. SQAH, Grades CCC. SQAV, Individual consideration. IB, 24 points.

Mod Information Systems/Wastes Management
UCAS Code: G5FX • Mode: 3 Years Major/Minor

Qualifications/Requirements: GCE, A/AS: 10-12 points. AGNVQ, Merit. ND/C, 5 Merits. SQAH, Grades CCC. SQAV, Individual consideration. IB, 24 points.

Mod Management Science/Information Systems
UCAS Code: G4G5 • Mode: 3 Years Major/Minor

Qualifications/Requirements: GCE, A/AS: 8 points. AGNVQ, Merit. ND/C, 5 Merits. SQAH, Grades CCC. SQAV, Individual consideration. IB, 24 points.

Mod Mathematics/Information Systems
UCAS Code: G1G5 • Mode: 3 Years Major/Minor

Qualifications/Requirements: GCE, A/AS: 8-10 points. Maths. AGNVQ, Individual consideration. ND/C, Individual consideration. SQAH, Grades CCC. SQAV, Individual consideration. IB, 24 points.

Mod Media & Popular Culture/Information Systems
UCAS Code: P4G5 • Mode: 3 Years Major/Minor

Qualifications/Requirements: GCE, A/AS: 14 points. AGNVQ, Merit. ND/C, Merit. SQAH, Grades BBC. SQAV, Individual consideration. IB, 24 points.

Mod Sociology/Information Systems
UCAS Code: L3G5 • Mode: 3 Years Major/Minor

Qualifications/Requirements: GCE, A/AS: 10-12 points. AGNVQ, Merit. ND/C, 5 Merits. SQAH, Grades CCC. SQAV, Individual consideration. IB, 24 points.

Mod Sport Studies/Information Systems
UCAS Code: B6G5 • Mode: 3 Years Major/Minor
Qualifications/Requirements: GCE, A/AS: 14 points. AGNVQ, Merit. ND/C, Merits and 2 Distinctions. SQAH, Grades BBB. IB, 24 points.

HND Business Information Technology
UCAS Code: 265G • Mode: 2 Years Single Subject
Qualifications/Requirements: GCE, A/AS: 2 points. AGNVQ, Pass. ND/C, 3 Merits. SQAH, Grades CC. SQAV, Individual consideration. IB, Individual consideration.

HND Computing
UCAS Code: 105G • Mode: 2 Years Single Subject
Qualifications/Requirements: GCE, A/AS: 2 points. AGNVQ, Pass. ND/C, 3 Merits. SQAH, Grades CC. SQAV, Individual consideration. IB, Individual consideration.

UNIVERSITY OF NEWCASTLE UPON TYNE N21

BSc Accounting and Computing Science (3 or 4 years)
UCAS Code: NG45 • Mode: 3 Years Equal Combination
Qualifications/Requirements: GCE, A/AS: 22 points. Maths at A level. AGNVQ, Individual consideration. ND/C, 4 Merits and 1 Distinction. SQAH, Grades AAABB. SQAV, Individual consideration. IB, 30 points.

BSc Computing Science and Surveying and Mapping Science
UCAS Code: GH5F • Mode: 3 Years Equal Combination
Qualifications/Requirements: GCE, A/AS: 18 points. AGNVQ, Individual consideration. ND/C, 4 Merits. SQAH, Grades AABB. SQAV, Individual consideration. IB, 28 points.

BSc Computing Science
UCAS Code: G500 • Mode: 3 Years Single Subject
Qualifications/Requirements: GCE, A/AS: 20 points. AGNVQ, Individual consideration. ND/C, Individual consideration. SQAH, Individual consideration. SQAV, Individual consideration. IB, Individual consideration.

BSc Computing Science (4 years)
UCAS Code: G501 • Mode: 4 Years Single Subject
Qualifications/Requirements: AGNVQ, Individual consideration. ND/C, Individual consideration. SQAH, Individual consideration. SQAV, Individual consideration. IB, Individual consideration.

BSc Computing Science and Economics (3 or 4 years)
UCAS Code: GL51 • Mode: 3 Years Equal Combination
Qualifications/Requirements: GCE, A/AS: 20 points. Maths at A level. AGNVQ, Individual consideration. ND/C, 4 Merits and 1 Distinction. SQAH, Grades AAABB. SQAV, Individual consideration. IB, 28 points.

BSc Computing Science and Physics (3 or 4 years)
UCAS Code: FG35 • Mode: 3 Years Equal Combination
Qualifications/Requirements: GCE, A/AS: 18 points. Maths and Physics at A level. AGNVQ, Individual consideration. ND/C, 4 Merits. SQAH, Grades AABB. SQAV, Individual consideration. IB, 28 points.

BSc Computing Science and Psychology (3 or 4 years)
UCAS Code: CG85 • Mode: 3 Years Equal Combination
Qualifications/Requirements: GCE, A/AS: 24 points. Maths. AGNVQ, Individual consideration. ND/C, Individual consideration. SQAH, Grades AAABB. SQAV, Individual consideration. IB, 30 points.

BSc Information Systems
UCAS Code: G520 • Mode: 3 Years Single Subject
Qualifications/Requirements: GCE, A/AS: 20 points. AGNVQ, Individual consideration. ND/C, Individual consideration. SQAH, Individual consideration. SQAV, Individual consideration. IB, Individual consideration.

NEWCASTLE COLLEGE N23

HND Computing
UCAS Code: 005G • Mode: 2 Years Single Subject
Qualifications/Requirements: Please refer to prospectus.

NEW COLLEGE DURHAM N28

HND Computing
UCAS Code: 005G • Mode: 2 Years Single Subject
Qualifications/Requirements: GCE, A/AS: 2 points. AGNVQ, Individual consideration. ND/C, National. SQAH, Individual consideration. SQAV, Individual consideration. IB, Individual consideration.

NEWMAN COLLEGE OF HIGHER EDUCATION N36

BA Information Technology
UCAS Code: XG65 • Mode: 3/4 Years Equal Combination
Qualifications/Requirements: GCE, A/AS: 12-14 points. AGNVQ, Merit with A/AS. ND/C, 3 Merits. SQAH, Grades CCC. SQAV, Individual consideration. IB, Pass in Diploma.

BA Information Technology (Secondary)
UCAS Code: GX57 • Mode: 3/4 Years Equal Combination
Qualifications/Requirements: GCE, A/AS: 12-14 points. AGNVQ, Merit. ND/C, 3 Merits. SQAH, Grades CCC. SQAV, Individual consideration. IB, Pass in Diploma.

BSc Information Technology (11-16 years)
UCAS Code: XG7N • Mode: 4 Years Equal Combination
Qualifications/Requirements: AGNVQ, Merit.

BEd Information Technology (11-16 years)
UCAS Code: XG7M • Mode: 3 Years Equal Combination

Qualifications/Requirements: AGNVQ, Merit.

UNIVERSITY OF WALES COLLEGE, NEWPORT N37

BA Information Technology and Archaeology
UCAS Code: GV56 • Mode: 3 Years Equal Combination

Qualifications/Requirements: GCE, A/AS: 10 points. AGNVQ, Distinction (in specific programmes). ND/C, Merits and Distinction. SQAH, Individual consideration. SQAV, Individual consideration. IB, Individual consideration.

BA Information Technology and English
UCAS Code: GQ53 • Mode: 3 Years Equal Combination

Qualifications/Requirements: GCE, A/AS: 10 points. AGNVQ, Distinction (in specific programmes). ND/C, Merits and Distinction. SQAH, Individual consideration. SQAV, Individual consideration. IB, Individual consideration.

BA Information Technology and Environmental Studies
UCAS Code: FG95 • Mode: 3 Years Equal Combination

Qualifications/Requirements: GCE, A/AS: 10 points. AGNVQ, Distinction (in specific programmes). ND/C, Merits and Distinction. SQAH, Individual consideration. SQAV, Individual consideration. IB, Individual consideration.

BA Information Technology and European Studies
UCAS Code: GT52 • Mode: 3 Years Equal Combination

Qualifications/Requirements: GCE, A/AS: 10 points. AGNVQ, Distinction (in specific programmes). ND/C, Merits and Distinction. SQAH, Individual consideration. SQAV, Individual consideration. IB, Individual consideration.

BA Information Technology and Geography
UCAS Code: GL58 • Mode: 3 Years Equal Combination

Qualifications/Requirements: GCE, A/AS: 10 points. AGNVQ, Distinction (in specific programmes). ND/C, Merits and Distinction. SQAH, Individual consideration. SQAV, Individual consideration. IB, Individual consideration.

BA Information Technology and History
UCAS Code: GV51 • Mode: 3 Years Equal Combination

Qualifications/Requirements: GCE, A/AS: 10 points. AGNVQ, Distinction (in specific programmes). ND/C, Merits and Distinction. SQAH, Individual consideration. SQAV, Individual consideration. IB, Individual consideration.

BA Sports Studies and Information Technology
UCAS Code: BG65 • Mode: 3 Years Equal Combination

Qualifications/Requirements: GCE, A/AS: 10 points. AGNVQ, Distinction (in specific programmes). ND/C, Merits and Distinction. SQAH, Individual consideration. SQAV, Individual consideration. IB, Individual consideration.

BSc Computing
UCAS Code: G501 • Mode: 3 Years Single Subject

Qualifications/Requirements: GCE, A/AS: 8-10 points. AGNVQ, Distinction (in specific programmes). ND/C, National. SQAH, Individual consideration. SQAV, Individual consideration. IB, Individual consideration.

BSc Industrial Information Technology
UCAS Code: G5H6 • Mode: 3 Years Major/Minor

Qualifications/Requirements: GCE, A/AS: 8-10 points. AGNVQ, Distinction (in specific programmes). ND/C, National. SQAH, Individual consideration. SQAV, Individual consideration. IB, Individual consideration.

Mod Information Technology
UCAS Code: G508 • Mode: 3/4 Years Single Subject

Qualifications/Requirements: AGNVQ, Pass.

HND Computing
UCAS Code: 105G • Mode: 2 Years Single Subject

Qualifications/Requirements: GCE, A/AS: 2 points. AGNVQ, Pass (in specific programmes). ND/C, National. SQAH, Individual consideration. SQAV, Individual consideration. IB, Individual consideration.

NORTHBROOK COLLEGE SUSSEX N41

BSc Computing and Information Systems for Business
UCAS Code: GG56 • Mode: 3 Years Equal Combination

Qualifications/Requirements: Please refer to prospectus.

BSc Information Systems and Business Management
UCAS Code: GN51 • Mode: 3 Years Equal Combination

Qualifications/Requirements: Please refer to prospectus.

NESCOT N49

BSc Computer Studies
UCAS Code: G500 • Mode: 4 Years Single Subject

Qualifications/Requirements: GCE, A level grades: DD. AGNVQ, Merit. ND/C, Merit overall. SQAH, Individual consideration. SQAV, National including specific subject(s). IB, Pass in Diploma.

BSc Computing and Business
UCAS Code: G508 • Mode: 5 Years Single Subject

Qualifications/Requirements: GCE, A level grades: E. AGNVQ, Pass. ND/C, National.

HND Business Information Technology
UCAS Code: 265G • Mode: 2 Years Single Subject
Qualifications/Requirements: GCE, A level grades: E. AGNVQ, Pass. ND/C, Merit overall. SQAH, Individual consideration. SQAV, National including specific subject(s). IB, Pass in Diploma.

HND Computing
UCAS Code: 105G • Mode: 2 Years Single Subject
Qualifications/Requirements: GCE, A level grades: E. AGNVQ, Pass. ND/C, Merit overall. SQAH, Individual consideration. SQAV, National including specific subject(s). IB, Pass in Diploma.

THE NORTH EAST WALES INSTITUTE OF HIGHER EDUCATION N56

BA Social Science and Information Technology
UCAS Code: LG35 • Mode: 3 Years Equal Combination
Qualifications/Requirements: GCE, A/AS: 6-12 points. AGNVQ, Merit (in specific programmes). ND/C, 3 to 4 Merits. SQAH, Individual consideration. SQAV, Individual consideration. IB, Individual consideration.

BSc Biology and Information Technology
UCAS Code: CG1M • Mode: 3 Years Equal Combination
Qualifications/Requirements: GCE, A/AS: 4-10 points. AGNVQ, Merit (in specific programmes). ND/C, 3 Merits. SQAH, Individual consideration. SQAV, Individual consideration. IB, Individual consideration.

BSc Business Information Systems
UCAS Code: G525 • Mode: 3 Years Single Subject
Qualifications/Requirements: GCE, A/AS: 4-8 points. AGNVQ, Merit (in specific programmes). ND/C, 4 Merits. SQAH, Grades CCC. SQAV, National including specific subject(s). IB, Individual consideration.

BSc Business Information Technology
UCAS Code: G501 • Mode: 3 Years Single Subject
Qualifications/Requirements: Please refer to prospectus.

BSc Business Information Technology with Accounting and Finance
UCAS Code: G5N4 • Mode: 3 Years Major/Minor
Qualifications/Requirements: GCE, A/AS: 4-10 points. AGNVQ, Merit (in specific programmes). ND/C, 3 to 4 Merits. SQAH, Individual consideration. SQAV, Individual consideration. IB, Individual consideration.

BSc Business Information Technology with Business Studies
UCAS Code: G5N1 • Mode: 3 Years Major/Minor
Qualifications/Requirements: GCE, A/AS: 4-10 points. AGNVQ, Merit (in specific programmes). ND/C, 3 to 4 Merits. SQAH, Individual consideration. SQAV, Individual consideration. IB, Individual consideration.

BSc Business Information Technology with Human Resource Management
UCAS Code: G5N6 • Mode: 3 Years Major/Minor
Qualifications/Requirements: GCE, A/AS: 4-10 points. AGNVQ, Merit (in specific programmes). ND/C, 3 to 4 Merits. SQAH, Individual consideration. SQAV, Individual consideration. IB, Individual consideration.

BSc Business Information Technology with Marketing
UCAS Code: G5N5 • Mode: 3 Years Major/Minor
Qualifications/Requirements: GCE, A/AS: 4-10 points. AGNVQ, Merit (in specific programmes). ND/C, 3 to 4 Merits. SQAH, Individual consideration. SQAV, Individual consideration. IB, Individual consideration.

BSc Computing
UCAS Code: G500 • Mode: 3 Years Single Subject
Qualifications/Requirements: GCE, A/AS: 6-10 points. AGNVQ, Merit (in specific programmes). ND/C, 3 Merits. SQAH, Grades CCC. SQAV, National including specific subject(s). IB, Individual consideration.

BSc Environmental Studies and Information Technology
UCAS Code: FG95 • Mode: 3 Years Equal Combination
Qualifications/Requirements: GCE, A/AS: 4-10 points. AGNVQ, Merit (in specific programmes). ND/C, 3 to 4 Merits. SQAH, Individual consideration. SQAV, Individual consideration. IB, Individual consideration.

BSc Geography and Information Technology
UCAS Code: FG8M • Mode: 3 Years Equal Combination
Qualifications/Requirements: GCE, A/AS: 4-10 points. AGNVQ, Merit (in specific programmes). ND/C, 3 to 4 Merits. SQAH, Individual consideration. SQAV, Individual consideration. IB, Individual consideration.

BSc Sports Science and Information Technology
UCAS Code: BG65 • Mode: 3 Years Equal Combination
Qualifications/Requirements: GCE, A/AS: 4-10 points. AGNVQ, Merit (in specific programmes). ND/C, 3 to 4 Merits. SQAH, Individual consideration. SQAV, Individual consideration. IB, Individual consideration.

HND Business Information Technology
UCAS Code: 1N5G • Mode: 2 Years Major/Minor
Qualifications/Requirements: GCE, A/AS: 2-4 points. AGNVQ, Pass (in specific programmes). ND/C, 2 Merits. SQAH, Grades CC. SQAV, National including specific subject(s). IB, Individual consideration.

HND Computing
UCAS Code: 005G • Mode: 2 Years Single Subject
Qualifications/Requirements: GCE, A/AS: 2-4 points. AGNVQ, Pass (in specific programmes). ND/C, 2 Merits. SQAH, Grades CC. SQAV, National including specific subject(s). IB, Individual consideration.

NORTH EAST WORCESTERSHIRE COLLEGE N58

HND Computing

UCAS Code: 005G • Mode: 2 Years Single Subject

Qualifications/Requirements: Please refer to prospectus.

UNIVERSITY OF NORTH LONDON N63

BSc Business Information Systems

UCAS Code: G5N1 • Mode: 3/4 Years Major/Minor

Qualifications/Requirements: GCE, A/AS: 10 points. A level: Maths, Physics, Computing, Statistics or Electronics. Specific GCSEs required. AGNVQ, Merit. ND/C, 4 Merits (in specific programmes). SQAH, Individual consideration. SQAV, Individual consideration. IB, Individual consideration.

BSc Computer Science (4 year sandwich)

UCAS Code: G500 • Mode: 3/4 Years Single Subject

Qualifications/Requirements: GCE, A/AS: 10 points. Any two from Maths,Computing,Physics,Statistics at A level. Specific GCSEs required. AGNVQ, Merit (in specific programmes). ND/C, 4 Merits (in specific programmes). SQAH, Grades CCCC including specific subject(s). SQAV, Individual consideration. IB, Individual consideration.

BSc Computing

UCAS Code: G501 • Mode: 3/4 Years Single Subject

Qualifications/Requirements: GCE, A/AS: 10 points. A level: Maths, Computing, Physics, Statistics or Electronics. Specific GCSEs required. AGNVQ, Merit (in specific programmes). ND/C, 4 Merits (in specific programmes). SQAH, Grades CCCC including specific subject(s). SQAV, Individual consideration. IB, Individual consideration.

BSc Computing/Mathematics Foundation

UCAS Code: GGN1 • Mode: 4/5 Years Equal Combination

Qualifications/Requirements: GCE, A/AS: 2 points. AGNVQ, Individual consideration. ND/C, National. SQAH, Individual consideration. SQAV, Individual consideration. IB, Individual consideration.

BSc Information Systems Design

UCAS Code: G520 • Mode: 3/4 Years Single Subject

Qualifications/Requirements: GCE, A/AS: 10 points. Science. AGNVQ, Merit (in specific programmes). ND/C, 4 Merits (in specific programmes). SQAH, Grades CCCC including specific subject(s). SQAV, Individual consideration. IB, Individual consideration.

Mod Business and Computing

UCAS Code: GN51 • Mode: 3/4 Years Equal Combination

Qualifications/Requirements: GCE, A/AS: 12 points. Maths, Computing, Physics, Electronics or Statistics at A level. AGNVQ, Distinction. ND/C, Merit overall and 3 Distinctions. SQAH, Individual consideration. SQAV, Individual consideration. IB, Individual consideration.

HND Computing

UCAS Code: 005G • Mode: 2 Years Single Subject

Qualifications/Requirements: GCE, A/AS: 4 points. A level: Maths, Computing, Physics, Electronics or Statistics. Specific GCSEs required. AGNVQ, Pass (in specific programmes). ND/C, 2 Merits (in specific programmes). SQAH, Grades CCC. SQAV, Individual consideration. IB, Individual consideration.

NORTH TYNESIDE COLLEGE N72

HND Computing

UCAS Code: 005G • Mode: 2 Years Single Subject

Qualifications/Requirements: ND/C, National.

UNIVERSITY OF NORTHUMBRIA AT NEWCASTLE N77

BA Art History and Information Studies

UCAS Code: W151 • Mode: 3 Years Single Subject

Qualifications/Requirements: GCE, A/AS: 18 points. English, History or Art at A level. AGNVQ, Pass. ND/C, Merit overall and 4 Distinctions. SQAH, Grades BBBCC. SQAV, Individual consideration. IB, 26 points.

BSc Business Information Systems

UCAS Code: G562 • Mode: 4 Years Single Subject

Qualifications/Requirements: GCE, A/AS: 16 points. AGNVQ, Distinction (in specific programmes) with 2 additional units or with A/AS. ND/C, Merit overall and 4 Distinctions. SQAH, Grades BBCCC. SQAV, Individual consideration. IB, 26 points.

BSc Business Information Technology

UCAS Code: G563 • Mode: 3 Years Single Subject

Qualifications/Requirements: GCE, A/AS: 12 points. AGNVQ, Distinction (in specific programmes) or Merit (in specific programmes) with 4 additional units or with A/AS. ND/C, Merit overall. SQAH, Grades BCCC. SQAV, Individual consideration. IB, 24 points.

BSc Computing for Business

UCAS Code: G5N1 • Mode: 4 Years Major/Minor

Qualifications/Requirements: GCE, A/AS: 14 points. AGNVQ, Merit. ND/C, 6 Merits. SQAH, Grades BCCC. SQAV, Individual consideration. IB, 26 points.

BSc Computing for Industry

UCAS Code: G501 • Mode: 4 Years Single Subject

Qualifications/Requirements: GCE, A/AS: 14 points. AGNVQ, Merit. ND/C, 6 Merits. SQAH, Grades BCCC. SQAV, Individual consideration. IB, 26 points.

BSc Computing Studies

UCAS Code: G502 • Mode: 4 Years Single Subject

Qualifications/Requirements: GCE, A/AS: 14 points. AGNVQ, Merit. ND/C, 6 Merits. SQAH, Grades BCCC. SQAV, Individual consideration. IB, 26 points.

BSc Computing Studies

UCAS Code: G508 • Mode: 5 Years Single Subject

Qualifications/Requirements: GCE, A/AS: 2-4 points. AGNVQ, Pass. ND/C, National. SQAH, Grades CCC. SQAV, Individual consideration. IB, 24 points.

BSc Computing with Cognitive Psychology

UCAS Code: G5C8 • Mode: 4 Years Major/Minor

Qualifications/Requirements: GCE, A/AS: 14 points. AGNVQ, Merit. ND/C, 6 Merits. SQAH, Grades BCCC. SQAV, Individual consideration. IB, 26 points.

BSc Information and Communication Management

UCAS Code: G560 • Mode: 3 Years Single Subject

Qualifications/Requirements: GCE, A/AS: 12 points. AGNVQ, Merit. ND/C, 3 Merits and 1 Distinction. SQAH, Grades CCCC. SQAV, Individual consideration. IB, 24 points.

HND Computing for Business

UCAS Code: 305G • Mode: 3 Years Single Subject

Qualifications/Requirements: GCE, A/AS: 4-6 points. AGNVQ, Pass. ND/C, 3 Merits. SQAH, Grades CCC. SQAV, Individual consideration. IB, 24 points.

HND Computing for Industry

UCAS Code: 205G • Mode: 3 Years Single Subject

Qualifications/Requirements: GCE, A/AS: 4-6 points. AGNVQ, Pass. ND/C, 3 Merits. SQAH, Grades CCC. SQAV, Individual consideration. IB, 24 points.

HND Computing Studies

UCAS Code: 105G • Mode: 3 Years Single Subject

Qualifications/Requirements: GCE, A/AS: 4-6 points. AGNVQ, Pass. ND/C, 3 Merits. SQAH, Grades CCC. SQAV, Individual consideration. IB, 24 points.

NORWICH: CITY COLLEGE N82

BSc Business Information Systems (17 mths)

UCAS Code: G562 • Mode: 1/2 Years Single Subject

Qualifications/Requirements: AGNVQ, Not normally sufficient. ND/C, Higher National. SQAH, Not normally sufficient. SQAV, Not normally sufficient. IB, Not normally sufficient.

HND Business Information Technology

UCAS Code: 265G • Mode: 2 Years Single Subject

Qualifications/Requirements: GCE, A level grades: EE. AGNVQ, Merit. ND/C, National. SQAH, Grades CC. SQAV, Individual consideration. IB, Individual consideration.

HND Computing

UCAS Code: 005G • Mode: 2 Years Single Subject

Qualifications/Requirements: GCE, A level grades: EE. AGNVQ, Merit. ND/C, National. SQAH, Grades CC. SQAV, Individual consideration. IB, Individual consideration.

THE UNIVERSITY OF NOTTINGHAM N84

BSc Computer Science

UCAS Code: G500 • Mode: 3 Years Single Subject

Qualifications/Requirements: GCE, A level grades: BBB-BCC. Maths at A level. AGNVQ, Pass. ND/C, Individual consideration. SQAH, Individual consideration. SQAV, Individual consideration. IB, Individual consideration.

BSc Computer Science and Management Studies

UCAS Code: GN51 • Mode: 3 Years Equal Combination

Qualifications/Requirements: GCE, A level grades: ABB-BBC. Maths. AGNVQ, Pass. ND/C, Individual consideration. SQAH, Individual consideration. SQAV, Individual consideration. IB, Individual consideration.

BSc Mathematics and Computer Science

UCAS Code: GG51 • Mode: 3 Years Equal Combination

Qualifications/Requirements: GCE, A level grades: AAB-ABB. Maths at A level. AGNVQ, Individual consideration. ND/C, Individual consideration. SQAH, Individual consideration. SQAV, Individual consideration. IB, Individual consideration.

THE NOTTINGHAM TRENT UNIVERSITY N91

BA Business Information Systems

UCAS Code: G522 • Mode: 4 Years Single Subject

Qualifications/Requirements: GCE, A/AS: 18 points. AGNVQ, Individual consideration. ND/C, Merits and Distinction (in specific programmes). SQAH, Individual consideration. SQAV, Individual consideration. IB, Individual consideration.

BSc Biology and Computing

UCAS Code: CG15 • Mode: 3 Years Equal Combination

Qualifications/Requirements: GCE, A/AS: 12 points. Biology. AGNVQ, Individual consideration. ND/C, Individual consideration. SQAH, Grades C. SQAV, Individual consideration. IB, Pass in Diploma.

BSc Business and Technology

UCAS Code: GN51 • Mode: 3 Years Equal Combination

Qualifications/Requirements: GCE, A/AS: 12-14 points. AGNVQ, Merit. ND/C, 4 Merits. SQAH, Grades CCCC. SQAV, National. IB, Individual consideration.

BSc Chemistry and Computing

UCAS Code: FG15 • Mode: 3 Years Equal Combination

Qualifications/Requirements: GCE, A/AS: 12 points. Chemistry. AGNVQ, Individual consideration. ND/C, Individual consideration. SQAH, Grades C. SQAV, Individual consideration. IB, Pass in Diploma.

BSc Computer Studies

UCAS Code: G501 • Mode: 3/4 Years Single Subject

Qualifications/Requirements: GCE, A/AS: 20 points. AGNVQ, Distinction. ND/C, Merits and Distinction. SQAH, Individual consideration. SQAV, Individual consideration. IB, 30 points.

BSc Computer Studies (Extended)
UCAS Code: G507 • Mode: 4/5 Years Single Subject

Qualifications/Requirements: AGNVQ, Individual consideration. ND/C, Individual consideration. SQAH, Individual consideration. SQAV, Individual consideration. IB, Individual consideration.

BSc Computing & Environmental Systems and Monitoring
UCAS Code: FGX5 • Mode: 3 Years Equal Combination

Qualifications/Requirements: GCE, A/AS: 12 points. Science. AGNVQ, Individual consideration. ND/C, Individual consideration. SQAH, Grades C. SQAV, Individual consideration. IB, Pass in Diploma.

BSc Computing and Physics
UCAS Code: FG35 • Mode: 3 Years Equal Combination

Qualifications/Requirements: GCE, A/AS: 12 points. Maths or Physics. AGNVQ, Individual consideration. ND/C, Individual consideration. SQAH, Grades C. SQAV, Individual consideration. IB, Pass in Diploma.

BSc Computing Systems
UCAS Code: G600 • Mode: 3/4 Years Single Subject

Qualifications/Requirements: GCE, A/AS: 20 points. AGNVQ, Distinction. ND/C, Merits and Distinction. SQAH, Individual consideration. SQAV, Individual consideration. IB, 30 points.

BSc Computing Systems (Extended)
UCAS Code: G508 • Mode: 4/5 Years Single Subject

Qualifications/Requirements: AGNVQ, Individual consideration. ND/C, Individual consideration. SQAH, Individual consideration. SQAV, Individual consideration. IB, Individual consideration.

BSc Information Technology for Science and Environmental Conservation & Management
UCAS Code: GF59 • Mode: 3 Years Equal Combination

Qualifications/Requirements: GCE, A/AS: 12 points. Science. AGNVQ, Individual consideration. ND/C, Individual consideration. SQAH, Grades C. SQAV, Individual consideration. IB, Pass in Diploma.

BSc Information Technology for Sciences and Biology
UCAS Code: GC51 • Mode: 3 Years Equal Combination

Qualifications/Requirements: GCE, A/AS: 12 points. Biology at A level. AGNVQ, Individual consideration. ND/C, Individual consideration. SQAH, Grades C. SQAV, Individual consideration. IB, Pass in Diploma.

BSc Information Technology for Sciences and Chemistry
UCAS Code: GF51 • Mode: 3 Years Equal Combination

Qualifications/Requirements: GCE, A/AS: 12 points. Chemistry at A level. AGNVQ, Individual consideration. ND/C, Individual consideration. SQAH, Grades C. SQAV, Individual consideration. IB, Pass in Diploma.

BSc Information Technology for Sciences and Maths
UCAS Code: GG51 • Mode: 3 Years Equal Combination

Qualifications/Requirements: GCE, A/AS: 12 points. Maths at A level. AGNVQ, Individual consideration. ND/C, Individual consideration. SQAH, Grades C. SQAV, Individual consideration. IB, Pass in Diploma.

BSc Information Technology for Sciences and Physics
UCAS Code: GF53 • Mode: 3 Years Equal Combination

Qualifications/Requirements: GCE, A/AS: 12 points. Maths or Physics. AGNVQ, Individual consideration. ND/C, Individual consideration. SQAH, Grades C. SQAV, Individual consideration. IB, Pass in Diploma.

BSc Mathematics and Computing
UCAS Code: GG15 • Mode: 3 Years Equal Combination

Qualifications/Requirements: GCE, A/AS: 12 points. Maths at A level. AGNVQ, Individual consideration. ND/C, Individual consideration. SQAH, Grades C. SQAV, Individual consideration. IB, Pass in Diploma.

BSc Sport & Exercise Science and Information Technology for Science
UCAS Code: BG6M • Mode: 3 Years Equal Combination

Qualifications/Requirements: GCE, A/AS: 16 points. Biology at A level and Physical Education or Sports Studies. AGNVQ, Individual consideration. ND/C, Individual consideration. SQAH, Grades B. SQAV, Individual consideration. IB, Pass in Diploma.

BEng Electronics and Computing
UCAS Code: GH56 • Mode: 3/4 Years Equal Combination

Qualifications/Requirements: GCE, A/AS: 18 points. AGNVQ, Individual consideration. ND/C, 4 Distinctions. SQAH, Grades BBCC. SQAV, Individual consideration. IB, Individual consideration.

BEng Electronics and Computing (Extended)
UCAS Code: GH5P • Mode: 4/5 Years Equal Combination

Qualifications/Requirements: AGNVQ, Individual consideration. ND/C, Individual consideration. SQAH, Individual consideration. SQAV, Individual consideration. IB, Individual consideration.

HND Computer Studies
UCAS Code: 105G • Mode: 2/3 Years Single Subject

Qualifications/Requirements: GCE, A/AS: 6 points. AGNVQ, Individual consideration. ND/C, 2 Merits. SQAH, Individual consideration. SQAV, Individual consideration. IB, Individual consideration.

HND Computing and Electronics
UCAS Code: 65HG • Mode: 2 Years Equal Combination

Qualifications/Requirements: GCE, A/AS: 4 points. AGNVQ, Individual consideration. ND/C, 4 Merits (in specific programmes). SQAH, Grades CC including specific subject(s). SQAV, Individual consideration. IB, Individual consideration.

OXFORD UNIVERSITY O33

BA Computation
UCAS Code: G500 • Mode: 3 Years Single Subject

Qualifications/Requirements: GCE, A level grades: AAB. Maths at A level. AGNVQ, Pass. ND/C, Distinction Overall. SQAH, Grades AAAAA. SQAV, Individual consideration. IB, 36 points.

BA Mathematics and Computation
UCAS Code: GG15 • Mode: 3 Years Equal Combination

Qualifications/Requirements: GCE, A level grades: AAB. Maths at A level. AGNVQ, Pass. ND/C, Distinction Overall. SQAH, Grades AAAAA. SQAV, Individual consideration. IB, 36 points.

MEng Engineering and Computing Science
UCAS Code: GH51 • Mode: 4 Years Equal Combination

Qualifications/Requirements: GCE, A level grades: AAB. Maths and Physics at A level. AGNVQ, Pass. ND/C, Distinction Overall. SQAH, Grades AAAAA. SQAV, Individual consideration. IB, 36 points.

OXFORD BROOKES UNIVERSITY O66

BA/BSc Computing Science/Spanish Studies
UCAS Code: GR5L • Mode: 3 Years Equal Combination

Qualifications/Requirements: Please refer to prospectus.

BA/BSc Computing/Spanish Studies
UCAS Code: GR5K • Mode: 3 Years Equal Combination

Qualifications/Requirements: Please refer to prospectus.

BA/BSc Information Systems/Spanish Studies
UCAS Code: RG45 • Mode: 3 Years Equal Combination

Qualifications/Requirements: Please refer to prospectus.

BSc Computing (FT or SW)
UCAS Code: G501 • Mode: 4 Years Single Subject

Qualifications/Requirements: GCE, A level grades: CCC. AGNVQ, Merit. ND/C, Individual consideration. SQAH, Individual consideration. SQAV, Individual consideration. IB, Individual consideration.

BSc Computing Science (FT or SW)
UCAS Code: G500 • Mode: 3/4 Years Single Subject

Qualifications/Requirements: GCE, A level grades: DDD-CD. AGNVQ, Merit. ND/C, Individual consideration. SQAH, Individual consideration. SQAV, Individual consideration. IB, Individual consideration.

BSc Information Systems (FT or SW)
UCAS Code: G520 • Mode: 4 Years Single Subject

Qualifications/Requirements: GCE, A level grades: DDD-CD. AGNVQ, Merit. ND/C, Individual consideration. SQAH, Individual consideration. SQAV, Individual consideration. IB, Individual consideration.

Mod Accounting and Finance/Computing
UCAS Code: GN54 • Mode: 3 Years Equal Combination

Qualifications/Requirements: GCE, A level grades: CDD-BCC. AGNVQ, Merit or Distinction with 3 additional units. ND/C, Individual consideration. SQAH, Individual consideration. SQAV, Individual consideration. IB, Individual consideration.

Mod Accounting and Finance/Computing Science
UCAS Code: NG45 • Mode: 3 Years Equal Combination

Qualifications/Requirements: GCE, A level grades: DDD-BBC. AGNVQ, Merit in Business with 4 additional units. ND/C, Individual consideration. SQAH, Individual consideration. SQAV, Individual consideration. IB, Individual consideration.

Mod Accounting and Finance/ Information Systems
UCAS Code: GNM4 • Mode: 3 Years Equal Combination

Qualifications/Requirements: GCE, A level grades: CDD-BCC. AGNVQ, Merit or Distinction with 3 additional units. ND/C, Individual consideration. SQAH, Individual consideration. SQAV, Individual consideration. IB, Individual consideration.

Mod Anthropology/Computing
UCAS Code: GL56 • Mode: 3 Years Equal Combination

Qualifications/Requirements: GCE, A level grades: CDD-BCC. AGNVQ, Merit with A/AS. ND/C, Individual consideration. SQAH, Individual consideration. SQAV, Individual consideration. IB, Individual consideration.

Mod Anthropology/Computing Science
UCAS Code: LG65 • Mode: 3 Years Equal Combination

Qualifications/Requirements: GCE, A level grades: DDD-BCC. AGNVQ, Merit with A/AS. ND/C, Individual consideration. SQAH, Individual consideration. SQAV, Individual consideration. IB, Individual consideration.

Mod Anthropology/Information Systems
UCAS Code: GLM6 • Mode: 3 Years Equal Combination

Qualifications/Requirements: GCE, A level grades: CDD-BCC. AGNVQ, Merit with A/AS. ND/C, Individual consideration. SQAH, Individual consideration. SQAV, Individual consideration. IB, Individual consideration.

Mod Biological Chemistry/Computing
UCAS Code: CG75 • Mode: 3 Years Equal Combination

Qualifications/Requirements: GCE, A level grades: DD-BC. Science. AGNVQ, Merit in Science. ND/C, Individual consideration. SQAH, Individual consideration. SQAV, Individual consideration. IB, Individual consideration.

Mod Biological Chemistry/Computing Science
UCAS Code: GC57 • Mode: 3 Years Equal Combination

Qualifications/Requirements: GCE, A level grades: DD-CCC. Science. AGNVQ, Merit in Science. ND/C, Individual consideration. SQAH, Individual consideration. SQAV, Individual consideration. IB, Individual consideration.

Mod Biological Chemistry/Information Systems

UCAS Code: CG7M • Mode: 3 Years Equal Combination

Qualifications/Requirements: GCE, A level grades: DD-CCC. Science. AGNVQ, Merit in Science. ND/C, Individual consideration. SQAH, Individual consideration. SQAV, Individual consideration. IB, Individual consideration.

Mod Biology/Computing

UCAS Code: CG15 • Mode: 2/3 Years Equal Combination

Qualifications/Requirements: GCE, A level grades: DD-BC. Science. AGNVQ, Merit in Science. ND/C, Individual consideration. SQAH, Individual consideration. SQAV, Individual consideration. IB, Individual consideration.

Mod Biology/Computing Science

UCAS Code: GC51 • Mode: 3 Years Equal Combination

Qualifications/Requirements: GCE, A level grades: DD-CCC. Science. AGNVQ, Merit in Science. ND/C, Individual consideration. SQAH, Individual consideration. SQAV, Individual consideration. IB, Individual consideration.

Mod Biology/Information Systems

UCAS Code: CG1M • Mode: 3 Years Equal Combination

Qualifications/Requirements: GCE, A level grades: DD-CCC. Science. AGNVQ, Merit in Science. ND/C, Individual consideration. SQAH, Individual consideration. SQAV, Individual consideration. IB, Individual consideration.

Mod Business Administration and Management/Software Engineering

UCAS Code: GN71 • Mode: 3 Years Equal Combination

Qualifications/Requirements: GCE, A level grades: CCC-BBC. AGNVQ, Merit or Merit in Business with 4 additional units. ND/C, Individual consideration. SQAH, Individual consideration. SQAV, Individual consideration. IB, Individual consideration.

Mod Business Administration and Management/Computing Science

UCAS Code: NG15 • Mode: 3 Years Equal Combination

Qualifications/Requirements: GCE, A level grades: CCC-BBB. AGNVQ, Merit in Business with 4 additional units. ND/C, Individual consideration. SQAH, Individual consideration. SQAV, Individual consideration. IB, Individual consideration.

Mod Business Statistics/Information Systems

UCAS Code: GG5K • Mode: 3 Years Equal Combination

Qualifications/Requirements: Please refer to prospectus.

Mod Cell Biology/Computer Systems

UCAS Code: CGC6 • Mode: 3 Years Equal Combination

Qualifications/Requirements: GCE, A level grades: DD-CCC. Science. AGNVQ, Merit in Science. ND/C, Individual consideration. SQAH, Individual consideration. SQAV, Individual consideration. IB, Individual consideration.

Mod Cell Biology/Computing

UCAS Code: CGC5 • Mode: 3 Years Equal Combination

Qualifications/Requirements: GCE, A level grades: DD-BCC. Science. AGNVQ, Merit in Science. ND/C, Individual consideration. SQAH, Individual consideration. SQAV, Individual consideration. IB, Individual consideration.

Mod Cell Biology/Computing Science

UCAS Code: GC5C • Mode: 3 Years Equal Combination

Qualifications/Requirements: GCE, A level grades: DD-CCC. Science. AGNVQ, Merit in Science. ND/C, Individual consideration. SQAH, Individual consideration. SQAV, Individual consideration. IB, Individual consideration.

Mod Cell Biology/Information Systems

UCAS Code: CGCM • Mode: 3 Years Equal Combination

Qualifications/Requirements: GCE, A level grades: DD-BC. Science. AGNVQ, Merit in Science. ND/C, Individual consideration. SQAH, Individual consideration. SQAV, Individual consideration. IB, Individual consideration.

Mod Cities and Society/Computing Science

UCAS Code: LG35 • Mode: 3 Years Equal Combination

Qualifications/Requirements: GCE, A level grades: DD-CCC. AGNVQ, Merit. ND/C, Individual consideration. SQAH, Individual consideration. SQAV, Individual consideration. IB, Individual consideration.

Mod Cities and Society/Information Systems

UCAS Code: LG3N • Mode: 3 Years Equal Combination

Qualifications/Requirements: GCE, A level grades: DD-CCC. AGNVQ, Merit. ND/C, Individual consideration. SQAH, Individual consideration. SQAV, Individual consideration. IB, Individual consideration.

Mod Combined Studies/Information Systems

UCAS Code: GYM4 • Mode: 3 Years

Qualifications/Requirements: AGNVQ, Not normally sufficient. ND/C, Individual consideration. SQAH, Not normally sufficient. SQAV, Individual consideration. IB, Not normally sufficient.

Mod Computing Mathematics/Information Systems

UCAS Code: GGM9 • Mode: 3 Years Equal Combination

Qualifications/Requirements: GCE, A level grades: CD-BC. AGNVQ, Merit. ND/C, Individual consideration. SQAH, Individual consideration. SQAV, Individual consideration. IB, Individual consideration.

Mod Computing Mathematics/Intelligent Systems

UCAS Code: GG89 • Mode: 3 Years Equal Combination

Qualifications/Requirements: GCE, A level grades: CD. AGNVQ, Merit. ND/C, Individual consideration. SQAH, Individual consideration. SQAV, Individual consideration. IB, Individual consideration.

Mod Computing Mathematics/ Multimedia Systems

UCAS Code: GGP9 • Mode: 3 Years Equal Combination

Qualifications/Requirements: Please refer to prospectus.

Mod Computing Mathematics/Software Engineering

UCAS Code: GG79 • Mode: 3 Years Equal Combination

Qualifications/Requirements: GCE, A level grades: CD-BC. AGNVQ, Merit. ND/C, Individual consideration. SQAH, Individual consideration. SQAV, Individual consideration. IB, Individual consideration.

Mod Computing Science/Ecology

UCAS Code: GC59 • Mode: 3 Years Equal Combination

Qualifications/Requirements: GCE, A level grades: DD-CCC. Science. AGNVQ, Merit in Science. ND/C, Individual consideration. SQAH, Individual consideration. SQAV, Individual consideration. IB, Individual consideration.

Mod Computing Science/Economics

UCAS Code: LG15 • Mode: 3 Years Equal Combination

Qualifications/Requirements: GCE, A level grades: DD-BCC. AGNVQ, Merit with 3 additional units. ND/C, Individual consideration. SQAH, Individual consideration. SQAV, Individual consideration. IB, Individual consideration.

Mod Computing Science/Educational Studies

UCAS Code: XG95 • Mode: 3 Years Equal Combination

Qualifications/Requirements: GCE, A level grades: DD-BCC. AGNVQ, Merit with 3 additional units. ND/C, Individual consideration. SQAH, Individual consideration. SQAV, Individual consideration. IB, Individual consideration.

Mod Computing Science/Electronics

UCAS Code: HG65 • Mode: 3 Years Equal Combination

Qualifications/Requirements: GCE, A level grades: DD-CCC. Science or Maths. AGNVQ, Merit (in specific programmes). ND/C, Individual consideration. SQAH, Individual consideration. SQAV, Individual consideration. IB, Individual consideration.

Mod Computing Science/English Studies

UCAS Code: QG35 • Mode: 3 Years Equal Combination

Qualifications/Requirements: GCE, A level grades: DD-BBC. AGNVQ, Merit with A/AS. ND/C, Individual consideration. SQAH, Individual consideration. SQAV, Individual consideration. IB, Individual consideration.

Mod Computing Science/Environmental Chemistry

UCAS Code: FG1N • Mode: 3 Years Equal Combination

Qualifications/Requirements: GCE, A level grades: DD-CCC. Science. AGNVQ, Merit in Science. ND/C, Individual consideration. SQAH, Individual consideration. SQAV, Individual consideration. IB, Individual consideration.

Mod Computing Science/Environmental Design & Conservation

UCAS Code: FG95 • Mode: 3 Years Equal Combination

Qualifications/Requirements: GCE, A level grades: DD-CCC. AGNVQ, Merit. ND/C, Individual consideration. SQAH, Individual consideration. SQAV, Individual consideration. IB, Individual consideration.

Mod Computing Science/Environmental Policy

UCAS Code: GK53 • Mode: 3 Years Equal Combination

Qualifications/Requirements: GCE, A level grades: DD-CCC. AGNVQ, Merit. ND/C, Individual consideration. SQAH, Individual consideration. SQAV, Individual consideration. IB, Individual consideration.

Mod Computing Science/Environmental Sciences

UCAS Code: FG9M • Mode: 3 Years Equal Combination

Qualifications/Requirements: GCE, A level grades: DD-CCC. Science. AGNVQ, Merit in Science. ND/C, Individual consideration. SQAH, Individual consideration. SQAV, Individual consideration. IB, Individual consideration.

Mod Computing Science/European Culture and Society

UCAS Code: TGGN • Mode: 3 Years Equal Combination

Qualifications/Requirements: GCE, A level grades: DD-CCC. AGNVQ, Merit with A/AS. ND/C, Individual consideration. SQAH, Individual consideration. SQAV, Individual consideration. IB, Individual consideration.

Mod Computing Science/Exercise and Health

UCAS Code: BG65 • Mode: 3 Years Equal Combination

Qualifications/Requirements: GCE, A level grades: DD-CCC. Science. AGNVQ, Merit in Science. ND/C, Individual consideration. SQAH, Individual consideration. SQAV, Individual consideration. IB, Individual consideration.

Mod Computing Science/Fine Art

UCAS Code: WG15 • Mode: 3 Years Equal Combination

Qualifications/Requirements: GCE, A level grades: DDD-BCC. Art. A portfolio of work is required. AGNVQ, Merit in Art & Design. ND/C, Individual consideration. SQAH, Individual consideration. SQAV, Individual consideration. IB, Individual consideration.

Mod Computing Science/Food Science and Nutrition

UCAS Code: GD54 • Mode: 3 Years Equal Combination

Qualifications/Requirements: GCE, A level grades: DD-CCC. Science. AGNVQ, Merit in Science. ND/C, Individual consideration. SQAH, Individual consideration. SQAV, Individual consideration. IB, Individual consideration.

Mod Computing Science/French Studies

UCAS Code: GRMD • Mode: 3 Years Equal Combination

Qualifications/Requirements: Please refer to prospectus.

Mod Computing Science/Geographic Information Science

UCAS Code: GFMV • Mode: 3 Years Equal Combination

Qualifications/Requirements: Please refer to prospectus.

Mod Computing Science/Geography

UCAS Code: GF5V • Mode: 3 Years Equal Combination

Qualifications/Requirements: GCE, A level grades: DDD-BCC. AGNVQ, Merit with A/AS. ND/C, Individual consideration. SQAH, Individual consideration. SQAV, Individual consideration. IB, Individual consideration.

Mod Computing Science/Geology

UCAS Code: GF56 • Mode: 3 Years Equal Combination

Qualifications/Requirements: GCE, A level grades: DD-CCC. Science or Maths. AGNVQ, Merit with A level. ND/C, Individual consideration. SQAH, Individual consideration. SQAV, Individual consideration. IB, Individual consideration.

Mod Computing Science/Geotechnics

UCAS Code: HG25 • Mode: 3 Years Equal Combination

Qualifications/Requirements: GCE, A level grades: DD-CCC. Science or Maths. AGNVQ, Merit with A level. ND/C, Individual consideration. SQAH, Individual consideration. SQAV, Individual consideration. IB, Individual consideration.

Mod Computing Science/German Studies

UCAS Code: RG25 • Mode: 3 Years Equal Combination

Qualifications/Requirements: GCE, A level grades: DD-CCC. German. AGNVQ, Merit with A level. ND/C, Individual consideration. SQAH, Individual consideration. SQAV, Individual consideration. IB, Individual consideration.

Mod Computing Science/History

UCAS Code: VG15 • Mode: 3 Years Equal Combination

Qualifications/Requirements: GCE, A level grades: DDD-BCC. AGNVQ, Merit with A/AS. ND/C, Individual consideration. SQAH, Individual consideration. SQAV, Individual consideration. IB, Individual consideration.

Mod Computing Science/History of Art

UCAS Code: VG45 • Mode: 3 Years Equal Combination

Qualifications/Requirements: GCE, A level grades: DDD-BCC. AGNVQ, Merit with A/AS. ND/C, Individual consideration. SQAH, Individual consideration. SQAV, Individual consideration. IB, Individual consideration.

Mod Computing Science/Hospitality Management Studies

UCAS Code: NG75 • Mode: 3 Years Equal Combination

Qualifications/Requirements: GCE, A level grades: DDD-BCC. AGNVQ, Merit with 3 additional units or Merit with A/AS. ND/C, Individual consideration. SQAH, Individual consideration. SQAV, Individual consideration. IB, Individual consideration.

Mod Computing Science/Human Biology

UCAS Code: GB51 • Mode: 3 Years Equal Combination

Qualifications/Requirements: GCE, A level grades: DD-CCC. Science. AGNVQ, Merit in Science. ND/C, Individual consideration. SQAH, Individual consideration. SQAV, Individual consideration. IB, Individual consideration.

Mod Computing Science/Information Systems

UCAS Code: G512 • Mode: 3 Years Single Subject

Qualifications/Requirements: GCE, A level grades: DD-CCC. AGNVQ, Merit. ND/C, Individual consideration. SQAH, Individual consideration. SQAV, Individual consideration. IB, Individual consideration.

Mod Computing Science/Intelligent Systems

UCAS Code: GG85 • Mode: 3 Years Equal Combination

Qualifications/Requirements: GCE, A level grades: DD-CCC. AGNVQ, Merit. ND/C, Individual consideration. SQAH, Individual consideration. SQAV, Individual consideration. IB, Individual consideration.

Mod Computing Science/Japanese Studies

UCAS Code: GT5K • Mode: 3 Years Equal Combination

Qualifications/Requirements: Please refer to prospectus.

Mod Computing Science/Law

UCAS Code: MG35 • Mode: 3 Years Equal Combination

Qualifications/Requirements: GCE, A level grades: CCC-ABB. AGNVQ, Distinction with 3 additional units or Distinction with A/AS. ND/C, Individual consideration. SQAH, Individual consideration. SQAV, Individual consideration. IB, Individual consideration.

Mod Computing Science/Leisure Planning

UCAS Code: GK5H • Mode: 3 Years Equal Combination

Qualifications/Requirements: GCE, A level grades: DD-CCC. AGNVQ, Merit. ND/C, Individual consideration. SQAH, Individual consideration. SQAV, Individual consideration. IB, Individual consideration.

Mod Computing Science/Marketing Management

UCAS Code: NG55 • Mode: 3 Years Equal Combination

Qualifications/Requirements: GCE, A level grades: DDD-BCC. AGNVQ, Merit with 3 additional units. ND/C, Individual consideration. SQAH, Individual consideration. SQAV, Individual consideration. IB, Individual consideration.

Mod Computing Science/Mathematics

UCAS Code: GG51 • Mode: 3 Years Equal Combination

Qualifications/Requirements: GCE, A level grades: DD-CCC. Maths. AGNVQ, Merit with A level. ND/C, Individual consideration. SQAH, Individual consideration. SQAV, Individual consideration. IB, Individual consideration.

Mod Computing Science/Multimedia Systems
UCAS Code: GGPM • Mode: 3 Years Equal Combination

Qualifications/Requirements: Please refer to prospectus.

Mod Computing Science/Music
UCAS Code: WG35 • Mode: 3 Years Equal Combination

Qualifications/Requirements: GCE, A level grades: DD-BCC. Music. AGNVQ, Merit with A level. ND/C, Individual consideration. SQAH, Individual consideration. SQAV, Individual consideration. IB, Individual consideration.

Mod Computing Science/Palliative Care
UCAS Code: GB57 • Mode: 3 Years Equal Combination

Qualifications/Requirements: Please refer to prospectus.

Mod Computing Science/Physical Geography
UCAS Code: GF5W • Mode: 3 Years Equal Combination

Qualifications/Requirements: GCE, A level grades: DDD-BCC. AGNVQ, Merit with A/AS. ND/C, Individual consideration. SQAH, Individual consideration. SQAV, Individual consideration. IB, Individual consideration.

Mod Computing Science/Planning Studies
UCAS Code: KG45 • Mode: 3 Years Equal Combination

Qualifications/Requirements: GCE, A level grades: DD-CCC. AGNVQ, Merit. ND/C, Individual consideration. SQAH, Individual consideration. SQAV, Individual consideration. IB, Individual consideration.

Mod Computing Science/Politics
UCAS Code: MG15 • Mode: 3 Years Equal Combination

Qualifications/Requirements: GCE, A level grades: DDD-BCC. AGNVQ, Merit with A/AS. ND/C, Individual consideration. SQAH, Individual consideration. SQAV, Individual consideration. IB, Individual consideration.

Mod Computing Science/Psychology
UCAS Code: GC58 • Mode: 3 Years Equal Combination

Qualifications/Requirements: GCE, A level grades: CCC-ABB. AGNVQ, Merit with A/AS. ND/C, Individual consideration. SQAH, Individual consideration. SQAV, Individual consideration. IB, Individual consideration.

Mod Computing Science/Publishing
UCAS Code: PG55 • Mode: 3 Years Equal Combination

Qualifications/Requirements: GCE, A level grades: DDD-BCC. AGNVQ, Merit with 3 additional units. ND/C, Individual consideration. SQAH, Individual consideration. SQAV, Individual consideration. IB, Individual consideration.

Mod Computing Science/Rehabilitation
UCAS Code: GB5R • Mode: 3 Years Equal Combination

Qualifications/Requirements: Please refer to prospectus.

Mod Computing Science/Retail Management
UCAS Code: NG5M • Mode: 3 Years Equal Combination

Qualifications/Requirements: GCE, A level grades: DDD-BCC. AGNVQ, Merit in Business with 4 additional units. ND/C, Individual consideration. SQAH, Individual consideration. SQAV, Individual consideration. IB, Individual consideration.

Mod Computing Science/Sociology
UCAS Code: LG3M • Mode: 3 Years Equal Combination

Qualifications/Requirements: GCE, A level grades: DDD-BCC. AGNVQ, Merit with A/AS. ND/C, Individual consideration. SQAH, Individual consideration. SQAV, Individual consideration. IB, Individual consideration.

Mod Computing Science/Software Engineering
UCAS Code: GG57 • Mode: 3 Years Equal Combination

Qualifications/Requirements: GCE, A level grades: DD-CCC. AGNVQ, Merit. ND/C, Individual consideration. SQAH, Individual consideration. SQAV, Individual consideration. IB, Individual consideration.

Mod Computing Science/Statistics
UCAS Code: GG54 • Mode: 3 Years Equal Combination

Qualifications/Requirements: GCE, A level grades: DD-CCC. AGNVQ, Merit with A level. ND/C, Individual consideration. SQAH, Individual consideration. SQAV, Individual consideration. IB, Individual consideration.

Mod Computing Science/ Telecommunications
UCAS Code: HG6M • Mode: 3 Years Equal Combination

Qualifications/Requirements: GCE, A level grades: DD-CCC. Science or Maths. AGNVQ, Merit (in specific programmes). ND/C, Individual consideration. SQAH, Individual consideration. SQAV, Individual consideration. IB, Individual consideration.

Mod Computing Science/Tourism
UCAS Code: PG75 • Mode: 3 Years Equal Combination

Qualifications/Requirements: GCE, A level grades: DD-CCC. AGNVQ, Merit with 3 additional units or Merit with A/AS. ND/C, Individual consideration. SQAH, Individual consideration. SQAV, Individual consideration. IB, Individual consideration.

Mod Computing Science/Transport and Travel
UCAS Code: NG95 • Mode: 3 Years Equal Combination

Qualifications/Requirements: GCE, A level grades: DD-CCC. AGNVQ, Merit. ND/C, Individual consideration. SQAH, Individual consideration. SQAV, Individual consideration. IB, Individual consideration.

Mod Computing Science/Water Resources
UCAS Code: HG2M • Mode: 3 Years Equal Combination

Qualifications/Requirements: GCE, A level grades: DD-CCC. AGNVQ, Merit. ND/C, Individual consideration. SQAH, Individual consideration. SQAV, Individual consideration. IB, Individual consideration.

Mod Computing/Computing Mathematics
UCAS Code: GG59 • Mode: 3 Years Equal Combination

Qualifications/Requirements: GCE, A level grades: CD-CDD. AGNVQ, Merit. ND/C, Individual consideration. SQAH, Individual consideration. SQAV, Individual consideration. IB, Individual consideration.

Mod Computing/Computing Science
UCAS Code: G511 • Mode: 3 Years Single Subject

Qualifications/Requirements: GCE, A level grades: DD-CCC. AGNVQ, Merit. ND/C, Individual consideration. SQAH, Individual consideration. SQAV, Individual consideration. IB, Individual consideration.

Mod Computing/Ecology
UCAS Code: CG95 • Mode: 3 Years Equal Combination

Qualifications/Requirements: GCE, A level grades: CD-BC. AGNVQ, Merit in Science. ND/C, Individual consideration. SQAH, Individual consideration. SQAV, Individual consideration. IB, Individual consideration.

Mod Computing/Economics
UCAS Code: GL51 • Mode: 3 Years Equal Combination

Qualifications/Requirements: GCE, A level grades: CDD-BB. AGNVQ, Merit or Merit with 3 additional units. ND/C, Individual consideration. SQAH, Individual consideration. SQAV, Individual consideration. IB, Individual consideration.

Mod Computing/Educational Studies
UCAS Code: GX59 • Mode: 3 Years Equal Combination

Qualifications/Requirements: GCE, A level grades: CC-CDD. AGNVQ, Merit or Merit with 3 additional units. ND/C, Individual consideration. SQAH, Individual consideration. SQAV, Individual consideration. IB, Individual consideration.

Mod Computing/Electronics
UCAS Code: GH56 • Mode: 2/3 Years Equal Combination

Qualifications/Requirements: GCE, A level grades: CDD-BB. Science or Maths at A level. AGNVQ, Merit (in specific programmes). ND/C, Individual consideration. SQAH, Individual consideration. SQAV, Individual consideration. IB, Individual consideration.

Mod Computing/English Studies
UCAS Code: GQ53 • Mode: 3 Years Equal Combination

Qualifications/Requirements: GCE, A level grades: CDD-AB. AGNVQ, Merit with A/AS. ND/C, Individual consideration. SQAH, Individual consideration. SQAV, Individual consideration. IB, Individual consideration.

Mod Computing/Environmental Chemistry
UCAS Code: GF51 • Mode: 3 Years Equal Combination

Qualifications/Requirements: GCE, A level grades: DD-CCC. Science. AGNVQ, Distinction in Science. ND/C, Individual consideration. SQAH, Individual consideration. SQAV, Individual consideration. IB, Individual consideration.

Mod Computing/Environmental Design and Conservation
UCAS Code: FG9N • Mode: 3 Years Equal Combination

Qualifications/Requirements: GCE, A level grades: DD-CCC. AGNVQ, Merit. ND/C, Individual consideration. SQAH, Individual consideration. SQAV, Individual consideration. IB, Individual consideration.

Mod Computing/Environmental Policy
UCAS Code: KG35 • Mode: 3 Years Equal Combination

Qualifications/Requirements: GCE, A level grades: DD-CCC. AGNVQ, Merit. ND/C, Individual consideration. SQAH, Individual consideration. SQAV, Individual consideration. IB, Individual consideration.

Mod Computing/Environmental Sciences
UCAS Code: FGX5 • Mode: 3 Years Equal Combination

Qualifications/Requirements: GCE, A level grades: CD-CCC. Science. AGNVQ, Merit or Distinction in Science. ND/C, Individual consideration. SQAH, Individual consideration. SQAV, Individual consideration. IB, Individual consideration.

Mod Computing/Exercise and Health
UCAS Code: GB56 • Mode: 3 Years Equal Combination

Qualifications/Requirements: GCE, A level grades: DD-BC. Science. AGNVQ, Merit in Science. ND/C, Individual consideration. SQAH, Individual consideration. SQAV, Individual consideration. IB, Individual consideration.

Mod Computing/Fine Art
UCAS Code: GW51 • Mode: 3 Years Equal Combination

Qualifications/Requirements: GCE, A level grades: CDD-BC. Art. A portfolio of work is required. AGNVQ, Merit in Art & Design with A/AS. ND/C, Individual consideration. SQAH, Individual consideration. SQAV, Individual consideration. IB, Individual consideration.

Mod Computing/Food Science and Nutrition
UCAS Code: DG45 • Mode: 3 Years Equal Combination

Qualifications/Requirements: GCE, A level grades: DD-BC. Science. AGNVQ, Merit in Science. ND/C, Individual consideration. SQAH, Individual consideration. SQAV, Individual consideration. IB, Individual consideration.

Mod Computing/French Studies
UCAS Code: GR5D • Mode: 3 Years Equal Combination

Qualifications/Requirements: Please refer to prospectus.

Mod Computing/Geographic Information Science
UCAS Code: GFM8 • Mode: 3 Years Equal Combination

Qualifications/Requirements: Please refer to prospectus.

Mod Computing/Geography
UCAS Code: GL58 • Mode: 3 Years Equal Combination

Qualifications/Requirements: GCE, A level grades: CDD-BB. AGNVQ, Merit. ND/C, Individual consideration. SQAH, Individual consideration. SQAV, Individual consideration. IB, Individual consideration.

Mod Computing/Geology
UCAS Code: FG65 • Mode: 3 Years Equal Combination

Qualifications/Requirements: GCE, A level grades: DD-BC. Science or Maths. AGNVQ, Pass in Science or Merit in Science. ND/C, Individual consideration. SQAH, Individual consideration. SQAV, Individual consideration. IB, Individual consideration.

Mod Computing/Geotechnics
UCAS Code: GH52 • Mode: 3 Years Equal Combination

Qualifications/Requirements: GCE, A level grades: CCC. Science, Maths, Design & Technology or Electronics. AGNVQ, Merit (in specific programmes). ND/C, Individual consideration. SQAH, Individual consideration. SQAV, Individual consideration. IB, Individual consideration.

Mod Computing/German Studies
UCAS Code: GR5G • Mode: 4 Years Equal Combination

Qualifications/Requirements: GCE, A level grades: DDD-CCC. German. AGNVQ, Merit with A level. ND/C, Individual consideration. SQAH, Individual consideration. SQAV, Individual consideration. IB, Individual consideration.

Mod Computing/History
UCAS Code: GV51 • Mode: 3 Years Equal Combination

Qualifications/Requirements: GCE, A level grades: CDD-BB. AGNVQ, Merit with A/AS. ND/C, Individual consideration. SQAH, Individual consideration. SQAV, Individual consideration. IB, Individual consideration.

Mod Computing/History of Art
UCAS Code: GV54 • Mode: 3 Years Equal Combination

Qualifications/Requirements: GCE, A level grades: CDD-BCC. AGNVQ, Merit with A/AS. ND/C, Individual consideration. SQAH, Individual consideration. SQAV, Individual consideration. IB, Individual consideration.

Mod Computing/Hospitality Management Studies
UCAS Code: GN57 • Mode: 3 Years Equal Combination

Qualifications/Requirements: GCE, A level grades: DDD-BC. AGNVQ, Merit or Merit with 3 additional units. ND/C, Individual consideration. SQAH, Individual consideration. SQAV, Individual consideration. IB, Individual consideration.

Mod Computing/Human Biology
UCAS Code: BG15 • Mode: 3 Years Equal Combination

Qualifications/Requirements: GCE, A level grades: DDD-BCC. Science. AGNVQ, Merit in Science. ND/C, Individual consideration. SQAH, Individual consideration. SQAV, Individual consideration. IB, Individual consideration.

Mod Computing/Information Systems
UCAS Code: G510 • Mode: 3 Years Single Subject

Qualifications/Requirements: GCE, A level grades: CDD-BC. AGNVQ, Merit. ND/C, Individual consideration. SQAH, Individual consideration. SQAV, Individual consideration. IB, Individual consideration.

Mod Computing/Intelligent Systems
UCAS Code: GG58 • Mode: 3 Years Equal Combination

Qualifications/Requirements: GCE, A level grades: BC-CDD. AGNVQ, Merit. ND/C, Individual consideration. SQAH, Individual consideration. SQAV, Individual consideration. IB, Individual consideration.

Mod Computing/Japanese Studies
UCAS Code: GT54 • Mode: 3 Years Equal Combination

Qualifications/Requirements: Please refer to prospectus.

Mod Computing/Law
UCAS Code: GM53 • Mode: 3 Years Equal Combination

Qualifications/Requirements: GCE, A level grades: CCC-ABB. AGNVQ, Merit or Distinction with 3 additional units. ND/C, Individual consideration. SQAH, Individual consideration. SQAV, Individual consideration. IB, Individual consideration.

Mod Computing/Leisure Planning
UCAS Code: KGH5 • Mode: 3 Years Equal Combination

Qualifications/Requirements: GCE, A level grades: DDD-BC. AGNVQ, Merit. ND/C, Individual consideration. SQAH, Individual consideration. SQAV, Individual consideration. IB, Individual consideration.

Mod Computing/Marketing Management
UCAS Code: GN5N • Mode: 3 Years Equal Combination

Qualifications/Requirements: GCE, A level grades: CDD-BCC. AGNVQ, Merit or Distinction with 3 additional units. ND/C, Individual consideration. SQAH, Individual consideration. SQAV, Individual consideration. IB, Individual consideration.

Mod Computing/Multimedia Systems
UCAS Code: GGP5 • Mode: 3 Years Equal Combination

Qualifications/Requirements: Please refer to prospectus.

Mod Computing/Music
UCAS Code: GW53 • Mode: 3 Years Equal Combination

Qualifications/Requirements: GCE, A level grades: DD-BC. Music. AGNVQ, Merit. ND/C, Individual consideration. SQAH, Individual consideration. SQAV, Individual consideration. IB, Individual consideration.

Mod Computing/Palliative Care
UCAS Code: BGR5 • Mode: 3 Years Equal Combination

Qualifications/Requirements: Please refer to prospectus.

Mod Computing/Physical Geography
UCAS Code: FGV5 • Mode: 3 Years Equal Combination

Qualifications/Requirements: GCE, A level grades: CCC-BB. AGNVQ, Merit. ND/C, Individual consideration. SQAH, Individual consideration. SQAV, Individual consideration. IB, Individual consideration.

Mod Computing/Planning Studies
UCAS Code: GK54 • Mode: 3 Years Equal Combination

Qualifications/Requirements: GCE, A level grades: DDD-BC. AGNVQ, Merit. ND/C, Individual consideration. SQAH, Individual consideration. SQAV, Individual consideration. IB, Individual consideration.

Mod Computing/Politics
UCAS Code: GM51 • Mode: 3 Years Equal Combination

Qualifications/Requirements: GCE, A level grades: CDD-AB. AGNVQ, Merit with A/AS. ND/C, Individual consideration. SQAH, Individual consideration. SQAV, Individual consideration. IB, Individual consideration.

Mod Computing/Psychology
UCAS Code: CG85 • Mode: 3 Years Equal Combination

Qualifications/Requirements: GCE, A level grades: CDD-BBC. AGNVQ, Merit with A/AS. ND/C, Individual consideration. SQAH, Individual consideration. SQAV, Individual consideration. IB, Individual consideration.

Mod Computing/Publishing
UCAS Code: GP55 • Mode: 3 Years Equal Combination

Qualifications/Requirements: GCE, A level grades: CDD-BB. AGNVQ, Merit or Merit (in specific programmes) with 3 additional units. ND/C, Individual consideration. SQAH, Individual consideration. SQAV, Individual consideration. IB, Individual consideration.

Mod Computing/Rehabilitation
UCAS Code: BGT5 • Mode: 3 Years Equal Combination

Qualifications/Requirements: Please refer to prospectus.

Mod Computing/Retail Management
UCAS Code: GN55 • Mode: 3 Years Equal Combination

Qualifications/Requirements: GCE, A level grades: CDD-CCD. AGNVQ, Merit with A/AS or Merit in Business with 4 additional units. ND/C, Individual consideration. SQAH, Individual consideration. SQAV, Individual consideration. IB, Individual consideration.

Mod Computing/Software Engineering
UCAS Code: GG75 • Mode: 3 Years Equal Combination

Qualifications/Requirements: GCE, A level grades: CDD-BCC. AGNVQ, Merit. ND/C, Individual consideration. SQAH, Individual consideration. SQAV, Individual consideration. IB, Individual consideration.

Mod Computing/Telecommunications
UCAS Code: GH5P • Mode: 3 Years Equal Combination

Qualifications/Requirements: GCE, A level grades: DD-CCC. Science or Maths. AGNVQ, Merit (in specific programmes). ND/C, Individual consideration. SQAH, Individual consideration. SQAV, Individual consideration. IB, Individual consideration.

Mod Computing/Tourism
UCAS Code: GP57 • Mode: 3 Years Equal Combination

Qualifications/Requirements: GCE, A level grades: CDD-BC. AGNVQ, Merit or Merit with 3 additional units. ND/C, Individual consideration. SQAH, Individual consideration. SQAV, Individual consideration. IB, Individual consideration.

Mod Computing/Transport and Travel
UCAS Code: GN59 • Mode: 3 Years Equal Combination

Qualifications/Requirements: GCE, A level grades: BC-DDD. AGNVQ, Merit. ND/C, Individual consideration. SQAH, Individual consideration. SQAV, Individual consideration. IB, Individual consideration.

Mod Computing/Water Resources
UCAS Code: GH5F • Mode: 3 Years Equal Combination

Qualifications/Requirements: GCE, A level grades: DD-CCC. AGNVQ, Merit. ND/C, Individual consideration. SQAH, Individual consideration. SQAV, Individual consideration. IB, Individual consideration.

Mod Ecology/Information Systems
UCAS Code: CG9M • Mode: 3 Years Equal Combination

Qualifications/Requirements: GCE, A level grades: CD-BC. AGNVQ, Merit in Science. ND/C, Individual consideration. SQAH, Individual consideration. SQAV, Individual consideration. IB, Individual consideration.

Mod Economics/Information Systems
UCAS Code: GLM1 • Mode: 3 Years Equal Combination

Qualifications/Requirements: GCE, A level grades: CDD-BB. AGNVQ, Merit or Merit with 3 additional units. ND/C, Individual consideration. SQAH, Individual consideration. SQAV, Individual consideration. IB, Individual consideration.

Mod Educational Studies/Information Systems
UCAS Code: GXM9 • Mode: 3 Years Equal Combination

Qualifications/Requirements: GCE, A level grades: CDD-BB. AGNVQ, Merit or Merit with 3 additional units. ND/C, Individual consideration. SQAH, Individual consideration. SQAV, Individual consideration. IB, Individual consideration.

Mod Educational Studies/Intelligent Systems
UCAS Code: GX89 • Mode: 3 Years Equal Combination

Qualifications/Requirements: GCE, A level grades: CD-CC. AGNVQ, Merit or Merit with 3 additional units. ND/C, Individual consideration. SQAH, Individual consideration. SQAV, Individual consideration. IB, Individual consideration.

Mod English Studies/Information Systems
UCAS Code: GQM3 • Mode: 3 Years Equal Combination

Qualifications/Requirements: GCE, A level grades: CDD-BCC. AGNVQ, Merit with A/AS. ND/C, Individual consideration. SQAH, Individual consideration. SQAV, Individual consideration. IB, Individual consideration.

Mod Environmental Design and Conservation/Information Systems
UCAS Code: FGXN • Mode: 3 Years Equal Combination

Qualifications/Requirements: GCE, A level grades: DD-CCC. AGNVQ, Merit. ND/C, Individual consideration. SQAH, Individual consideration. SQAV, Individual consideration. IB, Individual consideration.

Mod Environmental Chemistry/Information Systems
UCAS Code: GF5C • Mode: 3 Years Equal Combination

Qualifications/Requirements: GCE, A level grades: DD-BC. Science. AGNVQ, Merit in Science. ND/C, Individual consideration. SQAH, Individual consideration. SQAV, Individual consideration. IB, Individual consideration.

Mod Environmental Policy/Information Systems
UCAS Code: KG3M • Mode: 3 Years Equal Combination
Qualifications/Requirements: GCE, A level grades: DD-CCC. AGNVQ, Merit. ND/C, Individual consideration. SQAH, Individual consideration. SQAV, Individual consideration. IB, Individual consideration.

Mod Environmental Sciences/Information Systems
UCAS Code: FGXM • Mode: 3 Years Equal Combination
Qualifications/Requirements: GCE, A level grades: CD-BC. Science. AGNVQ, Merit or Distinction in Science. ND/C, Individual consideration. SQAH, Individual consideration. SQAV, Individual consideration. IB, Individual consideration.

Mod European Culture and Society/Information Systems
UCAS Code: TGGM • Mode: 3 Years Equal Combination
Qualifications/Requirements: GCE, A level grades: CD-CCC. AGNVQ, Merit with A/AS or Merit with 3 additional units. ND/C, Individual consideration. SQAH, Individual consideration. SQAV, Individual consideration. IB, Individual consideration.

Mod European Culture and Society/Intelligent Systems
UCAS Code: TGG8 • Mode: 3 Years Equal Combination
Qualifications/Requirements: GCE, A level grades: CD-CCC. AGNVQ, Merit with A/AS or Merit with 3 additional units. ND/C, Individual consideration. SQAH, Individual consideration. SQAV, Individual consideration. IB, Individual consideration.

Mod Exercise and Health/Information Systems
UCAS Code: GBM6 • Mode: 3 Years Equal Combination
Qualifications/Requirements: GCE, A level grades: DD-BCD. Science. AGNVQ, Merit in Science. ND/C, Individual consideration. SQAH, Individual consideration. SQAV, Individual consideration. IB, Individual consideration.

Mod Fine Art/Information Systems
UCAS Code: GWM1 • Mode: 3 Years Equal Combination
Qualifications/Requirements: GCE, A level grades: CDD-BC. Art. A portfolio of work is required. AGNVQ, Merit in Art & Design with A/AS. ND/C, Individual consideration. SQAH, Individual consideration. SQAV, Individual consideration. IB, Individual consideration.

Mod Food Science and Nutrition/ Information Systems
UCAS Code: DG4M • Mode: 3 Years Equal Combination
Qualifications/Requirements: GCE, A level grades: DD-BC. Science. AGNVQ, Merit in Science. ND/C, Individual consideration. SQAH, Individual consideration. SQAV, Individual consideration. IB, Individual consideration.

Mod Food Science and Nutrition/ Intelligent Systems
UCAS Code: DG48 • Mode: 3 Years Equal Combination
Qualifications/Requirements: GCE, A level grades: DD-CD. Science. AGNVQ, Merit in Science. ND/C, Individual consideration. SQAH, Individual consideration. SQAV, Individual consideration. IB, Individual consideration.

Mod Geographic Information Science/ Information Systems
UCAS Code: GFMW • Mode: 3 Years Equal Combination
Qualifications/Requirements: Please refer to prospectus.

Mod Geographic Information Science/ Mathematics
UCAS Code: GF18 • Mode: 3 Years Equal Combination
Qualifications/Requirements: Please refer to prospectus.

Mod Geographic Information Science/ Multimedia Systems
UCAS Code: GFP8 • Mode: 3 Years Equal Combination
Qualifications/Requirements: Please refer to prospectus.

Mod Geographic Information Science/Statistics
UCAS Code: FG8K • Mode: 3 Years Equal Combination
Qualifications/Requirements: Please refer to prospectus.

Mod Geography/Information Systems
UCAS Code: GLM8 • Mode: 3 Years Equal Combination
Qualifications/Requirements: GCE, A level grades: BC-BB. AGNVQ, Merit. ND/C, Individual consideration. SQAH, Individual consideration. SQAV, Individual consideration. IB, Individual consideration.

Mod Geology/Information Systems
UCAS Code: FG6N • Mode: 3 Years Equal Combination
Qualifications/Requirements: GCE, A level grades: DD-BC. Maths or Science. AGNVQ, Pass in Science or Merit. ND/C, Individual consideration. SQAH, Individual consideration. SQAV, Individual consideration. IB, Individual consideration.

Mod Geotechnics/Information Systems
UCAS Code: GHN2 • Mode: 3 Years Equal Combination
Qualifications/Requirements: GCE, A level grades: DD-BCD. Science, Maths, Design & Technology or Electronics. AGNVQ, Merit (in specific programmes). ND/C, Individual consideration. SQAH, Individual consideration. SQAV, Individual consideration. IB, Individual consideration.

Mod Hospitality Management Studies/Information Systems
UCAS Code: GNM7 • Mode: 3 Years Equal Combination
Qualifications/Requirements: GCE, A level grades: CC-BC. AGNVQ, Merit or Merit with 3 additional units. ND/C, Individual consideration. SQAH, Individual consideration. SQAV, Individual consideration. IB, Individual consideration.

Mod Human Biology/Information Systems

UCAS Code: BG1M • Mode: 3 Years Equal Combination

Qualifications/Requirements: GCE, A level grades: DD-BC. Science. AGNVQ, Merit in Science. ND/C, Individual consideration. SQAH, Individual consideration. SQAV, Individual consideration. IB, Individual consideration.

Mod Information Systems/Intelligent Systems

UCAS Code: GGM8 • Mode: 3 Years Equal Combination

Qualifications/Requirements: GCE, A level grades: CDD-BC. AGNVQ, Merit. ND/C, Individual consideration. SQAH, Individual consideration. SQAV, Individual consideration. IB, Individual consideration.

Mod Information Systems/Japanese Studies

UCAS Code: GT5L • Mode: 3 Years Equal Combination

Qualifications/Requirements: Please refer to prospectus.

Mod Information Systems/Law

UCAS Code: GMM3 • Mode: 3 Years Equal Combination

Qualifications/Requirements: GCE, A level grades: CCC-ABB. AGNVQ, Merit or Distinction with 3 additional units. ND/C, Individual consideration. SQAH, Individual consideration. SQAV, Individual consideration. IB, Individual consideration.

Mod Information Systems/Leisure Planning

UCAS Code: KGHM • Mode: 3 Years Equal Combination

Qualifications/Requirements: GCE, A level grades: DDD-BC. AGNVQ, Merit. ND/C, Individual consideration. SQAH, Individual consideration. SQAV, Individual consideration. IB, Individual consideration.

Mod Information Systems/Marketing Management

UCAS Code: GNMN • Mode: 3 Years Equal Combination

Qualifications/Requirements: GCE, A level grades: CDD-BCC. AGNVQ, Merit or Distinction with 3 additional units. ND/C, Individual consideration. SQAH, Individual consideration. SQAV, Individual consideration. IB, Individual consideration.

Mod Information Systems/Mathematics

UCAS Code: GGM1 • Mode: 3 Years Equal Combination

Qualifications/Requirements: GCE, A level grades: DD-BC. Maths. AGNVQ, Merit with A level. ND/C, Individual consideration. SQAH, Individual consideration. SQAV, Individual consideration. IB, Individual consideration.

Mod Information Systems/Multimedia Systems

UCAS Code: GGPN • Mode: 3 Years Equal Combination

Qualifications/Requirements: Please refer to prospectus.

Mod Information Systems/Music

UCAS Code: GWM3 • Mode: 3 Years Equal Combination

Qualifications/Requirements: GCE, A level grades: DD-BC. Music. AGNVQ, Merit. ND/C, Individual consideration. SQAH, Individual consideration. SQAV, Individual consideration. IB, Individual consideration.

Mod Information Systems/Palliative Care

UCAS Code: BGRM • Mode: 3 Years Equal Combination

Qualifications/Requirements: Please refer to prospectus.

Mod Information Systems/Physical Geography

UCAS Code: FGVM • Mode: 3 Years Equal Combination

Qualifications/Requirements: GCE, A level grades: CC-BBC. AGNVQ, Merit. ND/C, Individual consideration. SQAH, Individual consideration. SQAV, Individual consideration. IB, Individual consideration.

Mod Information Systems/Planning Studies

UCAS Code: GKM4 • Mode: 3 Years Equal Combination

Qualifications/Requirements: GCE, A level grades: DDD-BC. AGNVQ, Merit. ND/C, Individual consideration. SQAH, Individual consideration. SQAV, Individual consideration. IB, Individual consideration.

Mod Information Systems/Politics

UCAS Code: GMM1 • Mode: 3 Years Equal Combination

Qualifications/Requirements: GCE, A level grades: CDD-AB. AGNVQ, Merit with A/AS. ND/C, Individual consideration. SQAH, Individual consideration. SQAV, Individual consideration. IB, Individual consideration.

Mod Information Systems/Psychology

UCAS Code: CG8M • Mode: 3 Years Equal Combination

Qualifications/Requirements: GCE, A level grades: CCC-ABB. AGNVQ, Merit with A/AS. ND/C, Individual consideration. SQAH, Individual consideration. SQAV, Individual consideration. IB, Individual consideration.

Mod Information Systems/Publishing

UCAS Code: GPM5 • Mode: 3 Years Equal Combination

Qualifications/Requirements: GCE, A level grades: CDD-BB. AGNVQ, Merit or Merit (in specific programmes) with 3 additional units. ND/C, Individual consideration. SQAH, Individual consideration. SQAV, Individual consideration. IB, Individual consideration.

Mod Information Systems/Rehabilitation

UCAS Code: BGTM • Mode: 3 Years Equal Combination

Qualifications/Requirements: Please refer to prospectus.

Mod Information Systems/Retail Management

UCAS Code: GNM5 • Mode: 3 Years Equal Combination

Qualifications/Requirements: GCE, A level grades: DDD-BB. Science. AGNVQ, Merit with 3 additional units. ND/C, Individual consideration. SQAH, Individual consideration. SQAV, Individual consideration. IB, Individual consideration.

Mod Information Systems/Sociology

UCAS Code: GLM3 • Mode: 3 Years Equal Combination

Qualifications/Requirements: GCE, A level grades: CDD-BCC. AGNVQ, Merit with A/AS. ND/C, Individual consideration. SQAH, Individual consideration. SQAV, Individual consideration. IB, Individual consideration.

Mod Information Systems/Software Engineering

UCAS Code: GGM7 • Mode: 3 Years Equal Combination

Qualifications/Requirements: GCE, A level grades: CDD-BC. AGNVQ, Merit. ND/C, Individual consideration. SQAH, Individual consideration. SQAV, Individual consideration. IB, Individual consideration.

Mod Information Systems/Statistics

UCAS Code: GGM4 • Mode: 3 Years Equal Combination

Qualifications/Requirements: GCE, A level grades: DD-BC. AGNVQ, Merit. ND/C, Individual consideration. SQAH, Individual consideration. SQAV, Individual consideration. IB, Individual consideration.

Mod Information Systems/ Telecommunications

UCAS Code: GHMP • Mode: 3 Years Equal Combination

Qualifications/Requirements: GCE, A level grades: DD-CCC. Science or Maths. AGNVQ, Merit (in specific programmes). ND/C, Individual consideration. SQAH, Individual consideration. SQAV, Individual consideration. IB, Individual consideration.

Mod Information Systems/Tourism

UCAS Code: GPM7 • Mode: 3 Years Equal Combination

Qualifications/Requirements: GCE, A level grades: CDD-BC. AGNVQ, Merit or Merit with 3 additional units. ND/C, Individual consideration. SQAH, Individual consideration. SQAV, Individual consideration. IB, Individual consideration.

Mod Information Systems/Transport and Travel

UCAS Code: GNM9 • Mode: 3 Years Equal Combination

Qualifications/Requirements: GCE, A level grades: CC-BC. AGNVQ, Merit. ND/C, Individual consideration. SQAH, Individual consideration. SQAV, Individual consideration. IB, Individual consideration.

Mod Information Systems/Water Resources

UCAS Code: HGFM • Mode: 3 Years Equal Combination

Qualifications/Requirements: GCE, A level grades: DD-CCC. AGNVQ, Merit. ND/C, Individual consideration. SQAH, Individual consideration. SQAV, Individual consideration. IB, Individual consideration.

Mod Mapping and Cartography/ Computing

UCAS Code: FG85 • Mode: 3 Years Equal Combination

Qualifications/Requirements: GCE, A level grades: DDD-BC. AGNVQ, Merit. ND/C, Individual consideration. SQAH, Individual consideration. SQAV, Individual consideration. IB, Individual consideration.

Mod Mapping and Cartography/ Computing Science

UCAS Code: GF58 • Mode: 3 Years Equal Combination

Qualifications/Requirements: GCE, A level grades: DD-CCC. AGNVQ, Merit. ND/C, Individual consideration. SQAH, Individual consideration. SQAV, Individual consideration. IB, Individual consideration.

Mod Mapping and Cartography/ Information Systems

UCAS Code: FG8M • Mode: 3 Years Equal Combination

Qualifications/Requirements: GCE, A level grades: DDD-BC. AGNVQ, Merit. ND/C, Individual consideration. SQAH, Individual consideration. SQAV, Individual consideration. IB, Individual consideration.

UNIVERSITY OF PAISLEY P20

BSc Computing Science

UCAS Code: G500 • Mode: 3/4/5 Years Single Subject

Qualifications/Requirements: GCE, A level grades: CC. Maths. AGNVQ, Pass in Information Technology. ND/C, Individual consideration. SQAH, Grades BCCC including specific subject(s). SQAV, Higher National. IB, Pass in Diploma.

BSc Computing Science with French/ German/Spanish

UCAS Code: G5TF • Mode: 3/4/5 Years Major/Minor

Qualifications/Requirements: GCE, A level grades: CC. Maths. AGNVQ, Pass in Information Technology. ND/C, Individual consideration. SQAH, Grades BCCC including specific subject(s). SQAV, Higher National. IB, Pass in Diploma.

BSc Computing Science, Statistics and Operational Research

UCAS Code: GG54 • Mode: 3/4/5 Years Equal Combination

Qualifications/Requirements: GCE, A level grades: CC. Maths. AGNVQ, Pass in Information Technology. ND/C, Individual consideration. SQAH, Grades BCCC including specific subject(s). SQAV, Higher National. IB, Pass in Diploma.

BSc Computing Technology

UCAS Code: G600 • Mode: 3 Years Single Subject

Qualifications/Requirements: GCE, A level grades: DE. AGNVQ, Pass in Engineering. ND/C, Individual consideration. SQAH, Grades BCC including specific subject(s). SQAV, Higher National. IB, Pass in Diploma.

BSc Information Systems

UCAS Code: G520 • Mode: 3/4/5 Years Single Subject

Qualifications/Requirements: GCE, A level grades: CC. Maths. AGNVQ, Pass in Information Technology. ND/C, Individual consideration. SQAH, Grades BCCC including specific subject(s). SQAV, Higher National. IB, Pass in Diploma.

UNIVERSITY OF PLYMOUTH P60

BSc Business Information Management Systems
UCAS Code: G561 • Mode: 4 Years Single Subject

Qualifications/Requirements: GCE, A level grades: CDD-CCD. AGNVQ, Merit 12 additional units with A/AS. ND/C, Merit overall. SQAH, Individual consideration. SQAV, Individual consideration. IB, 24 points.

BSc Computing
UCAS Code: G500 • Mode: 3 Years Single Subject

Qualifications/Requirements: GCE, A/AS: 12 points. An approved subject from restricted list. AGNVQ, Distinction (in specific programmes). ND/C, Merit (in specific programmes). SQAH, Grades BBCC. SQAV, Individual consideration. IB, Individual consideration.

BSc Computing and Informatics
UCAS Code: G501 • Mode: 4 Years Single Subject

Qualifications/Requirements: GCE, A/AS: 16 points. An approved subject from restricted list. AGNVQ, Distinction (in specific programmes) with A/AS. ND/C, Distinction and Merit (in specific programmes). SQAH, Grades BBBC. SQAV, Individual consideration. IB, Individual consideration.

BEng Computer Engineering (MEng Option)
UCAS Code: G563 • Mode: 3/4/5 Years Single Subject

Qualifications/Requirements: GCE, A/AS: 18 points. An approved subject from restricted list at A level. AGNVQ, Merit in Engineering (in specific programmes). ND/C, 4 Merits (in specific programmes). SQAH, Individual consideration. SQAV, Individual consideration. IB, Individual consideration.

HND Business Information Technology
UCAS Code: 006G • Mode: 2 Years Single Subject

Qualifications/Requirements: GCE, A/AS: 2-4 points. AGNVQ, Pass. ND/C, National. SQAH, Individual consideration. SQAV, Individual consideration. IB, Individual consideration.

HND Business Information Technology (East Devon College)
UCAS Code: 165G • Mode: 2 Years Single Subject

Qualifications/Requirements: Please refer to prospectus.

HND Computing
UCAS Code: 105G • Mode: 2 Years Single Subject

Qualifications/Requirements: GCE, A/AS: 8 points. An approved subject from restricted list. AGNVQ, Merit (in specific programmes). ND/C, National (in specific programmes). SQAH, Grades CCCC. SQAV, Individual consideration. IB, Individual consideration.

UNIVERSITY OF PORTSMOUTH P80

BA Accounting and Business Information Systems (3 years or 4 years SW)
UCAS Code: NG45 • Mode: 3/4 Years Equal Combination

Qualifications/Requirements: GCE, A/AS: 18 points. AGNVQ, Distinction (in specific programmes) with 6 additional units or with A/AS. ND/C, 5 Merits and 1 Distinction. SQAH, Grades CCCCC. SQAV, Individual consideration. IB, Pass in Diploma.

BSc Business Information Systems
UCAS Code: G521 • Mode: 4 Years Single Subject

Qualifications/Requirements: GCE, A/AS: 20 points. AGNVQ, Distinction. ND/C, Merits and 2 Distinctions. SQAH, Grades BBBB. SQAV, Individual consideration. IB, 30 points.

BSc Business Information Technology
UCAS Code: G562 • Mode: 4 Years Single Subject

Qualifications/Requirements: GCE, A/AS: 20 points. AGNVQ, Distinction. ND/C, Merits and 2 Distinctions. SQAH, Grades BBBB. SQAV, Individual consideration. IB, 30 points.

BSc Computer Science
UCAS Code: G500 • Mode: 4 Years Single Subject

Qualifications/Requirements: GCE, A/AS: 20 points. AGNVQ, Distinction. ND/C, Merits and 2 Distinctions. SQAH, Grades BBBB. SQAV, Individual consideration. IB, 30 points.

BSc Computing
UCAS Code: GG57 • Mode: 3/4 Years Equal Combination

Qualifications/Requirements: GCE, A/AS: 20 points. AGNVQ, Distinction. ND/C, Merits and 2 Distinctions. SQAH, Grades BBBB. SQAV, Individual consideration. IB, 30 points.

BSc Decision Analysis and Information Technology
UCAS Code: G520 • Mode: 3 Years Single Subject

Qualifications/Requirements: GCE, A/AS: 20 points. AGNVQ, Distinction. ND/C, Merits and 2 Distinctions (in specific programmes). SQAH, Grades BBBB. SQAV, Individual consideration. IB, Individual consideration.

BSc Information Technology and Society
UCAS Code: GL53 • Mode: 4 Years Equal Combination

Qualifications/Requirements: GCE, A/AS: 20 points. AGNVQ, Distinction. ND/C, Merits and 2 Distinctions. SQAH, Grades BBBB. SQAV, Individual consideration. IB, 30 points.

BSc Technology Management with Computing
UCAS Code: N1G5 • Mode: 3/4 Years Major/Minor

Qualifications/Requirements: GCE, A/AS: 18 points. AGNVQ, Merit (in specific programmes). ND/C, 5 Merits. SQAH, Grades BBBB. SQAV, Individual consideration. IB, 30 points.

HND Computing

UCAS Code: 105G • Mode: 2 Years Single Subject

Qualifications/Requirements: GCE, A/AS: 6 points. AGNVQ, Merit. ND/C, 3 Merits. SQAH, Grades CCC. SQAV, Individual consideration. IB, Pass in Diploma.

QUEEN MARY AND WESTFIELD COLLEGE (UNIVERSITY OF LONDON) Q50

BSc Computer Science

UCAS Code: G500 • Mode: 3 Years Single Subject

Qualifications/Requirements: GCE, A level grades: BCC. AGNVQ, Distinction with A/AS. ND/C, 2 Merits and 3 Distinctions. SQAH, Grades BBBBB including specific subject(s). IB, 28 points.

BSc Computer Science with Business Studies

UCAS Code: G5N1 • Mode: 3 Years Major/Minor

Qualifications/Requirements: GCE, A level grades: BBC. AGNVQ, Distinction with A/AS. ND/C, 2 Merits and 3 Distinctions. SQAH, Grades BBBBB including specific subject(s). IB, 28 points.

BSc Mathematics and Computing

UCAS Code: GG51 • Mode: 3 Years Equal Combination

Qualifications/Requirements: Maths at A level. AGNVQ, Merit (in specific programmes) with A/AS. SQAH, Grades BBBCC including specific subject(s). IB, 30 points.

BSc Physics and Computer Science

UCAS Code: FG35 • Mode: 3 Years Equal Combination

Qualifications/Requirements: GCE, A/AS: 18 points. Maths and Physics at A level. AGNVQ, Merit in Science with A level. ND/C, Higher National (in specific programmes). SQAH, Grades BBBCC including specific subject(s). IB, 32 points.

THE QUEEN'S UNIVERSITY OF BELFAST Q75

BA Computer Science/Music

UCAS Code: GW53 • Mode: 3 Years Equal Combination

Qualifications/Requirements: GCE, A level grades: BCC. Maths, Computing or Physics at A level. AGNVQ, Distinction with A level. ND/C, Not normally sufficient. SQAH, Not normally sufficient. SQAV, Individual consideration. IB, 29 points.

BA Computer Science/Philosophy

UCAS Code: GV57 • Mode: 3 Years Equal Combination

Qualifications/Requirements: GCE, A level grades: BCC. Maths or Computing at A level. AGNVQ, Distinction with A level. ND/C, Not normally sufficient. SQAH, Not normally sufficient. SQAV, Individual consideration. IB, 29 points.

BSc Business Information Technology

UCAS Code: GN51 • Mode: 3 Years Equal Combination

Qualifications/Requirements: GCE, A level grades: BBC. AGNVQ, Individual consideration. ND/C, Individual consideration. SQAH, Individual consideration. SQAV, Individual consideration. IB, Individual consideration.

BSc Chemistry and Computer Science

UCAS Code: FG15 • Mode: 3/4 Years Equal Combination

Qualifications/Requirements: GCE, A level grades: BCC. Chemistry at A level. AGNVQ, Individual consideration. ND/C, Individual consideration. SQAH, Grades BBBC. SQAV, Individual consideration. IB, 29 points.

BSc Computer Science (Including Professional Experience) (4 years SW)

UCAS Code: G500 • Mode: 4/5 Years Single Subject

Qualifications/Requirements: GCE, A level grades: BCC. Maths, Computing, Chemistry or Physics. AGNVQ, Individual consideration. ND/C, Individual consideration. SQAH, Grades BBBC. SQAV, Individual consideration. IB, 29 points.

BSc Computer Science and Mathematics

UCAS Code: GG51 • Mode: 3/4 Years Equal Combination

Qualifications/Requirements: GCE, A level grades: BCC. Maths at A level. AGNVQ, Distinction with A level. ND/C, Individual consideration. SQAH, Grades BBBC. SQAV, Individual consideration. IB, 29 points.

BSc Computer Science and Physics

UCAS Code: GF53 • Mode: 3/4 Years Equal Combination

Qualifications/Requirements: GCE, A level grades: CCC. Physics and Maths at A level. AGNVQ, Individual consideration. ND/C, Individual consideration. SQAH, Grades BBBC. SQAV, Individual consideration. IB, 27 points.

BSc Computer Science with Business Administration

UCAS Code: G5ND • Mode: 3 Years Major/Minor

Qualifications/Requirements: GCE, A level grades: BCC. Maths, Computing, Physics or Chemistry. AGNVQ, Individual consideration. ND/C, Individual consideration. SQAH, Grades BBBC. SQAV, Individual consideration. IB, 29 points.

BSc Computer Science with Business Administration (including professional experience)

UCAS Code: GNMD • Mode: 4 Years Equal Combination

Qualifications/Requirements: GCE, A level grades: BCC. Maths, Computing, Physics or Chemistry. AGNVQ, Individual consideration. ND/C, Individual consideration. SQAH, Grades BBBC. SQAV, Individual consideration. IB, 29 points.

BSc Management and Information Systems

UCAS Code: NG15 • Mode: 3 Years Equal Combination

Qualifications/Requirements: Please refer to prospectus.

BEng Computer Science (including professional experience) (4 years SW)

UCAS Code: G506 • Mode: 4/5 Years Single Subject

Qualifications/Requirements: GCE, A level grades: BCC. Maths, Computing, Physics or Chemistry. AGNVQ, Individual consideration. ND/C, Individual consideration. SQAH, Grades BBBC. SQAV, Individual consideration. IB, 29 points.

MEng Computer Science

UCAS Code: G505 • Mode: 4 Years Single Subject

Qualifications/Requirements: GCE, A level grades: BBB. Maths, Computing, Physics or Chemistry. AGNVQ, Individual consideration. ND/C, Not normally sufficient. SQAH, Individual consideration. SQAV, Not normally sufficient. IB, 32 points.

MEng Computer Science

UCAS Code: G508 • Mode: 5 Years Single Subject

Qualifications/Requirements: GCE, A level grades: BBB. Maths, Computing, Physics or Chemistry. AGNVQ, Individual consideration. ND/C, Not normally sufficient. SQAH, Individual consideration. SQAV, Not normally sufficient. IB, 32 points.

READING COLLEGE AND SCHOOL OF ARTS & DESIGN R10

BSc Computing (Network Systems Support and Management)

UCAS Code: H640 • Mode: 1 Year Single Subject

Qualifications/Requirements: ND/C, Higher National (in specific programmes).

HND Computing

UCAS Code: 005G • Mode: 2 Years Single Subject

Qualifications/Requirements: Please refer to prospectus.

THE UNIVERSITY OF READING R12

BSc Analytical Computer Science

UCAS Code: G5G1 • Mode: 3 Years Major/Minor

Qualifications/Requirements: GCE, A/AS: 20 points. Maths at A level. AGNVQ, Distinction (in specific programmes) with A level. ND/C, 3 Merits and 2 Distinctions (in specific programmes). SQAH, Grades BBBB including specific subject(s). SQAV, Individual consideration. IB, 30 points.

BSc Applied Analytical Computer Science (4 years sandwich)

UCAS Code: G5GC • Mode: 4 Years Major/Minor

Qualifications/Requirements: GCE, A/AS: 20 points. Maths at A level. AGNVQ, Distinction (in specific programmes) with A level. ND/C, 3 Merits and 2 Distinctions (in specific programmes). SQAH, Grades BBBB including specific subject(s). SQAV, Individual consideration. IB, 30 points.

BSc Applied Computer Science

UCAS Code: G501 • Mode: 4 Years Single Subject

Qualifications/Requirements: GCE, A/AS: 20 points. AGNVQ, Distinction (in specific programmes) with 6 additional units or with A/AS. ND/C, 3 Merits and 2 Distinctions. SQAH, Grades BBBB. SQAV, Individual consideration. IB, 30 points.

BSc Computer Science

UCAS Code: G500 • Mode: 3 Years Single Subject

Qualifications/Requirements: GCE, A/AS: 20 points. AGNVQ, Distinction (in specific programmes) with 6 additional units or with A/AS. ND/C, 3 Merits and 2 Distinctions. SQAH, Grades BBBB. SQAV, Individual consideration. IB, 30 points.

BSc Computer Science and Cybernetics

UCAS Code: GH56 • Mode: 3 Years Equal Combination

Qualifications/Requirements: GCE, A/AS: 20 points. Maths or Science. AGNVQ, Distinction (in specific programmes) with 6 additional units or with A/AS. ND/C, 3 Merits and 2 Distinctions (in specific programmes). SQAH, Grades BBBB including specific subject(s). SQAV, Individual consideration. IB, 30 points.

BSc Computer Science with Philosophy

UCAS Code: G5V7 • Mode: 3 Years Major/Minor

Qualifications/Requirements: GCE, A/AS: 20 points. AGNVQ, Distinction (in specific programmes) with 6 additional units or with A/AS. ND/C, 3 Merits and 2 Distinctions. SQAH, Grades BBBB. SQAV, Individual consideration. IB, 30 points.

MEng Applied Computer Science

UCAS Code: G504 • Mode: 5 Years Single Subject

Qualifications/Requirements: GCE, A/AS: 24 points. AGNVQ, Distinction (in specific programmes) with 6 additional units or with A/AS. ND/C, 2 Merits and 3 Distinctions. SQAH, Grades ABBB. SQAV, Individual consideration. IB, 32 points.

MEng Computer Science

UCAS Code: G503 • Mode: 4 Years Single Subject

Qualifications/Requirements: GCE, A/AS: 24 points. AGNVQ, Distinction (in specific programmes) with 6 additional units or with A/AS. ND/C, 2 Merits and 3 Distinctions. SQAH, Grades ABBB. SQAV, Individual consideration. IB, 32 points.

MEng Computer Science and Cybernetics

UCAS Code: GHM6 • Mode: 4 Years Equal Combination

Qualifications/Requirements: GCE, A/AS: 24 points. Maths or Science. AGNVQ, Distinction (in specific programmes) with 6 additional units or with A/AS. ND/C, 2 Merits and 3 Distinctions (in specific programmes). SQAH, Grades ABBB including specific subject(s). SQAV, Individual consideration. IB, 32 points.

THE ROBERT GORDON UNIVERSITY R36

BSc Business Computing

UCAS Code: NG15 • Mode: 3/4 Years Equal Combination

Qualifications/Requirements: GCE, A level grades: DE. AGNVQ, Individual consideration. ND/C, National. SQAH, Grades CCC including specific subject(s). SQAV, Individual consideration. IB, Individual consideration.

BSc Computer Science

UCAS Code: G500 • Mode: 4/5 Years Single Subject

Qualifications/Requirements: GCE, A level grades: DE. Maths, Computing or Physics at A level. AGNVQ, Individual consideration. ND/C, National. SQAH, Grades CCC including specific subject(s). SQAV, Individual consideration. IB, Individual consideration.

BSc Computing

UCAS Code: G501 • Mode: 3/4 Years Single Subject

Qualifications/Requirements: GCE, A level grades: DE. AGNVQ, Individual consideration. ND/C, National. SQAH, Grades CCC including specific subject(s). SQAV, Individual consideration. IB, Individual consideration.

BSc Computing and Information

UCAS Code: G520 • Mode: 3/4 Years Single Subject

Qualifications/Requirements: GCE, A level grades: CE. AGNVQ, Individual consideration. ND/C, National. SQAH, Grades BBC including specific subject(s). SQAV, Individual consideration. IB, Individual consideration.

HND Computing: Information Technology

UCAS Code: 005G • Mode: 2 Years Single Subject

Qualifications/Requirements: GCE, A level grades: EE. AGNVQ, Individual consideration. ND/C, National. SQAH, Grades BC including specific subject(s). SQAV, Individual consideration. IB, Individual consideration.

ROYAL HOLLOWAY, UNIVERSITY OF LONDON R72

BSc Computer Science

UCAS Code: G500 • Mode: 3/4 Years Single Subject

Qualifications/Requirements: GCE, A level grades: BBB. Maths. AGNVQ, Distinction with A level. ND/C, Merit overall and 4 Distinctions. SQAH, Individual consideration. IB, 30 points.

BSc Computer Science and Mathematics

UCAS Code: GG51 • Mode: 3/4 Years Equal Combination

Qualifications/Requirements: GCE, A level grades: BBC-BBB. Maths at A level. AGNVQ, Distinction with A level. ND/C, Merit overall and 4 Distinctions. SQAH, Individual consideration. IB, Individual consideration.

BSc Computer Science and Physics

UCAS Code: GF53 • Mode: 3/4 Years Equal Combination

Qualifications/Requirements: GCE, A level grades: BBC-BBB. Maths and Physics. AGNVQ, Distinction with A level. ND/C, Individual consideration. SQAH, Grades BBBCC including specific subject(s). IB, 30 points.

BSc Computer Science with Artificial Intelligence

UCAS Code: G5G8 • Mode: 3 Years Major/Minor

Qualifications/Requirements: GCE, A level grades: BBB.

BSc Computer Science with Communications

UCAS Code: G560 • Mode: 3 Years Single Subject

Qualifications/Requirements: GCE, A level grades: BBB.

BSc Computer Science with Computer Architecture & Design

UCAS Code: G5G6 • Mode: 3 Years Major/Minor

Qualifications/Requirements: GCE, A level grades: BBB.

BSc Computer Science with French

UCAS Code: G5R1 • Mode: 3/4 Years Major/Minor

Qualifications/Requirements: GCE, A level grades: BBB. Maths and French at A level. AGNVQ, Distinction with A level. ND/C, Merit overall and 4 Distinctions. SQAH, Individual consideration. IB, 30 points.

BSc Computer Science with Management

UCAS Code: G5N1 • Mode: 3/4 Years Major/Minor

Qualifications/Requirements: GCE, A level grades: BBB. Maths. AGNVQ, Distinction with A level. ND/C, Merit overall and 4 Distinctions. SQAH, Individual consideration. IB, 30 points.

BSc Computer Science with Safety Critical Systems

UCAS Code: G590 • Mode: 3 Years Single Subject

Qualifications/Requirements: GCE, A level grades: BBB.

BSc Geology and Computing

UCAS Code: FG65 • Mode: 3 Years Equal Combination

Qualifications/Requirements: GCE, A level grades: BBC. Maths and Science. AGNVQ, Distinction with A level. ND/C, Individual consideration. SQAH, Individual consideration. IB, Individual consideration.

BSc Management and Information Systems

UCAS Code: NG15 • Mode: 3 Years Equal Combination

Qualifications/Requirements: GCE, A level grades: BBB. Maths. AGNVQ, Distinction with A/AS. ND/C, Individual consideration. SQAH, Individual consideration. IB, 30 points.

THE UNIVERSITY OF SALFORD S03

BSc Applied Computing and Mathematical Modelling with preliminary year

UCAS Code: G518 • Mode: 4 Years Single Subject

Qualifications/Requirements: GCE, A/AS: 16 points. Maths at A level. AGNVQ, Individual consideration. ND/C, Individual consideration. SQAH, Individual consideration. SQAV, Individual consideration. IB, Individual consideration.

BSc Biology and Information Technology

UCAS Code: GC51 • Mode: 3/4 Years Equal Combination

Qualifications/Requirements: GCE, A level grades: BCC-CCD. Biology or Italian at A level. AGNVQ, Individual consideration. ND/C, Individual consideration. SQAH, Individual consideration. SQAV, Individual consideration. IB, Individual consideration.

BSc Business Information Systems

UCAS Code: G520 • Mode: 3/4 Years Single Subject

Qualifications/Requirements: GCE, A/AS: 18 points. AGNVQ, Individual consideration. ND/C, Individual consideration. SQAH, Individual consideration. SQAV, Individual consideration. IB, Individual consideration.

BSc Chemistry and Information Technology

UCAS Code: FG15 • Mode: 3/4 Years Equal Combination

Qualifications/Requirements: GCE, A level grades: BCC-CCD. Chemistry or Italian at A level. AGNVQ, Individual consideration. ND/C, Individual consideration. SQAH, Individual consideration. SQAV, Individual consideration. IB, Individual consideration.

BSc Computer and Video Games

UCAS Code: G570 • Mode: 4 Years Single Subject

Qualifications/Requirements: AGNVQ, Individual consideration. ND/C, Individual consideration. SQAH, Individual consideration. SQAV, Individual consideration. IB, Individual consideration.

BSc Computer Science

UCAS Code: G500 • Mode: 3/4 Years Single Subject

Qualifications/Requirements: GCE, A/AS: 18 points. AGNVQ, Merit or Distinction. ND/C, Merits and Distinction. SQAH, Individual consideration. SQAV, Individual consideration. IB, Individual consideration.

BSc Computer Science and Applied Mathematics

UCAS Code: GG51 • Mode: 3/4 Years Equal Combination

Qualifications/Requirements: GCE, A/AS: 16 points. Maths at A level. AGNVQ, Individual consideration. ND/C, Individual consideration. SQAH, Individual consideration. SQAV, Individual consideration. IB, Individual consideration.

BSc Computer Science and Arabic

UCAS Code: GT56 • Mode: 4 Years Equal Combination

Qualifications/Requirements: AGNVQ, Individual consideration. ND/C, Individual consideration. SQAH, Individual consideration. SQAV, Individual consideration. IB, Individual consideration.

BSc Computer Science and French

UCAS Code: GR51 • Mode: 4 Years Equal Combination

Qualifications/Requirements: AGNVQ, Individual consideration. ND/C, Individual consideration. SQAH, Individual consideration. SQAV, Individual consideration. IB, Individual consideration.

BSc Computer Science and German

UCAS Code: GR52 • Mode: 4 Years Equal Combination

Qualifications/Requirements: AGNVQ, Individual consideration. ND/C, Individual consideration. SQAH, Individual consideration. SQAV, Individual consideration. IB, Individual consideration.

BSc Computer Science and Information Systems

UCAS Code: G505 • Mode: 4/5 Years Single Subject

Qualifications/Requirements: GCE, A/AS: 8 points. AGNVQ, Merit (in specific programmes). ND/C, National. SQAH, Individual consideration. SQAV, Individual consideration. IB, Individual consideration.

BSc Computer Science and Information Systems

UCAS Code: G506 • Mode: 3/4 Years Single Subject

Qualifications/Requirements: GCE, A/AS: 18 points. AGNVQ, Merit or Distinction. ND/C, Merits and Distinction. SQAH, Individual consideration. SQAV, Individual consideration. IB, Individual consideration.

BSc Computer Science and Italian

UCAS Code: GR53 • Mode: 4 Years Equal Combination

Qualifications/Requirements: AGNVQ, Individual consideration. ND/C, Individual consideration. SQAH, Individual consideration. SQAV, Individual consideration. IB, Individual consideration.

BSc Computer Science and Spanish

UCAS Code: GR54 • Mode: 4 Years Equal Combination

Qualifications/Requirements: AGNVQ, Individual consideration. ND/C, Individual consideration. SQAH, Individual consideration. SQAV, Individual consideration. IB, Individual consideration.

BSc Economics and Information Technology

UCAS Code: LG15 • Mode: 3/4 Years Equal Combination

Qualifications/Requirements: GCE, A level grades: BCC-CCD. Economics or Italian at A level. AGNVQ, Individual consideration. ND/C, Individual consideration. SQAH, Individual consideration. SQAV, Individual consideration. IB, Individual consideration.

BSc Geography and Information Technology

UCAS Code: GF58 • Mode: 3/4 Years Equal Combination

Qualifications/Requirements: GCE, A level grades: BCC-CCD. Geography or Italian at A level. AGNVQ, Individual consideration. ND/C, Individual consideration. SQAH, Individual consideration. SQAV, Individual consideration. IB, Individual consideration.

BSc Information Technology

UCAS Code: G5N1 • Mode: 3 Years Major/Minor

Qualifications/Requirements: GCE, A/AS: 18-20 points. AGNVQ, Distinction. ND/C, Merits and Distinction. SQAH, Individual consideration. SQAV, Individual consideration. IB, Individual consideration.

BSc Information Technology with English for Professional Purposes

UCAS Code: G5Q3 • Mode: 3 Years Major/Minor

Qualifications/Requirements: GCE, A/AS: 18-20 points. AGNVQ, Individual consideration. ND/C, Merits and Distinction. SQAH, Individual consideration. SQAV, Individual consideration. IB, Individual consideration.

BSc Information Technology with Language Training in French

UCAS Code: G5R1 • Mode: 3 Years Major/Minor

Qualifications/Requirements: GCE, A/AS: 18-20 points. AGNVQ, Distinction. ND/C, Merits and Distinction. SQAH, Individual consideration. SQAV, Individual consideration. IB, Individual consideration.

BSc Information Technology with Language Training in German

UCAS Code: G5R2 • Mode: 3 Years Major/Minor

Qualifications/Requirements: GCE, A/AS: 18-20 points. AGNVQ, Distinction. ND/C, Merits and Distinction. SQAH, Individual consideration. SQAV, Individual consideration. IB, Individual consideration.

BSc Information Technology with Studies in Japan

UCAS Code: G5T4 • Mode: 4 Years Major/Minor

Qualifications/Requirements: GCE, A/AS: 18-20 points. AGNVQ, Distinction. ND/C, Merits and Distinction. SQAH, Individual consideration. SQAV, Individual consideration. IB, Individual consideration.

BSc Management Science and Information Systems

UCAS Code: NG15 • Mode: 3/4 Years Equal Combination

Qualifications/Requirements: Please refer to prospectus.

BSc Management Science and Information Systems with Studies in North America

UCAS Code: NG1M • Mode: 3/4 Years Equal Combination

Qualifications/Requirements: Please refer to prospectus.

BSc Physics and Information Technology

UCAS Code: FG35 • Mode: 3/4 Years Equal Combination

Qualifications/Requirements: GCE, A level grades: BCC-CDD. Physics or Italian at A level. AGNVQ, Individual consideration. ND/C, Individual consideration. SQAH, Individual consideration. SQAV, Individual consideration. IB, Individual consideration.

BSc Physiology and Information Technology

UCAS Code: CG95 • Mode: 3/4 Years Equal Combination

Qualifications/Requirements: GCE, A level grades: BCC-CCD. Italian. AGNVQ, Individual consideration. ND/C, Individual consideration. SQAH, Individual consideration. SQAV, Individual consideration. IB, Individual consideration.

HND Business Studies/Computing in Business

UCAS Code: 51GN • Mode: 2 Years Equal Combination

Qualifications/Requirements: GCE, A/AS: 6 points. AGNVQ, Merit. ND/C, Individual consideration. SQAH, Individual consideration. SQAV, Individual consideration. IB, Individual consideration.

UNIVERSITY COLLEGE SCARBOROUGH S10

BA Business Management & Information Technology

UCAS Code: NG15 • Mode: 3 Years Equal Combination

Qualifications/Requirements: Please refer to prospectus.

BA Information & Communications Technology (with QTS)

UCAS Code: XG5C • Mode: 3 Years Equal Combination

Qualifications/Requirements: GCE, A/AS: 10 points. ND/C, 3 Merits. SQAH, Grades CCC. IB, 27 points.

THE UNIVERSITY OF SHEFFIELD S18

BSc Computer Science and French

UCAS Code: GR51 • Mode: 4 Years Equal Combination

Qualifications/Requirements: GCE, A/AS: 24 points. French and Maths at A level. AGNVQ, Distinction with A level. ND/C, 3 Merits and 3 Distinctions (in specific programmes). SQAH, Grades AABB including specific subject(s). SQAV, Individual consideration. IB, 32 points.

BSc Computer Science and German

UCAS Code: GR52 • Mode: 4 Years Equal Combination

Qualifications/Requirements: GCE, A/AS: 24 points. German and Maths at A level. AGNVQ, Distinction with A level. ND/C, 3 Merits and 3 Distinctions (in specific programmes). SQAH, Grades AABB including specific subject(s). SQAV, Individual consideration. IB, 32 points.

BSc Computer Science and Hispanic Studies

UCAS Code: GR54 • Mode: 4 Years Equal Combination

Qualifications/Requirements: GCE, A/AS: 24 points. Maths and a modern foreign language at A level. AGNVQ, Distinction with A level. ND/C, 3 Merits and 3 Distinctions (in specific programmes). SQAH, Grades AABB including specific subject(s). SQAV, Individual consideration. IB, 32 points.

BSc Computer Science and Mathematics (3/4 years)

UCAS Code: GG51 • Mode: 3 Years Equal Combination

Qualifications/Requirements: GCE, A/AS: 24 points. Maths at A level. AGNVQ, Distinction with A level. ND/C, 3 Merits and 3 Distinctions (in specific programmes). SQAH, Grades AABB including specific subject(s). SQAV, Individual consideration. IB, 32 points.

BSc Computer Science and Russian

UCAS Code: GR58 • Mode: 4 Years Equal Combination

Qualifications/Requirements: GCE, A/AS: 24 points. Maths and a modern foreign language at A level. AGNVQ, Distinction with A level. ND/C, 3 Merits and 3 Distinctions (in specific programmes). SQAH, Grades AABB including specific subject(s). SQAV, Individual consideration. IB, 32 points.

SHEFFIELD HALLAM UNIVERSITY S21

BA Accounting and Information Systems

UCAS Code: N4G5 • Mode: 3 Years Major/Minor

Qualifications/Requirements: GCE, A/AS: 14 points. AGNVQ, Individual consideration. ND/C, Merit overall and 3 Distinctions (in specific programmes). SQAH, Individual consideration. SQAV, Individual consideration. IB, Individual consideration.

BSc Business Information Systems

UCAS Code: G521 • Mode: 4 Years Single Subject

Qualifications/Requirements: GCE, A/AS: 12 points. AGNVQ, Merit. ND/C, 6 Merits. SQAH, Individual consideration. SQAV, Individual consideration. IB, Individual consideration.

BSc Computing & Management Science (Foundation Year)

UCAS Code: GN5D • Mode: 5 Years Equal Combination

Qualifications/Requirements: AGNVQ, Individual consideration. ND/C, Individual consideration. SQAH, Individual consideration. SQAV, Individual consideration. IB, Individual consideration.

BSc Computing (Visualisation)

UCAS Code: G5W2 • Mode: 3 Years Major/Minor

Qualifications/Requirements: GCE, A/AS: 12 points. AGNVQ, Merit. ND/C, Merits and Distinction. SQAH, Individual consideration. SQAV, Individual consideration. IB, Individual consideration.

BSc Computing and Management Sciences

UCAS Code: GN51 • Mode: 4 Years Equal Combination

Qualifications/Requirements: GCE, A/AS: 12 points. Maths at A level. AGNVQ, Merit. ND/C, 6 Merits. SQAH, Individual consideration. SQAV, Individual consideration. IB, Individual consideration.

BSc Mathematics and Technology

UCAS Code: GJ19 • Mode: 3/4 Years Equal Combination

Qualifications/Requirements: GCE, A/AS: 14 points. Maths, Statistics or Physics at A level. AGNVQ, Pass (in specific programmes). ND/C, National (in specific programmes). SQAH, Individual consideration. SQAV, Individual consideration. IB, Individual consideration.

HND Business Information Technology

UCAS Code: 165G • Mode: 3 Years Single Subject

Qualifications/Requirements: GCE, A/AS: 6 points. AGNVQ, Pass. ND/C, 2 Merits. SQAH, Individual consideration. SQAV, Individual consideration. IB, Individual consideration.

HND Computing

UCAS Code: 105G • Mode: 2/3 Years Single Subject

Qualifications/Requirements: GCE, A/AS: 6 points. AGNVQ, Pass. ND/C, 2 Merits. SQAH, Individual consideration. SQAV, Individual consideration. IB, Individual consideration.

SHEFFIELD COLLEGE S22

HND Computing

UCAS Code: 065G • Mode: 2 Years Single Subject

Qualifications/Requirements: AGNVQ, Individual consideration. ND/C, 3 Merits and 2 Distinctions. SQAH, Individual consideration. SQAV, Individual consideration. IB, Individual consideration.

SHREWSBURY COLLEGE OF ARTS AND TECHNOLOGY S23

HND Business Information Technology

UCAS Code: 265G • Mode: 2 Years Single Subject

Qualifications/Requirements: GCE, A/AS: 2 points. AGNVQ, Pass. SQAH, Individual consideration. SQAV, Individual consideration. IB, Individual consideration.

ST MARTIN'S COLLEGE, LANCASTER: AMBLESIDE: CARLISLE S24

BA Information & Communications Technology/Education (2+2, 5-11 years)

UCAS Code: X5G5 • Mode: 4 Years Major/Minor

Qualifications/Requirements: GCE, A level grades: CD-CEE. Italian at A level. AGNVQ, Merit in Information Technology. ND/C, Not normally sufficient. SQAH, Grades BCCC including specific subject(s). SQAV, Individual consideration. IB, 28 points.

BA Information & Communications Technology/Education (2+2, 7-14 years)

UCAS Code: X6GM • Mode: 4 Years Major/Minor

Qualifications/Requirements: GCE, A level grades: CD-CEE. Italian at A level. AGNVQ, Merit in Information Technology. ND/C, Not normally sufficient. SQAH, Grades BCCC including specific subject(s). SQAV, Individual consideration. IB, 28 points.

BA Information & Communications Technology/Education (nlp) (4 years)

UCAS Code: X2G5 • Mode: 4 Years Major/Minor

Qualifications/Requirements: GCE, A level grades: CD-CEE. Italian at A level. AGNVQ, Merit in Information Technology. ND/C, Not normally sufficient. SQAH, Grades BCCC including specific subject(s). SQAV, Individual consideration. IB, 28 points.

BA Information & Communications Technology/Education (upr)

UCAS Code: X4G5 • Mode: 4 Years Major/Minor

Qualifications/Requirements: GCE, A level grades: CD-CEE. Italian at A level. AGNVQ, Merit in Information Technology. ND/C, Not normally sufficient. SQAH, Grades BCCC including specific subject(s). SQAV, Individual consideration. IB, 28 points.

SOLIHULL COLLEGE S26

HND Computing

UCAS Code: 005G • Mode: 2 Years Single Subject

Qualifications/Requirements: GCE, A level grades: E. AGNVQ, Pass. ND/C, National. SQAH, Individual consideration. SQAV, Individual consideration. IB, Pass in Diploma.

UNIVERSITY OF SOUTHAMPTON S27

BSc Computer Science

UCAS Code: G500 • Mode: 3 Years Single Subject

Qualifications/Requirements: GCE, A/AS: 24 points. Maths at A level. AGNVQ, Distinction with A level. ND/C, Individual consideration. SQAH, CSYS required. SQAV, Individual consideration. IB, 32 points.

BSc Computer Science with Artificial Intelligence

UCAS Code: G5G8 • Mode: 3 Years Major/Minor

Qualifications/Requirements: GCE, A/AS: 24 points. Maths at A level. AGNVQ, Distinction with A level. ND/C, Individual consideration. SQAH, CSYS required. SQAV, Individual consideration. IB, 32 points.

BSc Computer Science with Distributed Systems and Network

UCAS Code: G5G6 • Mode: 3 Years Major/Minor

Qualifications/Requirements: GCE, A/AS: 24 points. Maths at A level. AGNVQ, Distinction with A level. ND/C, Individual consideration. SQAH, CSYS required. SQAV, Individual consideration. IB, 32 points.

BSc Computer Science with Image and Multimedia Systems

UCAS Code: G5GP • Mode: 3 Years Major/Minor

Qualifications/Requirements: GCE, A/AS: 24 points. Maths at A level. AGNVQ, Distinction with A level. ND/C, Individual consideration. SQAH, CSYS required. SQAV, Individual consideration. IB, 32 points.

BSc Computer Science with Parallel Computation

UCAS Code: G5GQ • Mode: 3 Years Major/Minor

Qualifications/Requirements: GCE, A/AS: 24 points. Maths at A level. AGNVQ, Distinction with A level. ND/C, Individual consideration. SQAH, CSYS required. SQAV, Individual consideration. IB, 32 points.

BSc Computer Science with Systems Integration

UCAS Code: G520 • Mode: 3 Years Single Subject

Qualifications/Requirements: GCE, A/AS: 24 points. Maths at A level. AGNVQ, Distinction with A level. ND/C, Individual consideration. SQAH, CSYS required. SQAV, Individual consideration. IB, 32 points.

BSc Mathematics and Information Technology

UCAS Code: GG15 • Mode: 3 Years Equal Combination

Qualifications/Requirements: GCE, A level grades: BBC. Maths at A level. AGNVQ, Individual consideration. ND/C, Individual consideration. SQAH, Grades ABBBB. SQAV, Individual consideration. IB, 30 points.

SOMERSET COLLEGE OF ARTS AND TECHNOLOGY S28

HND Computing

UCAS Code: 005G • Mode: 2 Years Single Subject

Qualifications/Requirements: GCE, A/AS: 2 points. AGNVQ, Merit. ND/C, National. SQAH, Individual consideration. SQAV, Individual consideration. IB, Individual consideration.

SOUTHAMPTON INSTITUTE S30

BSc Business Information Technology

UCAS Code: G562 • Mode: 4 Years Single Subject

Qualifications/Requirements: GCE, A/AS: 8 points. AGNVQ, Merit (in specific programmes). ND/C, Merit overall. SQAH, Grades CCCC. SQAV, National. IB, Pass in Diploma.

BSc Business Information Technology

UCAS Code: G563 • Mode: 5 Years Single Subject

Qualifications/Requirements: GCE, A/AS: 2 points. AGNVQ, Pass (in specific programmes). ND/C, National. SQAH, Grades CCCC. SQAV, National. IB, Pass in Diploma.

BSc Business Information Technology (with foundation)

UCAS Code: G568 • Mode: 5 Years Single Subject

Qualifications/Requirements: GCE, A/AS: 2-4 points. AGNVQ, Pass (in specific programmes). ND/C, National. SQAH, Grades CCCC. SQAV, National. IB, Pass in Diploma.

BSc Computer Studies

UCAS Code: G501 • Mode: 3 Years Single Subject

Qualifications/Requirements: GCE, A/AS: 8 points. AGNVQ, Merit (in specific programmes). ND/C, Merit overall. SQAH, Grades CCCC. SQAV, National. IB, Pass in Diploma.

BSc Computer Studies
UCAS Code: G503 • Mode: 4 Years Single Subject

Qualifications/Requirements: GCE, A/AS: 2 points. AGNVQ, Pass (in specific programmes). ND/C, National. SQAH, Grades CCCC. SQAV, National. IB, Pass in Diploma.

BSc Computer Studies (with foundation)
UCAS Code: G502 • Mode: 4 Years Single Subject

Qualifications/Requirements: GCE, A/AS: 2-4 points. AGNVQ, Pass (in specific programmes). ND/C, National. SQAH, Grades CCCC. SQAV, National. IB, Pass in Diploma.

HND Computing
UCAS Code: 105G • Mode: 2 Years Single Subject

Qualifications/Requirements: GCE, A/AS: 2 points. AGNVQ, Pass (in specific programmes). ND/C, National. SQAH, Grades CCCC. SQAV, National. IB, Pass in Diploma.

SOUTH DEVON COLLEGE S32

HND Computing
UCAS Code: 005G • Mode: 2 Years Single Subject

Qualifications/Requirements: GCE, A/AS: 4 points. AGNVQ, Merit. ND/C, Merit overall.

SOUTH BANK UNIVERSITY S33

BA/BSc Computing and Forensic Science
UCAS Code: BGC5 • Mode: 3 Years Equal Combination

Qualifications/Requirements: GCE, A/AS: 12-14 points. Maths and Science. AGNVQ, Merit. ND/C, Merit overall. SQAH, Individual consideration. SQAV, Individual consideration. IB, Individual consideration.

BA/BSc Computing and Music
UCAS Code: WG35 • Mode: 3 Years Equal Combination

Qualifications/Requirements: GCE, A/AS: 12-16 points. Maths and Music. AGNVQ, Merit. ND/C, Merit overall. SQAH, Individual consideration. SQAV, Individual consideration. IB, Individual consideration.

BSc Computing Studies
UCAS Code: G501 • Mode: 3/4 Years Single Subject

Qualifications/Requirements: GCE, A level grades: CD. Two Sciences at A level. AGNVQ, Merit. ND/C, 5 Merits (in specific programmes). SQAH, Individual consideration. SQAV, Individual consideration. IB, Individual consideration.

BSc Computing Studies
UCAS Code: G505 • Mode: 1/2 Years Single Subject

Qualifications/Requirements: ND/C, Higher National. SQAH, Individual consideration. SQAV, Individual consideration. IB, Individual consideration.

BSc Foundation Computing Studies
UCAS Code: G508 • Mode: 4 Years Single Subject

Qualifications/Requirements: GCE, A level grades: E. AGNVQ, Pass. ND/C, National. SQAH, Individual consideration. SQAV, Individual consideration. IB, Individual consideration.

Mod Accounting and Computing
UCAS Code: GN54 • Mode: 3 Years Equal Combination

Qualifications/Requirements: GCE, A/AS: 12-16 points. Accounting, Economics or Maths at A level. AGNVQ, Merit. ND/C, 4 Merits and 2 Distinctions. SQAH, Individual consideration. SQAV, Individual consideration. IB, Individual consideration.

Mod Computing and French
UCAS Code: GR51 • Mode: 3 Years Equal Combination

Qualifications/Requirements: GCE, A/AS: 12-16 points. Maths and French at A level. AGNVQ, Merit. ND/C, 4 Merits and 2 Distinctions. SQAH, Individual consideration. SQAV, Individual consideration. IB, Individual consideration.

Mod Computing and German
UCAS Code: GR5F • Mode: 3 Years Equal Combination

Qualifications/Requirements: GCE, A/AS: 14-18 points. Maths and German at A level. AGNVQ, Merit. ND/C, 2 Merits and 4 Distinctions. SQAH, Individual consideration. SQAV, Individual consideration. IB, Individual consideration.

Mod Computing and German ab initio
UCAS Code: GR52 • Mode: 3 Years Equal Combination

Qualifications/Requirements: GCE, A/AS: 12-16 points. Maths at A level. AGNVQ, Merit. ND/C, 4 Merits and 2 Distinctions. SQAH, Individual consideration. SQAV, Individual consideration. IB, Individual consideration.

Mod Computing and Health Studies
UCAS Code: GL54 • Mode: 3 Years Equal Combination

Qualifications/Requirements: GCE, A/AS: 12-16 points. Maths and Science at A level. AGNVQ, Merit. ND/C, 4 Merits and 2 Distinctions. SQAH, Individual consideration. SQAV, Individual consideration. IB, Individual consideration.

Mod Computing and Human Geography
UCAS Code: GL58 • Mode: 3 Years Equal Combination

Qualifications/Requirements: GCE, A/AS: 12-16 points. Maths and Geography at A level. AGNVQ, Merit. ND/C, 4 Merits and 2 Distinctions. SQAH, Individual consideration. SQAV, Individual consideration. IB, Individual consideration.

Mod Computing and Law
UCAS Code: GM53 • Mode: 3 Years Equal Combination

Qualifications/Requirements: GCE, A/AS: 14-18 points. Maths at A level. AGNVQ, Merit. ND/C, 2 Merits and 4 Distinctions. SQAH, Individual consideration. SQAV, Individual consideration. IB, Individual consideration.

Mod Computing and Marketing

UCAS Code: GN55 • Mode: 3 Years Equal Combination

Qualifications/Requirements: GCE, A/AS: 12 points. Maths at A level. AGNVQ, Merit. ND/C, 4 Merits and 2 Distinctions. SQAH, Individual consideration. SQAV, Individual consideration. IB, Individual consideration.

Mod Computing and Media Studies

UCAS Code: GP54 • Mode: 3 Years Equal Combination

Qualifications/Requirements: GCE, A/AS: 14-18 points. Maths and English at A level. AGNVQ, Distinction. ND/C, 2 Merits and 4 Distinctions. SQAH, Individual consideration. SQAV, Individual consideration. IB, Individual consideration.

Mod Computing and Product Design

UCAS Code: GH57 • Mode: 3 Years Equal Combination

Qualifications/Requirements: GCE, A/AS: 12-16 points. Maths and Art & Design at A level. AGNVQ, Merit. ND/C, 4 Merits and 2 Distinctions. SQAH, Individual consideration. SQAV, Individual consideration. IB, Individual consideration.

Mod Computing and Psychology

UCAS Code: CG85 • Mode: 3 Years Equal Combination

Qualifications/Requirements: GCE, A/AS: 14-18 points. Science and Maths at A level. AGNVQ, Merit. ND/C, 4 Merits and 2 Distinctions. SQAH, Individual consideration. SQAV, Individual consideration. IB, Individual consideration.

Mod Computing and Spanish

UCAS Code: GR5K • Mode: 3 Years Equal Combination

Qualifications/Requirements: GCE, A/AS: 14-18 points. Spanish and Maths at A level. AGNVQ, Merit. ND/C, 2 Merits and 4 Distinctions. SQAH, Individual consideration. SQAV, Individual consideration. IB, Individual consideration.

Mod Computing and Spanish ab initio

UCAS Code: GR54 • Mode: 3 Years Equal Combination

Qualifications/Requirements: GCE, A/AS: 12-16 points. Maths at A level. AGNVQ, Merit. ND/C, 4 Merits and 2 Distinctions. SQAH, Individual consideration. SQAV, Individual consideration. IB, Individual consideration.

Mod Computing and Sports Science

UCAS Code: BG65 • Mode: 3 Years Equal Combination

Qualifications/Requirements: GCE, A/AS: 12-16 points. Science and Maths at A level. AGNVQ, Merit. ND/C, 4 Merits and 2 Distinctions. SQAH, Individual consideration. SQAV, Individual consideration. IB, Individual consideration.

Mod Computing and Technology

UCAS Code: GJ59 • Mode: 3 Years Equal Combination

Qualifications/Requirements: GCE, A/AS: 12-16 points. Maths at A level. AGNVQ, Merit. ND/C, 4 Merits and 2 Distinctions. SQAH, Individual consideration. SQAV, Individual consideration. IB, Individual consideration.

Mod Computing and World Theatre

UCAS Code: GW54 • Mode: 3 Years Equal Combination

Qualifications/Requirements: GCE, A/AS: 14-18 points. Maths at A level. AGNVQ, Merit. ND/C, 2 Merits and 4 Distinctions. SQAH, Individual consideration. SQAV, Individual consideration. IB, Individual consideration.

Mod Economics and Computing

UCAS Code: GL51 • Mode: 3 Years Equal Combination

Qualifications/Requirements: GCE, A/AS: 12-16 points. Maths and Economics or Business Studies at A level. AGNVQ, Merit. ND/C, 4 Merits and 2 Distinctions. SQAH, Individual consideration. SQAV, Individual consideration. IB, Individual consideration.

Mod English Studies and Business Information Technology

UCAS Code: GQ73 • Mode: 3 Years Equal Combination

Qualifications/Requirements: GCE, A/AS: 14-18 points. Maths and English at A level. AGNVQ, Merit with A level. ND/C, Not normally sufficient. SQAH, Individual consideration. SQAV, Individual consideration. IB, Individual consideration.

HND Computing Studies

UCAS Code: 105G • Mode: 2 Years Single Subject

Qualifications/Requirements: GCE, A level grades: E. Science at A level. AGNVQ, Pass. ND/C, 3 Merits. SQAH, Individual consideration. SQAV, Individual consideration. IB, Individual consideration.

SOUTHPORT COLLEGE S35

HND Computing

UCAS Code: 005G • Mode: 2 Years Single Subject

Qualifications/Requirements: GCE, A/AS: 2 points. AGNVQ, Merit. ND/C, Merit overall.

UNIVERSITY OF ST ANDREWS S36

BSc Chemistry/Computer Science

UCAS Code: FG15 • Mode: 4 Years Equal Combination

Qualifications/Requirements: GCE, A level grades: CCC. Maths and Chemistry at A level. AGNVQ, Individual consideration. ND/C, Individual consideration. SQAH, Grades BBCC including specific subject(s). SQAV, Individual consideration. IB, 28 points.

BSc Computer Science

UCAS Code: G500 • Mode: 4 Years Single Subject

Qualifications/Requirements: GCE, A level grades: BCC. Maths at A level. AGNVQ, Individual consideration. ND/C, Individual consideration. SQAH, Grades BBBC including specific subject(s). SQAV, Individual consideration. IB, 28 points.

BSc Computer Science with French

UCAS Code: G5R1 • Mode: 4 Years Major/Minor

Qualifications/Requirements: GCE, A level grades: BBC. SQAH, Grades BBBC including specific subject(s). IB, 28 points.

BSc Computer Science with French (Integrated Year Abroad)

UCAS Code: G5RC • Mode: 4 Years Major/Minor

Qualifications/Requirements: GCE, A level grades: BBC. SQAH, Grades BBBC including specific subject(s). IB, 28 points.

BSc Computer Science with German

UCAS Code: G5R2 • Mode: 4 Years Major/Minor

Qualifications/Requirements: GCE, A level grades: BBC. SQAH, Grades BBBC including specific subject(s). IB, 28 points.

BSc Computer Science with German (Integrated Year Abroad)

UCAS Code: G5RF • Mode: 4 Years Major/Minor

Qualifications/Requirements: GCE, A level grades: BBC. SQAH, Grades BBB including specific subject(s). IB, 28 points.

BSc Computer Science/Geoscience

UCAS Code: FGP5 • Mode: 4 Years Equal Combination

Qualifications/Requirements: GCE, A level grades: BBC. SQAH, Grades BBBC including specific subject(s). IB, 28 points.

BSc Computer Science – Logic & Philosophy of Science

UCAS Code: GV5R • Mode: 3/4 Years Equal Combination

Qualifications/Requirements: GCE, A level grades: BBB. Maths at A level. AGNVQ, Individual consideration. ND/C, Individual consideration. SQAH, Grades BBCC including specific subject(s). SQAV, Individual consideration. IB, 28 points.

BSc Computer Science/Management

UCAS Code: GN5C • Mode: 4 Years Equal Combination

Qualifications/Requirements: GCE, A level grades: CCC. Maths at A level. AGNVQ, Individual consideration. ND/C, Not normally sufficient. SQAH, Grades BBBC including specific subject(s). SQAV, Individual consideration. IB, 28 points.

BSc Computer Science/Management Science

UCAS Code: GN51 • Mode: 4 Years Equal Combination

Qualifications/Requirements: GCE, A level grades: CCC. Maths at A level. AGNVQ, Individual consideration. ND/C, Individual consideration. SQAH, Grades BBBC including specific subject(s). SQAV, Individual consideration. IB, 28 points.

BSc Computer Science/Physics

UCAS Code: GF53 • Mode: 4 Years Equal Combination

Qualifications/Requirements: GCE, A level grades: BCC. Maths and Physics at A level. AGNVQ, Individual consideration. ND/C, Individual consideration. SQAH, Grades BBCC including specific subject(s). SQAV, Individual consideration. IB, 28 points.

SOUTHWARK COLLEGE S38

HND Computer Studies

UCAS Code: 005G • Mode: 2 Years Single Subject

Qualifications/Requirements: GCE, A level grades: E. Computing or Italian and Maths. AGNVQ, Merit in Information Technology. ND/C, National. SQAH, Individual consideration. SQAV, Individual consideration. IB, Individual consideration.

ST HELENS COLLEGE S51

HND Applied Computer Technology

UCAS Code: 015G • Mode: 2 Years Single Subject

Qualifications/Requirements: GCE, A/AS: 2-6 points. Maths, Physics, Computing or Italian at A level. AGNVQ, Pass in Engineering. ND/C, National. SQAH, Individual consideration. SQAV, Individual consideration. IB, Individual consideration.

HND Computer Security

UCAS Code: 095G • Mode: 2 Years Single Subject

Qualifications/Requirements: GCE, A/AS: 2 points. AGNVQ, Pass. ND/C, National. SQAH, Individual consideration. SQAV, Individual consideration. IB, Individual consideration.

HND Computing

UCAS Code: 005G • Mode: 2 Years Single Subject

Qualifications/Requirements: GCE, A/AS: 2 points. AGNVQ, Merit. ND/C, 3 Merits. SQAH, Individual consideration. SQAV, Individual consideration. IB, Individual consideration.

THE COLLEGE OF ST MARK AND ST JOHN S59

BA English (Literary Studies)/Information Technology

UCAS Code: Q3G5 • Mode: 3 Years Major/Minor

Qualifications/Requirements: GCE, A/AS: 14-16 points. English Literature. AGNVQ, Merit. ND/C, Individual consideration. SQAH, Individual consideration. SQAV, Individual consideration. IB, Individual consideration.

BA English Language Studies/Information Technology

UCAS Code: Q1G5 • Mode: 3 Years Major/Minor

Qualifications/Requirements: GCE, A/AS: 12 points. AGNVQ, Merit. ND/C, Merit overall. SQAH, Individual consideration. SQAV, Individual consideration. IB, Individual consideration.

BA Geography/Information Technology
UCAS Code: L8G5 • Mode: 3 Years Major/Minor

Qualifications/Requirements: GCE, A/AS: 8-10 points. Geography. AGNVQ, Merit. ND/C, Merit overall. SQAH, Individual consideration. SQAV, Individual consideration. IB, Individual consideration.

BA Information Technology/Education Studies
UCAS Code: G5X9 • Mode: 3 Years Major/Minor

Qualifications/Requirements: GCE, A/AS: 4 points. AGNVQ, Merit. ND/C, Merit overall. SQAH, Individual consideration. SQAV, Individual consideration. IB, Pass in Diploma.

BA Information Technology/English (Literary Studies)
UCAS Code: G5Q3 • Mode: 3 Years Major/Minor

Qualifications/Requirements: GCE, A/AS: 4 points. AGNVQ, Merit. ND/C, Merit overall. SQAH, Individual consideration. SQAV, Individual consideration. IB, Pass in Diploma.

BA Information Technology/English Language Studies
UCAS Code: G5Q1 • Mode: 3 Years Major/Minor

Qualifications/Requirements: GCE, A/AS: 4 points. AGNVQ, Merit. ND/C, Merit overall. SQAH, Individual consideration. SQAV, Individual consideration. IB, Pass in Diploma.

BA Information Technology/Geography
UCAS Code: G5L8 • Mode: 3 Years Major/Minor

Qualifications/Requirements: GCE, A/AS: 4 points. AGNVQ, Merit. ND/C, Merit overall. SQAH, Individual consideration. SQAV, Individual consideration. IB, Pass in Diploma.

BA Information Technology/Media Studies
UCAS Code: G5P4 • Mode: 3 Years Major/Minor

Qualifications/Requirements: GCE, A/AS: 4 points. AGNVQ, Merit. ND/C, Merit overall. SQAH, Individual consideration. SQAV, Individual consideration. IB, Pass in Diploma.

BA Information Technology/Public Relations
UCAS Code: G5P3 • Mode: 3 Years Major/Minor

Qualifications/Requirements: GCE, A/AS: 4 points. AGNVQ, Merit. ND/C, Merit overall. SQAH, Individual consideration. SQAV, Individual consideration. IB, Pass in Diploma.

BA Information Technology/Sports Science & Coaching
UCAS Code: G5B6 • Mode: 3 Years Major/Minor

Qualifications/Requirements: GCE, A/AS: 4 points. AGNVQ, Merit. ND/C, Merit overall. SQAH, Individual consideration. SQAV, Individual consideration. IB, Pass in Diploma.

BA Public Relations/Information Technology
UCAS Code: P3GN • Mode: 3 Years Major/Minor

Qualifications/Requirements: GCE, A/AS: 16 points. AGNVQ, Merit. ND/C, Merit overall. SQAH, Individual consideration. SQAV, Individual consideration. IB, Individual consideration.

BA Sports Science & Coaching/Information Technology
UCAS Code: B6G5 • Mode: 3 Years Major/Minor

Qualifications/Requirements: GCE, A/AS: 12-14 points. AGNVQ, Merit. ND/C, Merit overall. SQAH, Individual consideration. SQAV, Individual consideration. IB, Individual consideration.

BEd Information Technology/Primary Education
UCAS Code: X5G5 • Mode: 4 Years Major/Minor

Qualifications/Requirements: Please refer to prospectus.

STAFFORDSHIRE UNIVERSITY S72

BA Accounting Information Technology (3/4 years SW)
UCAS Code: N4G5 • Mode: 3/4 Years Major/Minor

Qualifications/Requirements: GCE, A/AS: 12 points. AGNVQ, Merit in Business or Merit in Information Technology. ND/C, Individual consideration. SQAH, Grades CCC. SQAV, Individual consideration. IB, 27 points.

BA Interactive Multimedia
UCAS Code: GP54 • Mode: 4 Years Equal Combination

Qualifications/Requirements: GCE, A/AS: 12 points. AGNVQ, Merit. ND/C, Individual consideration. SQAH, Grades CCC. SQAV, Individual consideration. IB, Individual consideration.

BSc Business Information Technology
UCAS Code: G562 • Mode: 4 Years Single Subject

Qualifications/Requirements: GCE, A/AS: 12 points. AGNVQ, Merit. ND/C, Individual consideration. SQAH, Grades CCC. SQAV, Individual consideration. IB, 27 points.

BSc Computer Graphics, Imaging and Visualisation
UCAS Code: GW5F • Mode: 4 Years Equal Combination

Qualifications/Requirements: GCE, A/AS: 12 points. AGNVQ, Merit. ND/C, Individual consideration. SQAH, Grades CCC. IB, 27 points.

BSc Computer Science
UCAS Code: G501 • Mode: 4 Years Single Subject

Qualifications/Requirements: GCE, A/AS: 12 points. AGNVQ, Merit. ND/C, Individual consideration. SQAH, Grades CCC. SQAV, Individual consideration. IB, 27 points.

BSc Computing and Electronics for Information Technology
UCAS Code: GH5Q • Mode: 3 Years Equal Combination

Qualifications/Requirements: GCE, A/AS: 12 points. AGNVQ, Merit. ND/C, 3 Merits (in specific programmes). SQAH, Grades CCC. SQAV, Individual consideration. IB, 24 points.

BSc Computing and Multimedia
UCAS Code: PG4M • Mode: 3 Years Equal Combination

Qualifications/Requirements: GCE, A/AS: 12 points. AGNVQ, Merit. ND/C, Individual consideration. SQAH, Grades CCC. IB, 27 points.

BSc Computing Science
UCAS Code: G500 • Mode: 4 Years Single Subject

Qualifications/Requirements: GCE, A/AS: 12 points. AGNVQ, Merit. ND/C, Individual consideration. SQAH, Grades CCC. SQAV, Individual consideration. IB, 27 points.

BSc Foundation Business Information Technology
UCAS Code: G561 • Mode: 5 Years Single Subject

Qualifications/Requirements: Please refer to prospectus.

BSc Foundation Computer Science
UCAS Code: G508 • Mode: 5 Years Single Subject

Qualifications/Requirements: Please refer to prospectus.

BSc Foundation Computer Systems
UCAS Code: G608 • Mode: 5 Years Single Subject

Qualifications/Requirements: Please refer to prospectus.

BSc Foundation Computing Science
UCAS Code: G509 • Mode: 5 Years Single Subject

Qualifications/Requirements: Please refer to prospectus.

BSc Foundation Computing with Applicable Mathematics
UCAS Code: G5GC • Mode: 5 Years Major/Minor

Qualifications/Requirements: Please refer to prospectus.

BSc Foundation Geology and Computing
UCAS Code: FG6M • Mode: 4 Years Equal Combination

Qualifications/Requirements: GCE, A/AS: 4 points. AGNVQ, Pass. ND/C, National. SQAH, Grades CCC. IB, 24 points.

BSc Foundation Information Systems
UCAS Code: G528 • Mode: 5 Years Single Subject

Qualifications/Requirements: Please refer to prospectus.

BSc Foundation Physics and Computing
UCAS Code: FG3N • Mode: 4 Years Equal Combination

Qualifications/Requirements: GCE, A/AS: 4 points. AGNVQ, Pass. ND/C, National. SQAH, Grades CCC. SQAV, Individual consideration. IB, 24 points.

BSc Foundation Technology Management
UCAS Code: GN51 • Mode: 5 Years Equal Combination

Qualifications/Requirements: Please refer to prospectus.

BSc Information Systems
UCAS Code: G521 • Mode: 4 Years Single Subject

Qualifications/Requirements: GCE, A/AS: 12 points. AGNVQ, Merit. ND/C, Individual consideration. SQAH, Grades CCC. SQAV, Individual consideration. IB, 27 points.

BSc Information Systems and Internet Commerce
UCAS Code: G5NC • Mode: 3 Years Major/Minor

Qualifications/Requirements: GCE, A/AS: 12 points. AGNVQ, Merit. ND/C, Individual consideration. SQAH, Grades CCC. IB, 27 points.

BSc Interactive Entertainment Technology
UCAS Code: GG57 • Mode: 3 Years Equal Combination

Qualifications/Requirements: GCE, A/AS: 12 points. AGNVQ, Merit. ND/C, 3 Merits. SQAH, Grades CCC. SQAV, Individual consideration. IB, 24 points.

BSc Management and Business Computing
UCAS Code: N111 • Mode: 3 Years Single Subject

Qualifications/Requirements: GCE, A/AS: 12 points. AGNVQ, Merit (in specific programmes). ND/C, Merit overall. SQAH, Grades BBC. SQAV, Individual consideration. IB, 24 points.

BSc Sport Sciences and Information Systems
UCAS Code: BG65 • Mode: 3 Years Equal Combination

Qualifications/Requirements: GCE, A/AS: 14 points. Science at A level. AGNVQ, Distinction. ND/C, 3 Merits and 3 Distinctions. SQAH, Grades BBC. SQAV, Individual consideration. IB, 28 points.

BEng Foundation Computer Science
UCAS Code: G507 • Mode: 5 Years Single Subject

Qualifications/Requirements: Please refer to prospectus.

BEng Foundation Computer Systems
UCAS Code: G605 • Mode: 5 Years Single Subject

Qualifications/Requirements: Please refer to prospectus.

BEng Foundation Computing Science
UCAS Code: G518 • Mode: 5 Years Single Subject

Qualifications/Requirements: Please refer to prospectus.

BEng Foundation Computing with Applicable Mathematics
UCAS Code: G5GD • Mode: 5 Years Major/Minor

Qualifications/Requirements: Please refer to prospectus.

BEng Foundation Information Systems
UCAS Code: G569 • Mode: 5 Years Single Subject

Qualifications/Requirements: Please refer to prospectus.

MEng Computer Science
UCAS Code: G502 • Mode: 4/5 Years Single Subject
Qualifications/Requirements: GCE, A/AS: 12 points. AGNVQ, Merit. ND/C, Individual consideration. SQAH, Grades CCC. SQAV, Individual consideration. IB, 27 points.

MEng Computing Science
UCAS Code: G503 • Mode: 4/5 Years Single Subject
Qualifications/Requirements: GCE, A/AS: 12 points. AGNVQ, Merit. ND/C, Individual consideration. SQAH, Grades CCC. SQAV, Individual consideration. IB, 27 points.

MEng Computing with Applicable Mathematics
UCAS Code: G5G1 • Mode: 4/5 Years Major/Minor
Qualifications/Requirements: GCE, A/AS: 12 points. AGNVQ, Merit. ND/C, Individual consideration. SQAH, Grades CCC. IB, 27 points.

MEng Information Systems
UCAS Code: G520 • Mode: 5 Years Single Subject
Qualifications/Requirements: GCE, A/AS: 12 points. AGNVQ, Merit. ND/C, Individual consideration. SQAH, Grades CCC. SQAV, Individual consideration. IB, 27 points.

Mod Biology/Computing
UCAS Code: CG15 • Mode: 3 Years Equal Combination
Qualifications/Requirements: GCE, A/AS: 12 points. Biology at A level. AGNVQ, Merit. ND/C, 4 Merits. SQAH, Grades BBC. SQAV, Individual consideration. IB, 26 points.

Mod Business Studies/Information Systems
UCAS Code: GN5C • Mode: 3 Years Equal Combination
Qualifications/Requirements: GCE, A/AS: 16 points. AGNVQ, Merit. ND/C, Individual consideration. SQAH, Grades BBB. SQAV, Individual consideration. IB, 24 points.

Mod Computing/Development Studies
UCAS Code: GMMY • Mode: 3 Years Equal Combination
Qualifications/Requirements: GCE, A/AS: 12 points. AGNVQ, Merit. ND/C, 3 Merits. SQAH, Grades BBB. SQAV, Individual consideration. IB, Individual consideration.

Mod Computing/Geography
UCAS Code: GL58 • Mode: 3 Years Equal Combination
Qualifications/Requirements: GCE, A/AS: 12 points. AGNVQ, Merit. ND/C, 3 Merits and 1 Distinction. SQAH, Grades BBC. SQAV, Individual consideration. IB, 26 points.

Mod Computing/Geology
UCAS Code: GF56 • Mode: 3 Years Equal Combination
Qualifications/Requirements: GCE, A/AS: 12 points. Science, Geography or Maths at A level. AGNVQ, Merit. ND/C, 4 Merits. SQAH, Grades BBC. SQAV, Individual consideration. IB, 26 points.

Mod Computing/Information Systems
UCAS Code: G529 • Mode: 3 Years Single Subject
Qualifications/Requirements: GCE, A/AS: 12 points. AGNVQ, Merit. ND/C, Individual consideration. SQAH, Grades CCC. SQAV, Individual consideration. IB, 27 points.

Mod Computing/Physics
UCAS Code: FG35 • Mode: 3 Years Equal Combination
Qualifications/Requirements: GCE, A/AS: 12 points. Physics at A level. AGNVQ, Merit. ND/C, 4 Merits. SQAH, Grades BBC. SQAV, Individual consideration. IB, 26 points.

Mod Computing/Psychology
UCAS Code: LG75 • Mode: 3 Years Equal Combination
Qualifications/Requirements: GCE, A/AS: 18 points. AGNVQ, Individual consideration. ND/C, 3 Merits and 3 Distinctions. SQAH, Grades BBB. SQAV, Individual consideration. IB, 27 points.

Mod Computing/Sociology
UCAS Code: LG35 • Mode: 3 Years Equal Combination
Qualifications/Requirements: GCE, A/AS: 12 points. AGNVQ, Merit. ND/C, 3 Merits. SQAH, Grades BBB. SQAV, Individual consideration. IB, 24 points.

Mod Environmental Studies/Information Systems
UCAS Code: GF59 • Mode: 3 Years Equal Combination
Qualifications/Requirements: GCE, A/AS: 12 points. AGNVQ, Merit. ND/C, Individual consideration. SQAH, Grades BBC. SQAV, Individual consideration. IB, 24 points.

Mod Geography/Information Systems
UCAS Code: GL5V • Mode: 3 Years Equal Combination
Qualifications/Requirements: GCE, A level grades: CC. AGNVQ, Merit. ND/C, 3 Merits and 1 Distinction. SQAH, Grades BBC. SQAV, Individual consideration. IB, 24 points.

Mod Information Systems/International Relations
UCAS Code: MG1M • Mode: 3 Years Equal Combination
Qualifications/Requirements: GCE, A/AS: 12 points. AGNVQ, Merit. ND/C, 3 Merits. SQAH, Grades BCC. SQAV, Individual consideration. IB, 24 points.

Mod Information Systems/Law
UCAS Code: MG35 • Mode: 3 Years Equal Combination
Qualifications/Requirements: GCE, A/AS: 14 points. AGNVQ, Merit with A/AS. ND/C, Higher National and 3 Distinctions. SQAH, Grades BBB. SQAV, Individual consideration. IB, 26 points.

Mod Information Systems/Legal Studies
UCAS Code: MG3M • Mode: 3 Years Equal Combination
Qualifications/Requirements: GCE, A/AS: 14 points. AGNVQ, Merit with A/AS. ND/C, 3 Distinctions. SQAH, Grades BBB. SQAV, Individual consideration. IB, 26 points.

Mod Information Systems/Politics
UCAS Code: MGC5 • Mode: 3 Years Equal Combination

Qualifications/Requirements: GCE, A/AS: 12 points. AGNVQ, Merit. ND/C, 3 Merits. SQAH, Grades BCC. SQAV, Individual consideration. IB, 24 points.

Mod Information Systems/Sociology
UCAS Code: LG3M • Mode: 3 Years Equal Combination

Qualifications/Requirements: GCE, A/AS: 12 points. AGNVQ, Merit. ND/C, 3 Merits. SQAH, Grades BBB. SQAV, Individual consideration. IB, 24 points.

HND Business Information Systems
UCAS Code: 225G • Mode: 2 Years Single Subject

Qualifications/Requirements: GCE, A/AS: 4 points. AGNVQ, Pass. ND/C, National. SQAH, Grades CC. SQAV, Individual consideration. IB, 24 points.

HND Business Information Technology
UCAS Code: 265G • Mode: 2 Years Single Subject

Qualifications/Requirements: GCE, A level grades: D. AGNVQ, Pass. ND/C, National. SQAH, Grades CC. SQAV, Individual consideration. IB, 24 points.

HND Computer Science
UCAS Code: 005G • Mode: 2 Years Single Subject

Qualifications/Requirements: GCE, A/AS: 4 points. AGNVQ, Pass. ND/C, National. SQAH, Grades CC. SQAV, Individual consideration. IB, Individual consideration.

HND Computing
UCAS Code: 105G • Mode: 2 Years Single Subject

Qualifications/Requirements: GCE, A/AS: 4 points. AGNVQ, Pass. ND/C, National. SQAH, Grades CC. SQAV, Individual consideration. IB, 24 points.

HND Information Systems
UCAS Code: 125G • Mode: 2 Years Single Subject

Qualifications/Requirements: GCE, A/AS: 4 points. AGNVQ, Pass. ND/C, National. SQAH, Grades CC. SQAV, Individual consideration. IB, Individual consideration.

THE UNIVERSITY OF STIRLING S75

BA Computing Science/Philosophy
UCAS Code: GV57 • Mode: 4 Years Equal Combination

Qualifications/Requirements: GCE, A level grades: CCD. AGNVQ, Merit (in specific programmes) with 2 additional units or with A/AS. ND/C, Merit overall. SQAH, Grades BBCC. SQAV, Higher National. IB, 28 points.

BA Computing Science/Sociology
UCAS Code: GL53 • Mode: 4 Years Equal Combination

Qualifications/Requirements: GCE, A level grades: CCD. AGNVQ, Merit (in specific programmes) with 2 additional units or with A/AS. ND/C, Merit overall. SQAH, Grades BBCC. SQAV, Higher National. IB, 28 points.

BA Finance/Computing Science
UCAS Code: NG35 • Mode: 4 Years Equal Combination

Qualifications/Requirements: GCE, A level grades: CCD. AGNVQ, Merit (in specific programmes) with 2 additional units or with A/AS. ND/C, Merit overall. SQAH, Grades BBCC. SQAV, Higher National. IB, 28 points.

BAcc Accountancy/Computing Science
UCAS Code: GN54 • Mode: 4 Years Equal Combination

Qualifications/Requirements: GCE, A level grades: BCC. AGNVQ, Distinction (in specific programmes) with 6 additional units or with A/AS. ND/C, Merit overall and 1 Distinction. SQAH, Grades BBBB. SQAV, Higher National. IB, 33 points.

BSc Biology/Computing Science
UCAS Code: CG15 • Mode: 4 Years Equal Combination

Qualifications/Requirements: GCE, A level grades: CCD. Science or Maths at A level. AGNVQ, Merit in Science with 3 additional units or with A/AS. ND/C, Merit overall and 2 Distinctions. SQAH, Grades BBCC including specific subject(s). SQAV, Higher National. IB, 28 points.

BSc Business Studies/Computing Science
UCAS Code: NG15 • Mode: 4 Years Equal Combination

Qualifications/Requirements: GCE, A level grades: CCC. AGNVQ, Merit (in specific programmes) with 2 additional units or with A/AS. ND/C, Merit overall and 1 Distinction. SQAH, Grades BBCC. SQAV, Higher National. IB, 28 points.

BSc Computing Science
UCAS Code: G500 • Mode: 4 Years Single Subject

Qualifications/Requirements: GCE, A level grades: CCD. AGNVQ, Merit (in specific programmes) with 2 additional units or with A/AS. ND/C, Merit overall. SQAH, Grades BBCC. SQAV, Higher National. IB, 28 points.

BSc Computing Science/Economics
UCAS Code: GL51 • Mode: 4 Years Equal Combination

Qualifications/Requirements: GCE, A level grades: CCD. AGNVQ, Merit (in specific programmes) with 2 additional units or with A/AS. ND/C, Merit overall. SQAH, Grades BBCC. SQAV, Higher National. IB, 28 points.

BSc Computing Science/Entrepreneurship
UCAS Code: GN5C • Mode: 4 Years Equal Combination

Qualifications/Requirements: GCE, A level grades: CCD. AGNVQ, Merit (in specific programmes) with 2 additional units or with A/AS. ND/C, Merit overall. SQAH, Grades BBCC. SQAV, Higher National. IB, 28 points.

BSc Computing Science/Film & Media Studies
UCAS Code: GW55 • Mode: 4 Years Equal Combination

Qualifications/Requirements: GCE, A level grades: BBC. AGNVQ, Distinction (in specific programmes) with 6 additional units or with A/AS. ND/C, Merit overall and 2 Distinctions. SQAH, Grades ABBB. SQAV, Higher National. IB, 35 points.

BSc Computing Science/French Language

UCAS Code: GR51 • Mode: 4 Years Equal Combination

Qualifications/Requirements: GCE, A level grades: CCD. AGNVQ, Merit (in specific programmes) with 2 additional units or with A/AS. ND/C, Merit overall. SQAH, Grades BBCC. SQAV, Higher National. IB, 28 points.

BSc Computing Science/German Language

UCAS Code: GR52 • Mode: 4 Years Equal Combination

Qualifications/Requirements: GCE, A level grades: CCD. AGNVQ, Merit (in specific programmes) with 2 additional units or with A/AS. ND/C, Merit overall. SQAH, Grades BBCC. SQAV, Higher National. IB, 28 points.

BSc Computing Science/Japanese Studies

UCAS Code: GT54 • Mode: 4 Years Equal Combination

Qualifications/Requirements: GCE, A level grades: CCD. AGNVQ, Merit (in specific programmes) with 2 additional units or with A/AS. ND/C, Merit overall. SQAH, Grades BBCC. SQAV, Higher National. IB, 28 points.

BSc Computing Science/Management Science

UCAS Code: GN51 • Mode: 4 Years Equal Combination

Qualifications/Requirements: GCE, A level grades: CCD. AGNVQ, Merit (in specific programmes) with 2 additional units or with A/AS. ND/C, Merit overall. SQAH, Grades BBCC. SQAV, Higher National. IB, 28 points.

BSc Computing Science/Marketing

UCAS Code: GN55 • Mode: 4 Years Equal Combination

Qualifications/Requirements: GCE, A level grades: CCD. AGNVQ, Merit (in specific programmes) with 2 additional units or with A/AS. ND/C, Merit overall. SQAH, Grades BBCC. SQAV, Higher National. IB, 28 points.

BSc Computing Science/Mathematics

UCAS Code: G5G1 • Mode: 4 Years Major/Minor

Qualifications/Requirements: GCE, A level grades: CCD. Maths at A level. AGNVQ, Merit (in specific programmes) with 2 additional units or with A/AS. ND/C, Merit overall. SQAH, Grades BBCC including specific subject(s). SQAV, Higher National. IB, 28 points.

BSc Computing Science/Psychology

UCAS Code: CG85 • Mode: 4 Years Equal Combination

Qualifications/Requirements: GCE, A level grades: CCD. AGNVQ, Merit (in specific programmes) with 2 additional units or with A/AS. ND/C, Merit overall. SQAH, Grades BBCC. SQAV, Higher National. IB, 28 points.

BSc Computing Science/Spanish Language

UCAS Code: GR54 • Mode: 4 Years Equal Combination

Qualifications/Requirements: GCE, A level grades: CCD. AGNVQ, Merit (in specific programmes) with 2 additional units or with A/AS. ND/C, Merit overall. SQAH, Grades BBCC. SQAV, Higher National. IB, 28 points.

BSc Education/Computing Science

UCAS Code: GX57 • Mode: 4/5 Years Equal Combination

Qualifications/Requirements: GCE, A level grades: CCD. AGNVQ, Merit (in specific programmes) with 6 additional units or with A/AS. ND/C, Merit overall. SQAH, Grades BBCC including specific subject(s). SQAV, Higher National. IB, 28 points.

STOCKPORT COLLEGE OF FURTHER & HIGHER EDUCATION S76

HND Business Information Technology

UCAS Code: 065G • Mode: 2 Years Single Subject

Qualifications/Requirements: English and Maths. AGNVQ, Pass in Business or in Information Technology. ND/C, National. SQAH, Individual consideration. SQAV, Individual consideration. IB, Individual consideration.

THE UNIVERSITY OF STRATHCLYDE S78

BSc Business Information Systems

UCAS Code: GN59 • Mode: 4 Years Equal Combination

Qualifications/Requirements: GCE, A level grades: BCC. Maths at A level. AGNVQ, Pass. ND/C, Higher National. SQAH, Grades BBBB including specific subject(s). SQAV, Higher National. IB, 28 points.

BSc Computer Science

UCAS Code: G500 • Mode: 4 Years Single Subject

Qualifications/Requirements: GCE, A level grades: BCC. Maths at A level. AGNVQ, Pass. ND/C, Individual consideration. SQAH, Grades BBBC including specific subject(s). SQAV, Individual consideration. IB, 28 points.

BSc Computer Science with Law

UCAS Code: G5M3 • Mode: 4 Years Major/Minor

Qualifications/Requirements: GCE, A level grades: BBB. Maths and English at A level. AGNVQ, Pass. ND/C, Individual consideration. SQAH, Grades AABB including specific subject(s). SQAV, Individual consideration. IB, 30 points.

BSc Mathematics and Computer Science

UCAS Code: GG15 • Mode: 4 Years Equal Combination

Qualifications/Requirements: GCE, A level grades: CD. Maths at A level. AGNVQ, Pass. ND/C, Individual consideration. SQAH, Grades BBBC including specific subject(s). SQAV, Individual consideration. IB, 32 points.

STRANMILLIS UNIVERSITY COLLEGE: A COLLEGE OF THE QUEEN'S UNIVERSITY OF BELFAST S79

BEd Information Technology with Education

UCAS Code: XG55 • Mode: 3 Years Equal Combination

Qualifications/Requirements: Please refer to prospectus.

SUFFOLK COLLEGE: AN ACCREDITED COLLEGE OF THE UNIVERSITY OF EAST ANGLIA S81

BA Fine & Applied Arts and Information Technology

UCAS Code: EG15 • Mode: 3 Years Equal Combination

Qualifications/Requirements: GCE, A level grades: CE. OTHER. AGNVQ, Pass (in specific programmes). ND/C, National (in specific programmes). SQAH, Individual consideration. SQAV, Individual consideration. IB, Individual consideration.

BA Fine & Applied Arts and Information Technology

UCAS Code: WG15 • Mode: 3 Years Equal Combination

Qualifications/Requirements: GCE, A level grades: CE. OTHER. AGNVQ, Pass (in specific programmes). ND/C, National (in specific programmes). SQAH, Individual consideration. SQAV, Individual consideration. IB, Individual consideration.

BA Graphic Design and Information Technology

UCAS Code: EG2N • Mode: 3 Years Equal Combination

Qualifications/Requirements: GCE, A level grades: CE. OTHER. AGNVQ, Pass (in specific programmes). ND/C, National (in specific programmes). SQAH, Individual consideration. SQAV, Individual consideration. IB, Individual consideration.

BA Graphic Design and Information Technology

UCAS Code: WG2N • Mode: 3 Years Equal Combination

Qualifications/Requirements: GCE, A level grades: CE. OTHER. AGNVQ, Pass (in specific programmes). ND/C, National (in specific programmes). SQAH, Individual consideration. SQAV, Individual consideration. IB, Individual consideration.

BA Illustration and Information Technology

UCAS Code: EG2M • Mode: 3 Years Equal Combination

Qualifications/Requirements: GCE, A level grades: CE. OTHER. AGNVQ, Pass (in specific programmes). ND/C, National (in specific programmes). SQAH, Individual consideration. SQAV, Individual consideration. IB, Individual consideration.

BA Illustration and Information Technology

UCAS Code: WG2M • Mode: 3 Years Equal Combination

Qualifications/Requirements: GCE, A level grades: CE. OTHER. AGNVQ, Pass (in specific programmes). ND/C, National (in specific programmes). SQAH, Individual consideration. SQAV, Individual consideration. IB, Individual consideration.

BA Literary Studies and Information Technology

UCAS Code: QG25 • Mode: 3 Years Equal Combination

Qualifications/Requirements: GCE, A level grades: CE. English. AGNVQ, Pass (in specific programmes). ND/C, National (in specific programmes). SQAH, Individual consideration. SQAV, Individual consideration. IB, Individual consideration.

BA Model Design and Information Technology

UCAS Code: GW5F • Mode: 3 Years Equal Combination

Qualifications/Requirements: GCE, A level grades: CE. OTHER. AGNVQ, Pass (in specific programmes). ND/C, National (in specific programmes). SQAH, Individual consideration. SQAV, Individual consideration. IB, Individual consideration.

BA Spatial Design and Information Technology

UCAS Code: GW5G • Mode: 3 Years Equal Combination

Qualifications/Requirements: GCE, A level grades: CE. OTHER. AGNVQ, Pass (in specific programmes). ND/C, National (in specific programmes). SQAH, Individual consideration. SQAV, Individual consideration. IB, Individual consideration.

BSc Animal Science & Welfare and Information Technology

UCAS Code: DGG5 • Mode: 3 Years Equal Combination

Qualifications/Requirements: GCE, A level grades: EE. Science. AGNVQ, Merit in Science or in Environmental Studies. ND/C, 3 Merits (in specific programmes). SQAH, Individual consideration. SQAV, Individual consideration. IB, Individual consideration.

BSc Applied Biology and Information Technology

UCAS Code: CG15 • Mode: 3 Years Equal Combination

Qualifications/Requirements: GCE, A level grades: EE. Science. AGNVQ, Pass in Science. ND/C, 3 Merits (in specific programmes). SQAH, Individual consideration. SQAV, Individual consideration. IB, Individual consideration.

BSc Behavioural Studies and Information Technology

UCAS Code: GL57 • Mode: 3 Years Equal Combination

Qualifications/Requirements: GCE, A level grades: DE. AGNVQ, Merit. ND/C, Merit overall. SQAH, Individual consideration. SQAV, Individual consideration. IB, Individual consideration.

BSc Business Information Technology

UCAS Code: G561 • Mode: 3 Years Single Subject

Qualifications/Requirements: GCE, A level grades: EE. AGNVQ, Pass. ND/C, National (in specific programmes). SQAH, Individual consideration. SQAV, Individual consideration. IB, Individual consideration.

BSc Environmental Studies and Information Technology

UCAS Code: FG95 • Mode: 3 Years Equal Combination

Qualifications/Requirements: GCE, A level grades: EE. Science or Geography. AGNVQ, Pass in Science or in Environmental Studies. ND/C, 3 Merits (in specific programmes). SQAH, Individual consideration. SQAV, Individual consideration. IB, Individual consideration.

BSc Food Production and Information Technology

UCAS Code: DG45 • Mode: 3 Years Equal Combination

Qualifications/Requirements: GCE, A level grades: EE. Science. AGNVQ, Merit in Environmental Studies or in Science. ND/C, 3 Merits (in specific programmes). SQAH, Individual consideration. SQAV, Individual consideration. IB, Individual consideration.

BSc Food Production with Information Technology

UCAS Code: D4G5 • Mode: 3 Years Major/Minor

Qualifications/Requirements: GCE, A level grades: EE. Science. AGNVQ, Merit in Environmental Studies or in Science. ND/C, 3 Merits (in specific programmes). SQAH, Individual consideration. SQAV, Individual consideration. IB, Individual consideration.

BSc Information Technology and Business Management

UCAS Code: GN51 • Mode: 3 Years Equal Combination

Qualifications/Requirements: GCE, A level grades: EE. AGNVQ, Pass (in specific programmes). ND/C, National (in specific programmes). SQAH, Individual consideration. SQAV, Individual consideration. IB, Individual consideration.

BSc Information Technology and Landscape and Garden Design

UCAS Code: GK53 • Mode: 3 Years Equal Combination

Qualifications/Requirements: GCE, A level grades: EE. Science or Geography. AGNVQ, Merit (in specific programmes). ND/C, Merit overall. SQAH, Individual consideration. SQAV, Individual consideration. IB, Individual consideration.

BSc Information Technology and Management

UCAS Code: GN5C • Mode: 3 Years Equal Combination

Qualifications/Requirements: GCE, A level grades: EE. AGNVQ, Pass (in specific programmes). ND/C, National (in specific programmes). SQAH, Individual consideration. SQAV, Individual consideration. IB, Individual consideration.

BSc Information Technology with Animal Science & Conservation

UCAS Code: G5DF • Mode: 3 Years Major/Minor

Qualifications/Requirements: GCE, A level grades: EE. AGNVQ, Pass (in specific programmes). ND/C, National (in specific programmes). SQAH, Individual consideration. SQAV, Individual consideration. IB, Individual consideration.

BSc Information Technology with Animal Science and Welfare

UCAS Code: G5DG • Mode: 3 Years Major/Minor

Qualifications/Requirements: GCE, A level grades: EE. AGNVQ, Pass (in specific programmes). ND/C, National (in specific programmes). SQAH, Individual consideration. SQAV, Individual consideration. IB, Individual consideration.

BSc Information Technology with Applied Biology

UCAS Code: G5C1 • Mode: 3 Years Major/Minor

Qualifications/Requirements: GCE, A level grades: EE. AGNVQ, Pass (in specific programmes). ND/C, National (in specific programmes). SQAH, Individual consideration. SQAV, Individual consideration. IB, Individual consideration.

BSc Information Technology with Behavioural Studies

UCAS Code: G5L7 • Mode: 3 Years Major/Minor

Qualifications/Requirements: GCE, A level grades: EE. AGNVQ, Pass (in specific programmes). ND/C, National (in specific programmes). SQAH, Individual consideration. SQAV, Individual consideration. IB, Individual consideration.

BSc Information Technology with Business Management

UCAS Code: G5NC • Mode: 3 Years Major/Minor

Qualifications/Requirements: GCE, A level grades: EE. AGNVQ, Pass (in specific programmes). ND/C, National (in specific programmes). SQAH, Individual consideration. SQAV, Individual consideration. IB, Individual consideration.

BSc Information Technology with Early Childhood Studies

UCAS Code: G5XX • Mode: 3 Years Major/Minor

Qualifications/Requirements: GCE, A level grades: EE. AGNVQ, Pass (in specific programmes). ND/C, National (in specific programmes). SQAH, Individual consideration. SQAV, Individual consideration. IB, Individual consideration.

BSc Information Technology with Environmental Studies

UCAS Code: G5F9 • Mode: 3 Years Major/Minor

Qualifications/Requirements: GCE, A level grades: EE. AGNVQ, Pass (in specific programmes). ND/C, National (in specific programmes). SQAH, Individual consideration. SQAV, Individual consideration. IB, Individual consideration.

BSc Information Technology with Fine & Applied Arts

UCAS Code: G5W1 • Mode: 3 Years Major/Minor

Qualifications/Requirements: GCE, A level grades: EE. AGNVQ, Pass (in specific programmes). ND/C, National (in specific programmes). SQAH, Individual consideration. SQAV, Individual consideration. IB, Individual consideration.

BSc Information Technology with Food Production

UCAS Code: G5D4 • Mode: 3 Years Major/Minor

Qualifications/Requirements: GCE, A level grades: EE. AGNVQ, Pass (in specific programmes). ND/C, National (in specific programmes). SQAH, Individual consideration. SQAV, Individual consideration. IB, Individual consideration.

BSc Information Technology with Garden Design

UCAS Code: G5K3 • Mode: 3 Years Major/Minor

Qualifications/Requirements: GCE, A level grades: EE. AGNVQ, Pass (in specific programmes). ND/C, National (in specific programmes). SQAH, Individual consideration. SQAV, Individual consideration. IB, Individual consideration.

BSc Information Technology with Graphic Design

UCAS Code: G5WF • Mode: 3 Years Major/Minor

Qualifications/Requirements: GCE, A level grades: EE. AGNVQ, Pass (in specific programmes). ND/C, National (in specific programmes). SQAH, Individual consideration. SQAV, Individual consideration. IB, Individual consideration.

BSc Information Technology with Human Biology

UCAS Code: G5B1 • Mode: 3 Years Major/Minor

Qualifications/Requirements: GCE, A level grades: EE. AGNVQ, Pass (in specific programmes). ND/C, National (in specific programmes). SQAH, Individual consideration. SQAV, Individual consideration. IB, Individual consideration.

BSc Information Technology with Illustration

UCAS Code: G5WG • Mode: 3 Years Major/Minor

Qualifications/Requirements: GCE, A level grades: EE. AGNVQ, Pass (in specific programmes). ND/C, National (in specific programmes). SQAH, Individual consideration. SQAV, Individual consideration. IB, Individual consideration.

BSc Information Technology with Management

UCAS Code: G5N1 • Mode: 3 Years Major/Minor

Qualifications/Requirements: GCE, A level grades: EE. AGNVQ, Pass (in specific programmes). ND/C, National (in specific programmes). SQAH, Individual consideration. SQAV, Individual consideration. IB, Individual consideration.

BSc Information Technology with Media Studies

UCAS Code: G5P4 • Mode: 3 Years Major/Minor

Qualifications/Requirements: GCE, A level grades: EE. AGNVQ, Pass (in specific programmes). ND/C, National (in specific programmes). SQAH, Individual consideration. SQAV, Individual consideration. IB, Individual consideration.

BSc Information Technology with Model Design

UCAS Code: GWNF • Mode: 3 Years Equal Combination

Qualifications/Requirements: GCE, A level grades: EE. AGNVQ, Pass (in specific programmes). ND/C, National (in specific programmes). SQAH, Individual consideration. SQAV, Individual consideration. IB, Individual consideration.

BSc Information Technology with Social Policy

UCAS Code: G5L4 • Mode: 3 Years Major/Minor

Qualifications/Requirements: GCE, A level grades: EE. AGNVQ, Pass (in specific programmes). ND/C, National (in specific programmes). SQAH, Individual consideration. SQAV, Individual consideration. IB, Individual consideration.

BSc Information Technology with Spatial Design

UCAS Code: GWNG • Mode: 3 Years Equal Combination

Qualifications/Requirements: GCE, A level grades: EE. AGNVQ, Pass (in specific programmes). ND/C, National (in specific programmes). SQAH, Individual consideration. SQAV, Individual consideration. IB, Individual consideration.

HND Business Information Technology

UCAS Code: 165G • Mode: 2 Years Single Subject

Qualifications/Requirements: GCE, A level grades: E. AGNVQ, Pass. ND/C, National. SQAH, Individual consideration. SQAV, Individual consideration. IB, Individual consideration.

UNIVERSITY OF SUNDERLAND S84

BA Accounting and Computing

UCAS Code: NG45 • Mode: 3/4 Years Equal Combination

Qualifications/Requirements: GCE, A/AS: 10 points. AGNVQ, Merit (in specific programmes). ND/C, 4 Merits (in specific programmes). SQAH, Grades CCCC. SQAV, National. IB, 24 points.

BA Business Computing

UCAS Code: G523 • Mode: 4 Years Single Subject

Qualifications/Requirements: GCE, A/AS: 10 points. AGNVQ, Merit. ND/C, 4 Merits (in specific programmes). SQAH, Grades CCCC. SQAV, National. IB, 24 points.

BA Business Computing (Foundation)

UCAS Code: G528 • Mode: 5 Years Single Subject

Qualifications/Requirements: AGNVQ, Pass.

BA Business Computing with French

UCAS Code: G5R1 • Mode: 4 Years Major/Minor

Qualifications/Requirements: GCE, A/AS: 10 points. AGNVQ, Merit. ND/C, 4 Merits (in specific programmes). SQAH, Grades CCCC. SQAV, National. IB, 24 points.

BA Business Computing with German
UCAS Code: G5R2 • Mode: 4 Years Major/Minor

Qualifications/Requirements: GCE, A/AS: 10 points. AGNVQ, Merit. ND/C, 4 Merits (in specific programmes). SQAH, Grades CCCC. SQAV, National. IB, 24 points.

BA Business Computing with Spanish
UCAS Code: G5R4 • Mode: 4 Years Major/Minor

Qualifications/Requirements: GCE, A/AS: 10 points. AGNVQ, Merit. ND/C, 4 Merits (in specific programmes). SQAH, Grades CCCC. SQAV, National. IB, 24 points.

BA Computer Studies (Top-up)
UCAS Code: G501 • Mode: 1 Year Single Subject

Qualifications/Requirements: Please refer to prospectus.

BA Information Technology Education (11-18 years)
UCAS Code: XG75 • Mode: 2 Years Equal Combination

Qualifications/Requirements: AGNVQ, Not normally sufficient. ND/C, Higher National (in specific programmes). SQAH, Not normally sufficient. SQAV, Higher National including specific subject(s). IB, Not normally sufficient.

BA Information Technology Education (11-18 years)
UCAS Code: XG7M • Mode: 3 Years Equal Combination

Qualifications/Requirements: GCE, A/AS: 12 points. Computing or Italian. AGNVQ, Merit (in specific programmes) with A level. ND/C, Higher National (in specific programmes). SQAH, Grades CCCCC including specific subject(s). SQAV, National including specific subject(s). IB, 24 points.

BA Information Technology Education (Key Stage 2/3) (11-18 years)
UCAS Code: XG6M • Mode: 3 Years Equal Combination

Qualifications/Requirements: GCE, A/AS: 12 points. Computing or Italian. AGNVQ, Merit (in specific programmes) with A level. ND/C, Higher National (in specific programmes). SQAH, Grades CCCCC including specific subject(s). SQAV, National including specific subject(s). IB, 24 points.

BA Interactive Media
UCAS Code: G700 • Mode: 3/4 Years Single Subject

Qualifications/Requirements: GCE, A/AS: 10 points. AGNVQ, Merit. ND/C, 4 Merits (in specific programmes). SQAH, Grades CCCC. SQAV, National. IB, 24 points.

BA/BSc Business Law and Computer Studies
UCAS Code: GM53 • Mode: 3 Years Equal Combination

Qualifications/Requirements: GCE, A/AS: 8 points. AGNVQ, Pass (in specific programmes). ND/C, 4 Merits. SQAH, Grades CCC. SQAV, Individual consideration. IB, Individual consideration.

BA/BSc Computer Studies and Human Resource Management
UCAS Code: GN56 • Mode: 3 Years Equal Combination

Qualifications/Requirements: GCE, A/AS: 8 points. AGNVQ, Pass (in specific programmes). ND/C, 4 Merits. SQAH, Grades CCC. SQAV, Individual consideration. IB, Individual consideration.

BA/BSc Computer Studies and Music
UCAS Code: GW53 • Mode: 3 Years Equal Combination

Qualifications/Requirements: Please refer to prospectus.

BSc Biochemistry and Computer Studies
UCAS Code: CG75 • Mode: 3 Years Equal Combination

Qualifications/Requirements: GCE, A/AS: 8 points. Biology or Chemistry at A level. AGNVQ, Merit (in specific programmes). ND/C, 4 Merits. SQAH, Grades CCC. SQAV, Individual consideration. IB, Individual consideration.

BSc Chemistry and Computer Studies
UCAS Code: FG15 • Mode: 3 Years Equal Combination

Qualifications/Requirements: GCE, A/AS: 8 points. Chemistry at A level. AGNVQ, Merit (in specific programmes). ND/C, 4 Merits. SQAH, Grades CCC. SQAV, Individual consideration. IB, Individual consideration.

BSc Computer Studies and Ecology
UCAS Code: CG95 • Mode: 3 Years Equal Combination

Qualifications/Requirements: GCE, A/AS: 8 points. Biology at A level. AGNVQ, Merit (in specific programmes). ND/C, 4 Merits. SQAH, Grades CCC. SQAV, Individual consideration. IB, Individual consideration.

BSc Computer Studies and Geology
UCAS Code: GF56 • Mode: 3 Years Equal Combination

Qualifications/Requirements: GCE, A/AS: 8 points. Geography or Geology at A level. AGNVQ, Merit. ND/C, 4 Merits. SQAH, Grades CCC. SQAV, Individual consideration. IB, Individual consideration.

BSc Computer Studies and Mathematical Sciences
UCAS Code: GG51 • Mode: 3 Years Equal Combination

Qualifications/Requirements: GCE, A/AS: 8 points. Maths at A level. AGNVQ, Pass (in specific programmes). ND/C, 4 Merits. SQAH, Grades CCC. SQAV, Individual consideration. IB, Individual consideration.

BSc Computer Studies and Microbiology
UCAS Code: CG55 • Mode: 3 Years Equal Combination

Qualifications/Requirements: GCE, A/AS: 8 points. Biology or Chemistry at A level. AGNVQ, Merit (in specific programmes). ND/C, 4 Merits. SQAH, Grades CCC. SQAV, Individual consideration. IB, Individual consideration.

BSc Computer Studies and Physiology

UCAS Code: GB51 • Mode: 3 Years Equal Combination

Qualifications/Requirements: GCE, A/AS: 8 points. Biology or Chemistry at A level. AGNVQ, Merit (in specific programmes). ND/C, 4 Merits. SQAH, Grades CCCC including specific subject(s). SQAV, Individual consideration. IB, 24 points.

BSc Computer Studies and Psychology

UCAS Code: GC58 • Mode: 3 Years Equal Combination

Qualifications/Requirements: GCE, A/AS: 12 points. AGNVQ, Merit with A/AS. ND/C, 5 Merits. SQAH, Grades BCCC including specific subject(s). SQAV, National. IB, 24 points.

BSc Computer Studies with American Studies

UCAS Code: G5Q4 • Mode: 3 Years Major/Minor

Qualifications/Requirements: GCE, A/AS: 8 points. AGNVQ, Pass. ND/C, 4 Merits. SQAH, Grades CCC. SQAV, Individual consideration. IB, Individual consideration.

BSc Computer Studies with Business Law

UCAS Code: G5M3 • Mode: 3 Years Major/Minor

Qualifications/Requirements: Please refer to prospectus.

BSc Computer Studies with Business Studies

UCAS Code: G5N1 • Mode: 3 Years Major/Minor

Qualifications/Requirements: GCE, A/AS: 8 points. AGNVQ, Pass. ND/C, 4 Merits. SQAH, Grades CCC. SQAV, Individual consideration. IB, Individual consideration.

BSc Computer Studies with Chemistry

UCAS Code: G5F1 • Mode: 3/4 Years Major/Minor

Qualifications/Requirements: GCE, A/AS: 8 points. AGNVQ, Merit. ND/C, 3 Merits. SQAH, Individual consideration. SQAV, Individual consideration. IB, Individual consideration.

BSc Computer Studies with Comparative Literature

UCAS Code: G5Q2 • Mode: 3 Years Major/Minor

Qualifications/Requirements: GCE, A/AS: 8 points. AGNVQ, Pass. ND/C, 4 Merits. SQAH, Grades CCC. SQAV, Individual consideration. IB, Individual consideration.

BSc Computer Studies with Criminology

UCAS Code: G5MH • Mode: 3 Years Major/Minor

Qualifications/Requirements: Please refer to prospectus.

BSc Computer Studies with Economics

UCAS Code: G5L1 • Mode: 3/4 Years Major/Minor

Qualifications/Requirements: GCE, A/AS: 8 points. AGNVQ, Merit. ND/C, 3 Merits. SQAH, Individual consideration. SQAV, Individual consideration. IB, Individual consideration.

BSc Computer Studies with English Studies

UCAS Code: G5Q3 • Mode: 3/4 Years Major/Minor

Qualifications/Requirements: GCE, A/AS: 10 points. AGNVQ, Merit. ND/C, 4 Merits. SQAH, Individual consideration. SQAV, Individual consideration. IB, Individual consideration.

BSc Computer Studies with European Studies

UCAS Code: G5T2 • Mode: 3 Years Major/Minor

Qualifications/Requirements: GCE, A/AS: 8 points. AGNVQ, Pass. ND/C, 4 Merits. SQAH, Grades CCC. SQAV, Individual consideration. IB, Individual consideration.

BSc Computer Studies with Geography

UCAS Code: G5L8 • Mode: 3/4 Years Major/Minor

Qualifications/Requirements: GCE, A/AS: 8 points. AGNVQ, Merit. ND/C, 3 Merits. SQAH, Individual consideration. SQAV, Individual consideration. IB, Individual consideration.

BSc Computer Studies with German Studies

UCAS Code: G5RF • Mode: 4 Years Major/Minor

Qualifications/Requirements: GCE, A/AS: 8 points. German at A level. AGNVQ, Pass (in specific programmes). ND/C, 4 Merits. SQAH, Grades CCC. SQAV, Individual consideration. IB, Individual consideration.

BSc Computer Studies with History

UCAS Code: G5V1 • Mode: 3/4 Years Major/Minor

Qualifications/Requirements: GCE, A/AS: 10 points. AGNVQ, Merit. ND/C, 4 Merits. SQAH, Individual consideration. SQAV, Individual consideration. IB, Individual consideration.

BSc Computer Studies with History of Art and Design

UCAS Code: G5V4 • Mode: 3/4 Years Major/Minor

Qualifications/Requirements: GCE, A/AS: 8 points. AGNVQ, Merit. ND/C, 3 Merits. SQAH, Individual consideration. SQAV, Individual consideration. IB, Individual consideration.

BSc Computer Studies with Human Resource Management

UCAS Code: G5N6 • Mode: 3 Years Major/Minor

Qualifications/Requirements: Please refer to prospectus.

BSc Computer Studies with Marketing

UCAS Code: G5N5 • Mode: 3 Years Major/Minor

Qualifications/Requirements: Please refer to prospectus.

BSc Computer Studies with Mathematical Sciences
UCAS Code: G5G1 • Mode: 3/4 Years Major/Minor

Qualifications/Requirements: GCE, A/AS: 8 points. AGNVQ, Merit. ND/C, 3 Merits. SQAH, Individual consideration. SQAV, Individual consideration. IB, Individual consideration.

BSc Computer Studies with Media Studies
UCAS Code: G5P4 • Mode: 3 Years Major/Minor

Qualifications/Requirements: GCE, A/AS: 24 points. AGNVQ, Individual consideration. ND/C, Individual consideration. SQAH, Individual consideration. SQAV, Individual consideration. IB, Individual consideration.

BSc Computer Studies with Music
UCAS Code: G5W3 • Mode: 3 Years Major/Minor

Qualifications/Requirements: GCE, A/AS: 8 points. AGNVQ, Pass. ND/C, 4 Merits. SQAH, Grades CCC. SQAV, Individual consideration. IB, Individual consideration.

BSc Computer Studies with Philosophy
UCAS Code: G5V7 • Mode: 3 Years Major/Minor

Qualifications/Requirements: GCE, A/AS: 8 points. AGNVQ, Pass. ND/C, 4 Merits. SQAH, Grades CCC. SQAV, Individual consideration. IB, Individual consideration.

BSc Computer Studies with Physiology
UCAS Code: G5B1 • Mode: 3/4 Years Major/Minor

Qualifications/Requirements: GCE, A/AS: 8 points. AGNVQ, Merit. ND/C, 3 Merits. SQAH, Individual consideration. SQAV, Individual consideration. IB, Individual consideration.

BSc Computer Studies with Politics
UCAS Code: G5M1 • Mode: 3 Years Major/Minor

Qualifications/Requirements: GCE, A/AS: 8 points. AGNVQ, Pass. ND/C, 4 Merits. SQAH, Grades CCC. SQAV, Individual consideration. IB, Individual consideration.

BSc Computer Studies with Psychology
UCAS Code: G5C8 • Mode: 3/4 Years Major/Minor

Qualifications/Requirements: GCE, A/AS: 10 points. AGNVQ, Merit with A/AS. ND/C, 4 Merits. SQAH, Individual consideration. SQAV, Individual consideration. IB, Individual consideration.

BSc Computer Studies with Religious Studies
UCAS Code: G5V8 • Mode: 3/4 Years Major/Minor

Qualifications/Requirements: GCE, A/AS: 8 points. AGNVQ, Merit. ND/C, 3 Merits. SQAH, Individual consideration. SQAV, Individual consideration. IB, Individual consideration.

BSc Computer Studies with Sociology
UCAS Code: G5L3 • Mode: 3/4 Years Major/Minor

Qualifications/Requirements: GCE, A/AS: 10 points. AGNVQ, Merit. ND/C, 4 Merits. SQAH, Individual consideration. SQAV, Individual consideration. IB, Individual consideration.

BSc Computer Studies with Spanish Studies
UCAS Code: G5RK • Mode: 4 Years Major/Minor

Qualifications/Requirements: GCE, A/AS: 8 points. Spanish at A level. AGNVQ, Pass. ND/C, 4 Merits. SQAH, Grades CCC. SQAV, Individual consideration. IB, Individual consideration.

BSc Computer Studies with Studies in Education & Training
UCAS Code: G5X9 • Mode: 3 Years Major/Minor

Qualifications/Requirements: Please refer to prospectus.

BSc Computing
UCAS Code: G500 • Mode: 3/4 Years Single Subject

Qualifications/Requirements: GCE, A/AS: 12 points. AGNVQ, Distinction. ND/C, 2 Merits and 2 Distinctions (in specific programmes). SQAH, Grades BCCC. SQAV, National. IB, 24 points.

BSc Information Technology
UCAS Code: G560 • Mode: 3/4 Years Single Subject

Qualifications/Requirements: GCE, A/AS: 12 points. AGNVQ, Distinction. ND/C, 2 Merits and 2 Distinctions (in specific programmes). SQAH, Grades BCCC. SQAV, National. IB, 24 points.

Mod American Studies and Computer Studies
UCAS Code: QG45 • Mode: 3 Years Equal Combination

Qualifications/Requirements: GCE, A/AS: 8 points. AGNVQ, Pass. ND/C, 4 Merits. SQAH, Grades CCC. SQAV, Individual consideration. IB, Individual consideration.

Mod Business Studies and Computer Studies
UCAS Code: NG15 • Mode: 3 Years Equal Combination

Qualifications/Requirements: GCE, A/AS: 10 points. AGNVQ, Pass (in specific programmes). ND/C, 4 Merits. SQAH, Grades CCC. SQAV, Individual consideration. IB, Individual consideration.

Mod Comparative Literature and Computer Studies
UCAS Code: QG25 • Mode: 3 Years Equal Combination

Qualifications/Requirements: GCE, A/AS: 8 points. AGNVQ, Pass (in specific programmes). ND/C, 4 Merits. SQAH, Grades CCC. SQAV, Individual consideration. IB, Individual consideration.

Mod Computer Studies and Economics
UCAS Code: GL51 • Mode: 3 Years Equal Combination
Qualifications/Requirements: GCE, A/AS: 8 points. Biology at A level. AGNVQ, Pass. ND/C, 4 Merits. SQAH, Grades CCC. SQAV, Individual consideration. IB, Individual consideration.

Mod Computer Studies and English Studies
UCAS Code: GQ53 • Mode: 3 Years Equal Combination
Qualifications/Requirements: GCE, A/AS: 10 points. English Literature at A level. AGNVQ, Merit. ND/C, 5 Merits. SQAH, Grades CCCC. SQAV, Individual consideration. IB, Individual consideration.

Mod Computer Studies and European Studies
UCAS Code: GT52 • Mode: 3 Years Equal Combination
Qualifications/Requirements: GCE, A/AS: 8 points. AGNVQ, Pass (in specific programmes). ND/C, 4 Merits. SQAH, Grades CCC. SQAV, Individual consideration. IB, Individual consideration.

Mod Computer Studies and French Studies
UCAS Code: GR51 • Mode: 4 Years Equal Combination
Qualifications/Requirements: GCE, A/AS: 8 points. French at A level. AGNVQ, Pass (in specific programmes). ND/C, 4 Merits. SQAH, Grades CCC. SQAV, Individual consideration. IB, Individual consideration.

Mod Computer Studies and Gender Studies
UCAS Code: GM59 • Mode: 3 Years Equal Combination
Qualifications/Requirements: GCE, A/AS: 8 points. AGNVQ, Pass (in specific programmes). ND/C, 4 Merits. SQAH, Grades CCC. SQAV, Individual consideration. IB, Individual consideration.

Mod Computer Studies and Geography
UCAS Code: GL58 • Mode: 3 Years Equal Combination
Qualifications/Requirements: GCE, A/AS: 10 points. Geography or Geology. AGNVQ, Merit. ND/C, National (in specific programmes). SQAH, Grades CCCC. SQAV, Individual consideration. IB, Individual consideration.

Mod Computer Studies and German Studies
UCAS Code: GR52 • Mode: 4 Years Equal Combination
Qualifications/Requirements: GCE, A/AS: 8 points. German at A level. AGNVQ, Pass (in specific programmes). ND/C, 4 Merits. SQAH, Grades CCC. SQAV, Individual consideration. IB, Individual consideration.

Mod Computer Studies and History
UCAS Code: GV51 • Mode: 3 Years Equal Combination
Qualifications/Requirements: GCE, A/AS: 10 points. History at A level. AGNVQ, Merit. ND/C, Individual consideration. SQAH, Grades CCCC. SQAV, Individual consideration. IB, 24 points.

Mod Computer Studies and History of Art and Design
UCAS Code: GV54 • Mode: 3/4 Years Equal Combination
Qualifications/Requirements: GCE, A/AS: 8 points. AGNVQ, Merit. ND/C, 4 Merits. SQAH, Grades CCC. SQAV, Individual consideration. IB, Individual consideration.

Mod Computer Studies and Media Studies
UCAS Code: GP54 • Mode: 3/4 Years Equal Combination
Qualifications/Requirements: GCE, A/AS: 24 points. AGNVQ, Distinction (in specific programmes). ND/C, Distinction Overall. SQAH, Grades BBCCC. SQAV, Individual consideration. IB, Individual consideration.

Mod Computer Studies and Philosophy
UCAS Code: GV57 • Mode: 3 Years Equal Combination
Qualifications/Requirements: GCE, A/AS: 8 points. AGNVQ, Pass (in specific programmes). ND/C, 4 Merits. SQAH, Grades CCC. SQAV, Individual consideration. IB, Individual consideration.

Mod Computer Studies and Politics
UCAS Code: GM51 • Mode: 3 Years Equal Combination
Qualifications/Requirements: GCE, A/AS: 8 points. AGNVQ, Pass (in specific programmes). ND/C, 4 Merits. SQAH, Grades CCC. SQAV, Individual consideration. IB, Individual consideration.

Mod Computer Studies and Religious Studies
UCAS Code: GV58 • Mode: 3 Years Equal Combination
Qualifications/Requirements: GCE, A/AS: 10 points. AGNVQ, Merit (in specific programmes). ND/C, 4 Merits. SQAH, Grades CCCC. SQAV, National. IB, 24 points.

Mod Computer Studies and Sociology
UCAS Code: GL53 • Mode: 3 Years Equal Combination
Qualifications/Requirements: GCE, A/AS: 10 points. AGNVQ, Merit (in specific programmes). ND/C, 5 Merits. SQAH, Grades CCCC. SQAV, Individual consideration. IB, Individual consideration.

Mod Computer Studies and Spanish Studies
UCAS Code: GR54 • Mode: 3/4 Years Equal Combination
Qualifications/Requirements: GCE, A/AS: 8 points. Spanish at A level. AGNVQ, Pass (in specific programmes). ND/C, 4 Merits. SQAH, Grades CCC. SQAV, Individual consideration. IB, Individual consideration.

HND Business Information Technology
UCAS Code: 065G • Mode: 2/3 Years Single Subject
Qualifications/Requirements: GCE, A/AS: 4-6 points. AGNVQ, Pass. ND/C, National. SQAH, Grades CC. SQAV, National. IB, Pass in Diploma.

HND Business Information Technology for Returners

UCAS Code: 165G • Mode: 2/3 Years Single Subject

Qualifications/Requirements: GCE, A/AS: 4-6 points. AGNVQ, Pass. ND/C, National. SQAH, Grades CC. SQAV, National. IB, Pass in Diploma.

HND Computing

UCAS Code: 105G • Mode: 2/3 Years Single Subject

Qualifications/Requirements: GCE, A/AS: 4-6 points. AGNVQ, Pass. ND/C, National. SQAH, Grades CC. SQAV, National. IB, Pass in Diploma.

HND Computing for Returners

UCAS Code: 005G • Mode: 2/3 Years Single Subject

Qualifications/Requirements: GCE, A/AS: 4-6 points. AGNVQ, Pass. ND/C, National. SQAH, Grades CC. SQAV, National. IB, Pass in Diploma.

THE UNIVERSITY OF SURREY S87

BSc Computing & Information Technology (h)

UCAS Code: G560 • Mode: 3 Years Single Subject

Qualifications/Requirements: GCE, A level grades: BBB-BCC. AGNVQ, Distinction (in specific programmes) with A/AS. ND/C, Merit overall and 4 Distinctions. SQAH, Individual consideration. SQAV, Individual consideration. IB, Individual consideration.

BSc Computing & Information Technology (h)

UCAS Code: G561 • Mode: 4 Years Single Subject

Qualifications/Requirements: GCE, A level grades: BBB-BCC. AGNVQ, Distinction (in specific programmes) with A/AS. ND/C, Merit overall and 4 Distinctions. SQAH, Individual consideration. SQAV, Individual consideration. IB, Individual consideration.

BSc Computing and German

UCAS Code: GR52 • Mode: 3/4 Years Equal Combination

Qualifications/Requirements: GCE, A level grades: BBC. Maths and German at A level. AGNVQ, Not normally sufficient. ND/C, Not normally sufficient. SQAH, Grades AABBB including specific subject(s). SQAV, Not normally sufficient. IB, 30 points.

BSc Computing Modelling and Simulation

UCAS Code: G510 • Mode: 3 Years Single Subject

Qualifications/Requirements: GCE, A/AS: 20-24 points. Maths at A level. AGNVQ, Individual consideration. ND/C, Individual consideration. SQAH, CSYS required. SQAV, Individual consideration. IB, Individual consideration.

BSc Computing Modelling and Simulation

UCAS Code: G511 • Mode: 4 Years Single Subject

Qualifications/Requirements: GCE, A/AS: 20-24 points. Maths at A level. AGNVQ, Individual consideration. ND/C, Individual consideration. SQAH, CSYS required. SQAV, Individual consideration. IB, Individual consideration.

BSc Mathematics and Computing Science with a European language (k)

UCAS Code: GGC5 • Mode: 3 Years Equal Combination

Qualifications/Requirements: GCE, A/AS: 18-20 points. Maths and a modern foreign language at A level. AGNVQ, Individual consideration. ND/C, Individual consideration. SQAH, CSYS required. SQAV, Individual consideration. IB, Individual consideration.

BSc Mathematics and Computing Science (i)

UCAS Code: GG1M • Mode: 3 Years Equal Combination

Qualifications/Requirements: GCE, A/AS: 18-20 points. Maths at A level. AGNVQ, Individual consideration. ND/C, Individual consideration. SQAH, CSYS required. SQAV, Individual consideration. IB, Individual consideration.

BSc Mathematics and Computing Science (i)

UCAS Code: GG15 • Mode: 4 Years Equal Combination

Qualifications/Requirements: GCE, A/AS: 18-20 points. Maths at A level. AGNVQ, Individual consideration. ND/C, Individual consideration. SQAH, CSYS required. SQAV, Individual consideration. IB, Individual consideration.

BSc Mathematics and Computing Science with a European Language (k)

UCAS Code: GG1N • Mode: 4 Years Equal Combination

Qualifications/Requirements: GCE, A/AS: 18-20 points. Maths and a modern foreign language at A level. AGNVQ, Individual consideration. ND/C, Individual consideration. SQAH, CSYS required. SQAV, Individual consideration. IB, Individual consideration.

BEng Computer Science and Engineering

UCAS Code: GH56 • Mode: 3 Years Equal Combination

Qualifications/Requirements: GCE, A/AS: 18-24 points. Maths at A level and Physics. AGNVQ, Distinction in Engineering with A/AS. ND/C, 5 Merits and 2 Distinctions (in specific programmes). SQAH, Grades BBBBC. SQAV, Not normally sufficient. IB, 28 points.

BEng Computer Science and Engineering

UCAS Code: GH5P • Mode: 4 Years Equal Combination

Qualifications/Requirements: GCE, A/AS: 18-24 points. Maths at A level and Physics. AGNVQ, Distinction in Engineering with A/AS. ND/C, 5 Merits and 2 Distinctions (in specific programmes). SQAH, Grades BBBBC. SQAV, Not normally sufficient. IB, 28 points.

MEng Computer Science and Engineering
UCAS Code: GH5Q • Mode: 4 Years Equal Combination

Qualifications/Requirements: GCE, A/AS: 24 points. Maths at A level and Physics. AGNVQ, Not normally sufficient. ND/C, Not normally sufficient. SQAH, Grades BBBBB. SQAV, Not normally sufficient. IB, 28 points.

MEng Computer Science and Engineering
UCAS Code: GHM6 • Mode: 5 Years Equal Combination

Qualifications/Requirements: GCE, A/AS: 24 points. Maths at A level and Physics. AGNVQ, Not normally sufficient. ND/C, Not normally sufficient. SQAH, Grades BBBBB. SQAV, Not normally sufficient. IB, 28 points.

MEng Computing and Information Technology
UCAS Code: G562 • Mode: 4 Years Single Subject

Qualifications/Requirements: GCE, A/AS: 24 points. AGNVQ, Distinction (in specific programmes) with A/AS. ND/C, Merit overall and 4 Distinctions. SQAH, Individual consideration. SQAV, Individual consideration. IB, Individual consideration.

MEng Computing and Information Technology
UCAS Code: G563 • Mode: 5 Years Single Subject

Qualifications/Requirements: GCE, A/AS: 24 points. AGNVQ, Distinction (in specific programmes) with A/AS. ND/C, Merit overall and 4 Distinctions. SQAH, Individual consideration. SQAV, Individual consideration. IB, Individual consideration.

UNIVERSITY OF SUSSEX S90

BSc Computer Science
UCAS Code: G500 • Mode: 3 Years Single Subject

Qualifications/Requirements: GCE, A level grades: BBB. Maths. AGNVQ, Merit with A level. ND/C, Merit overall (in specific programmes). SQAH, Individual consideration. SQAV, Individual consideration. IB, Individual consideration.

BSc Computer Science and Artificial Intelligence
UCAS Code: G575 • Mode: 3 Years Single Subject

Qualifications/Requirements: GCE, A level grades: BBB. AGNVQ, Merit with 6 additional units. ND/C, Merit overall (in specific programmes). SQAH, Individual consideration. SQAV, Individual consideration. IB, Individual consideration.

BSc Computer Science with European Studies
UCAS Code: G5T2 • Mode: 4 Years Major/Minor

Qualifications/Requirements: GCE, A level grades: BBB. Maths. AGNVQ, Merit with A level. ND/C, Merit overall (in specific programmes). SQAH, Individual consideration. SQAV, Individual consideration. IB, Individual consideration.

BSc Computer Science with Management Studies
UCAS Code: G5ND • Mode: 3 Years Major/Minor

Qualifications/Requirements: GCE, A level grades: BBB. Maths. AGNVQ, Merit with A level. ND/C, Merit overall (in specific programmes). SQAH, Individual consideration. SQAV, Individual consideration. IB, Individual consideration.

BSc Computing Sciences
UCAS Code: G502 • Mode: 4 Years Single Subject

Qualifications/Requirements: AGNVQ, Pass.

BSc Mathematics and Computer Science
UCAS Code: GG15 • Mode: 3 Years Equal Combination

Qualifications/Requirements: GCE, A level grades: BBC. Maths at A level. AGNVQ, Merit in Science with A level. ND/C, Merit overall (in specific programmes). SQAH, Individual consideration. SQAV, Individual consideration. IB, Individual consideration.

BEng Electronic Engineering and Computer Science
UCAS Code: HG65 • Mode: 3 Years Equal Combination

Qualifications/Requirements: GCE, A level grades: CCC. Maths at A level. AGNVQ, Merit in Science with A level. ND/C, Merit overall (in specific programmes). SQAH, Individual consideration. SQAV, Individual consideration. IB, Individual consideration.

UNIVERSITY OF WALES SWANSEA S93

BSc Computer Science
UCAS Code: G500 • Mode: 3 Years Single Subject

Qualifications/Requirements: GCE, A level grades: BBC. AGNVQ, Individual consideration. ND/C, 1 Merit and 5 Distinctions. SQAH, Grades ABBBB. SQAV, Individual consideration. IB, 30 points.

BSc Computer Science
UCAS Code: G501 • Mode: 4 Years Single Subject

Qualifications/Requirements: GCE, A level grades: CCC. AGNVQ, Individual consideration. ND/C, Individual consideration. SQAH, Individual consideration. SQAV, Individual consideration. IB, Individual consideration.

BSc Computer Science (with French)
UCAS Code: G5R1 • Mode: 4 Years Major/Minor

Qualifications/Requirements: GCE, A level grades: BBC. French at A level. AGNVQ, Individual consideration. ND/C, 1 Merit and 5 Distinctions (in specific programmes). SQAH, Grades BBBBB including specific subject(s). SQAV, Individual consideration. IB, 28 points.

BSc Computer Science (with German)
UCAS Code: G5R2 • Mode: 4 Years Major/Minor

Qualifications/Requirements: GCE, A level grades: BBC. German. AGNVQ, Individual consideration. ND/C, 1 Merit and 5 Distinctions (in specific programmes). SQAH, Grades BBBBB including specific subject(s). SQAV, Individual consideration. IB, 28 points.

BSc Computer Science (with Italian)

UCAS Code: G5R3 • Mode: 4 Years Major/Minor

Qualifications/Requirements: GCE, A level grades: BBC. A modern foreign language at A level. AGNVQ, Individual consideration. ND/C, 1 Merit and 5 Distinctions. SQAH, Grades BBBBB. SQAV, Individual consideration. IB, 28 points.

BSc Computer Science (with Russian)

UCAS Code: G5R8 • Mode: 4 Years Major/Minor

Qualifications/Requirements: GCE, A level grades: BBC. A modern foreign language at A level. AGNVQ, Individual consideration. ND/C, 1 Merit and 5 Distinctions. SQAH, Grades BBBBB. SQAV, Individual consideration. IB, 28 points.

BSc Computer Science (with Spanish)

UCAS Code: G5R4 • Mode: 4 Years Major/Minor

Qualifications/Requirements: GCE, A level grades: BBC. A modern foreign language at A level. AGNVQ, Individual consideration. ND/C, 1 Merit and 5 Distinctions. SQAH, Grades BBBBB. SQAV, Individual consideration. IB, 28 points.

BSc Computer Science (with Welsh)

UCAS Code: G5Q5 • Mode: 4 Years Major/Minor

Qualifications/Requirements: GCE, A level grades: BBC. AGNVQ, Individual consideration. ND/C, 1 Merit and 5 Distinctions. SQAH, Grades BBBBC. SQAV, Individual consideration. IB, 28 points.

BSc Computer Science and Physics

UCAS Code: FG35 • Mode: 3 Years Equal Combination

Qualifications/Requirements: GCE, A level grades: BBC-BCC. Maths and Physics at A level. AGNVQ, Individual consideration. ND/C, Individual consideration. SQAH, Individual consideration. SQAV, Individual consideration. IB, 28 points.

BSc Computer Science and Psychology

UCAS Code: CG85 • Mode: 3 Years Equal Combination

Qualifications/Requirements: GCE, A level grades: BBC. AGNVQ, Individual consideration. ND/C, 1 Merit and 5 Distinctions. SQAH, Grades BBBBB. SQAV, Individual consideration. IB, 30 points.

BSc Computer Science and Topographic Science

UCAS Code: GF58 • Mode: 3 Years Equal Combination

Qualifications/Requirements: GCE, A level grades: BBC. Maths and Geography at A level. AGNVQ, Individual consideration. ND/C, 1 Merit and 5 Distinctions (in specific programmes). SQAH, Grades ABBBB including specific subject(s). SQAV, Individual consideration. IB, 30 points.

BSc Computer Science with Electronics

UCAS Code: G5H6 • Mode: 3 Years Major/Minor

Qualifications/Requirements: GCE, A level grades: BBC. Maths. AGNVQ, Individual consideration. ND/C, 1 Merit and 5 Distinctions (in specific programmes). SQAH, Grades ABBBB including specific subject(s). SQAV, Individual consideration. IB, 30 points.

BSc Computing Mathematics

UCAS Code: G5G1 • Mode: 3 Years Major/Minor

Qualifications/Requirements: GCE, A level grades: BBC-BCC. Maths at A level. AGNVQ, Individual consideration. ND/C, 1 Merit and 5 Distinctions (in specific programmes). SQAH, Grades ABBBB including specific subject(s). SQAV, Individual consideration. IB, 30 points.

MEng Computing

UCAS Code: G503 • Mode: 4 Years Single Subject

Qualifications/Requirements: GCE, A level grades: BBC. Maths. AGNVQ, Distinction (in specific programmes). ND/C, 1 Merit and 5 Distinctions (in specific programmes). SQAH, Grades ABBBB. SQAV, Individual consideration. IB, 30 points.

SWANSEA INSTITUTE OF HIGHER EDUCATION S96

BA 3D Computer Animation

UCAS Code: G541 • Mode: 3 Years Single Subject

Qualifications/Requirements: Please refer to prospectus.

BA/BSc 3D Computer Animation

UCAS Code: G540 • Mode: 3 Years Single Subject

Qualifications/Requirements: Please refer to prospectus.

BSc Computing and Information Systems

UCAS Code: G520 • Mode: 3 Years Single Subject

Qualifications/Requirements: GCE, A/AS: 4 points. AGNVQ, Merit. ND/C, 2 Merits. SQAH, Grades CCCC. SQAV, Individual consideration. IB, 24 points.

HND Business Information Technology

UCAS Code: 265G • Mode: 2 Years Single Subject

Qualifications/Requirements: GCE, A/AS: 2-12 points. AGNVQ, Pass. ND/C, National. SQAH, Grades CC. SQAV, National. IB, Individual consideration.

HND Computing

UCAS Code: 105G • Mode: 2/3 Years Single Subject

Qualifications/Requirements: GCE, A/AS: 2-12 points. An approved subject from restricted list. AGNVQ, Pass. ND/C, National. SQAH, Grades CC. SQAV, National. IB, Individual consideration.

SWINDON COLLEGE S98

HND Computing

UCAS Code: 005G • Mode: 2 Years Single Subject

Qualifications/Requirements: GCE, A/AS: 4-12 points. AGNVQ, Merit (in specific programmes). ND/C, Individual consideration. SQAH, Individual consideration. SQAV, Individual consideration. IB, Individual consideration.

TAMESIDE COLLEGE T10

HND Computing
UCAS Code: 005G • Mode: 2 Years Single Subject

Qualifications/Requirements: GCE, A/AS: 2 points. AGNVQ, Pass. ND/C, National.

UNIVERSITY OF TEESSIDE T20

BA Computer Animation
UCAS Code: GW5F • Mode: 3/4 Years Equal Combination

Qualifications/Requirements: GCE, A/AS: 16-20 points. AGNVQ, Merit. ND/C, Individual consideration. SQAH, Grades CCCC. SQAV, Individual consideration. IB, Individual consideration.

BA Computer Games Design
UCAS Code: G570 • Mode: 3/4 Years Single Subject

Qualifications/Requirements: GCE, A/AS: 16-20 points. AGNVQ, Merit. ND/C, Individual consideration. SQAH, Grades CCCC. SQAV, Individual consideration. IB, Individual consideration.

BSc Chemical Systems Engineering with Information Technology
UCAS Code: H8G5 • Mode: 3 Years Major/Minor

Qualifications/Requirements: GCE, A/AS: 12-14 points. Two Sciences. AGNVQ, Merit. ND/C, 3 Merits (in specific programmes). SQAH, Grades CCC including specific subject(s). SQAV, National including specific subject(s). IB, Individual consideration.

BSc Computer Science
UCAS Code: G500 • Mode: 4 Years Single Subject

Qualifications/Requirements: GCE, A/AS: 12-16 points. AGNVQ, Merit. ND/C, Individual consideration. SQAH, Grades CCCC. SQAV, Individual consideration. IB, Individual consideration.

BSc Computer Studies
UCAS Code: G520 • Mode: 3/4 Years Single Subject

Qualifications/Requirements: GCE, A/AS: 12-16 points. AGNVQ, Merit. ND/C, Individual consideration. SQAH, Grades CCCC. SQAV, Individual consideration. IB, Individual consideration.

BSc Informatics
UCAS Code: G501 • Mode: 4 Years Single Subject

Qualifications/Requirements: GCE, A/AS: 12-16 points. AGNVQ, Merit. ND/C, Individual consideration. SQAH, Grades CCCC. SQAV, Individual consideration. IB, Individual consideration.

BSc Information Society
UCAS Code: GW52 • Mode: 3/4 Years Equal Combination

Qualifications/Requirements: GCE, A/AS: 12-16 points. AGNVQ, Merit. ND/C, Individual consideration. SQAH, Grades CCCC. SQAV, Individual consideration. IB, Individual consideration.

BSc Information Technology
UCAS Code: G560 • Mode: 4 Years Single Subject

Qualifications/Requirements: GCE, A/AS: 12-16 points. AGNVQ, Merit. ND/C, Individual consideration. SQAH, Grades CCCC. SQAV, Individual consideration. IB, Individual consideration.

BSc Interactive Computer Entertainment
UCAS Code: GG5R • Mode: 3/4 Years Equal Combination

Qualifications/Requirements: GCE, A/AS: 16-20 points. AGNVQ, Merit. ND/C, Individual consideration. SQAH, Grades CCCC. SQAV, Individual consideration. IB, Individual consideration.

HND Business Information Technology
UCAS Code: 265G • Mode: 2 Years Single Subject

Qualifications/Requirements: GCE, A/AS: 4 points. AGNVQ, Pass. ND/C, National. SQAH, Individual consideration. SQAV, Individual consideration. IB, Individual consideration.

HND Computing
UCAS Code: 005G • Mode: 2 Years Single Subject

Qualifications/Requirements: GCE, A/AS: 4 points. AGNVQ, Pass. ND/C, National. SQAH, Grades CC. SQAV, Individual consideration. IB, Individual consideration.

THAMES VALLEY UNIVERSITY T40

BA Digital Arts with Information Systems
UCAS Code: W9G5 • Mode: 3 Years Major/Minor

Qualifications/Requirements: GCE, A/AS: 8-12 points. AGNVQ, Merit. ND/C, Merit overall. SQAH, Grades CCC.

BA Information Systems with Finance
UCAS Code: G5N3 • Mode: 3 Years Major/Minor

Qualifications/Requirements: Please refer to prospectus.

BSc Information & Knowledge Management with Multimedia Computing
UCAS Code: P2GM • Mode: 2/3 Years Major/Minor

Qualifications/Requirements: GCE, A/AS: 8 points. AGNVQ, Merit. ND/C, Merit overall. SQAH, Grades CCCC. IB, 26 points.

BSc Information Systems
UCAS Code: G524 • Mode: 3 Years Single Subject

Qualifications/Requirements: GCE, A/AS: 8 points. AGNVQ, Merit. ND/C, Merit overall. SQAH, Grades CCCC. IB, 26 points.

BSc Information Systems with Multimedia Computing
UCAS Code: G523 • Mode: 2/3 Years Single Subject

Qualifications/Requirements: GCE, A/AS: 8 points. AGNVQ, Merit. ND/C, Merit overall. SQAH, Grades CCCC. IB, 26 points.

Mod Information Systems with Accounting

UCAS Code: G5NK • Mode: 2/3 Years Major/Minor

Qualifications/Requirements: GCE, A/AS: 8 points. AGNVQ, Merit. ND/C, Merit overall. SQAH, Grades CCCC. IB, 26 points.

Mod Information Systems with Business

UCAS Code: G5NC • Mode: 2/3 Years Major/Minor

Qualifications/Requirements: GCE, A/AS: 8 points. AGNVQ, Merit. ND/C, Merit overall. SQAH, Grades CCCC. IB, 26 points.

Mod Information Systems with Data Communications Support

UCAS Code: G5G6 • Mode: 2/3 Years Major/Minor

Qualifications/Requirements: GCE, A/AS: 8 points. AGNVQ, Merit. ND/C, Merit overall. SQAH, Grades CCCC. IB, 26 points.

Mod Information Systems with Electronic Publishing

UCAS Code: G5PM • Mode: 2/3 Years Major/Minor

Qualifications/Requirements: GCE, A/AS: 8 points. AGNVQ, Merit. ND/C, Merit overall. SQAH, Grades CCCC. IB, 26 points.

Mod Information Systems with Software Design

UCAS Code: G5G7 • Mode: 2/3 Years Major/Minor

Qualifications/Requirements: GCE, A/AS: 8 points. AGNVQ, Merit. ND/C, Merit overall. SQAH, Grades CCCC. IB, 26 points.

HND Business Information Technology

UCAS Code: 265G • Mode: 2 Years Single Subject

Qualifications/Requirements: GCE, A/AS: 4 points. AGNVQ, Pass. ND/C, National. SQAH, Grades CCCC. IB, 24 points.

HND Computing

UCAS Code: 017G • Mode: 2 Years Single Subject

Qualifications/Requirements: AGNVQ, Pass. ND/C, National. SQAH, Grades CCCC. IB, 24 points.

TRINITY COLLEGE CARMARTHEN T80

BSc Information Systems and Technologies

UCAS Code: G520 • Mode: 3 Years Single Subject

Qualifications/Requirements: GCE, A level grades: EE-CC. AGNVQ, Individual consideration. ND/C, Individual consideration. SQAH, Individual consideration. SQAV, Individual consideration. IB, Individual consideration.

UNIVERSITY OF ULSTER U20

BSc Computing and Linguistics

UCAS Code: GQ51 • Mode: 4 Years Equal Combination

Qualifications/Requirements: GCE, A level grades: BCC. AGNVQ, Distinction with 6 additional units or with A/AS. ND/C, Merit overall and 4 Distinctions. SQAH, Grades BBBC. SQAV, Individual consideration. IB, 30 points.

BSc Computing Science

UCAS Code: G502 • Mode: 4 Years Single Subject

Qualifications/Requirements: GCE, A level grades: CDD. AGNVQ, Distinction with 6 additional units or with A/AS. ND/C, Merit overall and 1 Distinction. SQAH, Grades CCCD. SQAV, Individual consideration. IB, 26 points.

BSc Computing Science (Hons)

UCAS Code: G500 • Mode: 4 Years Single Subject

Qualifications/Requirements: GCE, A level grades: BCC. AGNVQ, Distinction with 6 additional units or with A/AS. ND/C, Merit overall and 4 Distinctions. SQAH, Grades BBBC. SQAV, Individual consideration. IB, 30 points.

BSc Computing with Enterprise Studies

UCAS Code: G5N1 • Mode: 4 Years Major/Minor

Qualifications/Requirements: GCE, A level grades: BCC. AGNVQ, Distinction in Information Technology with 6 additional units or with A/AS. ND/C, Merit overall and 4 Distinctions. SQAH, Grades BBBC. SQAV, Individual consideration. IB, 30 points.

BSc Computing with Marketing

UCAS Code: G5N5 • Mode: 4 Years Major/Minor

Qualifications/Requirements: GCE, A level grades: BCC. AGNVQ, Distinction in Information Technology with 6 additional units or with A/AS. ND/C, Merit overall and 4 Distinctions. SQAH, Grades BBBC. SQAV, Individual consideration. IB, 30 points.

BSc Interactive Multimedia Design

UCAS Code: G532 • Mode: 4 Years Single Subject

Qualifications/Requirements: GCE, A level grades: BBC. AGNVQ, Distinction with 6 additional units or with A level. ND/C, Merit overall and 4 Distinctions. SQAH, Grades ABBB. SQAV, Individual consideration. IB, 32 points.

BSc Mathematics, Statistics and Computing

UCAS Code: G921 • Mode: 4 Years Single Subject

Qualifications/Requirements: GCE, A level grades: CDD. Maths. AGNVQ, Distinction with A level. ND/C, Merit overall and 1 Distinction. SQAH, Grades CCCD. SQAV, Individual consideration. IB, 26 points.

BSc Mathematics, Statistics and Computing (Hons)

UCAS Code: G920 • Mode: 4 Years Single Subject

Qualifications/Requirements: GCE, A level grades: BCC. Maths. AGNVQ, Distinction with A level. ND/C, Merit overall and 4 Distinctions. SQAH, Grades BBBC. SQAV, Individual consideration. IB, 30 points.

BEng Electronics and Computing

UCAS Code: HG67 • Mode: 4 Years Equal Combination

Qualifications/Requirements: GCE, A level grades: CDD. Science. AGNVQ, Distinction (in specific programmes). ND/C, Merit overall and 1 Distinction. SQAH, Grades CCCD. SQAV, Individual consideration. IB, 26 points.

BEng Electronics and Computing (Hons) with Diploma in Industrial Studies

UCAS Code: GH76 • Mode: 4 Years Equal Combination

Qualifications/Requirements: GCE, A level grades: CCD. Science. AGNVQ, Distinction (in specific programmes) with 4 additional units or with A level. ND/C, Merit overall and 3 Distinctions. SQAH, Grades BBCC. SQAV, Individual consideration. IB, 28 points.

HND Computing

UCAS Code: 105G • Mode: 2 Years Single Subject

Qualifications/Requirements: GCE, A level grades: CD. AGNVQ, Distinction in Information Technology. ND/C, Individual consideration. SQAH, Individual consideration. SQAV, Individual consideration. IB, 22 points.

HND Computing (includes Certificate in Industrial Studies) (3 year sandwich)

UCAS Code: 005G • Mode: 3 Years Single Subject

Qualifications/Requirements: GCE, A level grades: CD. AGNVQ, Distinction in Information Technology. ND/C, Individual consideration. SQAH, Individual consideration. SQAV, Individual consideration. IB, 22 points.

UNIVERSITY COLLEGE LONDON (UNIVERSITY OF LONDON) U80

BSc Computer Science

UCAS Code: G500 • Mode: 3 Years Single Subject

Qualifications/Requirements: GCE, A level grades: ABB. Maths at A level. AGNVQ, Individual consideration. ND/C, Merit overall and 2 Distinctions (in specific programmes). SQAH, Grades AABBB including specific subject(s). SQAV, Individual consideration. IB, 34 points.

BSc Computer Science with Cognitive Science

UCAS Code: G5C8 • Mode: 3 Years Major/Minor

Qualifications/Requirements: GCE, A level grades: ABB. Maths at A level. AGNVQ, Individual consideration. ND/C, Merit overall and 2 Distinctions (in specific programmes). SQAH, Grades AABBB including specific subject(s). SQAV, Individual consideration. IB, 34 points.

BSc Computer Science with Electronic Engineering

UCAS Code: G5H6 • Mode: 3 Years Major/Minor

Qualifications/Requirements: GCE, A level grades: ABB. Maths at A level. AGNVQ, Individual consideration. ND/C, Merit overall and 2 Distinctions (in specific programmes). SQAH, Grades AABBB including specific subject(s). SQAV, Individual consideration. IB, 34 points.

BSc Computer Science with Mathematics

UCAS Code: G5G1 • Mode: 3 Years Major/Minor

Qualifications/Requirements: GCE, A level grades: ABB. Maths at A level. AGNVQ, Individual consideration. ND/C, Merit overall and 2 Distinctions (in specific programmes). SQAH, Grades AABBB including specific subject(s). SQAV, Individual consideration. IB, 34 points.

BSc Information Management

UCAS Code: P2G5 • Mode: 3 Years Major/Minor

Qualifications/Requirements: GCE, A level grades: BBB. AGNVQ, Individual consideration. ND/C, Merit overall (in specific programmes). SQAH, Grades BBCCC including specific subject(s). SQAV, National including specific subject(s). IB, 30 points.

BSc Mathematics and Computer Science

UCAS Code: GG15 • Mode: 3 Years Equal Combination

Qualifications/Requirements: GCE, A level grades: ABB-AAA. Maths at A level. AGNVQ, Distinction with A level. ND/C, Distinction Overall (in specific programmes). SQAH, CSYS required. SQAV, National including specific subject(s). IB, 34 points.

MEng Electronic Engineering with Computer Science

UCAS Code: H6GN • Mode: 4 Years Major/Minor

Qualifications/Requirements: GCE, A level grades: ABB-AAB. Maths and Physics at A level. AGNVQ, Not normally sufficient. ND/C, 3 Merits and 4 Distinctions (in specific programmes). SQAH, Grades AAABB including specific subject(s). SQAV, Individual consideration. IB, 30 points.

MSci Computer Science

UCAS Code: G501 • Mode: 4 Years Single Subject

Qualifications/Requirements: GCE, A level grades: ABB. Maths at A level. AGNVQ, Individual consideration. ND/C, Merit overall and 2 Distinctions (in specific programmes). SQAH, Grades AABBB including specific subject(s). SQAV, Individual consideration. IB, 34 points.

MSci Computer Science with Cognitive Science

UCAS Code: G5CV • Mode: 4 Years Major/Minor

Qualifications/Requirements: GCE, A level grades: ABB. Maths at A level. AGNVQ, Individual consideration. ND/C, Merit overall and 2 Distinctions (in specific programmes). SQAH, Grades AABBB including specific subject(s). SQAV, Individual consideration. IB, 34 points.

MSci Computer Science with Electronic Engineering

UCAS Code: G5HP • Mode: 4 Years Major/Minor

Qualifications/Requirements: GCE, A level grades: ABB. Maths and Physics at A level. AGNVQ, Individual consideration. ND/C, Merit overall and 2 Distinctions (in specific programmes). SQAH, Grades AABBB including specific subject(s). SQAV, Individual consideration. IB, 34 points.

MSci Computer Science with Mathematics

UCAS Code: G5GC • Mode: 4 Years Major/Minor

Qualifications/Requirements: GCE, A level grades: ABB. Maths at A level. AGNVQ, Individual consideration. ND/C, Merit overall and 2 Distinctions (in specific programmes). SQAH, Grades AABBB including specific subject(s). SQAV, Individual consideration. IB, 34 points.

MSci Mathematics and Computer Science

UCAS Code: GG1M • Mode: 4 Years Equal Combination

Qualifications/Requirements: GCE, A level grades: ABB-AAA. Maths at A level. AGNVQ, Distinction with A level. ND/C, Distinction Overall (in specific programmes). SQAH, CSYS required. SQAV, National including specific subject(s). IB, 34 points.

UNIVERSITY COLLEGE WARRINGTON W17

BA Business Management and Information Technology

UCAS Code: N125 • Mode: 3 Years Single Subject

Qualifications/Requirements: GCE, A/AS: 14-16 points. AGNVQ, Individual consideration. ND/C, Individual consideration. SQAH, Individual consideration. SQAV, Individual consideration. IB, Individual consideration.

THE UNIVERSITY OF WARWICK W20

BSc Computer and Business Studies

UCAS Code: GN51 • Mode: 3 Years Equal Combination

Qualifications/Requirements: GCE, A level grades: AAB. AGNVQ, Distinction (in specific programmes) with A/AS. ND/C, Individual consideration. SQAH, CSYS required. IB, 36 points.

BSc Computer and Management Sciences (f)

UCAS Code: GN5C • Mode: 3 Years Equal Combination

Qualifications/Requirements: GCE, A level grades: AAB. Maths at A level. AGNVQ, Distinction (in specific programmes) with A/AS. ND/C, Individual consideration. SQAH, CSYS required. IB, 36 points.

BSc Computer Science (f)

UCAS Code: G500 • Mode: 3 Years Single Subject

Qualifications/Requirements: GCE, A level grades: AAB. Maths at A level. AGNVQ, Distinction (in specific programmes) with A/AS. ND/C, Individual consideration. SQAH, CSYS required. IB, 36 points.

MEng Computer Science

UCAS Code: G503 • Mode: 4 Years Single Subject

Qualifications/Requirements: GCE, A level grades: AAB. Maths at A level. AGNVQ, Distinction (in specific programmes) with A/AS. ND/C, Individual consideration. SQAH, CSYS required. IB, 36 points.

WARWICKSHIRE COLLEGE, ROYAL LEAMINGTON SPA AND MORETON MORRELL W25

HND Computing

UCAS Code: 005G • Mode: 2 Years Single Subject

Qualifications/Requirements: GCE, A/AS: 1 points. AGNVQ, Individual consideration. ND/C, Individual consideration. SQAH, Individual consideration. SQAV, Individual consideration. IB, Individual consideration.

HNC Computing

UCAS Code: 105G • Mode: 1 Year Single Subject

Qualifications/Requirements: GCE, A/AS: 1 points. AGNVQ, Individual consideration. ND/C, Individual consideration. SQAH, Individual consideration. SQAV, Individual consideration. IB, Individual consideration.

WEST HERTS COLLEGE, WATFORD W40

HND Business Information Technology

UCAS Code: 065G • Mode: 2 Years Single Subject

Qualifications/Requirements: GCE, A/AS: 2 points. AGNVQ, Pass. ND/C, National.

UNIVERSITY OF WESTMINSTER W50

BA Business Information Management and Finance

UCAS Code: NG15 • Mode: 4 Years Equal Combination

Qualifications/Requirements: GCE, A level grades: BBC. AGNVQ, Distinction. ND/C, Merit overall and 3 Distinctions. SQAH, Grades BBBB. SQAV, Individual consideration. IB, 28 points.

BA Management of Business Information

UCAS Code: NG17 • Mode: 3 Years Equal Combination

Qualifications/Requirements: GCE, A level grades: BC. AGNVQ, Distinction. ND/C, Merit overall and 2 Distinctions. SQAH, Grades BBBB. IB, 28 points.

BSc Computer Communications & Networks

UCAS Code: PG35 • Mode: 3 Years Equal Combination

Qualifications/Requirements: GCE, A level grades: CC. Maths, Physics or Computing. AGNVQ, Distinction. ND/C, 3 Merits. SQAH, Individual consideration. SQAV, Individual consideration. IB, Individual consideration.

BSc Computer Science
UCAS Code: G502 • Mode: 3 Years Single Subject

Qualifications/Requirements: GCE, A/AS: 12 points. AGNVQ, Distinction. ND/C, 3 Merits (in specific programmes).

BSc Computer Science (with foundation)
UCAS Code: G501 • Mode: 4 Years Single Subject

Qualifications/Requirements: GCE, A level grades: E. AGNVQ, Pass. ND/C, National.

BSc Computing
UCAS Code: G500 • Mode: 3 Years Single Subject

Qualifications/Requirements: GCE, A/AS: 12 points. AGNVQ, Distinction. ND/C, Merit overall (in specific programmes). SQAH, Individual consideration. SQAV, Individual consideration. IB, Individual consideration.

BSc Computing with International Foundation
UCAS Code: G5Q3 • Mode: 4 Years Major/Minor

Qualifications/Requirements: GCE, A level grades: D. AGNVQ, Pass. ND/C, National.

BSc Distributed and Network Computing
UCAS Code: G530 • Mode: 3 Years Single Subject

Qualifications/Requirements: GCE, A/AS: 12 points. AGNVQ, Distinction. ND/C, Merit overall. SQAH, Individual consideration. IB, Individual consideration.

BSc Information Systems Engineering
UCAS Code: G520 • Mode: 3 Years Single Subject

Qualifications/Requirements: GCE, A/AS: 14 points. AGNVQ, Merit (in specific programmes). ND/C, 4 Merits.

BSc Information Systems Engineering with International Foundation
UCAS Code: G5QM • Mode: 4 Years Major/Minor

Qualifications/Requirements: GCE, A level grades: D. AGNVQ, Pass. ND/C, National.

WESTMINSTER COLLEGE W52

HND Computer Studies
UCAS Code: 005G • Mode: 2 Years Single Subject

Qualifications/Requirements: GCE, A/AS: 8 points. An approved subject from restricted list. AGNVQ, Individual consideration. ND/C, Individual consideration. SQAH, Individual consideration. SQAV, Individual consideration. IB, 24 points.

WIGAN AND LEIGH COLLEGE W67

BSc Information Technology
UCAS Code: G560 • Mode: 3 Years Single Subject

Qualifications/Requirements: Please refer to prospectus.

HND Building Studies (Information Technology)
UCAS Code: 25KG • Mode: 2 Years Equal Combination

Qualifications/Requirements: GCE, A level grades: E. AGNVQ, Pass (in specific programmes). ND/C, National. SQAV, National. IB, Pass in Diploma.

HND Computing
UCAS Code: 005G • Mode: 2 Years Single Subject

Qualifications/Requirements: GCE, A level grades: DD. AGNVQ, Pass. ND/C, National.

WIRRAL METROPOLITAN COLLEGE W73

HND Computing
UCAS Code: 005G • Mode: 2 Years Single Subject

Qualifications/Requirements: AGNVQ, Individual consideration. ND/C, Individual consideration. SQAH, Individual consideration. SQAV, Individual consideration. IB, Individual consideration.

UNIVERSITY OF WOLVERHAMPTON W75

BSc Business Information Systems
UCAS Code: GN51 • Mode: 3/4 Years Equal Combination

Qualifications/Requirements: GCE, A/AS: 10 points. AGNVQ, Merit. ND/C, 4 Merits. SQAH, Grades BBBB. SQAV, Individual consideration. IB, 24 points.

BSc Computer Science
UCAS Code: G500 • Mode: 4 Years Single Subject

Qualifications/Requirements: GCE, A/AS: 10 points. AGNVQ, Merit. ND/C, 4 Merits. SQAH, Grades CCCC. SQAV, Individual consideration. IB, 24 points.

BSc Computer Science (Information Systems)
UCAS Code: G520 • Mode: 4 Years Single Subject

Qualifications/Requirements: GCE, A/AS: 10 points. AGNVQ, Merit. ND/C, 4 Merits. SQAH, Grades CCCC. SQAV, Individual consideration. IB, 24 points.

BSc Computer Science (Multimedia Technology) (4 years SW)
UCAS Code: G560 • Mode: 4 Years Single Subject

Qualifications/Requirements: GCE, A/AS: 10 points. AGNVQ, Merit. ND/C, 4 Merits. SQAH, Grades CCCC. SQAV, Individual consideration. IB, 24 points.

BSc Computer Science (Software Engineering)
UCAS Code: G700 • Mode: 4 Years Single Subject

Qualifications/Requirements: GCE, A/AS: 18 points. AGNVQ, Merit. ND/C, 4 Merits. SQAH, Grades CCCC. SQAV, Individual consideration. IB, 24 points.

BSc Computer Studies
UCAS Code: G502 • Mode: 1/2 Years Single Subject

Qualifications/Requirements: Please refer to prospectus.

BSc Computing

UCAS Code: G501 • Mode: 4 Years Single Subject

Qualifications/Requirements: GCE, A/AS: 10 points. AGNVQ, Merit. ND/C, 4 Merits. SQAH, Grades CCCC. SQAV, Individual consideration. IB, 24 points.

HND Computing

UCAS Code: 105G • Mode: 2 Years Single Subject

Qualifications/Requirements: GCE, A/AS: 4 points. AGNVQ, Merit. ND/C, 2 Merits. SQAH, Grades CCCC. SQAV, Individual consideration. IB, 24 points.

UNIVERSITY COLLEGE WORCESTER W80

Mod Art and Design/Information Technology

UCAS Code: WG95 • Mode: 3 Years Equal Combination

Qualifications/Requirements: GCE, A level grades: DD. Art. AGNVQ, Merit. ND/C, Individual consideration. SQAH, Individual consideration. SQAV, Individual consideration. IB, Individual consideration.

Mod Biological Science/Information Technology

UCAS Code: CG15 • Mode: 3 Years Equal Combination

Qualifications/Requirements: GCE, A level grades: DD. AGNVQ, Merit. ND/C, Individual consideration. SQAH, Individual consideration. SQAV, Individual consideration. IB, Individual consideration.

Mod Business Management/Information Technology

UCAS Code: NG15 • Mode: 3 Years Equal Combination

Qualifications/Requirements: GCE, A level grades: DD. AGNVQ, Merit. ND/C, Individual consideration. SQAH, Individual consideration. SQAV, Individual consideration. IB, Individual consideration.

Mod Drama/Information Technology

UCAS Code: WG45 • Mode: 3 Years Equal Combination

Qualifications/Requirements: GCE, A level grades: CD. AGNVQ, Merit. ND/C, Individual consideration. SQAH, Individual consideration. SQAV, Individual consideration. IB, Individual consideration.

Mod Education Studies/Information Technology

UCAS Code: XG95 • Mode: 3 Years Equal Combination

Qualifications/Requirements: GCE, A level grades: DD. AGNVQ, Merit. ND/C, Individual consideration. SQAH, Individual consideration. SQAV, Individual consideration. IB, Individual consideration.

Mod English and Literary Studies/Information Technology

UCAS Code: QG35 • Mode: 3 Years Equal Combination

Qualifications/Requirements: GCE, A level grades: CC. AGNVQ, Merit. ND/C, Individual consideration. SQAH, Individual consideration. SQAV, Individual consideration. IB, Individual consideration.

Mod Environmental Science/Information Technology

UCAS Code: FG95 • Mode: 3 Years Equal Combination

Qualifications/Requirements: GCE, A level grades: DD. AGNVQ, Merit. ND/C, Individual consideration. SQAH, Individual consideration. SQAV, Individual consideration. IB, Individual consideration.

Mod European Studies/Information Technology

UCAS Code: GT52 • Mode: 3 Years Equal Combination

Qualifications/Requirements: GCE, A level grades: DD. AGNVQ, Merit. ND/C, Individual consideration. SQAH, Individual consideration. SQAV, Individual consideration. IB, Individual consideration.

Mod Geography/Information Technology

UCAS Code: LG85 • Mode: 3 Years Equal Combination

Qualifications/Requirements: GCE, A level grades: DD. AGNVQ, Merit. ND/C, Individual consideration. SQAH, Individual consideration. SQAV, Individual consideration. IB, Individual consideration.

Mod Information Technology/Modern Foreign Languages

UCAS Code: GT59 • Mode: 3 Years Equal Combination

Qualifications/Requirements: GCE, A level grades: DD. A modern foreign language at A level. AGNVQ, Merit. ND/C, Individual consideration. SQAH, Individual consideration. SQAV, Individual consideration. IB, Individual consideration.

Mod Information Technology/Psychology

UCAS Code: GL57 • Mode: 3 Years Equal Combination

Qualifications/Requirements: GCE, A level grades: CC. AGNVQ, Merit. ND/C, Individual consideration. SQAH, Individual consideration. SQAV, Individual consideration. IB, Individual consideration.

Mod Information Technology/Sociology

UCAS Code: GL53 • Mode: 3 Years Equal Combination

Qualifications/Requirements: GCE, A level grades: DD. AGNVQ, Merit. ND/C, Individual consideration. SQAH, Individual consideration. SQAV, Individual consideration. IB, Individual consideration.

Mod Information Technology/Sports Studies

UCAS Code: GB56 • Mode: 3 Years Equal Combination

Qualifications/Requirements: GCE, A level grades: CC. AGNVQ, Merit. ND/C, Individual consideration. SQAH, Individual consideration. SQAV, Individual consideration. IB, Individual consideration.

Mod Information Technology/Women's Studies

UCAS Code: MG95 • Mode: 3 Years Equal Combination

Qualifications/Requirements: GCE, A level grades: DD. AGNVQ, Merit. ND/C, Individual consideration. SQAH, Individual consideration. SQAV, Individual consideration. IB, Individual consideration.

HND Computing and Information Technology
UCAS Code: 005G • Mode: 2 Years Single Subject

Qualifications/Requirements: GCE, A level grades: D. AGNVQ, Merit. ND/C, Individual consideration. SQAH, Individual consideration. SQAV, Individual consideration. IB, Individual consideration.

HND Information Technology Management
UCAS Code: 065G • Mode: 2 Years Single Subject

Qualifications/Requirements: GCE, A level grades: D. AGNVQ, Merit. ND/C, Individual consideration. SQAH, Individual consideration. SQAV, Individual consideration. IB, Individual consideration.

WORCESTER COLLEGE OF TECHNOLOGY W81

HND Computing
UCAS Code: 025G • Mode: 2 Years Single Subject

Qualifications/Requirements: GCE, A/AS: 2 points. AGNVQ, Pass. ND/C, National.

THE UNIVERSITY OF YORK Y50

BSc Computer Science/Mathematics
UCAS Code: GG51 • Mode: 3 Years Equal Combination

Qualifications/Requirements: GCE, A/AS: 26-28 points. Maths at A level. AGNVQ, Distinction (in specific programmes) with A level. ND/C, Higher National (in specific programmes). SQAH, CSYS required. SQAV, Higher National including specific subject(s). IB, 34 points.

BSc Computer Science/Mathematics (4 years SW)
UCAS Code: GG5C • Mode: 4 Years Equal Combination

Qualifications/Requirements: GCE, A/AS: 26-28 points. Maths at A level. AGNVQ, Distinction (in specific programmes) with A level. ND/C, Higher National (in specific programmes). SQAH, CSYS required. SQAV, Higher National including specific subject(s). IB, 34 points.

MMath Mathematics/Computer Science (4 years)
UCAS Code: GG15 • Mode: 4 Years Equal Combination

Qualifications/Requirements: GCE, A/AS: 26-28 points. Maths at A level. AGNVQ, Distinction (in specific programmes) with A level. ND/C, Higher National (in specific programmes). SQAH, CSYS required. SQAV, Higher National including specific subject(s). IB, 34 points.

Mix Computer Science
UCAS Code: G500 • Mode: 3 Years Single Subject

Qualifications/Requirements: GCE, A/AS: 26-28 points. Maths and any Physical Science or Biology at A level. AGNVQ, Distinction (in specific programmes) with A level. ND/C, Higher National (in specific programmes). SQAH, CSYS required. SQAV, Higher National including specific subject(s). IB, 34 points.

Mix Computer Science (4 years SW)
UCAS Code: G501 • Mode: 4 Years Single Subject

Qualifications/Requirements: GCE, A/AS: 26-28 points. Maths and any Physical Science or Biology at A level. AGNVQ, Distinction (in specific programmes) with A level. ND/C, Higher National (in specific programmes). SQAH, CSYS required. SQAV, Higher National including specific subject(s). IB, 34 points.

Mix Information Technology, Business Management and Language
UCAS Code: G5N1 • Mode: 3 Years Major/Minor

Qualifications/Requirements: GCE, A/AS: 26-28 points. AGNVQ, Distinction (in specific programmes) with A/AS. ND/C, Higher National. SQAH, Grades AAABBB. SQAV, Higher National. IB, 34 points.

Mix Information Technology, Business Management and Language (4 years SW)
UCAS Code: G5NC • Mode: 4 Years Major/Minor

Qualifications/Requirements: GCE, A/AS: 26-28 points. AGNVQ, Distinction (in specific programmes) with A/AS. ND/C, Higher National. SQAH, Grades AAABBB. SQAV, Higher National. IB, 34 points.

YORK COLLEGE OF FURTHER AND HIGHER EDUCATION Y70

HND Computing
UCAS Code: 005G • Mode: 2 Years Single Subject

Qualifications/Requirements: Please refer to prospectus.

YORKSHIRE COAST COLLEGE OF FURTHER AND HIGHER EDUCATION Y80

HND Computing
UCAS Code: 005G • Mode: 2 Years Single Subject

Qualifications/Requirements: GCE, A/AS: 2 points. AGNVQ, Pass (in specific programmes). ND/C, National. SQAH, Individual consideration. SQAV, Individual consideration. IB, Individual consideration.

THE UNIVERSITY OF ABERDEEN A20

BEng Engineering (Electronics and Computer Engineering)

UCAS Code: H6G6 • Mode: 4 Years Major/Minor

Qualifications/Requirements: GCE, A level grades: BBC. Maths and Physics or Design & Technology. AGNVQ, Merit in Science with A level. ND/C, Individual consideration. SQAH, Grades BBCC including specific subject(s). SQAV, Individual consideration. IB, 22 points.

BEng Engineering (Electronics and Software Engineering)

UCAS Code: H6G7 • Mode: 4 Years Major/Minor

Qualifications/Requirements: GCE, A level grades: BBC. Maths and Physics or Design & Technology. AGNVQ, Merit (in specific programmes). ND/C, Individual consideration. SQAH, Grades BBCC including specific subject(s). SQAV, Individual consideration. IB, 22 points.

MEng Electronic and Computer Engineering

UCAS Code: H6GP • Mode: 5 Years Major/Minor

Qualifications/Requirements: GCE, A level grades: BBB. Maths and Physics or Design & Technology. AGNVQ, Merit in Science with A level. ND/C, Individual consideration. SQAH, Grades ABBB including specific subject(s). SQAV, Individual consideration. IB, 24 points.

UNIVERSITY OF ABERTAY DUNDEE A30

BSc Internet & Communications Technology

UCAS Code: GH66 • Mode: 4/5 Years Equal Combination

Qualifications/Requirements: Please refer to prospectus.

ANGLIA POLYTECHNIC UNIVERSITY A60

BA/BSc Communication Studies and Internet Technology

UCAS Code: GP63 • Mode: 3 Years Equal Combination

Qualifications/Requirements: GCE, A/AS: 14 points. An approved subject from restricted list and Maths or Physics at A level. AGNVQ, Merit (in specific programmes) with A level. ND/C, 6 Merits. SQAH, Grades BBBC. SQAV, Individual consideration. IB, Pass in Diploma.

BA/BSc Internet Technology and Lens and Digital Media

UCAS Code: GW65 • Mode: 3 Years Equal Combination

Qualifications/Requirements: GCE, A/AS: 14 points. Art at A level. AGNVQ, Merit (in specific programmes). ND/C, 6 Merits. SQAH, Grades BCCC. SQAV, National. IB, Individual consideration.

BSc Audiotechnology and Internet Technology

UCAS Code: GH66 • Mode: 3 Years Equal Combination

Qualifications/Requirements: GCE, A/AS: 12 points. Maths or Physics at A level. AGNVQ, Merit (in specific programmes). ND/C, 4 Merits. SQAH, Grades BBCC. SQAV, National. IB, Pass in Diploma including specific subjects.

BSc Computer-Aided Product Design

UCAS Code: HW72 • Mode: 3 Years Equal Combination

Qualifications/Requirements: GCE, A/AS: 12 points. An approved subject from restricted list. AGNVQ, Merit (in specific programmes). ND/C, 6 Merits. SQAH, Grades CCC. SQAV, National. IB, Pass in Diploma.

BSc Electronics and Internet Technology

UCAS Code: GH6P • Mode: 3 Years Equal Combination

Qualifications/Requirements: GCE, A/AS: 12 points. Maths or any Physical Science. AGNVQ, Merit in Science. ND/C, 4 Merits. SQAH, Grades BBCC. SQAV, National. IB, Pass in Diploma.

BSc Imaging Science and Internet Technology

UCAS Code: BG86 • Mode: 3 Years Equal Combination

Qualifications/Requirements: GCE, A/AS: 12 points. Maths or Physics at A level. AGNVQ, Merit (in specific programmes). ND/C, 4 Merits. SQAH, Grades BBCC. SQAV, Individual consideration. IB, Pass in Diploma.

BSc Internet Technology

UCAS Code: G620 • Mode: 3 Years Single Subject

Qualifications/Requirements: GCE, A/AS: 12 points. AGNVQ, Merit (in specific programmes). ND/C, 4 Merits. SQAH, Grades BCCC. SQAV, National. IB, Individual consideration.

HND Computer-Aided Product Design

UCAS Code: 27WH • Mode: 2 Years Equal Combination

Qualifications/Requirements: GCE, A/AS: 6 points. An approved subject from restricted list. AGNVQ, Pass (in specific programmes). ND/C, 2 Merits. SQAH, Grades BCCC. SQAV, National. IB, Pass in Diploma.

ASTON UNIVERSITY A80

BEng Electronic Engineering and Computer Science

UCAS Code: GH56 • Mode: 3/4 Years Equal Combination

Qualifications/Requirements: GCE, A/AS: 20 points. Maths and any Physical Science, Computing or Electronics. AGNVQ, Merit (in specific programmes) with 6 additional units or with A level. ND/C, 2 Merits and 4 Distinctions (in specific programmes). SQAH, Grades BBBBB including specific subject(s). SQAV, Individual consideration. IB, 30 points.

UNIVERSITY OF WALES, BANGOR B06

BEng Computer-Aided Engineering

UCAS Code: H160 • Mode: 3 Years Single Subject

Qualifications/Requirements: GCE, A level grades: CCC. Maths and Physics, Electronics or Computing at A level. AGNVQ, Merit (in specific programmes) with 6 additional units or with A level. ND/C, 3 Merits (in specific programmes). SQAH, Grades BBCC including specific subject(s). SQAV, Individual consideration. IB, 28 points.

BEng Computer Systems Engineering

UCAS Code: H616 • Mode: 3 Years Single Subject

Qualifications/Requirements: GCE, A level grades: CCC. Maths and Physics, Electronics or Computing at A level. AGNVQ, Merit (in specific programmes) with 6 additional units or with A level. ND/C, 3 Merits (in specific programmes). SQAH, Grades CCCC including specific subject(s). SQAV, Individual consideration. IB, 26 points.

BEng Computer Systems Engineering (4 year Wide Entry)

UCAS Code: H615 • Mode: 4 Years Single Subject

Qualifications/Requirements: GCE, A level grades: DD. AGNVQ, Merit (in specific programmes) with 6 additional units or with A level. ND/C, 3 Merits. SQAH, Grades CCCC. SQAV, Individual consideration. IB, 26 points.

MEng Computer Systems Engineering

UCAS Code: H617 • Mode: 4 Years Single Subject

Qualifications/Requirements: GCE, A level grades: BBB. Maths and Physics, Electronics or Computing at A level. AGNVQ, Distinction (in specific programmes) with 6 additional units or with A level. ND/C, Individual consideration. SQAH, Grades BBBB including specific subject(s). SQAV, Individual consideration. IB, 30 points.

THE UNIVERSITY OF BIRMINGHAM B32

MEng Electronic and Computer Engineering (e)

UCAS Code: H610 • Mode: 3/4 Years Single Subject

Qualifications/Requirements: GCE, A level grades: BBB-BCC. Maths at A level and Physics. AGNVQ, Individual consideration. ND/C, 2 Merits and 4 Distinctions (in specific programmes). SQAH, Individual consideration. SQAV, Individual consideration. IB, 32 points.

BLACKPOOL AND THE FYLDE COLLEGE B41

HND Computer-Aided Engineering

UCAS Code: 061H • Mode: 2 Years Single Subject

Qualifications/Requirements: GCE, A/AS: 2 points. AGNVQ, Pass. ND/C, National. SQAH, Individual consideration. SQAV, Individual consideration. IB, Individual consideration.

HND Computing (Microelectronics)

UCAS Code: 66HG • Mode: 2 Years Equal Combination

Qualifications/Requirements: GCE, A/AS: 2 points. AGNVQ, Pass (in specific programmes). ND/C, National. SQAH, Individual consideration. SQAV, Individual consideration. IB, Individual consideration.

BOLTON INSTITUTE OF HIGHER EDUCATION B44

BSc Computer-Aided Product Design

UCAS Code: HW72 • Mode: 3 Years Equal Combination

Qualifications/Requirements: GCE, A/AS: 10 points. Maths, Science, Art & Design or Design & Technology. AGNVQ, Merit (in specific programmes). ND/C, 3 Merits. SQAH, Individual consideration. SQAV, Individual consideration. IB, Individual consideration.

BSc Computer-Aided Product Design

UCAS Code: HW7F • Mode: 4 Years Equal Combination

Qualifications/Requirements: GCE, A/AS: 4 points. Maths, Science, Art & Design or Design & Technology. AGNVQ, Pass (in specific programmes). ND/C, National. SQAH, Individual consideration. SQAV, Individual consideration. IB, Individual consideration.

BSc Internet Systems Development

UCAS Code: G610 • Mode: 3 Years Single Subject

Qualifications/Requirements: GCE, A level grades: CD. AGNVQ, Merit. ND/C, Merit overall. SQAH, Grades BBCC. SQAV, Individual consideration. IB, 24 points.

BEng Electronic and Computer Engineering

UCAS Code: GH66 • Mode: 3 Years Equal Combination

Qualifications/Requirements: GCE, A/AS: 12 points. Maths and Science. AGNVQ, Merit (in specific programmes). ND/C, 4 Merits. SQAH, Individual consideration. SQAV, Individual consideration. IB, Individual consideration.

BEng Electronic and Computer Engineering

UCAS Code: GH6Q • Mode: 4 Years Equal Combination

Qualifications/Requirements: GCE, A/AS: 4 points. Maths or Science. AGNVQ, Pass (in specific programmes). ND/C, National. SQAH, Individual consideration. SQAV, Individual consideration. IB, Individual consideration.

Mod Electronic/Electronic and Computer Engineering

UCAS Code: GH6P • Mode: 4/5 Years Equal Combination

Qualifications/Requirements: AGNVQ, Individual consideration. ND/C, Individual consideration. SQAH, Individual consideration. SQAV, Individual consideration. IB, Individual consideration.

Mod Motor Vehicle Studies and Simulation/Virtual Environment

UCAS Code: HGJ7 • Mode: 3 Years Equal Combination

Qualifications/Requirements: GCE, A/AS: 10 points. Maths or Science. AGNVQ, Merit. ND/C, 3 Merits.

BOURNEMOUTH UNIVERSITY B50

BEng Microelectronics and Computing (Extended)

UCAS Code: H618 • Mode: 4/5 Years Single Subject

Qualifications/Requirements: GCE, A level grades: DE. AGNVQ, Merit (in specific programmes). ND/C, Individual consideration. SQAH, Individual consideration. SQAV, Individual consideration. IB, Individual consideration.

HND Computer-Aided Design (Engineering)

UCAS Code: 061H • Mode: 2 Years Single Subject

Qualifications/Requirements: GCE, A/AS: 4-6 points. Design & Technology. AGNVQ, Individual consideration. ND/C, National. SQAH, Individual consideration. SQAV, Individual consideration. IB, Individual consideration.

HND Computer-Aided Design (Graphics Design)

UCAS Code: 52GE • Mode: 2 Years Equal Combination

Qualifications/Requirements: GCE, A/AS: 4-6 points. Design & Technology. AGNVQ, Individual consideration. ND/C, National. SQAH, Individual consideration. SQAV, Individual consideration. IB, Individual consideration.

HND Computer-Aided Design (Graphics Design)

UCAS Code: 52GW • Mode: 2 Years Equal Combination

Qualifications/Requirements: GCE, A/AS: 4-6 points. Design & Technology. AGNVQ, Individual consideration. ND/C, National. SQAH, Individual consideration. SQAV, Individual consideration. IB, Individual consideration.

HND Computer-Aided Design (Packaging Technology)

UCAS Code: 177H • Mode: 2 Years Single Subject

Qualifications/Requirements: GCE, A/AS: 4-6 points. Design & Technology. AGNVQ, Individual consideration. ND/C, National. SQAH, Individual consideration. SQAV, Individual consideration. IB, Individual consideration.

HND Computer-Aided Design (Products)

UCAS Code: 161H • Mode: 2 Years Single Subject

Qualifications/Requirements: GCE, A/AS: 4-6 points. Design & Technology. AGNVQ, Individual consideration. ND/C, National. SQAH, Individual consideration. SQAV, Individual consideration. IB, Individual consideration.

THE UNIVERSITY OF BRADFORD B56

BEng Computer-Aided Structural Engineering

UCAS Code: H240 • Mode: 3/4 Years Single Subject

Qualifications/Requirements: GCE, A level grades: CCC. Maths at A level. AGNVQ, Individual consideration. ND/C, 7 Merits (in specific programmes). SQAH, Individual consideration. SQAV, Individual consideration. IB, Individual consideration.

BEng Computer-Aided Structural Engineering

UCAS Code: H241 • Mode: 4/5 Years Single Subject

Qualifications/Requirements: GCE, A level grades: CCC. Maths at A level. AGNVQ, Individual consideration. ND/C, 7 Merits (in specific programmes). SQAH, Individual consideration. SQAV, Individual consideration. IB, Individual consideration.

BEng Electronic, Telecommunications and Computer Engineering

UCAS Code: H695 • Mode: 3 Years Single Subject

Qualifications/Requirements: GCE, A/AS: 14 points. Maths at A level. AGNVQ, Merit in Science with A level. ND/C, 5 Merits (in specific programmes). SQAH, Individual consideration. SQAV, Individual consideration. IB, Individual consideration.

BEng Electronic, Telecommunications and Computer Engineering (4 years)

UCAS Code: H690 • Mode: 4 Years Single Subject

Qualifications/Requirements: GCE, A/AS: 18 points. Maths at A level. AGNVQ, Merit in Science with A level. ND/C, 5 Merits (in specific programmes). SQAH, Individual consideration. SQAV, Individual consideration. IB, Individual consideration.

BEng Performance Engineering

UCAS Code: G740 • Mode: 3 Years Single Subject

Qualifications/Requirements: AGNVQ, Individual consideration. ND/C, Individual consideration. SQAH, Individual consideration. SQAV, Individual consideration. IB, Individual consideration.

BEng Performance Engineering

UCAS Code: G741 • Mode: 4 Years Single Subject

Qualifications/Requirements: AGNVQ, Individual consideration. ND/C, Individual consideration. SQAH, Individual consideration. SQAV, Individual consideration. IB, Individual consideration.

MEng Electronic, Telecommunications and Computer Engineering (4 years)

UCAS Code: H692 • Mode: 4 Years Single Subject

Qualifications/Requirements: GCE, A level grades: BBB. Maths and Physics at A level. AGNVQ, Distinction in Engineering with 6 additional units or with A level. ND/C, Merit overall and 3 Distinctions (in specific programmes). SQAH, Individual consideration. SQAV, Individual consideration. IB, Individual consideration.

UNIVERSITY OF BRIGHTON B72

MEng Electronic and Computer Engineering (Dip/BEng/MEng)

UCAS Code: HG66 • Mode: 3/4/5 Years Equal Combination

Qualifications/Requirements: GCE, A/AS: 10-24 points. Maths and Physics, Science or Design & Technology. AGNVQ, Individual consideration. ND/C, 7 Merits (in specific programmes). SQAH, Grades BBCC including specific subject(s). SQAV, Individual consideration. IB, 26 points.

UNIVERSITY OF BRISTOL B78

MEng Computer Systems Engineering

UCAS Code: H622 • Mode: 4 Years Single Subject

Qualifications/Requirements: GCE, A level grades: BBB. Maths and Physics at A level. AGNVQ, Distinction (in specific programmes) with A/AS. ND/C, Higher National (in specific programmes). SQAH, CSYS required. SQAV, Higher National including specific subject(s). IB, 33 points.

MEng Computer Systems Engineering with Study in Continental Europe

UCAS Code: H621 • Mode: 4 Years Single Subject

Qualifications/Requirements: GCE, A level grades: BBB. Maths and Physics at A level. AGNVQ, Distinction (in specific programmes) with A/AS. ND/C, Higher National (in specific programmes). SQAH, CSYS required. SQAV, Higher National including specific subject(s). IB, 33 points.

BRUNEL UNIVERSITY B84

BEng Computer Systems Engineering (MEng)

UCAS Code: G600 • Mode: 3/4 Years Single Subject

Qualifications/Requirements: GCE, A level grades: BCC. Maths and Science at A level. AGNVQ, Distinction in Engineering with A level. ND/C, 4 Merits and 1 Distinction (in specific programmes). SQAH, CSYS required. SQAV, Individual consideration. IB, 28 points.

BEng Computer Systems Engineering (thick SW)

UCAS Code: G602 • Mode: 4/5 Years Single Subject

Qualifications/Requirements: GCE, A level grades: BCC. Maths and Science at A level. AGNVQ, Distinction in Engineering with A level. ND/C, 4 Merits and 1 Distinction (in specific programmes). SQAH, CSYS required. SQAV, Individual consideration. IB, 28 points.

BEng Computer Systems Engineering (thin SW)

UCAS Code: G601 • Mode: 4/5 Years Single Subject

Qualifications/Requirements: GCE, A level grades: BCC. Maths and Science at A level. AGNVQ, Distinction in Engineering with A level. ND/C, 4 Merits and 1 Distinction (in specific programmes). SQAH, CSYS required. SQAV, Individual consideration. IB, 28 points.

BUCKINGHAMSHIRE CHILTERNS UNIVERSITY COLLEGE B94

BSc Computer-Aided Design

UCAS Code: H160 • Mode: 3 Years Single Subject

Qualifications/Requirements: GCE, A/AS: 8-10 points. AGNVQ, Merit. ND/C, Merit overall. SQAH, Grades CCCC. SQAV, Individual consideration. IB, Individual consideration.

BSc Computer Aided Design (1/2 years)

UCAS Code: HW72 • Mode: 1/2 Years Equal Combination

Qualifications/Requirements: ND/C, Higher National. SQAV, Higher National.

BSc Computer-Aided Design with Management

UCAS Code: H1N1 • Mode: 3 Years Major/Minor

Qualifications/Requirements: GCE, A/AS: 8-10 points. AGNVQ, Merit. ND/C, Merit overall. SQAH, Grades CCCC. SQAV, Individual consideration. IB, Individual consideration.

BSc Computer-Aided Design with Marketing

UCAS Code: H1N5 • Mode: 3 Years Major/Minor

Qualifications/Requirements: GCE, A/AS: 8-10 points. AGNVQ, Merit. ND/C, Merit overall. SQAH, Grades CCCC. SQAV, Individual consideration. IB, Individual consideration.

BSc Computer Engineering with Multimedia

UCAS Code: G6P4 • Mode: 3 Years Major/Minor

Qualifications/Requirements: GCE, A/AS: 8-10 points. AGNVQ, Merit. ND/C, Merit overall. SQAH, Grades CCCC. SQAV, Individual consideration. IB, Individual consideration.

CARDIFF UNIVERSITY C15

BEng Computer Systems Engineering

UCAS Code: HG66 • Mode: 3 Years Equal Combination

Qualifications/Requirements: GCE, A/AS: 20-22 points. Maths at A level. AGNVQ, Individual consideration. ND/C, 3 Merits and 2 Distinctions (in specific programmes). SQAH, Individual consideration. SQAV, Individual consideration. IB, 28 points.

CARMARTHENSHIRE COLLEGE C22

HND Computing (Network Administration)

UCAS Code: 016G • Mode: 2 Years Single Subject

Qualifications/Requirements: AGNVQ, Individual consideration. ND/C, Individual consideration. SQAH, Individual consideration. SQAV, Individual consideration. IB, Individual consideration.

UNIVERSITY OF CENTRAL ENGLAND IN BIRMINGHAM C25

BSc Business Systems Engineering

UCAS Code: G522 • Mode: 3/4 Years Single Subject

Qualifications/Requirements: GCE, A/AS: 16 points. AGNVQ, Merit. ND/C, 5 Merits. SQAH, Grades BBCCC. SQAV, Individual consideration. IB, 28 points.

BSc Business Systems Engineering Foundation Year

UCAS Code: G528 • Mode: 4/5 Years Single Subject

Qualifications/Requirements: GCE, A/AS: 4 points. AGNVQ, Pass. ND/C, 1 Merit. SQAH, Grades CC. SQAV, Individual consideration. IB, 24 points.

BSc Computer-Aided Design

UCAS Code: H161 • Mode: 3/4 Years Single Subject

Qualifications/Requirements: GCE, A/AS: 16 points. AGNVQ, Merit. ND/C, 5 Merits. SQAH, Grades BBCCC. SQAV, Individual consideration. IB, 28 points.

BSc Computer-Aided Design (Foundation Year)

UCAS Code: H169 • Mode: 4/5 Years Single Subject

Qualifications/Requirements: GCE, A/AS: 4 points. AGNVQ, Pass. ND/C, 1 Merit. SQAH, Grades CC. SQAV, Individual consideration. IB, 24 points.

BSc Computer Networks for Business

UCAS Code: G600 • Mode: 3/4 Years Single Subject

Qualifications/Requirements: GCE, A/AS: 16 points. AGNVQ, Merit. ND/C, 5 Merits. SQAH, Grades BBCCC. SQAV, Individual consideration. IB, 28 points.

BSc Computer Networks for Business Foundation Year

UCAS Code: G608 • Mode: 4/5 Years Single Subject

Qualifications/Requirements: GCE, A/AS: 4 points. AGNVQ, Pass. ND/C, 1 Merit. SQAH, Grades CC. SQAV, Individual consideration. IB, 24 points.

BEng Computer-Aided Design & Manufacture Foundation Year

UCAS Code: H168 • Mode: 4/5 Years Single Subject

Qualifications/Requirements: GCE, A/AS: 4 points. AGNVQ, Pass. ND/C, 1 Merit. SQAH, Grades CC. SQAV, Individual consideration. IB, 24 points.

BEng Computer-Aided Design and Manufacture

UCAS Code: H160 • Mode: 3/4 Years Single Subject

Qualifications/Requirements: GCE, A/AS: 16 points. Maths at A level. AGNVQ, Merit in Engineering with 1 additional unit. ND/C, 5 Merits (in specific programmes). SQAH, Grades BBCCC including specific subject(s). SQAV, Individual consideration. IB, 28 points.

HND Computer-Aided Design

UCAS Code: 161H • Mode: 2/3 Years Single Subject

Qualifications/Requirements: GCE, A/AS: 4 points. AGNVQ, Pass. ND/C, 1 Merit. SQAH, Grades CC. SQAV, Individual consideration. IB, 24 points.

HND Computer Networks for Business

UCAS Code: 006G • Mode: 2/3 Years Single Subject

Qualifications/Requirements: GCE, A/AS: 4 points. AGNVQ, Pass. ND/C, 1 Merit. SQAH, Grades CC. SQAV, Individual consideration. IB, 24 points.

UNIVERSITY OF CENTRAL LANCASHIRE C30

BSc Computer-Aided Technology (Year 0, Women)

UCAS Code: H161 • Mode: 4 Years Single Subject

Qualifications/Requirements: GCE, A/AS: points. AGNVQ, Individual consideration. ND/C, Individual consideration. SQAH, Individual consideration. SQAV, Individual consideration. IB, Individual consideration.

BSc Computer-Aided Engineering

UCAS Code: H160 • Mode: 3 Years Single Subject

Qualifications/Requirements: GCE, A/AS: 8 points. Maths and Science. AGNVQ, Merit (in specific programmes). ND/C, 3 Merits. SQAH, Grades CCC. SQAV, Individual consideration. IB, 24 points.

HND Computer-Aided Engineering

UCAS Code: 161H • Mode: 2 Years Single Subject

Qualifications/Requirements: GCE, A level grades: E. AGNVQ, Pass. ND/C, National. SQAH, Grades CC. IB, 24 points.

HND Engineering (Computer-Aided Engineering)

UCAS Code: 061H • Mode: 2/3 Years Single Subject

Qualifications/Requirements: GCE, A level grades: E. Maths. AGNVQ, Pass (in specific programmes). ND/C, National. SQAH, Grades CC. SQAV, Individual consideration. IB, 24 points.

HND Mechanical and Computer-Aided Engineering

UCAS Code: 31HH • Mode: 2 Years Equal Combination

Qualifications/Requirements: GCE, A level grades: E. AGNVQ, Pass. ND/C, National. SQAH, Grades CC. SQAV, Individual consideration. IB, 24 points.

CITY UNIVERSITY C60

BSc Computer Systems Engineering

UCAS Code: G656 • Mode: 3/4 Years Single Subject

Qualifications/Requirements: GCE, A level grades: DDE. Maths at A level and Physics or Computing. AGNVQ, Merit in Engineering with 6 additional units. ND/C, 5 Merits (in specific programmes). SQAH, Grades BBCCC including specific subject(s). SQAV, National including specific subject(s). IB, 24 points.

BEng Computer Systems Engineering (e)

UCAS Code: G600 • Mode: 3 Years Single Subject

Qualifications/Requirements: GCE, A level grades: CCD. Maths at A level and Physics or Computing. AGNVQ, Distinction in Engineering with A level. ND/C, 3 Merits and 2 Distinctions (in specific programmes). SQAH, Grades BBBBB including specific subject(s). SQAV, Higher National including specific subject(s). IB, 27 points.

BEng Computer Systems Engineering (e) (4 years SW)

UCAS Code: G601 • Mode: 4 Years Single Subject

Qualifications/Requirements: GCE, A level grades: CCD. Maths at A level and Physics or Computing. AGNVQ, Distinction in Engineering with A level. ND/C, 3 Merits and 2 Distinctions (in specific programmes). SQAH, Grades BBBBB including specific subject(s). SQAV, Higher National including specific subject(s). IB, 27 points.

BEng Computer Systems Engineering (Foundation)

UCAS Code: G608 • Mode: 4 Years Single Subject

Qualifications/Requirements: GCE, A level grades: EE. Maths at A level and Physics. AGNVQ, Merit in Engineering. ND/C, Individual consideration. SQAH, Grades CCCC including specific subject(s). SQAV, National including specific subject(s). IB, Individual consideration.

COVENTRY UNIVERSITY C85

BSc Computer Systems

UCAS Code: G502 • Mode: 3/4 Years Single Subject

Qualifications/Requirements: GCE, A/AS: 12 points. AGNVQ, Merit. ND/C, 4 Merits. SQAH, Grades CCCC. SQAV, Individual consideration. IB, Individual consideration.

HND Computer-Aided Engineering and Design

UCAS Code: 161H • Mode: 2 Years Single Subject

Qualifications/Requirements: GCE, A/AS: 2 points. Maths, Science or Design & Technology. AGNVQ, Pass. SQAH, Individual consideration. SQAV, Individual consideration. IB, Individual consideration.

UNIVERSITY OF DERBY D39

HND Computer-Aided Engineering Design

UCAS Code: 2RWH • Mode: 2 Years Equal Combination

Qualifications/Requirements: GCE, A/AS: 6 points. AGNVQ, Pass (in specific programmes). ND/C, 2 Merits. SQAH, Grades CCDD. SQAV, Individual consideration. IB, Pass in Diploma.

HND Computer-Aided Product Design

UCAS Code: 27WH • Mode: 2 Years Equal Combination

Qualifications/Requirements: GCE, A/AS: 6 points. AGNVQ, Pass (in specific programmes). ND/C, 2 Merits. SQAH, Grades CCDD. SQAV, Individual consideration. IB, Pass in Diploma including specific subjects.

DONCASTER COLLEGE D52

HND Electronic Engineering and Computing Technology

UCAS Code: 026H • Mode: 2 Years Single Subject

Qualifications/Requirements: Maths or Physics. AGNVQ, Pass in Engineering. ND/C, National. SQAH, Individual consideration. SQAV, Individual consideration. IB, Individual consideration.

UNIVERSITY OF DUNDEE D65

BEng Electronic Engineering and Microcomputer Systems

UCAS Code: HG66 • Mode: 4 Years Equal Combination

Qualifications/Requirements: GCE, A/AS: 14 points. Maths and Physics or Design & Technology at A level. AGNVQ, Merit in Manufacturing or in Engineering with A level. ND/C, 5 Merits (in specific programmes). SQAH, Grades BBBB including specific subject(s). SQAV, National including specific subject(s). IB, 26 points.

MEng Electronic Engineering and Microcomputer Systems

UCAS Code: GH66 • Mode: 5 Years Equal Combination

Qualifications/Requirements: GCE, A/AS: 24 points. Maths and Physics or Design & Technology at A level. AGNVQ, Merit in Manufacturing or in Engineering with A level. ND/C, 5 Merits (in specific programmes). SQAH, Grades ABBB including specific subject(s). SQAV, National including specific subject(s). IB, 26 points.

THE UNIVERSITY OF DURHAM D86

MEng Engineering (Information Systems)

UCAS Code: H610 • Mode: 4 Years Single Subject

Qualifications/Requirements: GCE, A/AS: 24 points. Maths at A level. AGNVQ, Individual consideration. ND/C, Individual consideration. SQAH, CSYS required. SQAV, Individual consideration. IB, 30 points.

UNIVERSITY OF EAST ANGLIA E14

BSc Computer Systems Engineering

UCAS Code: G600 • Mode: 3 Years Single Subject

Qualifications/Requirements: GCE, A level grades: CCC. Maths and Science at A level. AGNVQ, Merit with A level. ND/C, 4 Merits (in specific programmes). SQAH, Grades BBBC. SQAV, National including specific subject(s). IB, 28 points.

BSc Computer Systems Engineering with a Year in North America

UCAS Code: G601 • Mode: 3 Years Single Subject

Qualifications/Requirements: GCE, A level grades: BBB-BBC. Maths and Science at A level. AGNVQ, Distinction with A level. ND/C, 5 Distinctions (in specific programmes). SQAH, Grades ABBBB. SQAV, Higher National. IB, 30 points.

UNIVERSITY OF EAST LONDON E28

BSc Computer-Aided Engineering Design

UCAS Code: H160 • Mode: 3/4 Years Single Subject

Qualifications/Requirements: GCE, A/AS: 12 points. AGNVQ, Merit (in specific programmes). ND/C, 3 Merits. SQAH, Individual consideration. SQAV, Individual consideration. IB, Individual consideration.

THE UNIVERSITY OF EDINBURGH E56

BEng Electronics and Computer Science

UCAS Code: HG65 • Mode: 4 Years Equal Combination

Qualifications/Requirements: GCE, A level grades: CCC. Maths and Physics at A level. AGNVQ, Pass. ND/C, Merit overall (in specific programmes). SQAH, Grades BBBB including specific subject(s). SQAV, National including specific subject(s). IB, Pass in Diploma including specific subjects.

THE UNIVERSITY OF ESSEX E70

BSc Internet Computing

UCAS Code: G620 • Mode: 3 Years Single Subject

Qualifications/Requirements: GCE, A/AS: 20 points. AGNVQ, Distinction. ND/C, Merit overall and 2 Distinctions. SQAH, Grades BBBB. SQAV, Individual consideration. IB, 28 points.

BEng Computer Engineering

UCAS Code: H616 • Mode: 3/4 Years Single Subject

Qualifications/Requirements: GCE, A/AS: 14-18 points. Physics and Maths at A level. Entry point may vary according to qualifications held. AGNVQ, Distinction. ND/C, Merit overall (in specific programmes). SQAH, Grades BBBC. SQAV, Individual consideration. IB, 28 points.

BEng Information Systems Engineering

UCAS Code: H620 • Mode: 3/4 Years Single Subject

Qualifications/Requirements: GCE, A level grades: CCC. Physics and Maths at A level. Entry point may vary according to qualifications held. AGNVQ, Distinction. ND/C, Merit overall (in specific programmes). SQAH, Grades BBBC. SQAV, Individual consideration. IB, 28 points.

MEng Electronic & Computer Systems Engineering (MEng)

UCAS Code: H606 • Mode: 4 Years Single Subject

Qualifications/Requirements: GCE, A level grades: BBB. Maths and Computing, Electronics or Physics at A level. AGNVQ, Individual consideration. ND/C, Distinction Overall (in specific programmes). SQAH, CSYS required. SQAV, Individual consideration. IB, 32 points.

MEng Information Systems & Networks (MEng)

UCAS Code: HG6P • Mode: 4 Years Equal Combination

Qualifications/Requirements: GCE, A level grades: BBB. Maths and Computing, Electronics or Physics at A level. AGNVQ, Individual consideration. ND/C, Distinction Overall (in specific programmes). SQAH, CSYS required. SQAV, Individual consideration. IB, 32 points.

FARNBOROUGH COLLEGE OF TECHNOLOGY F66

BSc Computer Integrated Manufacturing

UCAS Code: H760 • Mode: 3 Years Single Subject

Qualifications/Requirements: GCE, A/AS: 10 points. Maths or Physics. AGNVQ, Merit. ND/C, Individual consideration. SQAH, Individual consideration. SQAV, Individual consideration. IB, Individual consideration.

HND Electronic and Computer Systems Engineering

UCAS Code: 66HG • Mode: 2 Years Equal Combination

Qualifications/Requirements: GCE, A level grades: E. AGNVQ, Pass. ND/C, National. SQAH, Individual consideration. SQAV, Individual consideration. IB, Individual consideration.

UNIVERSITY OF GLAMORGAN G14

BSc Electronics and IT Studies

UCAS Code: HG65 • Mode: 3 Years Equal Combination

Qualifications/Requirements: GCE, A level grades: DD. AGNVQ, Merit. ND/C, 3 Merits. SQAH, Individual consideration. SQAV, Individual consideration. IB, Individual consideration.

BSc Technical Marketing

UCAS Code: GH66 • Mode: 4 Years Equal Combination

Qualifications/Requirements: GCE, A level grades: CD. AGNVQ, Pass. ND/C, 3 Merits. SQAH, Individual consideration. SQAV, Individual consideration. IB, Individual consideration.

MEng Computer Systems Engineering

UCAS Code: G561 • Mode: 4/5 Years Single Subject

Qualifications/Requirements: GCE, A level grades: CC. Maths, Science or Computing. AGNVQ, Merit. ND/C, 3 Merits. SQAH, Individual consideration. SQAV, Individual consideration. IB, Individual consideration.

HND Computer-Aided Engineering

UCAS Code: 061H • Mode: 2/3 Years Single Subject

Qualifications/Requirements: GCE, A/AS: 2 points. Maths or Physics at A level. AGNVQ, Pass (in specific programmes). ND/C, National. SQAH, Individual consideration. SQAV, Individual consideration. IB, Individual consideration.

HND Computing (Network Administration)

UCAS Code: 026G • Mode: 2 Years Single Subject

Qualifications/Requirements: GCE, A/AS: 4 points. AGNVQ, Pass (in specific programmes). ND/C, 3 Merits. SQAH, Individual consideration. SQAV, Individual consideration. IB, Individual consideration.

UNIVERSITY OF GLASGOW G28

BEng Microcomputer Systems Engineering

UCAS Code: GH66 • Mode: 4 Years Equal Combination

Qualifications/Requirements: GCE, A level grades: BCC. Maths and Physics. AGNVQ, Merit in Engineering. ND/C, Merit overall. SQAH, Grades BBBBC including specific subject(s). SQAV, National including specific subject(s). IB, 24 points.

GLASGOW CALEDONIAN UNIVERSITY G42

BSc Computer-Aided Engineering

UCAS Code: H161 • Mode: 3 Years Single Subject

Qualifications/Requirements: GCE, A level grades: DEE. AGNVQ, Individual consideration. ND/C, Individual consideration. SQAH, Grades CCC including specific subject(s). SQAV, Individual consideration. IB, Individual consideration.

UNIVERSITY OF GREENWICH G70

BEng Computer Systems with Software Engineering

UCAS Code: G6G7 • Mode: 3/4 Years Major/Minor

Qualifications/Requirements: GCE, A/AS: 14 points. Science, Computing or Maths. AGNVQ, Merit (in specific programmes). ND/C, 4 Merits. SQAH, Grades CCCC. SQAV, Individual consideration. IB, 24 points.

BEng Computer Systems with Software Engineering (Extended)

UCAS Code: G6GR • Mode: 4/5 Years Major/Minor

Qualifications/Requirements: GCE, A/AS: 4 points. AGNVQ, Individual consideration. ND/C, Individual consideration. SQAH, Individual consideration. SQAV, Individual consideration. IB, Individual consideration.

HND Computer-Aided Manufacture

UCAS Code: 75HG • Mode: 3 Years Equal Combination

Qualifications/Requirements: Please refer to prospectus.

HND Computer Systems Engineering

UCAS Code: 006G • Mode: 3 Years Single Subject

Qualifications/Requirements: GCE, A/AS: 2 points. AGNVQ, Pass (in specific programmes). ND/C, 2 Merits. SQAH, Grades C. SQAV, Individual consideration. IB, Individual consideration.

HERIOT-WATT UNIVERSITY, EDINBURGH H24

BEng Electrical and Electronic Engineering (Computer Systems)

UCAS Code: HHM6 • Mode: 4 Years Equal Combination

Qualifications/Requirements: GCE, A level grades: CDD. Maths and Physics. AGNVQ, Merit (in specific programmes). ND/C, Merit overall. SQAH, Grades BBBC including specific subject(s). SQAV, Individual consideration. IB, 28 points.

BEng Mechanical Engineering (Computer-Aided Engineering

UCAS Code: H370 • Mode: 4 Years Single Subject

Qualifications/Requirements: GCE, A level grades: CDD. Maths and Physics. AGNVQ, Merit (in specific programmes) with A level. ND/C, Higher National. SQAH, Grades BBBC including specific subject(s). SQAV, Higher National. IB, 28 points.

UNIVERSITY OF HERTFORDSHIRE H36

BSc Electronic Music/Computing

UCAS Code: W3G5 • Mode: 3/4 Years Major/Minor

Qualifications/Requirements: GCE, A/AS: 14 points. Music. AGNVQ, Merit (in specific programmes) with A level. ND/C, Merit overall (in specific programmes). SQAH, Grades BCCC including specific subject(s). SQAV, Individual consideration. IB, 26 points.

BSc Electronics/Computing

UCAS Code: H6G5 • Mode: 3/4 Years Major/Minor

Qualifications/Requirements: GCE, A/AS: 14 points. AGNVQ, Merit (in specific programmes). ND/C, Merit overall (in specific programmes). SQAH, Grades BCCC including specific subject(s). SQAV, Individual consideration. IB, 24 points.

BEng Computer-Aided Engineering

UCAS Code: H160 • Mode: 3/4 Years Single Subject

Qualifications/Requirements: GCE, A/AS: 16 points. Maths, Biology, Chemistry, Physics or Computing. AGNVQ, Merit (in specific programmes). ND/C, 4 Merits (in specific programmes). SQAH, Individual consideration. SQAV, Individual consideration. IB, 26 points.

BEng Computer-Aided Engineering (Extended)

UCAS Code: H168 • Mode: 4/5 Years Single Subject

Qualifications/Requirements: GCE, A/AS: 2 points. AGNVQ, Individual consideration. ND/C, National. SQAH, Individual consideration. SQAV, Individual consideration. IB, Individual consideration.

MEng Computer-Aided Engineering
UCAS Code: H161 • Mode: 4/5 Years Single Subject
Qualifications/Requirements: GCE, A/AS: 24 points. Any three from Maths, Biology, Chemistry, Physics, Computing. AGNVQ, Not normally sufficient. ND/C, Not normally sufficient. SQAH, Individual consideration. SQAV, Individual consideration. IB, 30 points.

UHIp H49

HNC Computer-Aided Draughting & Design
UCAS Code: 161H • Mode: 1 Years Single Subject
Qualifications/Requirements: Please refer to prospectus.

THE UNIVERSITY OF HUDDERSFIELD H60

BSc Computer Integrated Manufacture
UCAS Code: H7H3 • Mode: 3/4 Years Major/Minor
Qualifications/Requirements: GCE, A/AS: 14 points. AGNVQ, Merit (in specific programmes). ND/C, 4 Merits. SQAH, Individual consideration. SQAV, Individual consideration. IB, Individual consideration.

BSc Computer-Aided Design
UCAS Code: H3H1 • Mode: 3/4 Years Major/Minor
Qualifications/Requirements: GCE, A/AS: 14 points. AGNVQ, Merit (in specific programmes). ND/C, 4 Merits. SQAH, Individual consideration. SQAV, Individual consideration. IB, Individual consideration.

BEng Computer Control Systems
UCAS Code: H660 • Mode: 4 Years Single Subject
Qualifications/Requirements: GCE, A/AS: 14 points. Maths and Science at A level. AGNVQ, Merit (in specific programmes). ND/C, 4 Merits (in specific programmes). SQAH, Grades BBB. SQAV, National including specific subject(s). IB, 26 points.

BEng Computer Control Systems (Extended)
UCAS Code: H668 • Mode: 5 Years Single Subject
Qualifications/Requirements: GCE, A/AS: 8 points. AGNVQ, Pass. ND/C, National. SQAH, Grades CCC. SQAV, Individual consideration. IB, Pass in Diploma.

BEng Computer-Aided Engineering
UCAS Code: H161 • Mode: 4 Years Single Subject
Qualifications/Requirements: GCE, A/AS: 10-14 points. Science at A level and Maths. AGNVQ, Merit in Engineering. ND/C, 3 Merits (in specific programmes). SQAH, Individual consideration. SQAV, Individual consideration. IB, Individual consideration.

BEng Computer-Aided Engineering (Extended)
UCAS Code: H168 • Mode: 5 Years Single Subject
Qualifications/Requirements: GCE, A/AS: 8 points. AGNVQ, Pass. ND/C, National. SQAH, Grades CCC. SQAV, Individual consideration. IB, Pass in Diploma.

BEng Electronic Engineering and Computer Systems
UCAS Code: GH56 • Mode: 4 Years Equal Combination
Qualifications/Requirements: GCE, A/AS: 14 points. Maths and Science at A level. AGNVQ, Merit (in specific programmes). ND/C, 4 Merits (in specific programmes). SQAH, Grades BBB. SQAV, National including specific subject(s). IB, 26 points.

BEng Electronic Engineering and Computer Systems (Extended)
UCAS Code: GH5P • Mode: 5 Years Equal Combination
Qualifications/Requirements: GCE, A/AS: 8 points. AGNVQ, Pass. ND/C, National. SQAH, Grades CCC. SQAV, Individual consideration. IB, Pass in Diploma.

MEng Computer Control Systems
UCAS Code: H661 • Mode: 5 Years Single Subject
Qualifications/Requirements: GCE, A level grades: CCC. Maths and Science at A level. AGNVQ, Merit (in specific programmes) with 3 additional units or with A level. ND/C, 3 Distinctions (in specific programmes). SQAH, Grades BBBB. SQAV, National including specific subject(s). IB, 30 points.

MEng Computer-Aided Engineering
UCAS Code: H162 • Mode: 5 Years Single Subject
Qualifications/Requirements: GCE, A level grades: CCC. Maths and Science at A level. AGNVQ, Distinction in Engineering. ND/C, 3 Distinctions. SQAH, Individual consideration. SQAV, Individual consideration. IB, Individual consideration.

MEng Electronic Engineering and Computer Systems
UCAS Code: GHM6 • Mode: 5 Years Equal Combination
Qualifications/Requirements: GCE, A level grades: CCC. Maths and Science at A level. AGNVQ, Merit (in specific programmes) with 3 additional units or with A level. ND/C, 3 Distinctions (in specific programmes). SQAH, Grades BBBB. SQAV, National including specific subject(s). IB, 30 points.

THE UNIVERSITY OF HULL H72

BEng Computer-Aided Engineering
UCAS Code: H160 • Mode: 3 Years Single Subject
Qualifications/Requirements: GCE, A level grades: CCC. Maths at A level. AGNVQ, Merit (in specific programmes) with A level. ND/C, Merit overall. SQAH, Grades BBBCC. SQAV, Individual consideration. IB, 26 points.

BEng Computer-Aided Engineering
UCAS Code: H161 • Mode: 4 Years Single Subject
Qualifications/Requirements: GCE, A level grades: CD. AGNVQ, Pass (in specific programmes). ND/C, National. SQAH, Grades CCCCD. SQAV, Individual consideration. IB, 24 points.

BEng Computer Systems Engineering

UCAS Code: G600 • Mode: 3 Years Single Subject

Qualifications/Requirements: GCE, A/AS: 18 points. Science and Maths at A level. AGNVQ, Merit in Engineering with A level. ND/C, Merit. SQAH, Grades BCCCC. SQAV, Individual consideration. IB, 26 points.

BEng Computer Systems Engineering (4 years)

UCAS Code: G601 • Mode: 4 Years Single Subject

Qualifications/Requirements: GCE, A/AS: 12 points. AGNVQ, Pass (in specific programmes). ND/C, National. SQAH, Grades CCCCD. SQAV, Individual consideration. IB, 24 points.

THE UNIVERSITY OF KENT AT CANTERBURY K24

BEng Computer Systems Engineering

UCAS Code: H610 • Mode: 3 Years Single Subject

Qualifications/Requirements: GCE, A level grades: BCC. Maths at A level and Physics. AGNVQ, Merit in Engineering with 3 additional units or with A level. ND/C, Merit overall (in specific programmes). SQAH, Individual consideration. SQAV, Individual consideration. IB, 26 points.

BEng Computer Systems Engineering (including a foundation year)

UCAS Code: H614 • Mode: 4 Years Single Subject

Qualifications/Requirements: AGNVQ, Pass. ND/C, National. SQAH, Individual consideration. SQAV, Individual consideration. IB, Pass in Diploma.

BEng Computer Systems Engineering with a year in industry

UCAS Code: H612 • Mode: 4 Years Single Subject

Qualifications/Requirements: GCE, A level grades: BCC. Maths at A level and Physics. AGNVQ, Merit in Engineering with 3 additional units or with A level. ND/C, Merit overall (in specific programmes). SQAH, Individual consideration. SQAV, Individual consideration. IB, 26 points.

MEng Computer Systems Engineering

UCAS Code: H611 • Mode: 4 Years Single Subject

Qualifications/Requirements: GCE, A/AS: 24 points. Maths at A level and Physics. AGNVQ, Merit in Engineering with 6 additional units or with A level. ND/C, Merit overall and 3 Distinctions (in specific programmes). SQAH, Individual consideration. SQAV, Individual consideration. IB, 30 points.

KING'S COLLEGE LONDON (UNIVERSITY OF LONDON) K60

BEng Computer-Aided Mechanical Engineering

UCAS Code: H303 • Mode: 3 Years Single Subject

Qualifications/Requirements: GCE, A level grades: BBC. Maths at A level and Physics. AGNVQ, Individual consideration. ND/C, Merit overall and 3 Distinctions. SQAH, Grades ABBBB. SQAV, Individual consideration. IB, 28 points.

MEng Computer-Aided Mechanical Engineering

UCAS Code: H304 • Mode: 4 Years Single Subject

Qualifications/Requirements: GCE, A level grades: BBC. Maths at A level and Physics. AGNVQ, Individual consideration. ND/C, Merit overall and 3 Distinctions. SQAH, Grades ABBBB. SQAV, Individual consideration. IB, 28 points.

LANCASTER UNIVERSITY L14

BEng Computer Systems Engineering

UCAS Code: GH66 • Mode: 3/4 Years Equal Combination

Qualifications/Requirements: GCE, A/AS: 20 points. Maths at A level and Physics, Computing, Design & Technology or Electronics. Specific GCSEs required. AGNVQ, Distinction with 6 additional units or with A/AS. ND/C, Merits and Distinction. SQAH, Grades BBBBB including specific subject(s). SQAV, Individual consideration. IB, 30 points.

MEng Computer Systems Engineering (4/5 years SW)

UCAS Code: GH6P • Mode: 4/5 Years Equal Combination

Qualifications/Requirements: GCE, A/AS: 24 points. Maths at A level and Physics, Computing, Design & Technology or Electronics. Specific GCSEs required. AGNVQ, Distinction with 6 additional units or with A/AS. ND/C, Merits and Distinction. SQAH, Grades AABBB including specific subject(s). SQAV, Individual consideration. IB, 33 points.

UNIVERSITY OF LEEDS L23

MEng Electronic and Computer Engineering

UCAS Code: H6G6 • Mode: 3/4 Years Major/Minor

Qualifications/Requirements: GCE, A level grades: BBC. Maths and Physics at A level. AGNVQ, Individual consideration. ND/C, 1 Merit and 5 Distinctions (in specific programmes). SQAH, Grades ABBBC. SQAV, Individual consideration. IB, 30 points.

UNIVERSITY OF LINCOLNSHIRE AND HUMBERSIDE L39

HND Computer-Aided Draughting and Design
UCAS Code: 061H • Mode: 2 Years Single Subject

Qualifications/Requirements: GCE, A/AS: 4 points. AGNVQ, Pass. ND/C, National. SQAH, Grades CC. SQAV, Individual consideration. IB, 24 points.

HND Engineering (Computer-Aided)
UCAS Code: 161H • Mode: 2/3 Years Single Subject

Qualifications/Requirements: GCE, A/AS: 2-4 points. Maths or Physics at A level. AGNVQ, Pass in Engineering.

HND Engineering (Electronics and Computers)
UCAS Code: 65HG • Mode: 2/3 Years Equal Combination

Qualifications/Requirements: GCE, A/AS: 4 points. Maths or Physics at A level. AGNVQ, Pass.

THE UNIVERSITY OF LIVERPOOL L41

BEng Computer and Microelectronic Systems
UCAS Code: GH66 • Mode: 3 Years Equal Combination

Qualifications/Requirements: GCE, A/AS: 22 points. Maths at A level and Physics or Computing. AGNVQ, Individual consideration. ND/C, 2 Merits and 3 Distinctions (in specific programmes). SQAH, Grades BBBBB including specific subject(s). SQAV, Individual consideration. IB, 30 points.

BEng Computer Electronics and Robotics
UCAS Code: H651 • Mode: 3 Years Single Subject

Qualifications/Requirements: GCE, A/AS: 22 points. Maths at A level and Physics. AGNVQ, Individual consideration. ND/C, 2 Merits and 3 Distinctions (in specific programmes). SQAH, Grades BBBBB including specific subject(s). SQAV, Individual consideration. IB, 30 points.

LIVERPOOL JOHN MOORES UNIVERSITY L51

BSc Applied Computer Technology
UCAS Code: G610 • Mode: 3/4 Years Single Subject

Qualifications/Requirements: GCE, A/AS: 12 points. Science at A level. AGNVQ, Merit (in specific programmes). ND/C, 5 Merits.

BEng Communications and Computer Engineering
UCAS Code: HG66 • Mode: 3 Years Equal Combination

Qualifications/Requirements: GCE, A/AS: 12 points. Maths and Physics at A level. AGNVQ, Merit (in specific programmes). ND/C, 5 Merits (in specific programmes).

BEng Computer-Aided Engineering
UCAS Code: H160 • Mode: 3/4 Years Single Subject

Qualifications/Requirements: GCE, A/AS: 10-14 points. Maths and Physics at A level. AGNVQ, Merit in Engineering with A level. ND/C, 3 Merits (in specific programmes).

BEng Computer Engineering
UCAS Code: G600 • Mode: 3 Years Single Subject

Qualifications/Requirements: GCE, A/AS: 12 points. Maths and Physics at A level. AGNVQ, Merit (in specific programmes). ND/C, 5 Merits (in specific programmes).

LLANDRILLO COLLEGE, NORTH WALES L53

HND Computer-Aided Engineering
UCAS Code: 061H • Mode: 2 Years Single Subject

Qualifications/Requirements: GCE, A/AS: 4 points. AGNVQ, Pass. ND/C, National. SQAH, Grades CCCC. SQAV, Individual consideration. IB, 26 points.

HND Network Administration
UCAS Code: 016G • Mode: 2 Years Single Subject

Qualifications/Requirements: Please refer to prospectus.

HND Telecommunications (Electronics and Computer Networks)
UCAS Code: 66HG • Mode: 2 Years Equal Combination

Qualifications/Requirements: GCE, A/AS: 4 points. AGNVQ, Pass. ND/C, National. SQAH, Grades CCCC. SQAV, Individual consideration. IB, 26 points.

LOUGHBOROUGH UNIVERSITY L79

BEng Computer Network and Internet Engineering
UCAS Code: GH66 • Mode: 3 Years Equal Combination

Qualifications/Requirements: Please refer to prospectus.

BEng Computer Network and Internet Engineering
UCAS Code: GH6P • Mode: 4 Years Equal Combination

Qualifications/Requirements: Please refer to prospectus.

BEng Electronic and Computer Systems Engineering
UCAS Code: H610 • Mode: 3 Years Single Subject

Qualifications/Requirements: GCE, A/AS: 18 points. Maths and Physics at A level. AGNVQ, Distinction in Engineering with 6 additional units or with A level. ND/C, 3 Merits and 2 Distinctions. SQAH, Individual consideration. SQAV, Individual consideration. IB, 28 points.

BEng Electronic and Computer Systems Engineering (4 years SW)
UCAS Code: H611 • Mode: 4 Years Single Subject

Qualifications/Requirements: GCE, A/AS: 18 points. Maths and Physics at A level. AGNVQ, Distinction in Engineering with 6 additional units or with A level. ND/C, 3 Merits and 2 Distinctions. SQAH, Individual consideration. SQAV, Individual consideration. IB, 28 points.

MEng Computer Network and Internet Engineering
UCAS Code: HG66 • Mode: 4 Years Equal Combination

Qualifications/Requirements: Please refer to prospectus.

MEng Computer Network and Internet Engineering
UCAS Code: HG6P • Mode: 5 Years Equal Combination

Qualifications/Requirements: Please refer to prospectus.

MEng Electronic and Computer Systems Engineering (4 years MEng)
UCAS Code: H613 • Mode: 4 Years Single Subject

Qualifications/Requirements: GCE, A/AS: 24 points. Maths and Physics at A level. AGNVQ, Distinction in Engineering with 6 additional units or with A level. ND/C, 3 Merits and 2 Distinctions. SQAH, Individual consideration. SQAV, Individual consideration. IB, 28 points.

MEng Electronic and Computer Systems Engineering (5 years MEng)
UCAS Code: H612 • Mode: 5 Years Single Subject

Qualifications/Requirements: GCE, A/AS: 24 points. Maths and Physics at A level. AGNVQ, Distinction in Engineering with 6 additional units or with A level. ND/C, 3 Merits and 2 Distinctions. SQAH, Individual consideration. SQAV, Individual consideration. IB, 28 points.

UNIVERSITY OF LUTON L93

BSc Communication System Design
UCAS Code: G610 • Mode: 3 Years Single Subject

Qualifications/Requirements: GCE, A/AS: 12 points. AGNVQ, Distinction or Merit with 6 additional units or with A/AS. ND/C, 5 Merits. SQAH, Grades BBCC. SQAV, Individual consideration. IB, 32 points.

BSc Computer-Aided Design and Manufacturing
UCAS Code: H765 • Mode: 3 Years Single Subject

Qualifications/Requirements: GCE, A/AS: 12 points. AGNVQ, Distinction or Merit with 6 additional units or with A/AS. ND/C, 5 Merits. SQAH, Grades BBCC. SQAV, Individual consideration. IB, 32 points.

BSc Computer-Aided Design Technology
UCAS Code: H160 • Mode: 3 Years Single Subject

Qualifications/Requirements: GCE, A/AS: 12 points. AGNVQ, Distinction or Merit with 6 additional units or with A/AS. ND/C, 5 Merits. SQAH, Grades BBCC. SQAV, Individual consideration. IB, 32 points.

BSc Computer Applications
UCAS Code: G611 • Mode: 3 Years Single Subject

Qualifications/Requirements: GCE, A/AS: 12 points. AGNVQ, Distinction or Merit with 6 additional units or with A/AS. ND/C, 5 Merits. SQAH, Grades BBCC. SQAV, Individual consideration. IB, 32 points.

BSc Computer System Engineering
UCAS Code: G601 • Mode: 3 Years Single Subject

Qualifications/Requirements: GCE, A/AS: 12 points. AGNVQ, Distinction or Merit with 6 additional units or with A/AS. ND/C, 5 Merits. SQAH, Grades BBCC. SQAV, Individual consideration. IB, 32 points.

HND Computer-Aided Design and Manufacturing
UCAS Code: 766H • Mode: 2 Years Single Subject

Qualifications/Requirements: GCE, A/AS: 12 points. IB, 32 points.

HND Computer Aided Design Technology
UCAS Code: 061H • Mode: 2 Years Single Subject

Qualifications/Requirements: GCE, A/AS: 6 points. AGNVQ, Merit or Pass with 6 additional units or with A/AS. ND/C, Merit. SQAH, Grades CCCC. SQAV, Individual consideration. IB, 26 points.

THE UNIVERSITY OF MANCHESTER M20

BSc Computer Engineering
UCAS Code: G600 • Mode: 3 Years Single Subject

Qualifications/Requirements: GCE, A level grades: BBB. Maths at A level. AGNVQ, Distinction in Science with A level. ND/C, Individual consideration. SQAH, Grades BBBBB including specific subject(s). SQAV, Individual consideration. IB, 30 points.

BSc Computer Engineering with Industrial Exp (4 years SW)
UCAS Code: G601 • Mode: 4 Years Single Subject

Qualifications/Requirements: GCE, A level grades: BBB. Maths at A level. AGNVQ, Distinction in Science with A level. ND/C, Individual consideration. SQAH, Grades BBBBB including specific subject(s). SQAV, Individual consideration. IB, 30 points.

THE UNIVERSITY OF MANCHESTER INSTITUTE OF SCIENCE AND TECHNOLOGY (UMIST) M25

BEng Computer Systems Engineering
UCAS Code: GH66 • Mode: 3 Years Equal Combination

Qualifications/Requirements: GCE, A level grades: BBC. Maths and Physics at A level. AGNVQ, Distinction in Engineering with A level. ND/C, Individual consideration. SQAH, Grades BBBBBC. SQAV, Individual consideration. IB, 30 points.

BEng Computer Systems Engineering with Industrial Experience
UCAS Code: GH6Q • Mode: 4 Years Equal Combination

Qualifications/Requirements: GCE, A level grades: BBC. Maths and Physics at A level. AGNVQ, Distinction in Engineering with A level. ND/C, Individual consideration. SQAH, Grades BBBBCC. SQAV, Individual consideration. IB, 30 points.

BEng Computing and Communications Systems Engineering

UCAS Code: H616 • Mode: 3 Years Single Subject

Qualifications/Requirements: GCE, A level grades: BBC. Maths at A level and Physics. AGNVQ, Individual consideration. ND/C, 4 Merits and 3 Distinctions. SQAH, Grades AABB including specific subject(s). SQAV, Individual consideration. IB, 30 points.

BEng Computing and Communications Systems Engineering with Industrial Experience

UCAS Code: H614 • Mode: 3 Years Single Subject

Qualifications/Requirements: Please refer to prospectus.

MEng Computer Systems Engineering (4 years)

UCAS Code: GH6P • Mode: 4 Years Equal Combination

Qualifications/Requirements: GCE, A level grades: BBC. Maths and Physics at A level. AGNVQ, Distinction in Engineering with A level. ND/C, Individual consideration. SQAH, Grades BBBBCC. SQAV, Individual consideration. IB, 30 points.

MEng Computing and Communications Systems Engineering

UCAS Code: H617 • Mode: 4 Years Single Subject

Qualifications/Requirements: GCE, A level grades: ABB-BBB. Maths at A level and Physics. AGNVQ, Individual consideration. ND/C, 2 Merits and 5 Distinctions. SQAH, Grades AABB including specific subject(s). SQAV, Individual consideration. IB, 30 points.

MEng Computing and Communications Systems Engineering with Industrial Experience

UCAS Code: H615 • Mode: 4 Years Single Subject

Qualifications/Requirements: Please refer to prospectus.

MChem Computer-Aided Chemistry

UCAS Code: F104 • Mode: 4 Years Single Subject

Qualifications/Requirements: GCE, A/AS: 18 points. Chemistry at A level. AGNVQ, Merit (in specific programmes) with 6 additional units or with A level. ND/C, 3 Merits and 1 Distinction. SQAH, CSYS required. SQAV, Individual consideration. IB, 26 points.

THE MANCHESTER METROPOLITAN UNIVERSITY M40

BSc Chemistry/Multimedia Technology

UCAS Code: FG16 • Mode: 3 Years Equal Combination

Qualifications/Requirements: GCE, A/AS: 12 points. Chemistry. AGNVQ, Merit (in specific programmes). ND/C, 1 Merit and 3 Distinctions (in specific programmes). SQAH, Grades BBBCC including specific subject(s). SQAV, Individual consideration. IB, 28 points.

BSc Computer Systems

UCAS Code: G600 • Mode: 4 Years Single Subject

Qualifications/Requirements: GCE, A/AS: 12-18 points. AGNVQ, Distinction (in specific programmes) or Merit (in specific programmes) with 4 additional units or with A/AS. ND/C, Merits and Distinction. SQAH, Grades BBCC. SQAV, Individual consideration. IB, 24 points.

BSc Management Systems and Multimedia Technology

UCAS Code: GN61 • Mode: 3 Years Equal Combination

Qualifications/Requirements: Please refer to prospectus.

BEng Computer and Electronic Engineering

UCAS Code: GH66 • Mode: 3/4 Years Equal Combination

Qualifications/Requirements: GCE, A/AS: 16 points. Maths and Physics. AGNVQ, Distinction with 1 additional unit. ND/C, Merit overall (in specific programmes). SQAH, Grades BBBB including specific subject(s). SQAV, Individual consideration. IB, 26 points.

BEng Computer and Electronic Engineering Foundation

UCAS Code: GH6Q • Mode: 4 Years Equal Combination

Qualifications/Requirements: GCE, A level grades: E. Maths or Physics. AGNVQ, Pass (in specific programmes). ND/C, 2 Merits (in specific programmes). SQAH, Individual consideration. SQAV, Individual consideration. IB, Individual consideration.

BEng Computer and Electronic Engineering with Study in Europe

UCAS Code: GH6P • Mode: 4 Years Equal Combination

Qualifications/Requirements: GCE, A/AS: 16 points. Maths and Physics. AGNVQ, Distinction with 1 additional unit. ND/C, Merit overall (in specific programmes). SQAH, Grades BBBB including specific subject(s). SQAV, Individual consideration. IB, 26 points.

MIDDLESEX UNIVERSITY M80

BSc Computer Communications plus minor subject

UCAS Code: G6Y4 • Mode: 3 Years Major/Minor

Qualifications/Requirements: GCE, A/AS: 12-16 points. AGNVQ, Merit (in specific programmes). ND/C, 5 Merits. SQAH, Grades CCCC. SQAV, Individual consideration. IB, 24 points.

BSc Computer Systems

UCAS Code: G500 • Mode: 3 Years Single Subject

Qualifications/Requirements: GCE, A/AS: 12 points. Science. AGNVQ, Merit (in specific programmes). ND/C, 5 Merits. SQAH, Individual consideration. SQAV, Individual consideration. IB, 24 points.

BEng Computer Systems Engineering
UCAS Code: H620 • Mode: 4 Years Single Subject

Qualifications/Requirements: GCE, A/AS: 12-16 points. Science. AGNVQ, Merit (in specific programmes). ND/C, 5 Merits (in specific programmes). SQAH, Grades CCCC. SQAV, Individual consideration. IB, 24 points.

NAPIER UNIVERSITY N07

BEng Automation Systems
UCAS Code: G610 • Mode: 3/4 Years Single Subject

Qualifications/Requirements: GCE, A level grades: CD. ND/C, Individual consideration. SQAH, Grades BBB. SQAV, Individual consideration.

BEng Computer Based Systems Engineering
UCAS Code: G600 • Mode: 3/4 Years Single Subject

Qualifications/Requirements: GCE, A level grades: CD. ND/C, Individual consideration. SQAH, Grades BBB. SQAV, Individual consideration.

BEng Electronic and Computer Engineering
UCAS Code: GH56 • Mode: 4 Years Equal Combination

Qualifications/Requirements: GCE, A level grades: DD. AGNVQ, Individual consideration. ND/C, Individual consideration. SQAH, Grades BBC. SQAV, Individual consideration. IB, Individual consideration.

BEng Systems Integration
UCAS Code: G601 • Mode: 3/4 Years Single Subject

Qualifications/Requirements: GCE, A level grades: CD. ND/C, Individual consideration. SQAH, Grades BBB. SQAV, Individual consideration.

UNIVERSITY COLLEGE NORTHAMPTON N14

HND Computer Systems
UCAS Code: 005G • Mode: 2 Years Single Subject

Qualifications/Requirements: GCE, A/AS: 2 points. AGNVQ, Pass. ND/C, 2 Merits. SQAH, Grades CC. SQAV, Individual consideration. IB, Individual consideration.

UNIVERSITY OF NEWCASTLE UPON TYNE N21

BSc Computer Systems Engineering
UCAS Code: G600 • Mode: 3 Years Single Subject

Qualifications/Requirements: GCE, A/AS: 20 points. Maths at A level and Science. AGNVQ, Individual consideration. ND/C, Individual consideration. SQAH, CSYS required. SQAV, Individual consideration. IB, Individual consideration.

BSc Computer Systems Engineering (foundation)
UCAS Code: G601 • Mode: 4 Years Single Subject

Qualifications/Requirements: AGNVQ, Individual consideration. ND/C, Individual consideration. SQAH, Individual consideration. SQAV, Individual consideration. IB, Individual consideration.

THE NORTH EAST WALES INSTITUTE OF HIGHER EDUCATION N56

HND Computer Engineering
UCAS Code: 006H • Mode: 2 Years Single Subject

Qualifications/Requirements: GCE, A/AS: 4-8 points. AGNVQ, Pass (in specific programmes). ND/C, 3 Merits. SQAH, Individual consideration. SQAV, Individual consideration. IB, Individual consideration.

HND Electronics and Microcomputing
UCAS Code: 016H • Mode: 2 Years Single Subject

Qualifications/Requirements: GCE, A/AS: 2-6 points. AGNVQ, Pass (in specific programmes). ND/C, 2 Merits. SQAH, Grades CC. SQAV, National including specific subject(s). IB, Individual consideration.

UNIVERSITY OF NORTH LONDON N63

BSc Microcomputer Systems Technology
UCAS Code: H610 • Mode: 3/4 Years Single Subject

Qualifications/Requirements: GCE, A/AS: 8 points. Maths and Electronics, Computing, Design & Technology at A level or Physics. Specific GCSEs required. AGNVQ, Merit (in specific programmes). ND/C, 3 Merits (in specific programmes). SQAH, Grades CCCC including specific subject(s). SQAV, Individual consideration. IB, Individual consideration.

Mod Microcomputer Systems Technology and Marketing
UCAS Code: HN65 • Mode: 3 Years Equal Combination

Qualifications/Requirements: GCE, A/AS: 14 points. Any Physical Science, Electronics, Computing or Maths. AGNVQ, Individual consideration. ND/C, Individual consideration. SQAH, Individual consideration. SQAV, Individual consideration. IB, Individual consideration.

HND Microcomputer Systems Technology
UCAS Code: 016H • Mode: 2 Years Single Subject

Qualifications/Requirements: GCE, A/AS: 4 points. Maths, Physics, Computing or Electronics. AGNVQ, Pass in Engineering. ND/C, 1 Merit (in specific programmes). SQAH, Grades CCC including specific subject(s). SQAV, Individual consideration. IB, Individual consideration.

UNIVERSITY OF NORTHUMBRIA AT NEWCASTLE N77

BSc Computer and Network Technology
UCAS Code: G621 • Mode: 3/4 Years Single Subject

Qualifications/Requirements: GCE, A/AS: 10 points. AGNVQ, Merit. ND/C, 3 Merits. SQAH, Grades BCC. SQAV, Individual consideration. IB, 24 points.

BEng Computer-Aided Engineering
UCAS Code: H160 • Mode: 3/4 Years Single Subject

Qualifications/Requirements: GCE, A/AS: 10-18 points. Maths and Physics or Science at A level. AGNVQ, Merit in Engineering. ND/C, Individual consideration. SQAH, Individual consideration. SQAV, Individual consideration. IB, Individual consideration.

BEng Computer-Aided Engineering
UCAS Code: H168 • Mode: 5 Years Single Subject

Qualifications/Requirements: GCE, A/AS: 2-4 points. AGNVQ, Pass. ND/C, National. SQAH, Individual consideration. SQAV, Individual consideration. IB, Individual consideration.

NORWICH: CITY COLLEGE N82

HND Engineering (Computer-Aided) (SW)
UCAS Code: 061H • Mode: 3 Years Single Subject

Qualifications/Requirements: GCE, A level grades: E. AGNVQ, Pass (in specific programmes) with 6 additional units. ND/C, Individual consideration. SQAH, Individual consideration. SQAV, Individual consideration. IB, Individual consideration.

THE UNIVERSITY OF NOTTINGHAM N84

BEng Electronic and Computer Engineering
UCAS Code: H610 • Mode: 3 Years Single Subject

Qualifications/Requirements: GCE, A level grades: BBC-CCE. Maths and Physics at A level. AGNVQ, Individual consideration. ND/C, Individual consideration. SQAH, CSYS required. SQAV, Individual consideration. IB, 27 points.

THE NOTTINGHAM TRENT UNIVERSITY N91

BEng Electronics and Computing
UCAS Code: GH56 • Mode: 3/4 Years Equal Combination

Qualifications/Requirements: GCE, A/AS: 18 points. AGNVQ, Individual consideration. ND/C, 4 Distinctions. SQAH, Grades BBCC. SQAV, Individual consideration. IB, Individual consideration.

BEng Electronics and Computing (Extended)
UCAS Code: GH5P • Mode: 4/5 Years Equal Combination

Qualifications/Requirements: AGNVQ, Individual consideration. ND/C, Individual consideration. SQAH, Individual consideration. SQAV, Individual consideration. IB, Individual consideration.

BEng Engineering Systems and Computing (Extended)
UCAS Code: H618 • Mode: 4/5 Years Single Subject

Qualifications/Requirements: AGNVQ, Individual consideration. ND/C, Individual consideration. SQAH, Individual consideration. SQAV, Individual consideration. IB, Individual consideration.

OXFORD BROOKES UNIVERSITY O66

BA/BSc Computer Systems/Spanish Studies
UCAS Code: GR54 • Mode: 3 Years Equal Combination

Qualifications/Requirements: Please refer to prospectus.

BSc Computer-Aided Product Design
UCAS Code: H160 • Mode: 3 Years Single Subject

Qualifications/Requirements: GCE, A/AS: 14 points. Design & Technology at A level. AGNVQ, Merit (in specific programmes). ND/C, Individual consideration. SQAH, Individual consideration. SQAV, Individual consideration. IB, Individual consideration.

BSc Computer Systems (FT or SW)
UCAS Code: G600 • Mode: 4 Years Single Subject

Qualifications/Requirements: GCE, A level grades: CCC. AGNVQ, Merit. ND/C, Individual consideration. SQAH, Individual consideration. SQAV, Individual consideration. IB, Individual consideration.

Mod Accounting and Finance/Computer Systems
UCAS Code: GN64 • Mode: 3 Years Equal Combination

Qualifications/Requirements: GCE, A level grades: BC-BCC. AGNVQ, Merit or Distinction with 3 additional units. ND/C, Individual consideration. SQAH, Individual consideration. SQAV, Individual consideration. IB, Individual consideration.

Mod Anthropology/Computer Systems
UCAS Code: GL66 • Mode: 3 Years Equal Combination

Qualifications/Requirements: GCE, A level grades: BC-BCC. AGNVQ, Merit with A/AS. ND/C, Individual consideration. SQAH, Individual consideration. SQAV, Individual consideration. IB, Individual consideration.

Mod Biological Chemistry/Computer Systems
UCAS Code: CG76 • Mode: 3 Years Equal Combination

Qualifications/Requirements: GCE, A level grades: DD-CCC. Science. AGNVQ, Merit in Science. ND/C, Individual consideration. SQAH, Individual consideration. SQAV, Individual consideration. IB, Individual consideration.

Mod Biology/Computer Systems
UCAS Code: CG16 • Mode: 3 Years Equal Combination

Qualifications/Requirements: GCE, A level grades: DD-CDD. Science. AGNVQ, Merit in Science. ND/C, Individual consideration. SQAH, Individual consideration. SQAV, Individual consideration. IB, Individual consideration.

Mod Business Administration and Management/Computer Systems
UCAS Code: GN61 • Mode: 3 Years Equal Combination

Qualifications/Requirements: GCE, A level grades: BC-BBC. AGNVQ, Merit in Business or Merit in Business with 4 additional units. ND/C, Individual consideration. SQAH, Individual consideration. SQAV, Individual consideration. IB, Individual consideration.

Mod Cities and Society/Computer Systems
UCAS Code: LG36 • Mode: 3 Years Equal Combination

Qualifications/Requirements: GCE, A level grades: DD-CCC. AGNVQ, Merit. ND/C, Individual consideration. SQAH, Individual consideration. SQAV, Individual consideration. IB, Individual consideration.

Mod Combined Studies/Computer Systems

UCAS Code: GY64 • Mode: 3 Years

Qualifications/Requirements: AGNVQ, Not normally sufficient. ND/C, Individual consideration. SQAH, Not normally sufficient. SQAV, Individual consideration. IB, Not normally sufficient.

Mod Complementary Therapies – Aromatherapy/Computer Systems

UCAS Code: GW68 • Mode: 3 Years Equal Combination

Qualifications/Requirements: Please refer to prospectus.

Mod Computer Systems/Computing

UCAS Code: GG65 • Mode: 3 Years Equal Combination

Qualifications/Requirements: GCE, A level grades: CDD-BC. AGNVQ, Merit. ND/C, Individual consideration. SQAH, Individual consideration. SQAV, Individual consideration. IB, Individual consideration.

Mod Computer Systems/Computing Mathematics

UCAS Code: GG69 • Mode: 3 Years Equal Combination

Qualifications/Requirements: GCE, A level grades: CD-BC. AGNVQ, Merit. ND/C, Individual consideration. SQAH, Individual consideration. SQAV, Individual consideration. IB, Individual consideration.

Mod Computer Systems/Computing Science

UCAS Code: GG56 • Mode: 3 Years Equal Combination

Qualifications/Requirements: GCE, A level grades: DD-CCC. AGNVQ, Merit. ND/C, Individual consideration. SQAH, Individual consideration. SQAV, Individual consideration. IB, Individual consideration.

Mod Computer Systems/Ecology

UCAS Code: CG96 • Mode: 3 Years Equal Combination

Qualifications/Requirements: GCE, A level grades: CD-BC. AGNVQ, Merit in Science. ND/C, Individual consideration. SQAH, Individual consideration. SQAV, Individual consideration. IB, Individual consideration.

Mod Computer Systems/Economics

UCAS Code: GL61 • Mode: 3 Years Equal Combination

Qualifications/Requirements: GCE, A level grades: CDD-BB. AGNVQ, Merit or Merit with 3 additional units. ND/C, Individual consideration. SQAH, Individual consideration. SQAV, Individual consideration. IB, Individual consideration.

Mod Computer Systems/Educational Studies

UCAS Code: GX69 • Mode: 3 Years Equal Combination

Qualifications/Requirements: GCE, A level grades: DDD-BC. AGNVQ, Merit or Merit with 3 additional units. ND/C, Individual consideration. SQAH, Individual consideration. SQAV, Individual consideration. IB, Individual consideration.

Mod Computer Systems/Electronics

UCAS Code: GH66 • Mode: 3 Years Equal Combination

Qualifications/Requirements: GCE, A level grades: CC-BC. Science or Maths at A level. AGNVQ, Merit (in specific programmes). ND/C, Individual consideration. SQAH, Individual consideration. SQAV, Individual consideration. IB, Individual consideration.

Mod Computer Systems/English Studies

UCAS Code: GQ63 • Mode: 3 Years Equal Combination

Qualifications/Requirements: GCE, A level grades: CCC-BBB. AGNVQ, Merit with A/AS. ND/C, Individual consideration. SQAH, Individual consideration. SQAV, Individual consideration. IB, Individual consideration.

Mod Computer Systems/Environmental Chemistry

UCAS Code: GF61 • Mode: 3 Years Equal Combination

Qualifications/Requirements: GCE, A level grades: DD-CCC. Science. AGNVQ, Distinction in Science. ND/C, Individual consideration. SQAH, Individual consideration. SQAV, Individual consideration. IB, Individual consideration.

Mod Computer Systems/Environmental Design and Conservation

UCAS Code: FG96 • Mode: 3 Years Equal Combination

Qualifications/Requirements: GCE, A level grades: DD-CCC. AGNVQ, Merit. ND/C, Individual consideration. SQAH, Individual consideration. SQAV, Individual consideration. IB, Individual consideration.

Mod Computer Systems/Environmental Policy

UCAS Code: KG36 • Mode: 3 Years Equal Combination

Qualifications/Requirements: GCE, A level grades: DD-CCC. AGNVQ, Merit. ND/C, Individual consideration. SQAH, Individual consideration. SQAV, Individual consideration. IB, Individual consideration.

Mod Computer Systems/Environmental Sciences

UCAS Code: FGX6 • Mode: 3 Years Equal Combination

Qualifications/Requirements: GCE, A level grades: CD-BC. Science. AGNVQ, Merit or Distinction in Science. ND/C, Individual consideration. SQAH, Individual consideration. SQAV, Individual consideration. IB, Individual consideration.

Mod Computer Systems/European Culture and Society

UCAS Code: TGG6 • Mode: 3 Years Equal Combination

Qualifications/Requirements: GCE, A level grades: DD-CCC. AGNVQ, Merit with A/AS. ND/C, Individual consideration. SQAH, Individual consideration. SQAV, Individual consideration. IB, Individual consideration.

Mod Computer Systems/Exercise and Health
UCAS Code: GB66 • Mode: 3 Years Equal Combination

Qualifications/Requirements: GCE, A level grades: DD-BC. Science. AGNVQ, Merit in Science. ND/C, Individual consideration. SQAH, Individual consideration. SQAV, Individual consideration. IB, Individual consideration.

Mod Computer Systems/Fine Art
UCAS Code: GW61 • Mode: 3 Years Equal Combination

Qualifications/Requirements: GCE, A level grades: CDD-BC. Art. A portfolio of work is required. AGNVQ, Merit in Art & Design with A/AS. ND/C, Individual consideration. SQAH, Individual consideration. SQAV, Individual consideration. IB, Individual consideration.

Mod Computer Systems/Food Science and Nutrition
UCAS Code: DG46 • Mode: 3 Years Equal Combination

Qualifications/Requirements: GCE, A level grades: DD-BC. Science. AGNVQ, Merit in Science. ND/C, Individual consideration. SQAH, Individual consideration. SQAV, Individual consideration. IB, Individual consideration.

Mod Computer Systems/French Studies
UCAS Code: RG16 • Mode: 3 Years Equal Combination

Qualifications/Requirements: Please refer to prospectus.

Mod Computer Systems/Geographic Information Science
UCAS Code: FG8P • Mode: 3 Years Equal Combination

Qualifications/Requirements: Please refer to prospectus.

Mod Computer Systems/Geography
UCAS Code: GL68 • Mode: 3 Years Equal Combination

Qualifications/Requirements: GCE, A level grades: CDD-BB. AGNVQ, Merit. ND/C, Individual consideration. SQAH, Individual consideration. SQAV, Individual consideration. IB, Individual consideration.

Mod Computer Systems/Geology
UCAS Code: FG66 • Mode: 3 Years Equal Combination

Qualifications/Requirements: GCE, A level grades: DD-BC. AGNVQ, Pass in Science or Merit in Science. ND/C, Individual consideration. SQAH, Individual consideration. SQAV, Individual consideration. IB, Individual consideration.

Mod Computer Systems/Geotechnics
UCAS Code: GH62 • Mode: 3 Years Equal Combination

Qualifications/Requirements: GCE, A level grades: DD-BC. Science, Maths, Design & Technology or Electronics. AGNVQ, Merit (in specific programmes). ND/C, Individual consideration. SQAH, Individual consideration. SQAV, Individual consideration. IB, Individual consideration.

Mod Computer Systems/German Studies
UCAS Code: GR6G • Mode: 4 Years Equal Combination

Qualifications/Requirements: GCE, A level grades: DDD-CCC. German. AGNVQ, Merit with A level. ND/C, Individual consideration. SQAH, Individual consideration. SQAV, Individual consideration. IB, Individual consideration.

Mod Computer Systems/History
UCAS Code: GV61 • Mode: 3 Years Equal Combination

Qualifications/Requirements: GCE, A level grades: CDD-BB. AGNVQ, Merit with A/AS. ND/C, Individual consideration. SQAH, Individual consideration. SQAV, Individual consideration. IB, Individual consideration.

Mod Computer Systems/History of Art
UCAS Code: GV64 • Mode: 3 Years Equal Combination

Qualifications/Requirements: GCE, A level grades: CDD-BCC. AGNVQ, Merit with A/AS. ND/C, Individual consideration. SQAH, Individual consideration. SQAV, Individual consideration. IB, Individual consideration.

Mod Computer Systems/Hospitality Management Studies
UCAS Code: GN67 • Mode: 3 Years Equal Combination

Qualifications/Requirements: GCE, A level grades: DDD-BC. AGNVQ, Merit or Merit with 3 additional units. ND/C, Individual consideration. SQAH, Individual consideration. SQAV, Individual consideration. IB, Individual consideration.

Mod Computer Systems/Human Biology
UCAS Code: BG16 • Mode: 3 Years Equal Combination

Qualifications/Requirements: GCE, A level grades: DDD-BCC. Science. AGNVQ, Merit in Science. ND/C, Individual consideration. SQAH, Individual consideration. SQAV, Individual consideration. IB, Individual consideration.

Mod Computer Systems/Information Systems
UCAS Code: GG6M • Mode: 3 Years Equal Combination

Qualifications/Requirements: GCE, A level grades: CDD-BC. AGNVQ, Merit. ND/C, Individual consideration. SQAH, Individual consideration. SQAV, Individual consideration. IB, Individual consideration.

Mod Computer Systems/Intelligent Systems
UCAS Code: GG68 • Mode: 3 Years Equal Combination

Qualifications/Requirements: GCE, A level grades: CDD-BC. AGNVQ, Merit. ND/C, Individual consideration. SQAH, Individual consideration. SQAV, Individual consideration. IB, Individual consideration.

Mod Computer Systems/Japanese Studies
UCAS Code: GT64 • Mode: 3 Years Equal Combination

Qualifications/Requirements: Please refer to prospectus.

Mod Computer Systems/Law
UCAS Code: GM63 • Mode: 3 Years Equal Combination

Qualifications/Requirements: GCE, A level grades: CCC-ABB. AGNVQ, Merit or Distinction with 3 additional units. ND/C, Individual consideration. SQAH, Individual consideration. SQAV, Individual consideration. IB, Individual consideration.

Mod Computer Systems/Leisure Planning
UCAS Code: KGH6 • Mode: 3 Years Equal Combination

Qualifications/Requirements: GCE, A level grades: DDD-BC. AGNVQ, Merit. ND/C, Individual consideration. SQAH, Individual consideration. SQAV, Individual consideration. IB, Individual consideration.

Mod Computer Systems/Marketing Management
UCAS Code: GN6N • Mode: 3 Years Equal Combination

Qualifications/Requirements: GCE, A level grades: CDD-BCC. AGNVQ, Merit or Distinction with 3 additional units. ND/C, Individual consideration. SQAH, Individual consideration. SQAV, Individual consideration. IB, Individual consideration.

Mod Computer Systems/Mathematics
UCAS Code: GG61 • Mode: 3 Years Equal Combination

Qualifications/Requirements: GCE, A level grades: DD-BC. Maths. AGNVQ, Merit with A level. ND/C, Individual consideration. SQAH, Individual consideration. SQAV, Individual consideration. IB, Individual consideration.

Mod Computer Systems/Multimedia Systems
UCAS Code: G610 • Mode: 3 Years Single Subject

Qualifications/Requirements: Please refer to prospectus.

Mod Computer Systems/Music
UCAS Code: GW63 • Mode: 3 Years Equal Combination

Qualifications/Requirements: GCE, A level grades: DD-BC. Music. AGNVQ, Merit. ND/C, Individual consideration. SQAH, Individual consideration. SQAV, Individual consideration. IB, Individual consideration.

Mod Computer Systems/Palliative Care
UCAS Code: BGR6 • Mode: 3 Years Equal Combination

Qualifications/Requirements: Please refer to prospectus.

Mod Computer Systems/Physical Geography
UCAS Code: FGV6 • Mode: 3 Years Equal Combination

Qualifications/Requirements: GCE, A level grades: CDD-BB. AGNVQ, Merit. ND/C, Individual consideration. SQAH, Individual consideration. SQAV, Individual consideration. IB, Individual consideration.

Mod Computer Systems/Planning Studies
UCAS Code: GK64 • Mode: 3 Years Equal Combination

Qualifications/Requirements: GCE, A level grades: DDD-BC. AGNVQ, Merit. ND/C, Individual consideration. SQAH, Individual consideration. SQAV, Individual consideration. IB, Individual consideration.

Mod Computer Systems/Politics
UCAS Code: GM61 • Mode: 3 Years Equal Combination

Qualifications/Requirements: GCE, A level grades: BC-AB. AGNVQ, Merit with A/AS. ND/C, Individual consideration. SQAH, Individual consideration. SQAV, Individual consideration. IB, Individual consideration.

Mod Computer Systems/Psychology
UCAS Code: CG86 • Mode: 3 Years Equal Combination

Qualifications/Requirements: GCE, A level grades: CCC-ABB. AGNVQ, Merit with A/AS. ND/C, Individual consideration. SQAH, Individual consideration. SQAV, Individual consideration. IB, Individual consideration.

Mod Computer Systems/Publishing
UCAS Code: GP65 • Mode: 3 Years Equal Combination

Qualifications/Requirements: GCE, A level grades: CDD-CCC. AGNVQ, Merit or Merit (in specific programmes) with 3 additional units. ND/C, Individual consideration. SQAH, Individual consideration. SQAV, Individual consideration. IB, Individual consideration.

Mod Computer Systems/Rehabilitation
UCAS Code: BGT6 • Mode: 3 Years Equal Combination

Qualifications/Requirements: Please refer to prospectus.

Mod Computer Systems/Retail Management
UCAS Code: GN65 • Mode: 3 Years Equal Combination

Qualifications/Requirements: GCE, A level grades: CDD-BBC. AGNVQ, Merit with A/AS or Merit in Business with 4 additional units. ND/C, Individual consideration. SQAH, Individual consideration. SQAV, Individual consideration. IB, Individual consideration.

Mod Computer Systems/Sociology
UCAS Code: GL63 • Mode: 3 Years Equal Combination

Qualifications/Requirements: GCE, A level grades: CDD-BCC. AGNVQ, Merit with A/AS. ND/C, Individual consideration. SQAH, Individual consideration. SQAV, Individual consideration. IB, Individual consideration.

Mod Computer Systems/Software Engineering
UCAS Code: GG67 • Mode: 3 Years Equal Combination

Qualifications/Requirements: GCE, A level grades: CDD-BC. AGNVQ, Merit. ND/C, Individual consideration. SQAH, Individual consideration. SQAV, Individual consideration. IB, Individual consideration.

Mod Computer Systems/Statistics
UCAS Code: GG64 • Mode: 3 Years Equal Combination

Qualifications/Requirements: GCE, A level grades: DD-BC. AGNVQ, Merit. ND/C, Individual consideration. SQAH, Individual consideration. SQAV, Individual consideration. IB, Individual consideration.

Mod Computer Systems/Telecommunications
UCAS Code: GH6P • Mode: 3 Years Equal Combination

Qualifications/Requirements: GCE, A level grades: DD-CCC. Science or Maths. AGNVQ, Merit (in specific programmes). ND/C, Individual consideration. SQAH, Individual consideration. SQAV, Individual consideration. IB, Individual consideration.

Mod Computer Systems/Tourism
UCAS Code: GP67 • Mode: 3 Years Equal Combination

Qualifications/Requirements: GCE, A level grades: CDD-BC. AGNVQ, Merit or Merit with 3 additional units. ND/C, Individual consideration. SQAH, Individual consideration. SQAV, Individual consideration. IB, Individual consideration.

Mod Computer Systems/Transport and Travel
UCAS Code: GN69 • Mode: 3 Years Equal Combination

Qualifications/Requirements: GCE, A level grades: CDD-BC. AGNVQ, Merit. ND/C, Individual consideration. SQAH, Individual consideration. SQAV, Individual consideration. IB, Individual consideration.

Mod Computer Systems/Water Resources
UCAS Code: GH6F • Mode: 3 Years Equal Combination

Qualifications/Requirements: GCE, A level grades: DD-CCC. AGNVQ, Merit. ND/C, Individual consideration. SQAH, Individual consideration. SQAV, Individual consideration. IB, Individual consideration.

Mod Electronics/Information Systems
UCAS Code: GHM6 • Mode: 3 Years Equal Combination

Qualifications/Requirements: GCE, A level grades: CC-BCC. Science or Maths at A level. AGNVQ, Merit (in specific programmes). ND/C, Individual consideration. SQAH, Individual consideration. SQAV, Individual consideration. IB, Individual consideration.

Mod Electronics/Intelligent Systems
UCAS Code: GH86 • Mode: 3 Years Equal Combination

Qualifications/Requirements: GCE, A level grades: CD-CCC. Science or Maths at A level. AGNVQ, Merit (in specific programmes). ND/C, Individual consideration. SQAH, Individual consideration. SQAV, Individual consideration. IB, Individual consideration.

Mod Electronics/Multimedia Systems
UCAS Code: GHP6 • Mode: 3 Years Equal Combination

Qualifications/Requirements: Please refer to prospectus.

Mod Mapping and Cartography/ Computer Systems
UCAS Code: FG86 • Mode: 3 Years Equal Combination

Qualifications/Requirements: GCE, A level grades: DDD-BC. AGNVQ, Merit. ND/C, Individual consideration. SQAH, Individual consideration. SQAV, Individual consideration. IB, Individual consideration.

HND Computer-Aided Product Design
UCAS Code: 27WH • Mode: 2 Years Equal Combination

Qualifications/Requirements: AGNVQ, Individual consideration. ND/C, Individual consideration. SQAH, Individual consideration. SQAV, Individual consideration. IB, Individual consideration.

HND Computer Systems Engineering
UCAS Code: 006G • Mode: 2 Years Single Subject

Qualifications/Requirements: AGNVQ, Individual consideration. ND/C, Individual consideration. SQAH, Individual consideration. SQAV, Individual consideration. IB, Individual consideration.

UNIVERSITY OF PAISLEY P20

BSc Computer-Aided Design
UCAS Code: H160 • Mode: 1 Years Single Subject

Qualifications/Requirements: AGNVQ, Not normally sufficient. ND/C, Higher National. SQAH, Not normally sufficient. SQAV, Higher National. IB, Not normally sufficient.

BSc Computer Networking
UCAS Code: G601 • Mode: 3 Years Single Subject

Qualifications/Requirements: GCE, A level grades: DE. AGNVQ, Pass in Engineering. ND/C, Individual consideration. SQAH, Grades BCC including specific subject(s). SQAV, Higher National. IB, Pass in Diploma.

BEng Computer Engineering
UCAS Code: GH66 • Mode: 3/4/5 Years Equal Combination

Qualifications/Requirements: GCE, A level grades: CD. Maths and Physics. AGNVQ, Pass in Engineering. ND/C, Individual consideration. SQAH, Grades BBBC including specific subject(s). SQAV, Higher National. IB, Pass in Diploma.

UNIVERSITY OF PLYMOUTH P60

BSc Civil Engineering and Computer-Aided Design
UCAS Code: K2G5 • Mode: 3/4 Years Major/Minor

Qualifications/Requirements: GCE, A/AS: 14 points. An approved subject from restricted list at A level. AGNVQ, Merit (in specific programmes). ND/C, 4 Merits and 1 Distinction (in specific programmes). SQAH, Individual consideration. SQAV, Individual consideration. IB, Individual consideration.

BSc Computer Systems Engineering
UCAS Code: G601 • Mode: 3/4 Years Single Subject

Qualifications/Requirements: GCE, A/AS: 10 points. An approved subject from restricted list at A level. AGNVQ, Merit (in specific programmes). ND/C, 2 Merits. SQAH, Individual consideration. SQAV, Individual consideration. IB, Individual consideration.

HND Computer Systems Engineering
UCAS Code: 106G • Mode: 2 Years Single Subject

Qualifications/Requirements: GCE, A/AS: 4 points. An approved subject from restricted list at A level. AGNVQ, Pass (in specific programmes). ND/C, National. SQAH, Individual consideration. SQAV, Individual consideration. IB, Individual consideration.

UNIVERSITY OF PORTSMOUTH P80

BSc Computer Systems Engineering
UCAS Code: G602 • Mode: 3/4 Years Single Subject

Qualifications/Requirements: GCE, A/AS: 8 points. Any two from Maths,Physics,Electronics,Computing. AGNVQ, Pass. ND/C, 2 Merits. SQAH, Grades BCC. SQAV, Individual consideration. IB, Pass in Diploma.

BSc Internet Technology
UCAS Code: HG66 • Mode: 3/4 Years Equal Combination

Qualifications/Requirements: GCE, A/AS: 18 points. AGNVQ, Distinction (in specific programmes). ND/C, 6 Merits (in specific programmes). SQAH, Grades BBCC. SQAV, Individual consideration. IB, 26 points.

BEng Computer Engineering
UCAS Code: G601 • Mode: 3/4 Years Single Subject

Qualifications/Requirements: GCE, A/AS: 18 points. Any two from Maths, Physics, Computing, Electronics. AGNVQ, Merit (in specific programmes) with 3 additional units or with A/AS. ND/C, 6 Merits (in specific programmes). SQAH, Grades BBCC. SQAV, Individual consideration. IB, 26 points.

BEng Computer Technology (ext route available)
UCAS Code: G600 • Mode: 3/4 Years Single Subject

Qualifications/Requirements: GCE, A/AS: 12 points. Any two from Maths, Physics, Computing, Electronics, Design & Technology. AGNVQ, Merit (in specific programmes) with 2 additional units or with A/AS. ND/C, 4 Merits (in specific programmes). SQAH, Grades BCCC. SQAV, Individual consideration. IB, 24 points.

BEng Electronic and Computer Engineering
UCAS Code: GH66 • Mode: 3/4 Years Equal Combination

Qualifications/Requirements: GCE, A/AS: 18 points. Any two from Maths, Physics, Computing, Electronics. AGNVQ, Merit (in specific programmes) with 3 additional units or with A/AS. ND/C, 6 Merits (in specific programmes). SQAH, Grades BBCC. SQAV, Individual consideration. IB, 26 points.

MEng Computer Engineering
UCAS Code: G603 • Mode: 4/5 Years Single Subject

Qualifications/Requirements: GCE, A/AS: 22 points. Any two from Maths, Physics, Computing, Electronics. AGNVQ, Distinction (in specific programmes) with 6 additional units or with A/AS. ND/C, 4 Merits and 3 Distinctions (in specific programmes). SQAH, Grades BBBB. SQAV, Individual consideration. IB, 30 points.

MEng Electronic and Computer Engineering
UCAS Code: GH6P • Mode: 4/5 Years Equal Combination

Qualifications/Requirements: GCE, A/AS: 22 points. Any two from Maths,Physics,Computing,Electronics. AGNVQ, Distinction (in specific programmes) with 6 additional units or with A level. ND/C, 4 Merits and 3 Distinctions (in specific programmes). SQAH, Grades BBBB. SQAV, Individual consideration. IB, 30 points.

HND Computer Systems Engineering
UCAS Code: 106G • Mode: 2/3 Years Single Subject

Qualifications/Requirements: GCE, A/AS: 4 points. Maths, Physics, Electronics or Computing. AGNVQ, Pass (in specific programmes). ND/C, 1 Merit (in specific programmes). SQAH, Grades CC. SQAV, Individual consideration. IB, Pass in Diploma.

QUEEN MARY AND WESTFIELD COLLEGE (UNIVERSITY OF LONDON) Q50

BSc Computer Systems and Digital Electronics
UCAS Code: G521 • Mode: 3 Years Single Subject

Qualifications/Requirements: GCE, A level grades: BCC. Maths and Physics or Electronics at A level. AGNVQ, Distinction with A/AS. ND/C, 2 Merits and 3 Distinctions. SQAH, Grades BBBBB including specific subject(s). IB, 28 points.

BEng Computer Engineering
UCAS Code: H610 • Mode: 3 Years Single Subject

Qualifications/Requirements: GCE, A level grades: BBC. Maths at A level. AGNVQ, Merit in Engineering or in Science with A level. ND/C, 3 Merits and 4 Distinctions (in specific programmes). SQAH, Grades BBBCC including specific subject(s). IB, 26 points.

MEng Computer Engineering
UCAS Code: H611 • Mode: 4 Years Single Subject

Qualifications/Requirements: GCE, A level grades: AAB. Maths at A level. AGNVQ, Distinction in Engineering or in Science with A level. ND/C, 2 Merits and 5 Distinctions (in specific programmes). SQAH, Grades BBBCC including specific subject(s). IB, 30 points.

THE UNIVERSITY OF READING R12

BSc Applied Computer Science and Cybernetics
UCAS Code: GH5P • Mode: 4 Years Equal Combination

Qualifications/Requirements: GCE, A/AS: 20 points. Maths or Science. AGNVQ, Distinction (in specific programmes) with 6 additional units or with A/AS. ND/C, 3 Merits and 2 Distinctions (in specific programmes). SQAH, Grades BBBB. SQAV, Individual consideration. IB, 30 points.

MEng Applied Computer Science and Cybernetics
UCAS Code: GH5Q • Mode: 5 Years Equal Combination

Qualifications/Requirements: GCE, A/AS: 24 points. Maths or Science. AGNVQ, Distinction (in specific programmes) with 6 additional units or with A/AS. ND/C, 2 Merits and 3 Distinctions (in specific programmes). SQAH, Grades ABBB including specific subject(s). SQAV, Individual consideration. IB, 32 points.

THE ROBERT GORDON UNIVERSITY R36

BSc Computer Network Management and Design

UCAS Code: G620 • Mode: 3/4 Years Single Subject

Qualifications/Requirements: GCE, A level grades: DE. AGNVQ, Individual consideration. ND/C, Individual consideration. SQAH, Grades BCC including specific subject(s). SQAV, Individual consideration. IB, Individual consideration.

BEng Electronic and Computer Engineering

UCAS Code: GH66 • Mode: 4 Years Equal Combination

Qualifications/Requirements: GCE, A level grades: CCC. Maths and Science at A level. AGNVQ, Individual consideration. ND/C, National. SQAH, Grades BBCC including specific subject(s). SQAV, Individual consideration. IB, Individual consideration.

THE UNIVERSITY OF SALFORD S03

BSc Computer Systems and Telecommunications Engineering

UCAS Code: H610 • Mode: 3/4 Years Single Subject

Qualifications/Requirements: GCE, A/AS: 16 points. Maths and Physics. AGNVQ, Merit with A/AS. ND/C, 5 Merits. SQAH, Individual consideration. SQAV, Individual consideration. IB, Individual consideration.

SANDWELL COLLEGE S08

HND Computer-Aided Engineering

UCAS Code: 061H • Mode: 2 Years Single Subject

Qualifications/Requirements: GCE, A level grades: DD-EE. Computing, Design & Technology, Maths or Physics. AGNVQ, Pass (in specific programmes). ND/C, National. SQAH, Individual consideration. SQAV, Individual consideration. IB, Individual consideration.

THE UNIVERSITY OF SHEFFIELD S18

MEng Computer Systems Engineering

UCAS Code: G600 • Mode: 3/4 Years Single Subject

Qualifications/Requirements: GCE, A level grades: BCC. Maths at A level and Physics or Electronics. AGNVQ, Merit or Distinction with A level. ND/C, 5 Merits and 1 Distinction (in specific programmes). SQAH, Grades ABBB including specific subject(s). SQAV, Individual consideration. IB, 30 points.

MEng Electronic Engineering (Computing)

UCAS Code: H611 • Mode: 4 Years Single Subject

Qualifications/Requirements: GCE, A level grades: ABB. Maths and Physics at A level. AGNVQ, Distinction in Science or in Engineering with A level. ND/C, 3 Merits and 3 Distinctions (in specific programmes). SQAH, Grades AAAB including specific subject(s). SQAV, Individual consideration. IB, 33 points.

SHEFFIELD HALLAM UNIVERSITY S21

BSc Applied Computing

UCAS Code: G610 • Mode: 1 Years Single Subject

Qualifications/Requirements: AGNVQ, Not normally sufficient. ND/C, Higher National. SQAH, Not normally sufficient. SQAV, Not normally sufficient. IB, Not normally sufficient.

BSc Computer and Network Engineering

UCAS Code: GG67 • Mode: 3/4 Years Equal Combination

Qualifications/Requirements: GCE, A/AS: 14 points. Science at A level. AGNVQ, Merit (in specific programmes). ND/C, 3 Merits (in specific programmes). SQAH, Individual consideration. SQAV, Individual consideration. IB, Individual consideration.

BSc Computing (Networks and Communication)

UCAS Code: G600 • Mode: 4 Years Single Subject

Qualifications/Requirements: GCE, A/AS: 12 points. AGNVQ, Merit. ND/C, 6 Merits. SQAH, Individual consideration. SQAV, Individual consideration. IB, Individual consideration.

BSc Electronics and Information Technology

UCAS Code: GG65 • Mode: 1 Year Equal Combination

Qualifications/Requirements: AGNVQ, Not normally sufficient. ND/C, Higher National (in specific programmes). SQAH, Not normally sufficient. SQAV, Not normally sufficient. IB, Not normally sufficient.

BSc Information Engineering and Technology Management

UCAS Code: G560 • Mode: 3 Years Single Subject

Qualifications/Requirements: GCE, A/AS: 8 points. Science. AGNVQ, Merit. ND/C, 3 Merits. SQAH, Individual consideration. SQAV, Individual consideration. IB, Individual consideration.

BEng Computer-Aided Engineering and Design

UCAS Code: H161 • Mode: 3/4 Years Single Subject

Qualifications/Requirements: GCE, A/AS: 10 points. Maths, Physics, Chemistry or any Physical Science at A level. AGNVQ, Merit (in specific programmes). ND/C, 4 Merits (in specific programmes). SQAH, Individual consideration. SQAV, Individual consideration. IB, Individual consideration.

BEng Computer Engineering

UCAS Code: G601 • Mode: 3 Years Single Subject

Qualifications/Requirements: GCE, A/AS: 10 points. Maths or Science at A level. AGNVQ, Merit (in specific programmes). ND/C, 4 Merits (in specific programmes). SQAH, Individual consideration. SQAV, Individual consideration. IB, Individual consideration.

BEng Mechanical and Computer-Aided Engineering
UCAS Code: HH31 • Mode: 3 Years Equal Combination

Qualifications/Requirements: GCE, A/AS: 18 points. Two Sciences at A level. AGNVQ, Merit (in specific programmes) with 3 additional units. ND/C, 7 Merits (in specific programmes). SQAH, Individual consideration. SQAV, Individual consideration. IB, Individual consideration.

HND Computer Aided Engineering and Design
UCAS Code: 161H • Mode: 2 Years Single Subject

Qualifications/Requirements: GCE, A level grades: E. Maths or Science at A level. AGNVQ, Pass. ND/C, National. SQAH, Individual consideration. SQAV, Individual consideration. IB, Individual consideration.

HND Computer and Network Engineering
UCAS Code: 006G • Mode: 2 Years Single Subject

Qualifications/Requirements: GCE, A/AS: 6 points. Maths, Physics, Chemistry or Science at A level. AGNVQ, Pass (in specific programmes). ND/C, 2 Merits (in specific programmes). SQAH, Individual consideration. SQAV, Individual consideration. IB, Individual consideration.

HND Computer Engineering
UCAS Code: 106G • Mode: 2 Years Single Subject

Qualifications/Requirements: GCE, A level grades: E. Maths or Science at A level. AGNVQ, Pass. ND/C, National. SQAH, Individual consideration. SQAV, Individual consideration. IB, Individual consideration.

HND Computer Networks
UCAS Code: 206G • Mode: 2/3 Years Single Subject

Qualifications/Requirements: GCE, A/AS: 6 points. AGNVQ, Pass. ND/C, 2 Merits. SQAH, Individual consideration. SQAV, Individual consideration. IB, Individual consideration.

HND Information Engineering and Technology Management
UCAS Code: 065G • Mode: 2 Years Single Subject

Qualifications/Requirements: GCE, A/AS: 12 points. AGNVQ, Pass. SQAH, Individual consideration. SQAV, Individual consideration. IB, Individual consideration.

UNIVERSITY OF SOUTHAMPTON S27

BEng Computer Engineering
UCAS Code: H614 • Mode: 3 Years Single Subject

Qualifications/Requirements: GCE, A/AS: 24 points. Maths at A level and Science. AGNVQ, Distinction with A level. ND/C, Individual consideration. SQAH, CSYS required. SQAV, Individual consideration. IB, 32 points.

MEng Computer Engineering
UCAS Code: H615 • Mode: 4 Years Single Subject

Qualifications/Requirements: GCE, A/AS: 26 points. Maths at A level and Science. AGNVQ, Distinction with A level. ND/C, Individual consideration. SQAH, CSYS required. SQAV, Individual consideration. IB, 34 points.

SOUTHAMPTON INSTITUTE S30

BSc Computer Network Communications
UCAS Code: G602 • Mode: 3 Years Single Subject

Qualifications/Requirements: GCE, A/AS: 10 points. AGNVQ, Merit (in specific programmes). ND/C, Merit overall. SQAH, Grades CCCC. SQAV, National. IB, Pass in Diploma.

BSc Computer Network Communications
UCAS Code: G607 • Mode: 4 Years Single Subject

Qualifications/Requirements: GCE, A/AS: 2-4 points. AGNVQ, Pass (in specific programmes). ND/C, National. SQAH, Grades CCCC. SQAV, National. IB, Pass in Diploma.

BSc Computer Network Management
UCAS Code: G601 • Mode: 3 Years Single Subject

Qualifications/Requirements: GCE, A/AS: 10 points. AGNVQ, Merit (in specific programmes). ND/C, Merit overall. SQAH, Grades CCCC. SQAV, National. IB, Pass in Diploma.

BSc Computer Network Management (with foundation)
UCAS Code: G609 • Mode: 4 Years Single Subject

Qualifications/Requirements: GCE, A/AS: 2-4 points. AGNVQ, Pass (in specific programmes). ND/C, National. SQAH, Grades CCCC. SQAV, National. IB, Pass in Diploma.

SOUTH BANK UNIVERSITY S33

BSc Computer-Aided Engineering
UCAS Code: H161 • Mode: 3/4 Years Single Subject

Qualifications/Requirements: GCE, A level grades: EE. Maths, Physics, Chemistry, Computing or Design & Technology at A level. AGNVQ, Merit. ND/C, 5 Merits. SQAH, Individual consideration. SQAV, Individual consideration. IB, Individual consideration.

BSc Computer-Aided Engineering
UCAS Code: H165 • Mode: 1/2 Years Single Subject

Qualifications/Requirements: ND/C, Higher National. SQAH, Individual consideration. SQAV, Individual consideration. IB, Individual consideration.

BSc Internet Computing
UCAS Code: G610 • Mode: 3/4 Years Single Subject

Qualifications/Requirements: GCE, A/AS: 10 points. Computing, Italian or Science. AGNVQ, Merit (in specific programmes). ND/C, Merit overall. SQAH, Individual consideration. SQAV, Individual consideration. IB, Individual consideration.

BSc Internet Computing (Top-Up)
UCAS Code: G611 • Mode: 1 Years Single Subject

Qualifications/Requirements: ND/C, Higher National.

BSc Telecommunications and Computer Networks Engineering (Top-up)
UCAS Code: G6H6 • Mode: 1/2 Years Major/Minor

Qualifications/Requirements: ND/C, Higher National.

BEng Internet Engineering
UCAS Code: G620 • Mode: 3/4 Years Single Subject

Qualifications/Requirements: GCE, A level grades: DD. Maths and Science. AGNVQ, Merit in Engineering. ND/C, 5 Merits (in specific programmes). SQAH, Individual consideration. SQAV, Individual consideration. IB, Individual consideration.

HND Computer Aided Engineering
UCAS Code: 061H • Mode: 2/3 Years Single Subject

Qualifications/Requirements: GCE, A level grades: E. Maths, Physics, Chemistry, Computing or Design & Technology at A level. AGNVQ, Pass in Engineering or in Manufacturing or in Science. ND/C, (in specific programmes). SQAH, Individual consideration. SQAV, Individual consideration. IB, Individual consideration.

HND Internet Computing
UCAS Code: 016G • Mode: 2 Years Single Subject

Qualifications/Requirements: GCE, A level grades: E. Computing, Italian or Science. AGNVQ, Pass (in specific programmes). ND/C, National. SQAH, Individual consideration. SQAV, Individual consideration. IB, Individual consideration.

SOUTHPORT COLLEGE S35

HND Computer-Aided Design
UCAS Code: 061H • Mode: 2 Years Single Subject

Qualifications/Requirements: GCE, A/AS: 2 points. AGNVQ, Merit. ND/C, Merit overall.

ST HELENS COLLEGE S51

HND Computer Games Production
UCAS Code: 016G • Mode: 2 Years Single Subject

Qualifications/Requirements: GCE, A/AS: 2-6 points. Maths, Physics, Electronics or Computing at A level. AGNVQ, Pass (in specific programmes). ND/C, National. SQAH, Individual consideration. SQAV, Individual consideration. IB, Individual consideration.

HND Computer Networking
UCAS Code: 026G • Mode: 2 Years Single Subject

Qualifications/Requirements: GCE, A/AS: 2-6 points. Maths, Physics, Electronics or Computing at A level. AGNVQ, Pass (in specific programmes). ND/C, National. SQAH, Individual consideration. SQAV, Individual consideration. IB, Individual consideration.

STAFFORDSHIRE UNIVERSITY S72

BSc Computer Systems
UCAS Code: G601 • Mode: 4 Years Single Subject

Qualifications/Requirements: GCE, A/AS: 12 points. AGNVQ, Merit. ND/C, Individual consideration. SQAH, Grades CCC. SQAV, Individual consideration. IB, 27 points.

BSc Foundation Internet Technology
UCAS Code: G607 • Mode: 5 Years Single Subject

Qualifications/Requirements: Please refer to prospectus.

BSc Internet Technology
UCAS Code: G611 • Mode: 4 Years Single Subject

Qualifications/Requirements: GCE, A/AS: 12 points. AGNVQ, Merit. ND/C, Individual consideration. SQAH, Grades CCC. IB, 27 points.

BEng Foundation Internet Technology
UCAS Code: G606 • Mode: 5 Years Single Subject

Qualifications/Requirements: Please refer to prospectus.

MEng Computer-Aided Engineering
UCAS Code: H110 • Mode: 4/5 Years Single Subject

Qualifications/Requirements: GCE, A/AS: 12 points. Maths or Physics. AGNVQ, Merit (in specific programmes). ND/C, 3 Merits. SQAH, Grades CCC including specific subject(s). SQAV, Individual consideration. IB, 24 points.

MEng Extended Computer-Aided Engineering
UCAS Code: H169 • Mode: 5/6 Years Single Subject

Qualifications/Requirements: GCE, A/AS: 4 points. AGNVQ, Merit. ND/C, National. SQAH, Grades CCC. SQAV, Individual consideration. IB, 24 points.

MEng Microelectronics and Computer Engineering
UCAS Code: GH66 • Mode: 4/5 Years Equal Combination

Qualifications/Requirements: GCE, A/AS: 12 points. Maths or Physics. AGNVQ, Merit (in specific programmes). ND/C, 3 Merits. SQAH, Grades CCC including specific subject(s). SQAV, Individual consideration. IB, 24 points.

MEng Microelectronics and Computer Engineering
UCAS Code: HH6P • Mode: 4 Years Equal Combination

Qualifications/Requirements: GCE, A/AS: 4 points. AGNVQ, Pass. ND/C, 4 Merits. SQAH, Grades CCC. SQAV, Individual consideration. IB, 24 points.

HND Computer-Aided Engineering
UCAS Code: 061H • Mode: 2 Years Single Subject

Qualifications/Requirements: GCE, A/AS: 2 points. Maths or Physics. AGNVQ, Pass in Engineering or in Science. ND/C, National. SQAH, Grades DDD. SQAV, Individual consideration. IB, 24 points.

HND Extended Computer-Aided Engineering
UCAS Code: 961H • Mode: 3 Years Single Subject

Qualifications/Requirements: GCE, A/AS: 2 points. AGNVQ, Pass (in specific programmes). ND/C, National. SQAH, Grades DDD. SQAV, Individual consideration. IB, 24 points.

THE UNIVERSITY OF STRATHCLYDE S78

BEng Computer and Electronic Systems (BEng)
UCAS Code: GH56 • Mode: 4 Years Equal Combination

Qualifications/Requirements: GCE, A level grades: CCD. Maths and Physics at A level. AGNVQ, Pass. ND/C, Higher National. SQAH, Grades ABBB including specific subject(s). SQAV, Higher National. IB, 30 points.

MEng Computer and Electronic Systems
UCAS Code: GH5P • Mode: 5 Years Equal Combination

Qualifications/Requirements: GCE, A level grades: BBB. Maths and Physics at A level. ND/C, Individual consideration. SQAH, Grades AAAA including specific subject(s). SQAV, Individual consideration. IB, 38 points.

UNIVERSITY OF SUNDERLAND S84

BSc Internet Information Systems
UCAS Code: G621 • Mode: 3/4 Years Single Subject

Qualifications/Requirements: GCE, A/AS: 10 points. AGNVQ, Merit. ND/C, 4 Merits (in specific programmes). SQAH, Grades CCCC. SQAV, National. IB, 24 points.

BEng Computer Systems Engineering
UCAS Code: H611 • Mode: 3/4 Years Single Subject

Qualifications/Requirements: GCE, A level grades: CC. Maths and Physics. AGNVQ, Merit in Engineering. ND/C, 3 Merits (in specific programmes). SQAH, Grades CCC including specific subject(s). SQAV, National including specific subject(s). IB, 24 points.

BEng Computer Systems Engineering (Foundation)
UCAS Code: H618 • Mode: 4/5 Years Single Subject

Qualifications/Requirements: GCE, A/AS: 4 points. AGNVQ, Pass. ND/C, National. SQAH, Grades CC. SQAV, National. IB, 24 points.

THE UNIVERSITY OF SURREY S87

BEng Electronic and Information Systems Engineering with Foundation year
UCAS Code: HHP5 • Mode: 4/5 Years Equal Combination

Qualifications/Requirements: GCE, A/AS: 16-20 points. AGNVQ, Individual consideration. ND/C, Individual consideration. SQAH, Individual consideration. SQAV, Individual consideration. IB, Individual consideration.

BEng Information Systems Engineering (c)
UCAS Code: H632 • Mode: 3 Years Single Subject

Qualifications/Requirements: GCE, A/AS: 18-24 points. Maths at A level and Physics. AGNVQ, Distinction in Engineering with A level. ND/C, 5 Merits and 2 Distinctions (in specific programmes). SQAH, Grades BBBBC. SQAV, Not normally sufficient. IB, 28 points.

BEng Information Systems Engineering (c)
UCAS Code: H630 • Mode: 4 Years Single Subject

Qualifications/Requirements: GCE, A/AS: 18-24 points. Maths at A level and Physics. AGNVQ, Distinction in Engineering with A level. ND/C, 5 Merits and 2 Distinctions (in specific programmes). SQAH, Grades BBBBC. SQAV, Not normally sufficient. IB, 28 points.

MEng Information Systems Engineering (c)
UCAS Code: H633 • Mode: 4 Years Single Subject

Qualifications/Requirements: GCE, A/AS: 24 points. Maths at A level and Physics. AGNVQ, Not normally sufficient. ND/C, Not normally sufficient. SQAH, Grades BBBBB. SQAV, Not normally sufficient. IB, 28 points.

MEng Information Systems Engineering (c)
UCAS Code: H631 • Mode: 5 Years Single Subject

Qualifications/Requirements: GCE, A/AS: 24 points. Maths at A level and Physics. AGNVQ, Pass. ND/C, Not normally sufficient. SQAH, Grades BBBBB. SQAV, Not normally sufficient. IB, 28 points.

MChem Computer-Aided Chemistry
UCAS Code: F152 • Mode: 4 Years Single Subject

Qualifications/Requirements: GCE, A/AS: 20 points. Chemistry at A level and two Sciences. ND/C, Individual consideration. SQAH, Individual consideration. SQAV, Individual consideration. IB, Individual consideration.

UNIVERSITY OF SUSSEX S90

BEng Computer Systems Engineering
UCAS Code: H6G5 • Mode: 3 Years Major/Minor

Qualifications/Requirements: GCE, A level grades: BCC. Maths at A level. AGNVQ, Merit in Science with A level. ND/C, Merit overall (in specific programmes). SQAH, Individual consideration. SQAV, Individual consideration. IB, Individual consideration.

BEng Computer Systems Engineering
UCAS Code: H618 • Mode: 4 Years Single Subject

Qualifications/Requirements: GCE, A level grades: DD. Two Sciences. AGNVQ, Pass in Science. ND/C, National. SQAH, Individual consideration. SQAV, Individual consideration. IB, Individual consideration.

BEng Electronic Engineering and Computer Science

UCAS Code: HG65 • Mode: 3 Years Equal Combination

Qualifications/Requirements: GCE, A level grades: CCC. Maths at A level. AGNVQ, Merit in Science with A level. ND/C, Merit overall (in specific programmes). SQAH, Individual consideration. SQAV, Individual consideration. IB, Individual consideration.

MEng Computer Systems Engineering

UCAS Code: H6GM • Mode: 4 Years Major/Minor

Qualifications/Requirements: GCE, A level grades: BBB. Maths at A level. AGNVQ, Merit in Science with A level. ND/C, Merit overall (in specific programmes). SQAH, Individual consideration. SQAV, Individual consideration. IB, Individual consideration.

SWANSEA INSTITUTE OF HIGHER EDUCATION S96

BEng Computer Systems (Electronics)

UCAS Code: H600 • Mode: 3 Years Single Subject

Qualifications/Requirements: GCE, A/AS: 4 points. AGNVQ, Pass. ND/C, National. SQAH, Grades CCCC. SQAV, Individual consideration. IB, 24 points.

BEng Computer Systems (Electronics)

UCAS Code: H608 • Mode: 4 Years Single Subject

Qualifications/Requirements: AGNVQ, Pass.

BEng Computer Systems (Laser Technology)

UCAS Code: H635 • Mode: 3 Years Single Subject

Qualifications/Requirements: GCE, A/AS: 4 points. AGNVQ, Pass. ND/C, National. SQAH, Grades CCCC. SQAV, Individual consideration. IB, 24 points.

BEng Computer Systems (Laser Technology)

UCAS Code: H638 • Mode: 4 Years Single Subject

Qualifications/Requirements: Please refer to prospectus.

BEng Computer Systems (Multimedia)

UCAS Code: G500 • Mode: 3 Years Single Subject

Qualifications/Requirements: GCE, A/AS: 4 points. AGNVQ, Pass. ND/C, National. SQAH, Grades CCCC. SQAV, Individual consideration. IB, 24 points.

BEng Computer Systems (Multimedia)

UCAS Code: G508 • Mode: 4 Years Single Subject

Qualifications/Requirements: Please refer to prospectus.

UNIVERSITY OF TEESSIDE T20

BSc Computer-Aided Design Engineering

UCAS Code: H7G5 • Mode: 3 Years Major/Minor

Qualifications/Requirements: GCE, A/AS: 12-14 points. AGNVQ, Merit. ND/C, 3 Merits. SQAH, Grades CCC including specific subject(s). SQAV, National including specific subject(s). IB, Individual consideration.

BSc Computer Engineering & Microelectronics

UCAS Code: GH66 • Mode: 3/4 Years Equal Combination

Qualifications/Requirements: GCE, A/AS: 10-12 points. Two Sciences. AGNVQ, Merit. ND/C, 3 Merits. SQAH, Grades CCC including specific subject(s). SQAV, National including specific subject(s). IB, Individual consideration.

BSc Electronic Systems Engineering with Information Technology

UCAS Code: H6GM • Mode: 3/4 Years Major/Minor

Qualifications/Requirements: GCE, A/AS: 12-14 points. Two Sciences. AGNVQ, Merit. ND/C, 3 Merits (in specific programmes). SQAH, Grades CCC including specific subject(s). SQAV, National including specific subject(s). IB, Individual consideration.

BSc Materials Systems Engineering with Information Technology

UCAS Code: J5GM • Mode: 3/4 Years Major/Minor

Qualifications/Requirements: GCE, A/AS: 12-14 points. Two Sciences. AGNVQ, Merit. ND/C, 3 Merits (in specific programmes). SQAH, Grades CCC including specific subject(s). SQAV, National including specific subject(s). IB, Individual consideration.

HND Computer-Aided Engineering

UCAS Code: 161H • Mode: 2 Years Single Subject

Qualifications/Requirements: GCE, A/AS: 4 points. Maths or Science. AGNVQ, Pass. ND/C, National. SQAH, Grades CC including specific subject(s). SQAV, National including specific subject(s). IB, Individual consideration.

HND Electronic and Computer Engineering

UCAS Code: 116H • Mode: 2 Years Single Subject

Qualifications/Requirements: GCE, A/AS: 4 points. Maths or Physics. AGNVQ, Pass. ND/C, National. SQAH, Grades CC including specific subject(s). SQAV, National. IB, Individual consideration.

THAMES VALLEY UNIVERSITY T40

Mod Information and Knowledge Management with Data Communications Support

UCAS Code: P2G6 • Mode: 2/3 Years Major/Minor

Qualifications/Requirements: GCE, A/AS: 8 points. AGNVQ, Merit. ND/C, Merit overall. SQAH, Grades CCCC. IB, 26 points.

UNIVERSITY OF ULSTER U20

BEng Electronics and Computing

UCAS Code: HG67 • Mode: 4 Years Equal Combination

Qualifications/Requirements: GCE, A level grades: CDD. Science. AGNVQ, Distinction (in specific programmes). ND/C, Merit overall and 1 Distinction. SQAH, Grades CCCD. SQAV, Individual consideration. IB, 26 points.

BEng Electronics and Computing (Hons) with Diploma in Industrial Studies

UCAS Code: GH76 • Mode: 4 Years Equal Combination

Qualifications/Requirements: GCE, A level grades: CCD. Science. AGNVQ, Distinction (in specific programmes) with 4 additional units or with A level. ND/C, Merit overall and 3 Distinctions. SQAH, Grades BBCC. SQAV, Individual consideration. IB, 28 points.

UNIVERSITY COLLEGE LONDON (UNIVERSITY OF LONDON) U80

MEng Electronic Engineering with Computer Science

UCAS Code: H6GN • Mode: 4 Years Major/Minor

Qualifications/Requirements: GCE, A level grades: ABB-AAB. Maths and Physics at A level. AGNVQ, Not normally sufficient. ND/C, 3 Merits and 4 Distinctions (in specific programmes). SQAH, Grades AAABB including specific subject(s). SQAV, Individual consideration. IB, 30 points.

THE UNIVERSITY OF WARWICK W20

BEng Computer Systems Engineering (b)

UCAS Code: GH66 • Mode: 3 Years Equal Combination

Qualifications/Requirements: GCE, A level grades: ABB-BBB. Maths at A level. AGNVQ, Distinction in Science with A/AS. ND/C, Individual consideration. SQAH, Grades AAABB including specific subject(s). IB, 32 points.

MEng Computer Systems Engineering (b)

UCAS Code: GH6P • Mode: 4 Years Equal Combination

Qualifications/Requirements: GCE, A level grades: ABB-BBB. Maths at A level. AGNVQ, Distinction in Science with A/AS. ND/C, Individual consideration. SQAH, Grades AAABB including specific subject(s). IB, 32 points.

WARWICKSHIRE COLLEGE, ROYAL LEAMINGTON SPA AND MORETON MORRELL W25

HND Engineering (Computer-Aided)

UCAS Code: 061H • Mode: 2 Years Single Subject

Qualifications/Requirements: GCE, A/AS: 1 points. AGNVQ, Individual consideration. ND/C, Individual consideration. SQAH, Individual consideration. SQAV, Individual consideration. IB, Individual consideration.

UNIVERSITY OF WESTMINSTER W50

BSc Computer Systems Technology

UCAS Code: G600 • Mode: 3 Years Single Subject

Qualifications/Requirements: GCE, A level grades: CC. Maths, Science or Computing. AGNVQ, Distinction. ND/C, 5 Merits. SQAH, Individual consideration. SQAV, Individual consideration. IB, Individual consideration.

BEng Control & Computer Engineering

UCAS Code: H640 • Mode: 3 Years Single Subject

Qualifications/Requirements: GCE, A/AS: 18 points. Maths and Physics at A level. AGNVQ, Merit (in specific programmes). ND/C, 3 Merits and 1 Distinction (in specific programmes). SQAH, Grades CCCC including specific subject(s). SQAV, Individual consideration. IB, Individual consideration.

BEng Control & Computer Engineering (Ext)

UCAS Code: H648 • Mode: 4 Years Single Subject

Qualifications/Requirements: GCE, A level grades: D-E. AGNVQ, Pass. ND/C, National. SQAH, Individual consideration. SQAV, Individual consideration. IB, Individual consideration.

BEng Control and Computer Engineering with International Foundation

UCAS Code: H6QJ • Mode: 4 Years Major/Minor

Qualifications/Requirements: GCE, A level grades: D. AGNVQ, Pass. ND/C, National.

HND Mechanical, Manufacturing & Computer-Aided Engineering

UCAS Code: 73GH • Mode: 2 Years Equal Combination

Qualifications/Requirements: GCE, A level grades: E. Science. AGNVQ, Pass (in specific programmes). ND/C, National (in specific programmes).

UNIVERSITY OF WOLVERHAMPTON W75

BSc Computer-Aided Design & Construction

UCAS Code: H1K2 • Mode: 3/4 Years Major/Minor

Qualifications/Requirements: GCE, A/AS: 16 points. AGNVQ, Merit. ND/C, 3 Merits. SQAH, Grades BBBB. SQAV, Individual consideration. IB, 28 points.

BSc Computer-Aided Engineering Design

UCAS Code: H160 • Mode: 3/4 Years Single Subject

Qualifications/Requirements: GCE, A/AS: 8 points. AGNVQ, Merit. ND/C, 3 Merits. SQAH, Grades CCCC. SQAV, Individual consideration. IB, 24 points.

BSc Computer-Aided Industrial Design

UCAS Code: E230 • Mode: 3/4 Years Single Subject

Qualifications/Requirements: GCE, A/AS: 8 points. AGNVQ, Merit. ND/C, 3 Merits. SQAH, Grades CCCC. SQAV, Individual consideration. IB, 24 points.

BSc Computer-Aided Industrial Design

UCAS Code: H770 • Mode: 3/4 Years Single Subject

Qualifications/Requirements: GCE, A/AS: 8 points. AGNVQ, Merit. ND/C, 3 Merits. SQAH, Grades CCCC. SQAV, Individual consideration. IB, 24 points.

BSc Computer-Aided Industrial Design

UCAS Code: W230 • Mode: 3/4 Years Single Subject

Qualifications/Requirements: GCE, A/AS: 8 points. AGNVQ, Merit. ND/C, 3 Merits. SQAH, Grades CCCC. SQAV, Individual consideration. IB, 24 points.

BSc Computer-Aided Product Design

UCAS Code: EW7F • Mode: 3/4 Years Equal Combination

Qualifications/Requirements: GCE, A/AS: 8 points. AGNVQ, Pass. ND/C, 3 Merits. SQAH, Individual consideration. SQAV, Individual consideration. IB, Individual consideration.

BSc Computer-Aided Product Design

UCAS Code: HW72 • Mode: 3/4 Years Equal Combination

Qualifications/Requirements: GCE, A/AS: 8 points. AGNVQ, Pass. ND/C, 3 Merits. SQAH, Individual consideration. SQAV, Individual consideration. IB, Individual consideration.

BSc Computer-Aided Product Design

UCAS Code: HW7F • Mode: 3/4 Years Equal Combination

Qualifications/Requirements: GCE, A/AS: 8 points. AGNVQ, Pass. ND/C, 3 Merits. SQAH, Individual consideration. SQAV, Individual consideration. IB, Individual consideration.

HND Computer-Aided Design

UCAS Code: 061H • Mode: 2/3 Years Single Subject

Qualifications/Requirements: GCE, A/AS: 8 points. AGNVQ, Pass. ND/C, National. SQAH, Grades CCCC. SQAV, Individual consideration. IB, 24 points.

THE UNIVERSITY OF YORK Y50

BEng Electronic and Computer Engineering

UCAS Code: H633 • Mode: 3 Years Single Subject

Qualifications/Requirements: GCE, A/AS: 26 points. Maths and Science at A level. AGNVQ, Distinction (in specific programmes) with A level. ND/C, Individual consideration. SQAH, CSYS required. SQAV, Individual consideration. IB, 30 points.

BEng Electronic and Computer Engineering

UCAS Code: H634 • Mode: 4 Years Single Subject

Qualifications/Requirements: GCE, A/AS: 26 points. Maths and Science at A level. AGNVQ, Distinction (in specific programmes) with A level. ND/C, Individual consideration. SQAH, CSYS required. SQAV, Individual consideration. IB, 30 points.

MEng Computer Systems and Software Engineering

UCAS Code: GG67 • Mode: 4 Years Equal Combination

Qualifications/Requirements: GCE, A/AS: 26-28 points. Maths and any Physical Science or Biology at A level. AGNVQ, Distinction (in specific programmes) with A level. ND/C, Higher National (in specific programmes). SQAH, CSYS required. SQAV, Higher National including specific subject(s). IB, 34 points.

MEng Electronic and Computer Engineering

UCAS Code: H636 • Mode: 4/5 Years Single Subject

Qualifications/Requirements: GCE, A/AS: 26 points. Maths and Science at A level. AGNVQ, Distinction (in specific programmes) with A level. ND/C, Individual consideration. SQAH, CSYS required. SQAV, Individual consideration. IB, 30 points.

COMPUTING WITH LANGUAGES

THE UNIVERSITY OF ABERDEEN A20

BSc Computing Science with French

UCAS Code: G5R1 • Mode: 4 Years Major/Minor

Qualifications/Requirements: GCE, A level grades: CCD. Two Sciences or Science and Maths. AGNVQ, Merit in Science. ND/C, Individual consideration. SQAH, Grades BBBC including specific subject(s). SQAV, Individual consideration. IB, 26 points.

BSc Computing Science with French with Industrial Placement

UCAS Code: G5RC • Mode: 5 Years Major/Minor

Qualifications/Requirements: GCE, A level grades: CCD. Two Sciences or Science and Maths. AGNVQ, Merit in Science. ND/C, Individual consideration. SQAH, Grades BBBC including specific subject(s). SQAV, Individual consideration. IB, 26 points.

BSc Computing Science with Spanish

UCAS Code: G5R4 • Mode: 4 Years Major/Minor

Qualifications/Requirements: GCE, A level grades: CCD. Two Sciences or Science and Maths. AGNVQ, Merit in Science. ND/C, Individual consideration. SQAH, Grades BBBC including specific subject(s). SQAV, Individual consideration. IB, 26 points.

BSc Computing Science with Spanish with Industrial Placement

UCAS Code: G5RK • Mode: 5 Years Major/Minor

Qualifications/Requirements: GCE, A level grades: CCD. Two Sciences or Science and Maths. AGNVQ, Merit in Science. ND/C, Individual consideration. SQAH, Grades BBBC including specific subject(s). SQAV, Individual consideration. IB, 26 points.

THE UNIVERSITY OF WALES, ABERYSTWYTH A40

BSc Computer Science with German

UCAS Code: G5R2 • Mode: 4 Years Major/Minor

Qualifications/Requirements: GCE, A/AS: 20 points. German at A level. AGNVQ, Merit with A level. ND/C, Not normally sufficient. SQAH, Grades BBBCC including specific subject(s). SQAV, Individual consideration. IB, 30 points.

BSc Computer Science with Italian

UCAS Code: G5R3 • Mode: 4 Years Major/Minor

Qualifications/Requirements: GCE, A/AS: 20 points. A modern foreign language at A level. AGNVQ, Merit with A level. ND/C, Not normally sufficient. SQAH, Grades BBBCC including specific subject(s). SQAV, Individual consideration. IB, 30 points.

BSc Computer Science with Spanish

UCAS Code: G5R4 • Mode: 4 Years Major/Minor

Qualifications/Requirements: GCE, A/AS: 20 points. Science at A level. AGNVQ, Merit with A level. ND/C, Not normally sufficient. SQAH, Grades BBBCC including specific subject(s). SQAV, Individual consideration. IB, 30 points.

BSc Cymraeg gyda Ffrangeg

UCAS Code: G5R1 • Mode: 4 Years Major/Minor

Qualifications/Requirements: GCE, A/AS: 20 points. French at A level. AGNVQ, Merit with A level. ND/C, Not normally sufficient. SQAH, Grades BBBCC including specific subject(s). SQAV, Individual consideration. IB, 30 points.

ANGLIA POLYTECHNIC UNIVERSITY A60

Mod Computer Science and French

UCAS Code: GR51 • Mode: 3/4 Years Equal Combination

Qualifications/Requirements: GCE, A/AS: 12 points. AGNVQ, Merit (in specific programmes). ND/C, 4 Merits. SQAH, Grades BBCC. SQAV, Individual consideration. IB, Pass in Diploma.

Mod Computer Science and German

UCAS Code: GR52 • Mode: 3/4 Years Equal Combination

Qualifications/Requirements: GCE, A/AS: 12 points. AGNVQ, Merit (in specific programmes). ND/C, 4 Merits. SQAH, Grades BBCC. SQAV, National. IB, Pass in Diploma.

Mod Computer Science and Italian

UCAS Code: GR53 • Mode: 3/4 Years Equal Combination

Qualifications/Requirements: GCE, A/AS: 12 points. AGNVQ, Merit (in specific programmes). ND/C, 4 Merits. SQAH, Grades BBCC. SQAV, National. IB, Pass in Diploma.

Mod Computer Science and Spanish

UCAS Code: GR54 • Mode: 3/4 Years Equal Combination

Qualifications/Requirements: GCE, A/AS: 10 points. AGNVQ, Merit (in specific programmes). ND/C, 3 Merits. SQAH, Grades BBCC. SQAV, National. IB, Pass in Diploma.

ASTON UNIVERSITY A80

BSc Computer Science/French

UCAS Code: GR51 • Mode: 4 Years Equal Combination

Qualifications/Requirements: GCE, A/AS: 20 points. French. AGNVQ, Distinction (in specific programmes) with 6 additional units or with A level. ND/C, Not normally sufficient. SQAH, Grades BBBBB including specific subject(s). SQAV, Individual consideration. IB, 30 points.

UNIVERSITY OF WALES, BANGOR B06

BSc Computer Systems with French

UCAS Code: H6R1 • Mode: 4 Years Major/Minor

Qualifications/Requirements: GCE, A level grades: CCD. French at A level. AGNVQ, Merit (in specific programmes) with A level. ND/C, 3 Merits (in specific programmes). SQAH, Grades BBCC including specific subject(s). SQAV, Not normally sufficient. IB, 26 points.

BSc Computer Systems with German

UCAS Code: H6R2 • Mode: 4 Years Major/Minor

Qualifications/Requirements: GCE, A level grades: CCD. German at A level. AGNVQ, Merit (in specific programmes) with A level. ND/C, 3 Merits (in specific programmes). SQAH, Grades BBCC. SQAV, Individual consideration. IB, 26 points.

THE UNIVERSITY OF BIRMINGHAM B32

BA Computer Studies/French Studies

UCAS Code: GR51 • Mode: 4 Years Equal Combination

Qualifications/Requirements: GCE, A level grades: BBB. French at A level. AGNVQ, Distinction with A level. ND/C, Individual consideration. SQAH, Grades ABBBB. SQAV, Individual consideration. IB, 32 points.

BA Computer Studies/German Studies

UCAS Code: GR52 • Mode: 4 Years Equal Combination

Qualifications/Requirements: GCE, A level grades: BBB. German. AGNVQ, Distinction with A level. ND/C, Individual consideration. SQAH, Grades ABBBB. SQAV, Individual consideration. IB, 32 points.

BA Computer Studies/Hispanic Studies

UCAS Code: GR54 • Mode: 4 Years Equal Combination

Qualifications/Requirements: GCE, A level grades: BBB. AGNVQ, Distinction with A/AS. ND/C, Individual consideration. SQAH, Grades ABBBB. SQAV, Individual consideration. IB, 32 points.

BA Computer Studies/Italian

UCAS Code: GR53 • Mode: 4 Years Equal Combination

Qualifications/Requirements: GCE, A level grades: BBB. AGNVQ, Distinction with A/AS. ND/C, Individual consideration. SQAH, Grades ABBBB. SQAV, Individual consideration. IB, 32 points.

BA Computer Studies/Latin

UCAS Code: GQ56 • Mode: 3 Years Equal Combination

Qualifications/Requirements: GCE, A level grades: BBB. Latin. AGNVQ, Distinction with A level. ND/C, Individual consideration. SQAH, Grades ABBBB. SQAV, Individual consideration. IB, 32 points.

BA Computer Studies/Modern Greek Studies

UCAS Code: GT52 • Mode: 4 Years Equal Combination

Qualifications/Requirements: GCE, A level grades: BBB. AGNVQ, Distinction with A/AS. ND/C, Individual consideration. SQAH, Grades ABBBB. SQAV, Individual consideration. IB, 32 points.

BA Computer Studies/Portuguese

UCAS Code: GR55 • Mode: 4 Years Equal Combination

Qualifications/Requirements: GCE, A level grades: BBB. AGNVQ, Distinction with A/AS. ND/C, Individual consideration. SQAH, Grades ABBBB. SQAV, Individual consideration. IB, 32 points.

BA Computer Studies/Russian

UCAS Code: GR58 • Mode: 4 Years Equal Combination

Qualifications/Requirements: GCE, A level grades: BBB. AGNVQ, Distinction with A/AS. ND/C, Individual consideration. SQAH, Grades ABBBB. SQAV, Individual consideration. IB, 32 points.

BOLTON INSTITUTE OF HIGHER EDUCATION B44

Mod Business Information Systems and French

UCAS Code: GR5C • Mode: 3 Years Equal Combination

Qualifications/Requirements: GCE, A level grades: CD. French at A level. AGNVQ, Individual consideration. ND/C, Individual consideration. SQAH, Grades BBCC. SQAV, Individual consideration. IB, 24 points.

Mod Business Information Systems and German

UCAS Code: GR5F • Mode: 3 Years Equal Combination

Qualifications/Requirements: GCE, A level grades: CD. German at A level. AGNVQ, Individual consideration. ND/C, Individual consideration. SQAH, Grades BBCC. SQAV, Individual consideration. IB, 24 points.

Mod Computing and European Studies

UCAS Code: GLM3 • Mode: 3 Years Equal Combination

Qualifications/Requirements: GCE, A level grades: CD. AGNVQ, Merit. ND/C, Merit overall. SQAH, Grades BBCC. SQAV, Individual consideration. IB, 24 points.

Mod Computing and French

UCAS Code: GR51 • Mode: 3 Years Equal Combination

Qualifications/Requirements: GCE, A level grades: CD. French at A level. AGNVQ, Individual consideration. ND/C, Individual consideration. SQAH, Grades BBCC. SQAV, Individual consideration. IB, 24 points.

Mod Computing and German

UCAS Code: GR52 • Mode: 3 Years Equal Combination

Qualifications/Requirements: GCE, A level grades: CD. German at A level. AGNVQ, Individual consideration. ND/C, Individual consideration. SQAH, Grades BBCC. SQAV, Individual consideration. IB, 24 points.

Mod Computing and Language Studies

UCAS Code: GQ51 • Mode: 3 Years Equal Combination

Qualifications/Requirements: GCE, A level grades: CD. AGNVQ, Merit. ND/C, Merit overall. SQAH, Grades BBCC. SQAV, Individual consideration. IB, 24 points.

Mod French and Leisure Computing Technology

UCAS Code: GR71 • Mode: 3 Years Equal Combination

Qualifications/Requirements: GCE, A/AS: 10 points. French at A level. AGNVQ, Merit (in specific programmes). ND/C, 3 Merits. SQAH, Individual consideration. SQAV, Individual consideration. IB, Individual consideration.

Mod French and Simulation/Virtual Environment

UCAS Code: RG17 • Mode: 3 Years Equal Combination

Qualifications/Requirements: GCE, A/AS: 10 points. French at A level. AGNVQ, Individual consideration. ND/C, 3 Merits. SQAH, Individual consideration. SQAV, Individual consideration. IB, Individual consideration.

Mod German and Leisure Computing Technology

UCAS Code: GR72 • Mode: 3 Years Equal Combination

Qualifications/Requirements: GCE, A level grades: CD. German at A level. AGNVQ, Individual consideration. ND/C, Individual consideration. SQAH, Individual consideration. SQAV, Individual consideration. IB, Individual consideration.

Mod German and Simulation/Virtual Environment

UCAS Code: RG27 • Mode: 3 Years Equal Combination

Qualifications/Requirements: GCE, A level grades: CD. German at A level. AGNVQ, Individual consideration. ND/C, Individual consideration. SQAH, Individual consideration. SQAV, Individual consideration. IB, Individual consideration.

UNIVERSITY OF THE WEST OF ENGLAND, BRISTOL B80

BA EFL and Information Systems

UCAS Code: QG35 • Mode: 3 Years Equal Combination

Qualifications/Requirements: GCE, A/AS: 18-22 points. AGNVQ, Merit with A level. ND/C, 5 Merits and 1 Distinction. SQAH, Grades BBBC including specific subject(s). SQAV, Individual consideration. IB, 26 points.

BA French and Information Systems

UCAS Code: RG15 • Mode: 3 Years Equal Combination

Qualifications/Requirements: GCE, A/AS: 18-22 points. French at A level. AGNVQ, Merit with A level. ND/C, 4 to 5 Merits. SQAH, Grades BBBC including specific subject(s). SQAV, Individual consideration. IB, 28 points.

BA German and Information Systems

UCAS Code: RG25 • Mode: 3 Years Equal Combination

Qualifications/Requirements: GCE, A/AS: 18-22 points. AGNVQ, Merit with A level. ND/C, 4 to 5 Merits. SQAH, Grades BBBC. SQAV, Individual consideration. IB, 28 points.

BA Information Systems, EFL and French

UCAS Code: G5Q3 • Mode: 3 Years Major/Minor

Qualifications/Requirements: GCE, A/AS: 18-22 points. AGNVQ, Merit with A level. ND/C, 4 to 5 Merits. SQAH, Grades BBBC. SQAV, Individual consideration. IB, 28 points.

BA Information Systems, EFL and German

UCAS Code: G5QH • Mode: 3 Years Major/Minor

Qualifications/Requirements: GCE, A/AS: 18-22 points. AGNVQ, Merit with A level. ND/C, 4 to 5 Merits. SQAH, Grades BBBC including specific subject(s). SQAV, Individual consideration. IB, 28 points.

BA Information Systems, EFL and Spanish

UCAS Code: G5QJ • Mode: 3 Years Major/Minor

Qualifications/Requirements: GCE, A/AS: 18-22 points. AGNVQ, Merit with A level. ND/C, 4 to 5 Merits. SQAH, Grades BBBC. SQAV, Individual consideration. IB, 28 points.

BA Information Systems, French and German

UCAS Code: G5T9 • Mode: 3 Years Major/Minor

Qualifications/Requirements: GCE, A/AS: 18-22 points. French or German at A level. AGNVQ, Merit with A level. ND/C, 4 to 5 Merits. SQAH, Grades BBBC. SQAV, Individual consideration. IB, 28 points.

BA Information Systems, French and Spanish

UCAS Code: G5TX • Mode: 3 Years Major/Minor

Qualifications/Requirements: GCE, A/AS: 18-22 points. French or Spanish at A level. AGNVQ, Merit with A level. ND/C, 4 to 5 Merits. SQAH, Grades BBBC. SQAV, Individual consideration. IB, 28 points.

BA Information Systems, German and Spanish

UCAS Code: G5TY • Mode: 3 Years Major/Minor

Qualifications/Requirements: GCE, A/AS: 18-22 points. German or Spanish at A level. AGNVQ, Merit with A level. ND/C, 4 to 5 Merits. SQAH, Grades BBBC including specific subject(s). SQAV, Individual consideration. IB, 28 points.

BA Spanish and Information Systems

UCAS Code: RG45 • Mode: 3 Years Equal Combination

Qualifications/Requirements: GCE, A/AS: 18-22 points. AGNVQ, Merit with A level. ND/C, 4 to 5 Merits. SQAH, Grades BBBC. SQAV, Individual consideration. IB, 28 points.

THE UNIVERSITY OF BUCKINGHAM B90

BSc Information Systems with French

UCAS Code: G5R1 • Mode: 2 Years Major/Minor

Qualifications/Requirements: GCE, A/AS: 12 points. AGNVQ, Merit. ND/C, 5 Merits. SQAH, Grades CCCC. SQAV, Individual consideration. IB, 24 points.

BSc Information Systems with Spanish

UCAS Code: G5R4 • Mode: 2 Years Major/Minor

Qualifications/Requirements: GCE, A/AS: 12 points. AGNVQ, Merit. ND/C, 5 Merits. SQAH, Grades CCCC. SQAV, Individual consideration. IB, 24 points.

CANTERBURY CHRIST CHURCH UNIVERSITY COLLEGE C10

Mod Information Technology with French

UCAS Code: G5R1 • Mode: 3 Years Major/Minor

Qualifications/Requirements: GCE, A level grades: DD. French at A level. AGNVQ, Merit with A level. ND/C, Merit overall. SQAH, Individual consideration. SQAV, Individual consideration. IB, 24 points.

Mix French and Information Technology

UCAS Code: GR51 • Mode: 3 Years Equal Combination

Qualifications/Requirements: GCE, A level grades: DD. French at A level. AGNVQ, Merit with A level. ND/C, Merit overall. SQAH, Individual consideration. SQAV, Individual consideration. IB, 24 points.

CHELTENHAM & GLOUCESTER COLLEGE OF HIGHER EDUCATION C50

BSc Business Computer Systems with Modern Languages

UCAS Code: G5T9 • Mode: 3 Years Major/Minor

Qualifications/Requirements: GCE, A/AS: 8-12 points. AGNVQ, Merit. ND/C, Merit overall. SQAH, Grades CCCC. SQAV, Individual consideration. IB, 24 points.

CHESTER: A COLLEGE OF THE UNIVERSITY OF LIVERPOOL C55

BA Computer Science/IS and French

UCAS Code: GR51 • Mode: 3 Years Equal Combination

Qualifications/Requirements: GCE, A/AS: 12 points. AGNVQ, Merit (in specific programmes). ND/C, Merit overall. SQAH, Grades CCCC. IB, 24 points.

BA Computer Science/IS and German

UCAS Code: GR52 • Mode: 3 Years Equal Combination

Qualifications/Requirements: GCE, A/AS: 12 points. AGNVQ, Merit (in specific programmes). ND/C, Merit overall. SQAH, Grades CCCC. IB, 24 points.

BA Computer Science/IS and Spanish

UCAS Code: GR54 • Mode: 3 Years Equal Combination

Qualifications/Requirements: GCE, A/AS: 12 points. AGNVQ, Merit (in specific programmes). ND/C, Merit overall. SQAH, Grades CCCC. IB, 24 points.

BA Computer Science/IS with French

UCAS Code: G5R1 • Mode: 3 Years Major/Minor

Qualifications/Requirements: GCE, A/AS: 12 points. AGNVQ, Merit. ND/C, Merit overall. SQAH, Grades CCCC. SQAV, National. IB, 24 points.

BA Computer Science/IS with German

UCAS Code: G5R2 • Mode: 3 Years Major/Minor

Qualifications/Requirements: GCE, A/AS: 12 points. AGNVQ, Merit (in specific programmes). ND/C, Merit overall. SQAH, Grades CCCC. SQAV, National. IB, 24 points.

BA Computer Science/IS with Spanish

UCAS Code: G5R4 • Mode: 3 Years Major/Minor

Qualifications/Requirements: GCE, A/AS: 12 points. AGNVQ, Merit (in specific programmes). ND/C, Merit overall. SQAH, Grades CCCC. IB, 24 points.

DE MONTFORT UNIVERSITY D26

Mod Computing and French

UCAS Code: GR51 • Mode: 3 Years Equal Combination

Qualifications/Requirements: GCE, A/AS: 8-14 points. AGNVQ, Merit. ND/C, 4 Merits (in specific programmes). SQAH, Grades BBCC including specific subject(s). SQAV, Individual consideration. IB, 28 points.

Mod Computing and German

UCAS Code: GR52 • Mode: 3 Years Equal Combination

Qualifications/Requirements: GCE, A/AS: 8-14 points. AGNVQ, Merit. ND/C, 4 Merits (in specific programmes). SQAH, Grades BBCC including specific subject(s). SQAV, Individual consideration. IB, 28 points.

Mod Computing and Spanish

UCAS Code: GR54 • Mode: 3 Years Equal Combination

Qualifications/Requirements: GCE, A/AS: 8-14 points. AGNVQ, Merit. ND/C, 4 Merits (in specific programmes). SQAH, Grades BBBC including specific subject(s). SQAV, Individual consideration. IB, 28 points.

UNIVERSITY OF EAST LONDON E28

BA Computing & Business Information Systems/French

UCAS Code: GR5C • Mode: 3 Years Equal Combination

Qualifications/Requirements: GCE, A/AS: 12 points. AGNVQ, Merit. ND/C, Merit overall. SQAH, Individual consideration. SQAV, Individual consideration. IB, Individual consideration.

BA Computing & Business Information Systems/German

UCAS Code: GR5F • Mode: 3 Years Equal Combination

Qualifications/Requirements: GCE, A/AS: 12 points. AGNVQ, Merit. ND/C, Merit overall. SQAH, Individual consideration. SQAV, Individual consideration. IB, Individual consideration.

BA Computing & Business Information Systems/Italian

UCAS Code: G5RH • Mode: 3 Years Major/Minor

Qualifications/Requirements: GCE, A/AS: 12 points. AGNVQ, Merit. ND/C, Merit overall. SQAH, Individual consideration. SQAV, Individual consideration. IB, Individual consideration.

BA Computing & Business Information Systems/Spanish
UCAS Code: GR5K • Mode: 3 Years Equal Combination

Qualifications/Requirements: GCE, A/AS: 12 points. AGNVQ, Merit. ND/C, Merit overall. SQAH, Individual consideration. SQAV, Individual consideration. IB, Individual consideration.

BA French/Information Technology
UCAS Code: GR51 • Mode: 3 Years Equal Combination

Qualifications/Requirements: GCE, A/AS: 12 points. AGNVQ, Merit (in specific programmes). ND/C, Merit overall. SQAH, Individual consideration. SQAV, Individual consideration. IB, Individual consideration.

BA German/Information Technology
UCAS Code: GR52 • Mode: 3 Years Equal Combination

Qualifications/Requirements: GCE, A/AS: 12 points. AGNVQ, Merit (in specific programmes). ND/C, Merit overall. SQAH, Individual consideration. SQAV, Individual consideration. IB, Individual consideration.

BA Information Technology/Linguistics
UCAS Code: GQ51 • Mode: 3 Years Equal Combination

Qualifications/Requirements: GCE, A/AS: 12 points. AGNVQ, Not normally sufficient. ND/C, Merit overall. SQAH, Individual consideration. SQAV, Individual consideration. IB, Individual consideration.

BA Information Technology/Spanish
UCAS Code: GR54 • Mode: 3 Years Equal Combination

Qualifications/Requirements: GCE, A/AS: 12 points. AGNVQ, Merit with A/AS. ND/C, Merit overall. SQAH, Individual consideration. SQAV, Individual consideration. IB, Individual consideration.

BSc Information Technology/European Studies
UCAS Code: G5T2 • Mode: 3 Years Major/Minor

Qualifications/Requirements: GCE, A/AS: 12 points. AGNVQ, Merit. ND/C, Merit overall. SQAH, Individual consideration. SQAV, Individual consideration. IB, Individual consideration.

BSc Information Technology/Italian
UCAS Code: G5R3 • Mode: 3 Years Major/Minor

Qualifications/Requirements: GCE, A/AS: 12 points. AGNVQ, Merit. ND/C, Merit overall. SQAH, Individual consideration. SQAV, Individual consideration. IB, Individual consideration.

UNIVERSITY OF EXETER E84

BSc Computer Science with European Study
UCAS Code: G5T2 • Mode: 4 Years Major/Minor

Qualifications/Requirements: GCE, A/AS: 20 points. AGNVQ, Merit or Distinction (in specific programmes). ND/C, Merit overall and 2 Distinctions. SQAH, Grades BBBBB. SQAV, Individual consideration. IB, 29 points.

UNIVERSITY OF GLASGOW G28

MA Computing/Czech
UCAS Code: GT51 • Mode: 5 Years Equal Combination

Qualifications/Requirements: GCE, A level grades: BBC. AGNVQ, Merit. ND/C, Higher National. SQAH, Grades BBBB. SQAV, Higher National. IB, 30 points.

MA Computing/French
UCAS Code: GR51 • Mode: 5 Years Equal Combination

Qualifications/Requirements: GCE, A level grades: BBC. AGNVQ, Merit. ND/C, Higher National. SQAH, Grades BBBB. SQAV, Higher National. IB, 30 points.

MA Computing/Greek
UCAS Code: GQ57 • Mode: 4 Years Equal Combination

Qualifications/Requirements: GCE, A level grades: BBC. AGNVQ, Merit. ND/C, Higher National. SQAH, Grades BBBB. SQAV, Higher National. IB, 30 points.

MA Computing/Russian
UCAS Code: GR58 • Mode: 5 Years Equal Combination

Qualifications/Requirements: GCE, A level grades: BBC. AGNVQ, Merit. ND/C, Higher National. SQAH, Grades BBBB. SQAV, Higher National. IB, 30 points.

UNIVERSITY OF GREENWICH G70

BSc Computing with French
UCAS Code: G5R1 • Mode: 3/4 Years Major/Minor

Qualifications/Requirements: GCE, A/AS: 8-10 points. AGNVQ, Merit. ND/C, 8 Merits. SQAH, Grades CCC. SQAV, Individual consideration. IB, Individual consideration.

BSc Computing with German
UCAS Code: G5R2 • Mode: 3/4 Years Major/Minor

Qualifications/Requirements: GCE, A/AS: 8-10 points. AGNVQ, Merit. ND/C, 8 Merits. SQAH, Grades CCC. SQAV, Individual consideration. IB, Individual consideration.

BSc Computing with Italian
UCAS Code: G5R3 • Mode: 3 Years Major/Minor

Qualifications/Requirements: Please refer to prospectus.

BSc Computing with Spanish
UCAS Code: G5R4 • Mode: 3/4 Years Major/Minor

Qualifications/Requirements: GCE, A/AS: 8-10 points. AGNVQ, Merit. ND/C, 8 Merits. SQAH, Grades CCC. SQAV, Individual consideration. IB, Individual consideration.

BSc Information Systems with French
UCAS Code: G5RC • Mode: 3/4 Years Major/Minor

Qualifications/Requirements: GCE, A/AS: 8-10 points. AGNVQ, Pass (in specific programmes). ND/C, 4 Merits. SQAH, Grades CCC. SQAV, Individual consideration. IB, Individual consideration.

BSc Information Systems with German

UCAS Code: G5RF • Mode: 3/4 Years Major/Minor

Qualifications/Requirements: GCE, A/AS: 8-10 points. AGNVQ, Pass (in specific programmes). ND/C, 4 Merits. SQAH, Grades CCC. SQAV, Individual consideration. IB, Individual consideration.

BSc Information Systems with Spanish

UCAS Code: G5RK • Mode: 3/4 Years Major/Minor

Qualifications/Requirements: GCE, A/AS: 8-10 points. AGNVQ, Pass (in specific programmes). ND/C, 4 Merits. SQAH, Grades CCC. SQAV, Individual consideration. IB, Individual consideration.

UNIVERSITY OF HERTFORDSHIRE H36

BSc Computing/Linguistic Sciences

UCAS Code: G5Q1 • Mode: 3/4 Years Major/Minor

Qualifications/Requirements: GCE, A/AS: 14 points. AGNVQ, Merit (in specific programmes). ND/C, Merit overall. SQAH, Grades BCCC. SQAV, Individual consideration. IB, 24 points.

BSc European Studies/Computing

UCAS Code: T2G5 • Mode: 3/4 Years Major/Minor

Qualifications/Requirements: GCE, A/AS: 14 points. AGNVQ, Merit (in specific programmes). ND/C, Merit overall. SQAH, Grades BCCC. SQAV, Individual consideration. IB, 24 points.

BSc Linguistic Sciences/Computing

UCAS Code: Q1G5 • Mode: 3/4 Years Major/Minor

Qualifications/Requirements: GCE, A/AS: 14 points. AGNVQ, Merit (in specific programmes). ND/C, Merit overall. SQAH, Grades BCCC. SQAV, Individual consideration. IB, 24 points.

THE UNIVERSITY OF HUDDERSFIELD H60

BA Business Computing with French

UCAS Code: G5R1 • Mode: 4 Years Major/Minor

Qualifications/Requirements: GCE, A/AS: 16 points. AGNVQ, Merit. ND/C, Merit overall. SQAH, Grades BBBB. SQAV, Individual consideration. IB, Individual consideration.

BA Business Computing with German

UCAS Code: G5R2 • Mode: 4 Years Major/Minor

Qualifications/Requirements: GCE, A/AS: 16 points. AGNVQ, Merit. ND/C, Merit overall. SQAH, Grades BBBB. SQAV, Individual consideration. IB, Individual consideration.

IMPERIAL COLLEGE OF SCIENCE, TECHNOLOGY AND MEDICINE (UNIVERSITY OF LONDON) I50

MEng Computing (European Programme of Study) (4 years) (h)

UCAS Code: G502 • Mode: 4 Years Single Subject

Qualifications/Requirements: GCE, A level grades: AAB. Maths at A level. AGNVQ, Pass. ND/C, Individual consideration. SQAH, Individual consideration. SQAV, Individual consideration. IB, Individual consideration.

KEELE UNIVERSITY K12

Mod Computer Science and German

UCAS Code: GR52 • Mode: 3 Years Equal Combination

Qualifications/Requirements: GCE, A level grades: BCC-CCC. German at A level and Maths or Science. AGNVQ, Distinction (in specific programmes) with A level. ND/C, Individual consideration. SQAH, CSYS required. SQAV, Individual consideration. IB, 28 points.

Mod Computer Science and Latin

UCAS Code: GQ56 • Mode: 3 Years Equal Combination

Qualifications/Requirements: GCE, A level grades: BCC-CCC. Latin at A level and Maths or Science. AGNVQ, Distinction (in specific programmes) with A level. ND/C, Individual consideration. SQAH, CSYS required. SQAV, Individual consideration. IB, 28 points.

Mod Computer Science and Russian

UCAS Code: GR58 • Mode: 3 Years Equal Combination

Qualifications/Requirements: GCE, A level grades: BCC-CCC. Russian at A level and Maths or Science. AGNVQ, Distinction (in specific programmes) with A level. ND/C, Individual consideration. SQAH, CSYS required. SQAV, Individual consideration. IB, 28 points.

Mod Computer Science and Russian Studies

UCAS Code: GRM8 • Mode: 3 Years Equal Combination

Qualifications/Requirements: GCE, A level grades: BCC-CCC. Maths or Science. AGNVQ, Distinction (in specific programmes) with A level. ND/C, Individual consideration. SQAH, CSYS required. SQAV, Individual consideration. IB, 28 points.

THE UNIVERSITY OF KENT AT CANTERBURY K24

BA European Studies/Computing

UCAS Code: TG25 • Mode: 4 Years Equal Combination

Qualifications/Requirements: GCE, A/AS: 20 points. AGNVQ, Individual consideration. ND/C, 3 Merits and 3 Distinctions. SQAH, Individual consideration. SQAV, Individual consideration. IB, 28 points.

BA French/Computing
UCAS Code: RG15 • Mode: 4 Years Equal Combination

Qualifications/Requirements: GCE, A/AS: 20 points. French. AGNVQ, Individual consideration. ND/C, 3 Merits and 3 Distinctions. SQAH, Individual consideration. SQAV, Individual consideration. IB, 28 points.

BA German/Computing
UCAS Code: RG25 • Mode: 4 Years Equal Combination

Qualifications/Requirements: GCE, A/AS: 20 points. German. AGNVQ, Individual consideration. ND/C, 3 Merits and 3 Distinctions. SQAH, Individual consideration. SQAV, Individual consideration. IB, 28 points.

BA Italian/Computing
UCAS Code: RG35 • Mode: 4 Years Equal Combination

Qualifications/Requirements: GCE, A/AS: 20 points. AGNVQ, Individual consideration. ND/C, 3 Merits and 3 Distinctions. SQAH, Individual consideration. SQAV, Individual consideration. IB, 28 points.

BA Spanish/Computing
UCAS Code: GR54 • Mode: 4 Years Equal Combination

Qualifications/Requirements: GCE, A/AS: 20 points. AGNVQ, Individual consideration. ND/C, 3 Merits and 3 Distinctions. SQAH, Individual consideration. SQAV, Individual consideration. IB, 28 points.

THE UNIVERSITY OF WALES, LAMPETER L07

BA Information Technology and French
UCAS Code: GR51 • Mode: 4 Years Equal Combination

Qualifications/Requirements: GCE, A/AS: 14-16 points. French at A level. AGNVQ, Individual consideration. ND/C, Individual consideration. SQAH, Individual consideration. SQAV, Individual consideration. IB, Individual consideration.

BA Information Technology and Greek
UCAS Code: GQ57 • Mode: 3 Years Equal Combination

Qualifications/Requirements: GCE, A/AS: 14-16 points. AGNVQ, Individual consideration. ND/C, Individual consideration. SQAH, Individual consideration. SQAV, Individual consideration. IB, Individual consideration.

BA Islamic Studies and Information Technology
UCAS Code: GT56 • Mode: 3 Years Equal Combination

Qualifications/Requirements: GCE, A/AS: 14 points. AGNVQ, Individual consideration. ND/C, Individual consideration. SQAH, Individual consideration. SQAV, Individual consideration. IB, Individual consideration.

BA Jewish Studies and Information Technology
UCAS Code: GQ59 • Mode: 3 Years Equal Combination

Qualifications/Requirements: Please refer to prospectus.

BA Latin and Information Technology
UCAS Code: GQ56 • Mode: 3 Years Equal Combination

Qualifications/Requirements: GCE, A/AS: 16 points. AGNVQ, Individual consideration. ND/C, Individual consideration. SQAH, Individual consideration. SQAV, Individual consideration. IB, Individual consideration.

BA Welsh and Information Technology
UCAS Code: GQ55 • Mode: 3/4 Years Equal Combination

Qualifications/Requirements: GCE, A/AS: 14 points. Welsh at A level. AGNVQ, Individual consideration. ND/C, Individual consideration. SQAH, Individual consideration. SQAV, Individual consideration. IB, Individual consideration.

BA Welsh Studies and Information Technology
UCAS Code: GQ5N • Mode: 3 Years Equal Combination

Qualifications/Requirements: GCE, A/AS: 14 points. AGNVQ, Individual consideration. ND/C, Individual consideration. SQAH, Individual consideration. SQAV, Individual consideration. IB, Individual consideration.

DipHE French/Information Technology
UCAS Code: RG15 • Mode: 2 Years Equal Combination

Qualifications/Requirements: Please refer to prospectus.

LANCASTER UNIVERSITY L14

BSc Computing and European Languages
UCAS Code: GT52 • Mode: 4 Years Equal Combination

Qualifications/Requirements: GCE, A level grades: BCC. Two modern foreign languages. AGNVQ, Distinction with 6 additional units or with A/AS. ND/C, Individual consideration. SQAH, Grades BBBBB including specific subject(s). SQAV, Individual consideration. IB, 30 points.

BSc French Studies and Computing
UCAS Code: GR51 • Mode: 4 Years Equal Combination

Qualifications/Requirements: GCE, A level grades: BCC. French at A level. AGNVQ, Distinction with 6 additional units or with A/AS. ND/C, Individual consideration. SQAH, Grades BBBBB including specific subject(s). SQAV, Individual consideration. IB, 30 points.

BSc German Studies and Computing
UCAS Code: GR52 • Mode: 4 Years Equal Combination

Qualifications/Requirements: GCE, A level grades: BCC. German or a modern foreign language at A level. AGNVQ, Distinction with 6 additional units or with A/AS. ND/C, Individual consideration. SQAH, Grades BBBBB including specific subject(s). SQAV, Individual consideration. IB, 30 points.

BSc Italian Studies and Computing
UCAS Code: GR53 • Mode: 4 Years Equal Combination

Qualifications/Requirements: GCE, A level grades: BCC. Italian or a modern foreign language at A level. AGNVQ, Distinction with 6 additional units or with A/AS. ND/C, Individual consideration. SQAH, Grades BBBBB including specific subject(s). SQAV, Individual consideration. IB, 30 points.

BSc Spanish Studies and Computing
UCAS Code: GR54 • Mode: 4 Years Equal Combination

Qualifications/Requirements: GCE, A level grades: BCC. Spanish or a modern foreign language at A level. AGNVQ, Distinction with 6 additional units or with A/AS. ND/C, Individual consideration. SQAH, Grades BBBBB including specific subject(s). SQAV, Individual consideration. IB, 30 points.

UNIVERSITY OF LEEDS L23

BA Computing/Russian Civilisation
UCAS Code: RG85 • Mode: 3 Years Equal Combination

Qualifications/Requirements: GCE, A level grades: BBC. AGNVQ, Individual consideration. ND/C, Individual consideration. SQAH, Individual consideration. SQAV, Individual consideration. IB, Individual consideration.

BSc Computing/French
UCAS Code: GR51 • Mode: 4 Years Equal Combination

Qualifications/Requirements: GCE, A level grades: BBC. French at A level. AGNVQ, Individual consideration. SQAH, Individual consideration. SQAV, Individual consideration. IB, 30 points.

BSc Computing/German
UCAS Code: GR52 • Mode: 4 Years Equal Combination

Qualifications/Requirements: GCE, A level grades: BBC. German at A level. AGNVQ, Individual consideration. SQAH, Individual consideration. SQAV, Individual consideration. IB, 30 points.

UNIVERSITY OF LINCOLNSHIRE AND HUMBERSIDE L39

BA/BSc Computing and European Studies
UCAS Code: GT52 • Mode: 3 Years Equal Combination

Qualifications/Requirements: GCE, A/AS: 12 points. AGNVQ, Merit. ND/C, 3 Merits and 1 Distinction. SQAH, Grades CCCC. SQAV, Individual consideration. IB, 24 points.

BA/BSc Computing and Modern Languages
UCAS Code: GT5X • Mode: 3 Years Equal Combination

Qualifications/Requirements: GCE, A/AS: 12 points. A modern foreign language. AGNVQ, Merit. ND/C, 3 Merits and 1 Distinction. SQAH, Grades CCCC. SQAV, Individual consideration. IB, 24 points.

THE UNIVERSITY OF LIVERPOOL L41

BSc Computer Science with a European Language
UCAS Code: G5T2 • Mode: 4 Years Major/Minor

Qualifications/Requirements: GCE, A/AS: 20 points. Maths at A level. AGNVQ, Individual consideration. ND/C, Merit overall and 4 Distinctions (in specific programmes). SQAH, Grades BBBBC including specific subject(s). SQAV, Individual consideration. IB, 30 points.

LIVERPOOL HOPE L46

BA Information Technology/European Studies
UCAS Code: TG25 • Mode: 3 Years Equal Combination

Qualifications/Requirements: GCE, A/AS: 12 points. AGNVQ, Merit. ND/C, 8 Merits. SQAH, Individual consideration. SQAV, Individual consideration. IB, Individual consideration.

BA Information Technology/French
UCAS Code: GR51 • Mode: 3 Years Equal Combination

Qualifications/Requirements: GCE, A/AS: 12 points. French at A level. AGNVQ, Pass with A level. ND/C, 8 Merits. SQAH, Individual consideration. SQAV, Individual consideration. IB, Individual consideration.

LONDON GUILDHALL UNIVERSITY L55

Mod Computing & European Studies
UCAS Code: GT52 • Mode: 3 Years Equal Combination

Qualifications/Requirements: GCE, A level grades: DD. AGNVQ, Merit (in specific programmes). ND/C, Merit overall. SQAH, Individual consideration. SQAV, Individual consideration. IB, 24 points.

Mod Computing & French
UCAS Code: GR51 • Mode: 4 Years Equal Combination

Qualifications/Requirements: GCE, A level grades: DD. AGNVQ, Merit (in specific programmes). ND/C, Merit overall. SQAH, Individual consideration. SQAV, Individual consideration. IB, 24 points.

Mod Computing & German
UCAS Code: GR52 • Mode: 4 Years Equal Combination

Qualifications/Requirements: GCE, A level grades: DD. AGNVQ, Merit (in specific programmes). ND/C, Merit overall. SQAH, Individual consideration. SQAV, Individual consideration. IB, 24 points.

Mod Computing & Spanish
UCAS Code: GR54 • Mode: 4 Years Equal Combination

Qualifications/Requirements: GCE, A level grades: DD. AGNVQ, Merit (in specific programmes). ND/C, Merit overall. SQAH, Individual consideration. SQAV, Individual consideration. IB, 24 points.

THE UNIVERSITY OF MANCHESTER INSTITUTE OF SCIENCE AND TECHNOLOGY (UMIST) M25

BSc Computational Linguistics
UCAS Code: G5Q1 • Mode: 3 Years Major/Minor

Qualifications/Requirements: GCE, A/AS: 20 points. AGNVQ, Merit in Information Technology. ND/C, Merit overall and 4 Distinctions. SQAH, Grades ABBBC. SQAV, Individual consideration. IB, 28 points.

THE MANCHESTER METROPOLITAN UNIVERSITY M40

BSc Applicable Mathematics/Multimedia Technology

UCAS Code: GG16 • Mode: 3 Years Equal Combination

Qualifications/Requirements: GCE, A/AS: 16 points. Maths. AGNVQ, Merit (in specific programmes). ND/C, 1 Merit and 3 Distinctions (in specific programmes). SQAH, Grades BBBCC including specific subject(s). SQAV, Individual consideration. IB, 28 points.

UNIVERSITY COLLEGE NORTHAMPTON N14

Mod French/Information Systems

UCAS Code: R1G5 • Mode: 3 Years Major/Minor

Qualifications/Requirements: GCE, A/AS: 10 points. French. AGNVQ, Individual consideration. ND/C, 5 Merits. SQAH, Grades CCC. SQAV, Individual consideration. IB, 24 points.

Mod Information Systems/French

UCAS Code: G5R1 • Mode: 3 Years Major/Minor

Qualifications/Requirements: GCE, A/AS: 10-12 points. French. AGNVQ, Merit. ND/C, 5 Merits. SQAH, Grades CCC. SQAV, Individual consideration. IB, 24 points.

Mod Information Systems/Italian

UCAS Code: G5R3 • Mode: 3 Years Major/Minor

Qualifications/Requirements: GCE, A/AS: 10-12 points. AGNVQ, Merit. ND/C, 5 Merits. SQAH, Grades CCC. SQAV, Individual consideration. IB, 24 points.

Mod Information Systems/Spanish

UCAS Code: G5R4 • Mode: 3 Years Major/Minor

Qualifications/Requirements: GCE, A/AS: 10-12 points. AGNVQ, Merit. ND/C, 5 Merits. SQAH, Grades CCC. SQAV, Individual consideration. IB, 24 points.

OXFORD BROOKES UNIVERSITY O66

BA/BSc Computing Science/Spanish Studies

UCAS Code: GR5L • Mode: 3 Years Equal Combination

Qualifications/Requirements: Please refer to prospectus.

BA/BSc Computing/Spanish Studies

UCAS Code: GR5K • Mode: 3 Years Equal Combination

Qualifications/Requirements: Please refer to prospectus.

BA/BSc Information Systems/Spanish Studies

UCAS Code: RG45 • Mode: 3 Years Equal Combination

Qualifications/Requirements: Please refer to prospectus.

Mod Computing Science/French Studies

UCAS Code: GRMD • Mode: 3 Years Equal Combination

Qualifications/Requirements: Please refer to prospectus.

Mod Computing Science/German Studies

UCAS Code: RG25 • Mode: 3 Years Equal Combination

Qualifications/Requirements: GCE, A level grades: DD-CCC. German. AGNVQ, Merit with A level. ND/C, Individual consideration. SQAH, Individual consideration. SQAV, Individual consideration. IB, Individual consideration.

Mod Computing Science/Japanese Studies

UCAS Code: GT5K • Mode: 3 Years Equal Combination

Qualifications/Requirements: Please refer to prospectus.

Mod Computing/French Studies

UCAS Code: GR5D • Mode: 3 Years Equal Combination

Qualifications/Requirements: Please refer to prospectus.

Mod Computing/German Studies

UCAS Code: GR5G • Mode: 4 Years Equal Combination

Qualifications/Requirements: GCE, A level grades: DDD-CCC. German. AGNVQ, Merit with A level. ND/C, Individual consideration. SQAH, Individual consideration. SQAV, Individual consideration. IB, Individual consideration.

Mod Computing/Japanese Studies

UCAS Code: GT54 • Mode: 3 Years Equal Combination

Qualifications/Requirements: Please refer to prospectus.

Mod French Studies/Information Systems

UCAS Code: GRN1 • Mode: 3 Years Equal Combination

Qualifications/Requirements: Please refer to prospectus.

Mod French Studies/Multimedia Systems

UCAS Code: RG1P • Mode: 3 Years Equal Combination

Qualifications/Requirements: Please refer to prospectus.

Mod French Studies/Statistics

UCAS Code: RG1K • Mode: 3 Years Equal Combination

Qualifications/Requirements: Please refer to prospectus.

Mod German Studies/Information Systems

UCAS Code: GRMG • Mode: 4 Years Equal Combination

Qualifications/Requirements: GCE, A level grades: CD-BC. German. AGNVQ, Merit with A level. ND/C, Individual consideration. SQAH, Individual consideration. SQAV, Individual consideration. IB, Individual consideration.

Mod German Studies/Multimedia Systems

UCAS Code: GRP2 • Mode: 3 Years Equal Combination

Qualifications/Requirements: Please refer to prospectus.

Mod Information Systems/Japanese Studies

UCAS Code: GT5L • Mode: 3 Years Equal Combination

Qualifications/Requirements: Please refer to prospectus.

Mod Information Systems/Transport and Travel

UCAS Code: GNM9 • Mode: 3 Years Equal Combination

Qualifications/Requirements: GCE, A level grades: CC-BC. AGNVQ, Merit. ND/C, Individual consideration. SQAH, Individual consideration. SQAV, Individual consideration. IB, Individual consideration.

Mod Japanese Studies/Multimedia Systems

UCAS Code: GT6K • Mode: 3 Years Equal Combination

Qualifications/Requirements: Please refer to prospectus.

Mod Japanese Studies/Software Engineering

UCAS Code: GT74 • Mode: 3 Years Equal Combination

Qualifications/Requirements: Please refer to prospectus.

UNIVERSITY OF PAISLEY P20

BSc Computing Science with French/German/Spanish

UCAS Code: G5TF • Mode: 3/4/5 Years Major/Minor

Qualifications/Requirements: GCE, A level grades: CC. Maths. AGNVQ, Pass in Information Technology. ND/C, Individual consideration. SQAH, Grades BCCC including specific subject(s). SQAV, Higher National. IB, Pass in Diploma.

QUEEN MARY AND WESTFIELD COLLEGE (UNIVERSITY OF LONDON) Q50

BA French, Linguistics and Computer Science

UCAS Code: GR51 • Mode: 4 Years Equal Combination

Qualifications/Requirements: GCE, A level grades: BBC-BCD. French at A level. AGNVQ, Distinction with A/AS. ND/C, 3 Distinctions (in specific programmes). SQAH, Grades BBBCC including specific subject(s). IB, 28 points.

BA German, Linguistics and Computer Science

UCAS Code: GR52 • Mode: 4 Years Equal Combination

Qualifications/Requirements: GCE, A level grades: BBC-BCD. A modern foreign language. AGNVQ, Distinction with A/AS. ND/C, 3 Distinctions (in specific programmes). SQAH, Grades BBCCC including specific subject(s). IB, 28 points.

BA Hispanic Studies, Linguistics and Computer Science

UCAS Code: GR5K • Mode: 4 Years Equal Combination

Qualifications/Requirements: GCE, A level grades: BBC-BCD. A modern foreign language. AGNVQ, Distinction with A/AS. ND/C, 3 Distinctions (in specific programmes). SQAH, Grades BBCCC including specific subject(s). IB, 28 points.

BA Russian, Linguistics and Computer Science

UCAS Code: GR58 • Mode: 4 Years Equal Combination

Qualifications/Requirements: GCE, A level grades: BBC-BCD. A modern foreign language. AGNVQ, Distinction with A/AS. ND/C, 3 Distinctions (in specific programmes). SQAH, Grades BBCCC including specific subject(s). IB, 28 points.

THE UNIVERSITY OF READING R12

MEng German Informatics

UCAS Code: G5R2 • Mode: 4 Years Major/Minor

Qualifications/Requirements: GCE, A/AS: 24 points. German at A level. AGNVQ, Distinction (in specific programmes) with A level. ND/C, 2 Merits and 3 Distinctions (in specific programmes). SQAH, Grades ABBB including specific subject(s). SQAV, Individual consideration. IB, 32 points.

ROYAL HOLLOWAY, UNIVERSITY OF LONDON R72

BSc Computer Science with French

UCAS Code: G5R1 • Mode: 3/4 Years Major/Minor

Qualifications/Requirements: GCE, A level grades: BBB. Maths and French at A level. AGNVQ, Distinction with A level. ND/C, Merit overall and 4 Distinctions. SQAH, Individual consideration. IB, 30 points.

THE UNIVERSITY OF SALFORD S03

BSc Computer Science and Arabic

UCAS Code: GT56 • Mode: 4 Years Equal Combination

Qualifications/Requirements: AGNVQ, Individual consideration. ND/C, Individual consideration. SQAH, Individual consideration. SQAV, Individual consideration. IB, Individual consideration.

BSc Computer Science and French

UCAS Code: GR51 • Mode: 4 Years Equal Combination

Qualifications/Requirements: AGNVQ, Individual consideration. ND/C, Individual consideration. SQAH, Individual consideration. SQAV, Individual consideration. IB, Individual consideration.

BSc Computer Science and German

UCAS Code: GR52 • Mode: 4 Years Equal Combination

Qualifications/Requirements: AGNVQ, Individual consideration. ND/C, Individual consideration. SQAH, Individual consideration. SQAV, Individual consideration. IB, Individual consideration.

BSc Computer Science and Italian

UCAS Code: GR53 • Mode: 4 Years Equal Combination

Qualifications/Requirements: AGNVQ, Individual consideration. ND/C, Individual consideration. SQAH, Individual consideration. SQAV, Individual consideration. IB, Individual consideration.

BSc Computer Science and Spanish
UCAS Code: GR54 • Mode: 4 Years Equal Combination

Qualifications/Requirements: AGNVQ, Individual consideration. ND/C, Individual consideration. SQAH, Individual consideration. SQAV, Individual consideration. IB, Individual consideration.

BSc Information Technology with Language Training in French
UCAS Code: G5R1 • Mode: 3 Years Major/Minor

Qualifications/Requirements: GCE, A/AS: 18-20 points. AGNVQ, Distinction. ND/C, Merits and Distinction. SQAH, Individual consideration. SQAV, Individual consideration. IB, Individual consideration.

BSc Information Technology with Language Training in German
UCAS Code: G5R2 • Mode: 3 Years Major/Minor

Qualifications/Requirements: GCE, A/AS: 18-20 points. AGNVQ, Distinction. ND/C, Merits and Distinction. SQAH, Individual consideration. SQAV, Individual consideration. IB, Individual consideration.

THE UNIVERSITY OF SHEFFIELD S18

BSc Computer Science and French
UCAS Code: GR51 • Mode: 4 Years Equal Combination

Qualifications/Requirements: GCE, A/AS: 24 points. French and Maths at A level. AGNVQ, Distinction with A level. ND/C, 3 Merits and 3 Distinctions (in specific programmes). SQAH, Grades AABB including specific subject(s). SQAV, Individual consideration. IB, 32 points.

BSc Computer Science and German
UCAS Code: GR52 • Mode: 4 Years Equal Combination

Qualifications/Requirements: GCE, A/AS: 24 points. German and Maths at A level. AGNVQ, Distinction with A level. ND/C, 3 Merits and 3 Distinctions (in specific programmes). SQAH, Grades AABB including specific subject(s). SQAV, Individual consideration. IB, 32 points.

BSc Computer Science and Hispanic Studies
UCAS Code: GR54 • Mode: 4 Years Equal Combination

Qualifications/Requirements: GCE, A/AS: 24 points. Maths and a modern foreign language at A level. AGNVQ, Distinction with A level. ND/C, 3 Merits and 3 Distinctions (in specific programmes). SQAH, Grades AABB including specific subject(s). SQAV, Individual consideration. IB, 32 points.

BSc Computer Science and Russian
UCAS Code: GR58 • Mode: 4 Years Equal Combination

Qualifications/Requirements: GCE, A/AS: 24 points. Maths and a modern foreign language at A level. AGNVQ, Distinction with A level. ND/C, 3 Merits and 3 Distinctions (in specific programmes). SQAH, Grades AABB including specific subject(s). SQAV, Individual consideration. IB, 32 points.

SOUTH BANK UNIVERSITY S33

Mod Computing and French
UCAS Code: GR51 • Mode: 3 Years Equal Combination

Qualifications/Requirements: GCE, A/AS: 12-16 points. Maths and French at A level. AGNVQ, Merit. ND/C, 4 Merits and 2 Distinctions. SQAH, Individual consideration. SQAV, Individual consideration. IB, Individual consideration.

Mod Computing and German
UCAS Code: GR5F • Mode: 3 Years Equal Combination

Qualifications/Requirements: GCE, A/AS: 14-18 points. Maths and German at A level. AGNVQ, Merit. ND/C, 2 Merits and 4 Distinctions. SQAH, Individual consideration. SQAV, Individual consideration. IB, Individual consideration.

Mod Computing and German ab initio
UCAS Code: GR52 • Mode: 3 Years Equal Combination

Qualifications/Requirements: GCE, A/AS: 12-16 points. Maths at A level. AGNVQ, Merit. ND/C, 4 Merits and 2 Distinctions. SQAH, Individual consideration. SQAV, Individual consideration. IB, Individual consideration.

Mod Computing and Spanish
UCAS Code: GR5K • Mode: 3 Years Equal Combination

Qualifications/Requirements: GCE, A/AS: 14-18 points. Spanish and Maths at A level. AGNVQ, Merit. ND/C, 2 Merits and 4 Distinctions. SQAH, Individual consideration. SQAV, Individual consideration. IB, Individual consideration.

Mod Computing and Spanish ab initio
UCAS Code: GR54 • Mode: 3 Years Equal Combination

Qualifications/Requirements: GCE, A/AS: 12-16 points. Maths at A level. AGNVQ, Merit. ND/C, 4 Merits and 2 Distinctions. SQAH, Individual consideration. SQAV, Individual consideration. IB, Individual consideration.

UNIVERSITY OF ST ANDREWS S36

BSc Computer Science with French
UCAS Code: G5R1 • Mode: 4 Years Major/Minor

Qualifications/Requirements: GCE, A level grades: BBC. SQAH, Grades BBBC including specific subject(s). IB, 28 points.

BSc Computer Science with French (Integrated Year Abroad)
UCAS Code: G5RC • Mode: 4 Years Major/Minor

Qualifications/Requirements: GCE, A level grades: BBC. SQAH, Grades BBBC including specific subject(s). IB, 28 points.

BSc Computer Science with German
UCAS Code: G5R2 • Mode: 4 Years Major/Minor

Qualifications/Requirements: GCE, A level grades: BBC. SQAH, Grades BBBC including specific subject(s). IB, 28 points.

BSc Computer Science with German (Integrated Year Abroad)
UCAS Code: G5RF • Mode: 4 Years Major/Minor
Qualifications/Requirements: GCE, A level grades: BBC. SQAH, Grades BBB including specific subject(s). IB, 28 points.

THE UNIVERSITY OF STIRLING S75

BSc Computing Science/French Language
UCAS Code: GR51 • Mode: 4 Years Equal Combination
Qualifications/Requirements: GCE, A level grades: CCD. AGNVQ, Merit (in specific programmes) with 2 additional units or with A/AS. ND/C, Merit overall. SQAH, Grades BBCC. SQAV, Higher National. IB, 28 points.

BSc Computing Science/German Language
UCAS Code: GR52 • Mode: 4 Years Equal Combination
Qualifications/Requirements: GCE, A level grades: CCD. AGNVQ, Merit (in specific programmes) with 2 additional units or with A/AS. ND/C, Merit overall. SQAH, Grades BBCC. SQAV, Higher National. IB, 28 points.

BSc Computing Science/Japanese Studies
UCAS Code: GT54 • Mode: 4 Years Equal Combination
Qualifications/Requirements: GCE, A level grades: CCD. AGNVQ, Merit (in specific programmes) with 2 additional units or with A/AS. ND/C, Merit overall. SQAH, Grades BBCC. SQAV, Higher National. IB, 28 points.

BSc Computing Science/Spanish Language
UCAS Code: GR54 • Mode: 4 Years Equal Combination
Qualifications/Requirements: GCE, A level grades: CCD. AGNVQ, Merit (in specific programmes) with 2 additional units or with A/AS. ND/C, Merit overall. SQAH, Grades BBCC. SQAV, Higher National. IB, 28 points.

UNIVERSITY OF SUNDERLAND S84

BA Business Computing with French
UCAS Code: G5R1 • Mode: 4 Years Major/Minor
Qualifications/Requirements: GCE, A/AS: 10 points. AGNVQ, Merit. ND/C, 4 Merits (in specific programmes). SQAH, Grades CCCC. SQAV, National. IB, 24 points.

BA Business Computing with German
UCAS Code: G5R2 • Mode: 4 Years Major/Minor
Qualifications/Requirements: GCE, A/AS: 10 points. AGNVQ, Merit. ND/C, 4 Merits (in specific programmes). SQAH, Grades CCCC. SQAV, National. IB, 24 points.

BA Business Computing with Spanish
UCAS Code: G5R4 • Mode: 4 Years Major/Minor
Qualifications/Requirements: GCE, A/AS: 10 points. AGNVQ, Merit. ND/C, 4 Merits (in specific programmes). SQAH, Grades CCCC. SQAV, National. IB, 24 points.

BSc Computer Studies with American Studies
UCAS Code: G5Q4 • Mode: 3 Years Major/Minor
Qualifications/Requirements: GCE, A/AS: 8 points. AGNVQ, Pass. ND/C, 4 Merits. SQAH, Grades CCC. SQAV, Individual consideration. IB, Individual consideration.

BSc Computer Studies with European Studies
UCAS Code: G5T2 • Mode: 3 Years Major/Minor
Qualifications/Requirements: GCE, A/AS: 8 points. AGNVQ, Pass. ND/C, 4 Merits. SQAH, Grades CCC. SQAV, Individual consideration. IB, Individual consideration.

BSc Computer Studies with German Studies
UCAS Code: G5RF • Mode: 4 Years Major/Minor
Qualifications/Requirements: GCE, A/AS: 8 points. German at A level. AGNVQ, Pass (in specific programmes). ND/C, 4 Merits. SQAH, Grades CCC. SQAV, Individual consideration. IB, Individual consideration.

BSc Computer Studies with Spanish Studies
UCAS Code: G5RK • Mode: 4 Years Major/Minor
Qualifications/Requirements: GCE, A/AS: 8 points. Spanish at A level. AGNVQ, Pass. ND/C, 4 Merits. SQAH, Grades CCC. SQAV, Individual consideration. IB, Individual consideration.

Mod Computer Studies and French Studies
UCAS Code: GR51 • Mode: 4 Years Equal Combination
Qualifications/Requirements: GCE, A/AS: 8 points. French at A level. AGNVQ, Pass (in specific programmes). ND/C, 4 Merits. SQAH, Grades CCC. SQAV, Individual consideration. IB, Individual consideration.

Mod Computer Studies and German Studies
UCAS Code: GR52 • Mode: 4 Years Equal Combination
Qualifications/Requirements: GCE, A/AS: 8 points. German at A level. AGNVQ, Pass (in specific programmes). ND/C, 4 Merits. SQAH, Grades CCC. SQAV, Individual consideration. IB, Individual consideration.

Mod Computer Studies and Spanish Studies
UCAS Code: GR54 • Mode: 3/4 Years Equal Combination
Qualifications/Requirements: GCE, A/AS: 8 points. Spanish at A level. AGNVQ, Pass (in specific programmes). ND/C, 4 Merits. SQAH, Grades CCC. SQAV, Individual consideration. IB, Individual consideration.

THE UNIVERSITY OF SURREY S87

BSc Computing and German
UCAS Code: GR52 • Mode: 3/4 Years Equal Combination

Qualifications/Requirements: GCE, A level grades: BBC. Maths and German at A level. AGNVQ, Not normally sufficient. ND/C, Not normally sufficient. SQAH, Grades AABBB including specific subject(s). SQAV, Not normally sufficient. IB, 30 points.

UNIVERSITY OF SUSSEX S90

BA Linguistics: School of Cognitive and Computing Sciences
UCAS Code: Q1G5 • Mode: 3 Years Major/Minor

Qualifications/Requirements: GCE, A level grades: BBC. AGNVQ, Merit with 6 additional units. ND/C, Merit overall. SQAH, Individual consideration. SQAV, Individual consideration. IB, Individual consideration.

BSc Computer Science with European Studies
UCAS Code: G5T2 • Mode: 4 Years Major/Minor

Qualifications/Requirements: GCE, A level grades: BBB. Maths. AGNVQ, Merit with A level. ND/C, Merit overall (in specific programmes). SQAH, Individual consideration. SQAV, Individual consideration. IB, Individual consideration.

UNIVERSITY OF WALES SWANSEA S93

BA French/Russian (with Computer Studies)
UCAS Code: RR1W • Mode: 4 Years Equal Combination

Qualifications/Requirements: GCE, A level grades: BBC-BCC. French at A level. AGNVQ, Individual consideration. ND/C, 1 Merit and 5 Distinctions (in specific programmes). SQAH, Grades BBBBC including specific subject(s). SQAV, Individual consideration. IB, 28 points.

BA Welsh (with Computer Studies)
UCAS Code: Q5G5 • Mode: 3/4 Years Major/Minor

Qualifications/Requirements: GCE, A level grades: BCC-CCC. Welsh at A level. AGNVQ, Not normally sufficient. ND/C, 1 Merit and 4 Distinctions (in specific programmes). SQAH, Individual consideration. SQAV, Individual consideration. IB, Individual consideration.

BSc Computer Science (with French)
UCAS Code: G5R1 • Mode: 4 Years Major/Minor

Qualifications/Requirements: GCE, A level grades: BBC. French at A level. AGNVQ, Individual consideration. ND/C, 1 Merit and 5 Distinctions (in specific programmes). SQAH, Grades BBBBB including specific subject(s). SQAV, Individual consideration. IB, 28 points.

BSc Computer Science (with German)
UCAS Code: G5R2 • Mode: 4 Years Major/Minor

Qualifications/Requirements: GCE, A level grades: BBC. German. AGNVQ, Individual consideration. ND/C, 1 Merit and 5 Distinctions (in specific programmes). SQAH, Grades BBBBB including specific subject(s). SQAV, Individual consideration. IB, 28 points.

BSc Computer Science (with Italian)
UCAS Code: G5R3 • Mode: 4 Years Major/Minor

Qualifications/Requirements: GCE, A level grades: BBC. A modern foreign language at A level. AGNVQ, Individual consideration. ND/C, 1 Merit and 5 Distinctions. SQAH, Grades BBBBB. SQAV, Individual consideration. IB, 28 points.

BSc Computer Science (with Russian)
UCAS Code: G5R8 • Mode: 4 Years Major/Minor

Qualifications/Requirements: GCE, A level grades: BBC. A modern foreign language at A level. AGNVQ, Individual consideration. ND/C, 1 Merit and 5 Distinctions. SQAH, Grades BBBBB. SQAV, Individual consideration. IB, 28 points.

BSc Computer Science (with Spanish)
UCAS Code: G5R4 • Mode: 4 Years Major/Minor

Qualifications/Requirements: GCE, A level grades: BBC. A modern foreign language at A level. AGNVQ, Individual consideration. ND/C, 1 Merit and 5 Distinctions. SQAH, Grades BBBBB. SQAV, Individual consideration. IB, 28 points.

BSc Computer Science (with Welsh)
UCAS Code: G5Q5 • Mode: 4 Years Major/Minor

Qualifications/Requirements: GCE, A level grades: BBC. AGNVQ, Individual consideration. ND/C, 1 Merit and 5 Distinctions. SQAH, Grades BBBBC. SQAV, Individual consideration. IB, 28 points.

UNIVERSITY OF ULSTER U20

BSc Computing and Linguistics
UCAS Code: GQ51 • Mode: 4 Years Equal Combination

Qualifications/Requirements: GCE, A level grades: BCC. AGNVQ, Distinction with 6 additional units or with A/AS. ND/C, Merit overall and 4 Distinctions. SQAH, Grades BBBC. SQAV, Individual consideration. IB, 30 points.

UNIVERSITY COLLEGE WORCESTER W80

Mod Information Technology/Modern Foreign Languages
UCAS Code: GT59 • Mode: 3 Years Equal Combination

Qualifications/Requirements: GCE, A level grades: DD. A modern foreign language at A level. AGNVQ, Merit. ND/C, Individual consideration. SQAH, Individual consideration. SQAV, Individual consideration. IB, Individual consideration.

THE UNIVERSITY OF WALES, ABERYSTWYTH A40

BEng Interactive Multimedia Engineering (including integegrated individual & professional training)

UCAS Code: GG65•Mode: 4 Years Equal Combination

Qualifications/Requirements: GCE, A/AS: 20 points. AGNVQ, Merit. ND/C, 3 Merits and 2 Distinctions. SQAH, Grades BBBCC. SQAV, Individual consideration. IB, 30 points.

ANGLIA POLYTECHNIC UNIVERSITY A60

BSc Information Systems and Multimedia

UCAS Code: G524•Mode: 3 Years Single Subject

Qualifications/Requirements: GCE, A/AS: 12 points. AGNVQ, Merit. ND/C, 4 Merits. SQAH, Grades CCCC. SQAV, National. IB, Pass in Diploma.

BSc Multimedia and Product Design

UCAS Code: HH76•Mode: 3 Years Equal Combination

Qualifications/Requirements: GCE, A/AS: 12 points. AGNVQ, Merit (in specific programmes). ND/C, 4 Merits. SQAH, Grades BBCC. SQAV, Individual consideration. IB, Pass in Diploma.

BSc Multimedia Computing

UCAS Code: G611•Mode: 3 Years Single Subject

Qualifications/Requirements: GCE, A/AS: 14 points. AGNVQ, Merit (in specific programmes). ND/C, 6 Merits. SQAH, Grades BBCC. SQAV, Individual consideration. IB, Pass in Diploma.

BSc Multimedia Systems

UCAS Code: G610•Mode: 3 Years Single Subject

Qualifications/Requirements: GCE, A/AS: 14 points. AGNVQ, Merit (in specific programmes). ND/C, 6 Merits. SQAH, Grades BBCC. SQAV, Individual consideration. IB, Pass in Diploma.

Mod Business Administration and Multimedia

UCAS Code: HN61•Mode: 3 Years Equal Combination

Qualifications/Requirements: GCE, A/AS: 12 points. AGNVQ, Merit. ND/C, 4 Merits. SQAH, Grades BBCC. SQAV, Individual consideration. IB, Pass in Diploma.

Mod Lens and Digital Media and Real Time Computer Systems

UCAS Code: GW5N•Mode: 3 Years Equal Combination

Qualifications/Requirements: GCE, A/AS: 14 points. Art at A level. AGNVQ, Merit in Art & Design. ND/C, 6 Merits. SQAH, Grades BBBC. SQAV, Individual consideration. IB, Pass in Diploma.

HND Multimedia Computing

UCAS Code: 045G•Mode: 2 Years Single Subject

Qualifications/Requirements: GCE, A/AS: 12 points. Science at A level. AGNVQ, Merit. ND/C, 4 Merits. SQAH, Grades BBCC. SQAV, National. IB, Pass in Diploma.

BARKING COLLEGE B11

HND Multimedia

UCAS Code: 52GE•Mode: 2 Years Equal Combination

Qualifications/Requirements: Please refer to prospectus.

HND Multimedia

UCAS Code: 52GW•Mode: 2 Years Equal Combination

Qualifications/Requirements: Please refer to prospectus.

BARNSLEY COLLEGE B13

BSc Multimedia Technology

UCAS Code: HP64•Mode: 3 Years Equal Combination

Qualifications/Requirements: GCE, A/AS: 8 points. AGNVQ, Merit. ND/C, Individual consideration. SQAH, Individual consideration. SQAV, Individual consideration. IB, Individual consideration.

HND Multimedia Design and Production

UCAS Code: 042E•Mode: 2 Years Single Subject

Qualifications/Requirements: GCE, A/AS: 2 points. OTHER. AGNVQ, Pass (in specific programmes). ND/C, Individual consideration.

HND Multimedia Design and Production

UCAS Code: 042W•Mode: 2 Years Single Subject

Qualifications/Requirements: GCE, A/AS: 2 points. OTHER. AGNVQ, Pass (in specific programmes). ND/C, Individual consideration.

HND Multimedia Design Technology

UCAS Code: 46PH•Mode: 2 Years Equal Combination

Qualifications/Requirements: GCE, A/AS: 2 points. AGNVQ, Pass (in specific programmes). ND/C, Individual consideration. SQAH, Individual consideration. SQAV, Individual consideration. IB, Individual consideration.

UNIVERSITY OF BATH B16

HND Computing and Multimedia

UCAS Code: 035G•Mode: 2 Years Single Subject

Qualifications/Requirements: Please refer to prospectus.

HND Multimedia

UCAS Code: 045G•Mode: 2 Years Single Subject

Qualifications/Requirements: Please refer to prospectus.

BOURNEMOUTH UNIVERSITY B50

BA Multimedia Journalism

UCAS Code: P600•Mode: 3 Years Single Subject

Qualifications/Requirements: GCE, A level grades: BBB. AGNVQ, Distinction (in specific programmes) with A level. ND/C, Distinction Overall. SQAH, Grades AABBB. SQAV, Individual consideration. IB, 32 points.

BSc Multimedia Communications
UCAS Code: G524•Mode: 4 Years Single Subject

Qualifications/Requirements: GCE, A/AS: 12 points. AGNVQ, Merit. ND/C, 3 Merits. SQAH, Individual consideration. SQAV, Individual consideration. IB, 28 points.

THE ARTS INSTITUTE AT BOURNEMOUTH B53

HND Multimedia
UCAS Code: 52WE•Mode: 2 Years Equal Combination

Qualifications/Requirements: GCE, A level grades: E. OTHER. AGNVQ, Pass (in specific programmes). ND/C, National (in specific programmes). SQAH, Individual consideration. SQAV, National including specific subject(s). IB, Individual consideration.

HND Multimedia
UCAS Code: 52WW•Mode: 2 Years Equal Combination

Qualifications/Requirements: GCE, A level grades: E. OTHER. AGNVQ, Pass (in specific programmes). ND/C, National (in specific programmes). SQAH, Individual consideration. SQAV, National including specific subject(s). IB, Individual consideration.

THE UNIVERSITY OF BRADFORD B56

BSc Multimedia Computing
UCAS Code: G540•Mode: 3 Years Single Subject

Qualifications/Requirements: GCE, A/AS: 18 points. AGNVQ, Distinction (in specific programmes) with 4 additional units or with A/AS. ND/C, 3 Merits and 2 Distinctions. SQAH, Individual consideration. SQAV, Individual consideration. IB, Individual consideration.

BSc Multimedia Computing
UCAS Code: G541•Mode: 4 Years Single Subject

Qualifications/Requirements: GCE, A/AS: 18 points. AGNVQ, Distinction (in specific programmes) with 4 additional units or with A/AS. ND/C, 3 Merits and 2 Distinctions. SQAH, Individual consideration. SQAV, Individual consideration. IB, Individual consideration.

BEng Telecommunications and Multimedia Systems
UCAS Code: H620•Mode: 3 Years Single Subject

Qualifications/Requirements: GCE, A/AS: 18 points. Maths at A level. AGNVQ, Merit (in specific programmes) with A level. ND/C, Individual consideration. SQAH, Individual consideration. SQAV, Individual consideration. IB, Individual consideration.

BEng Telecommunications and Multimedia Systems
UCAS Code: H621•Mode: 4 Years Single Subject

Qualifications/Requirements: GCE, A/AS: 18 points. Maths at A level. AGNVQ, Merit (in specific programmes) with A level. ND/C, Individual consideration. SQAH, Individual consideration. SQAV, Individual consideration. IB, Individual consideration.

MEng Telecommunications and Multimedia Systems
UCAS Code: H622•Mode: 4 Years Single Subject

Qualifications/Requirements: GCE, A level grades: BBB. Maths and Physics at A level. AGNVQ, Distinction in Science with A level. ND/C, 5 Distinctions (in specific programmes). SQAH, Individual consideration. SQAV, Individual consideration. IB, Individual consideration.

UNIVERSITY OF BRIGHTON B72

BSc Computing and Media
UCAS Code: GP54•Mode: 3/4 Years Equal Combination

Qualifications/Requirements: GCE, A/AS: 18 points. AGNVQ, Merit with 6 additional units or with A/AS. ND/C, Distinction Overall. SQAH, Grades BBBC. SQAV, Individual consideration. IB, 27 points.

BRUNEL UNIVERSITY B84

HND Computing (Multimedia Production)
UCAS Code: 45PG•Mode: 2 Years Equal Combination

Qualifications/Requirements: GCE, A/AS: 8 points. AGNVQ, Merit. ND/C, Merit overall (in specific programmes). SQAH, Grades BCCC including specific subject(s). SQAV, Individual consideration. IB, 24 points.

BUCKINGHAMSHIRE CHILTERNS UNIVERSITY COLLEGE B94

BA Multimedia Production
UCAS Code: P410•Mode: 3 Years Single Subject

Qualifications/Requirements: GCE, A/AS: 8-12 points. AGNVQ, Merit. ND/C, Merit overall. SQAH, Grades CCCC. SQAV, Individual consideration. IB, 27 points.

BSc Computer Engineering with Multimedia
UCAS Code: G6P4•Mode: 3 Years Major/Minor

Qualifications/Requirements: GCE, A/AS: 8-10 points. AGNVQ, Merit. ND/C, Merit overall. SQAH, Grades CCCC. SQAV, Individual consideration. IB, Individual consideration.

BSc Computing with Multimedia
UCAS Code: G5P4•Mode: 3 Years Major/Minor

Qualifications/Requirements: GCE, A/AS: 8-10 points. AGNVQ, Merit. ND/C, Merit overall. SQAH, Grades CCCC. SQAV, Individual consideration. IB, Individual consideration.

BSc Creative Multimedia with Video
UCAS Code: G5W5•Mode: 3 Years Major/Minor

Qualifications/Requirements: GCE, A/AS: 8-10 points. AGNVQ, Merit. ND/C, Merit overall. SQAH, Grades CCCC. SQAV, Individual consideration. IB, 27 points.

BSc Multimedia Technology

UCAS Code: P411•Mode: 3 Years Single Subject

Qualifications/Requirements: GCE, A/AS: 8-10 points. AGNVQ, Merit. ND/C, Merit overall. SQAH, Grades CCCC. SQAV, Individual consideration. IB, 27 points.

CANTERBURY CHRIST CHURCH UNIVERSITY COLLEGE C10

Mod Information Technology with Media & Cultural Studies

UCAS Code: G5P4•Mode: 3 Years Major/Minor

Qualifications/Requirements: GCE, A level grades: CC. AGNVQ, Merit. ND/C, Merit overall. SQAH, Individual consideration. SQAV, Individual consideration. IB, 24 points.

Mix Information Technology and Media & Cultural Studies

UCAS Code: GP54•Mode: 3 Years Equal Combination

Qualifications/Requirements: GCE, A level grades: CC. AGNVQ, Merit. ND/C, Merit overall. SQAH, Individual consideration. SQAV, Individual consideration. IB, 24 points.

HND Multimedia Computing

UCAS Code: 105G•Mode: 2 Years Single Subject

Qualifications/Requirements: GCE, A level grades: D. AGNVQ, Pass. ND/C, Merit overall. SQAH, Individual consideration. SQAV, Individual consideration. IB, 24 points.

CARMARTHENSHIRE COLLEGE C22

HND Multimedia Computing

UCAS Code: 065G•Mode: 2 Years Single Subject

Qualifications/Requirements: AGNVQ, Individual consideration. ND/C, Individual consideration. SQAH, Individual consideration. SQAV, Individual consideration. IB, Individual consideration.

UNIVERSITY OF CENTRAL ENGLAND IN BIRMINGHAM C25

BA Multimedia, Communication and Culture

UCAS Code: PP24•Mode: 3 Years Equal Combination

Qualifications/Requirements: GCE, A/AS: 12 points. AGNVQ, Merit. ND/C, Merit overall. SQAH, Individual consideration. IB, 24 points.

BSc Multimedia Technology

UCAS Code: P410•Mode: 3/4 Years Single Subject

Qualifications/Requirements: GCE, A/AS: 18 points. AGNVQ, Merit. ND/C, 6 Merits. SQAH, Grades BBBCC. SQAV, Individual consideration. IB, 30 points.

BSc Multimedia Technology Foundation Year

UCAS Code: P418•Mode: 4/5 Years Single Subject

Qualifications/Requirements: GCE, A/AS: 4 points. AGNVQ, Pass. ND/C, 1 Merit. SQAH, Grades CC. SQAV, Individual consideration. IB, 24 points.

HND Multimedia Technology

UCAS Code: 014P•Mode: 2/3 Years Single Subject

Qualifications/Requirements: GCE, A/AS: 4 points. AGNVQ, Pass. ND/C, 1 Merit. SQAH, Grades CC. SQAV, Individual consideration. IB, 24 points.

UNIVERSITY OF CENTRAL LANCASHIRE C30

BSc Computing and Media Technology

UCAS Code: GP5K•Mode: 3 Years Equal Combination

Qualifications/Requirements: GCE, A/AS: 14 points. AGNVQ, Distinction in Science with 6 additional units or with A/AS. ND/C, Merit overall. SQAH, Grades BBBC. SQAV, Individual consideration. IB, 26 points.

BSc Media Technology and Computing

UCAS Code: GP54•Mode: 3 Years Equal Combination

Qualifications/Requirements: GCE, A/AS: 14 points. AGNVQ, Distinction (in specific programmes) with 6 additional units or with A level. ND/C, Merit overall. SQAH, Grades BBBC. SQAV, Individual consideration. IB, 26 points.

BSc Multimedia System Design

UCAS Code: GG56•Mode: 3 Years Equal Combination

Qualifications/Requirements: GCE, A/AS: 14 points. AGNVQ, Distinction (in specific programmes) with 6 additional units or with A level. ND/C, Merit overall. SQAH, Grades BBBC. SQAV, Individual consideration. IB, 26 points.

BSc Web and Multimedia

UCAS Code: G620•Mode: 3 Years Single Subject

Qualifications/Requirements: GCE, A/AS: 18 points. AGNVQ, Distinction (in specific programmes). ND/C, Merit overall. SQAH, Grades BBBB. SQAV, Individual consideration. IB, 28 points.

CHELTENHAM & GLOUCESTER COLLEGE OF HIGHER EDUCATION C50

BA Business Management with Multimedia (4 years SW)

UCAS Code: NGCN•Mode: 4 Years Equal Combination

Qualifications/Requirements: GCE, A/AS: 12 points. AGNVQ, Merit in Business with 3 additional units. ND/C, Merit overall and 3 Distinctions. SQAH, Grades CCCC. SQAV, Individual consideration. IB, 26 points.

BA English Studies with Multimedia

UCAS Code: Q3G5•Mode: 3 Years Major/Minor

Qualifications/Requirements: GCE, A/AS: 12 points. English. AGNVQ, Merit with A level. ND/C, Merit overall and 3 Distinctions. SQAH, Grades CCCC. SQAV, Individual consideration. IB, 26 points.

BA Film Studies with Multimedia
UCAS Code: W5PK•Mode: 3 Years Major/Minor

Qualifications/Requirements: GCE, A/AS: 12 points. AGNVQ, Merit. ND/C, Merit overall. SQAH, Grades CCCC. SQAV, Individual consideration. IB, 24 points.

BA Financial Services Management with Multimedia (4 years SW)
UCAS Code: NGJN•Mode: 4 Years Equal Combination

Qualifications/Requirements: GCE, A/AS: 8-12 points. AGNVQ, Merit with 3 additional units. ND/C, Merit overall and 2 Distinctions. SQAH, Grades CCCC. SQAV, Individual consideration. IB, 26 points.

BA Garden Design with Multimedia
UCAS Code: W2GM•Mode: 3 Years Major/Minor

Qualifications/Requirements: GCE, A/AS: 8-12 points. AGNVQ, Merit. ND/C, Merit overall. SQAH, Grades CCCC. SQAV, Individual consideration. IB, 24 points.

BA Heritage Management with Multimedia
UCAS Code: K9G5•Mode: 3 Years Major/Minor

Qualifications/Requirements: GCE, A/AS: 8-12 points. AGNVQ, Merit. ND/C, Merit overall. SQAH, Grades CCCC. SQAV, Individual consideration. IB, 24 points.

BA Human Resource Management with Multimedia (4 years SW)
UCAS Code: NGDM•Mode: 4 Years Equal Combination

Qualifications/Requirements: GCE, A/AS: 12 points. AGNVQ, Merit in Business with 3 additional units. ND/C, Merit overall and 2 Distinctions. SQAH, Grades CCCC. SQAV, Individual consideration. IB, 26 points.

BA International Business Management with Multimedia
UCAS Code: N1P4•Mode: 4 Years Major/Minor

Qualifications/Requirements: GCE, A/AS: 12 points. AGNVQ, Merit. ND/C, Merit overall and 3 Distinctions. SQAH, Grades CCCC. SQAV, Individual consideration. IB, 26 points.

BA International Marketing Management with Multimedia
UCAS Code: N5P4•Mode: 4 Years Major/Minor

Qualifications/Requirements: GCE, A/AS: 12 points. AGNVQ, Merit. ND/C, Merit overall and 3 Distinctions. SQAH, Grades CCCC. SQAV, Individual consideration. IB, 26 points.

BA Leisure Management with Multimedia (4 years SW)
UCAS Code: N790•Mode: 4 Years Single Subject

Qualifications/Requirements: GCE, A/AS: 12-14 points. AGNVQ, Merit in Leisure and Tourism with 3 additional units. ND/C, Merit overall and 2 Distinctions. SQAH, Grades CCCC. SQAV, Individual consideration. IB, 26 points.

BA Marketing Management with Multimedia (4 years SW)
UCAS Code: NGN5•Mode: 4 Years Equal Combination

Qualifications/Requirements: GCE, A/AS: 12 points. AGNVQ, Merit in Business with 3 additional units. ND/C, Merit overall and 2 Distinctions. SQAH, Grades CCCC. SQAV, Individual consideration. IB, 26 points.

BA Media Communications with Multimedia
UCAS Code: PGK5•Mode: 3 Years Equal Combination

Qualifications/Requirements: GCE, A/AS: 12 points. AGNVQ, Merit in Media with 3 additional units. ND/C, Merit overall and 2 Distinctions. SQAH, Grades CCCC. SQAV, Individual consideration. IB, 26 points.

BA Tourism Management with Multimedia (4 years SW)
UCAS Code: GP5R•Mode: 4 Years Equal Combination

Qualifications/Requirements: GCE, A/AS: 12 points. AGNVQ, Merit in Leisure and Tourism with 3 additional units. ND/C, Merit overall. SQAH, Grades CCCC. SQAV, Individual consideration. IB, 26 points.

BA Visual Arts with Multimedia
UCAS Code: W1G5•Mode: 3 Years Major/Minor

Qualifications/Requirements: GCE, A/AS: 10-14 points. Art. AGNVQ, Merit in Art & Design with 3 additional units. ND/C, Merit overall and 2 Distinctions. SQAH, Grades CCCC. SQAV, Individual consideration. IB, 26 points.

BSc Business Computer Systems and Multimedia
UCAS Code: G520•Mode: 3 Years Single Subject

Qualifications/Requirements: GCE, A/AS: 8-12 points. AGNVQ, Merit. ND/C, Merit overall. SQAH, Grades CCCC. SQAV, Individual consideration. IB, 24 points.

BSc Business Computer Systems with Media Communications
UCAS Code: G5P4•Mode: 3 Years Major/Minor

Qualifications/Requirements: GCE, A/AS: 8-12 points. AGNVQ, Merit. ND/C, Merit overall. SQAH, Grades CCCC. SQAV, Individual consideration. IB, 24 points.

BSc Business Computer Systems with Multimedia
UCAS Code: G521•Mode: 3 Years Single Subject

Qualifications/Requirements: GCE, A/AS: 8 points. AGNVQ, Merit. ND/C, Merit overall. SQAH, Grades CCCC. SQAV, Individual consideration. IB, 26 points.

BSc Computing and Multimedia
UCAS Code: G561•Mode: 3 Years Single Subject

Qualifications/Requirements: GCE, A/AS: 8-12 points. AGNVQ, Merit. ND/C, Merit overall. SQAH, Grades CCCC. SQAV, Individual consideration. IB, 26 points.

BSc Computing with Media Communications
UCAS Code: G5PL•Mode: 3 Years Major/Minor
Qualifications/Requirements: GCE, A/AS: 8-12 points. AGNVQ, Merit. ND/C, Merit overall. SQAH, Grades CCCC. SQAV, Individual consideration. IB, 24 points.

BSc Computing with Multimedia
UCAS Code: G505•Mode: 3 Years Single Subject
Qualifications/Requirements: GCE, A/AS: 8 points. AGNVQ, Merit. ND/C, Merit overall. SQAH, Grades CCCC. SQAV, Individual consideration. IB, 26 points.

BSc Geology and Multimedia
UCAS Code: FG6M•Mode: 3 Years Equal Combination
Qualifications/Requirements: GCE, A/AS: 8-12 points. AGNVQ, Merit with 3 additional units with A level. ND/C, Merit overall. SQAH, Grades CCCC. SQAV, Individual consideration. IB, 24 points.

BSc Geology with Multimedia
UCAS Code: FGPM•Mode: 3 Years Equal Combination
Qualifications/Requirements: GCE, A/AS: 8-12 points. AGNVQ, Merit with 3 additional units with A level. ND/C, Merit overall. SQAH, Grades CCCC. SQAV, Individual consideration. IB, 24 points.

BSc Information Technology and Multimedia
UCAS Code: G560•Mode: 3 Years Single Subject
Qualifications/Requirements: GCE, A/AS: 8-12 points. AGNVQ, Merit with 3 additional units. ND/C, Merit overall. SQAH, Grades CCCC. SQAV, Individual consideration. IB, 26 points.

BSc Information Technology with Media Communications
UCAS Code: G5PK•Mode: 3 Years Major/Minor
Qualifications/Requirements: GCE, A/AS: 8-12 points. AGNVQ, Merit with 3 additional units. ND/C, Merit overall. SQAH, Grades CCCC. SQAV, Individual consideration. IB, 24 points.

BSc Information Technology with Multimedia
UCAS Code: G562•Mode: 3 Years Single Subject
Qualifications/Requirements: GCE, A/AS: 8 points. AGNVQ, Merit with 3 additional units. ND/C, Merit overall. SQAH, Grades CCCC. SQAV, Individual consideration. IB, 26 points.

BSc Multimedia and Physical Geography
UCAS Code: GF5V•Mode: 3 Years Equal Combination
Qualifications/Requirements: GCE, A/AS: 8-12 points. AGNVQ, Merit with 3 additional units. ND/C, Merit overall and 2 Distinctions. SQAH, Grades CCCC. SQAV, Individual consideration. IB, 26 points.

BSc Multimedia and Psychology
UCAS Code: GL57•Mode: 3 Years Equal Combination
Qualifications/Requirements: GCE, A/AS: 12 points. AGNVQ, Merit in Information Technology with 3 additional units. ND/C, Merit overall and 2 Distinctions. SQAH, Grades CCCC. SQAV, Individual consideration. IB, 26 points.

BSc Multimedia Computing
UCAS Code: G540•Mode: 3 Years Single Subject
Qualifications/Requirements: GCE, A/AS: 8-12 points. AGNVQ, Merit with 3 additional units. ND/C, Merit overall. SQAH, Grades CCCC. SQAV, Individual consideration. IB, 24 points.

BSc Multimedia Management
UCAS Code: G542•Mode: 3 Years Single Subject
Qualifications/Requirements: GCE, A/AS: 8-12 points. AGNVQ, Merit with 3 additional units. ND/C, Merit overall. SQAH, Grades CCCC. SQAV, Individual consideration. IB, 24 points.

BSc Multimedia Marketing
UCAS Code: G543•Mode: 3 Years Single Subject
Qualifications/Requirements: GCE, A/AS: 8-12 points. AGNVQ, Merit with 3 additional units. ND/C, Merit overall. SQAH, Grades CCCC. SQAV, Individual consideration. IB, 24 points.

BSc Multimedia Systems Design
UCAS Code: G545•Mode: 3 Years Single Subject
Qualifications/Requirements: GCE, A/AS: 8-12 points. AGNVQ, Merit with 3 additional units. ND/C, Merit overall. SQAH, Grades CCCC. SQAV, Individual consideration. IB, 24 points.

BSc Multimedia with Accounting & Financial Management
UCAS Code: G565•Mode: 4 Years Single Subject
Qualifications/Requirements: GCE, A/AS: 8-12 points. AGNVQ, Merit in Information Technology with 3 additional units. ND/C, Merit overall. SQAH, Grades CCCC. SQAV, Individual consideration. IB, 24 points.

BSc Multimedia with Accounting & Financial Management
UCAS Code: NGHN•Mode: 3 Years Equal Combination
Qualifications/Requirements: GCE, A/AS: 8-12 points. AGNVQ, Merit in Information Technology with 3 additional units. ND/C, Merit overall. SQAH, Grades CCCC. SQAV, Individual consideration. IB, 26 points.

BSc Multimedia with Business Computer Systems
UCAS Code: G528•Mode: 3 Years Single Subject
Qualifications/Requirements: GCE, A/AS: 8-12 points. AGNVQ, Merit in Information Technology with 3 additional units. ND/C, Merit overall. SQAH, Grades CCCC. SQAV, Individual consideration. IB, 26 points.

BSc Multimedia with Business Computer Systems
UCAS Code: G569•Mode: 4 Years Single Subject

Qualifications/Requirements: GCE, A/AS: 8-12 points. AGNVQ, Merit in Information Technology with 3 additional units. ND/C, Merit overall. SQAH, Grades CCCC. SQAV, Individual consideration. IB, 24 points.

BSc Multimedia with Business Management
UCAS Code: G568•Mode: 4 Years Single Subject

Qualifications/Requirements: GCE, A/AS: 8-12 points. AGNVQ, Merit in Information Technology with 3 additional units. ND/C, Merit overall. SQAH, Grades CCCC. SQAV, Individual consideration. IB, 24 points.

BSc Multimedia with Business Management
UCAS Code: NGDN•Mode: 3 Years Equal Combination

Qualifications/Requirements: GCE, A/AS: 12 points. AGNVQ, Merit in Information Technology with 3 additional units. ND/C, Merit overall and 2 Distinctions. SQAH, Grades CCCC. SQAV, Individual consideration. IB, 26 points.

BSc Multimedia with Combined Arts
UCAS Code: G5Y3•Mode: 3 Years Major/Minor

Qualifications/Requirements: GCE, A/AS: 8-12 points. AGNVQ, Merit in Information Technology with 3 additional units. ND/C, Merit overall. SQAH, Grades CCCC. SQAV, Individual consideration. IB, 24 points.

BSc Multimedia with Combined Arts
UCAS Code: G5YH•Mode: 4 Years Major/Minor

Qualifications/Requirements: GCE, A/AS: 8-12 points. AGNVQ, Merit in Information Technology with 3 additional units. ND/C, Merit overall. SQAH, Grades CCCC. SQAV, Individual consideration. IB, 26 points.

BSc Multimedia with Computing
UCAS Code: G564•Mode: 3 Years Single Subject

Qualifications/Requirements: GCE, A/AS: 8-12 points. AGNVQ, Merit in Information Technology with 3 additional units. ND/C, Merit overall. SQAH, Grades CCCC. SQAV, Individual consideration. IB, 26 points.

BSc Multimedia with Computing
UCAS Code: G573•Mode: 4 Years Single Subject

Qualifications/Requirements: GCE, A/AS: 8-12 points. AGNVQ, Merit in Information Technology with 3 additional units. ND/C, Merit overall. SQAH, Grades CCCC. SQAV, Individual consideration. IB, 26 points.

BSc Multimedia with English Studies
UCAS Code: G5Q3•Mode: 3 Years Major/Minor

Qualifications/Requirements: GCE, A/AS: 8-12 points. AGNVQ, Merit in Information Technology with 3 additional units. ND/C, Merit overall. SQAH, Grades CCCC. SQAV, Individual consideration. IB, 26 points.

BSc Multimedia with English Studies
UCAS Code: G5QH•Mode: 4 Years Major/Minor

Qualifications/Requirements: GCE, A/AS: 8-12 points. AGNVQ, Merit in Information Technology with 3 additional units. ND/C, Merit overall. SQAH, Grades CCCC. SQAV, Individual consideration. IB, 24 points.

BSc Multimedia with Financial Services Management
UCAS Code: G570•Mode: 4 Years Single Subject

Qualifications/Requirements: GCE, A/AS: 8-12 points. AGNVQ, Merit in Information Technology with 3 additional units. ND/C, Merit overall. SQAH, Grades CCCC. SQAV, Individual consideration. IB, 24 points.

BSc Multimedia with Financial Services Management
UCAS Code: NGJM•Mode: 3 Years Equal Combination

Qualifications/Requirements: GCE, A/AS: 8-12 points. AGNVQ, Merit in Information Technology with 3 additional units. ND/C, Merit overall. SQAH, Grades CCCC. SQAV, Individual consideration. IB, 26 points.

BSc Multimedia with Garden Design
UCAS Code: G5DF•Mode: 3 Years Major/Minor

Qualifications/Requirements: GCE, A/AS: 8-12 points. AGNVQ, Merit. ND/C, Merit overall. SQAH, Grades CCCC. SQAV, Individual consideration. IB, 24 points.

BSc Multimedia with Garden Design (4 years SW)
UCAS Code: G5DG•Mode: 4 Years Major/Minor

Qualifications/Requirements: GCE, A/AS: 8-12 points. AGNVQ, Merit. ND/C, Merit overall. SQAH, Grades CCCC. SQAV, Individual consideration. IB, 24 points.

BSc Multimedia with Geology
UCAS Code: GFMP•Mode: 3 Years Equal Combination

Qualifications/Requirements: GCE, A/AS: 8-12 points. AGNVQ, Merit in Information Technology with 3 additional units. ND/C, Merit overall. SQAH, Grades CCCC. SQAV, Individual consideration. IB, 24 points.

BSc Multimedia with Geology
UCAS Code: GFMQ•Mode: 4 Years Equal Combination

Qualifications/Requirements: GCE, A/AS: 8-12 points. AGNVQ, Merit in Information Technology with 3 additional units. ND/C, Merit overall. SQAH, Grades CCCC. SQAV, Individual consideration. IB, 24 points.

BSc Multimedia with Heritage Management
UCAS Code: G5KY•Mode: 3 Years Major/Minor

Qualifications/Requirements: GCE, A/AS: 8-12 points. AGNVQ, Merit. ND/C, Merit overall. SQAH, Grades CCCC. SQAV, Individual consideration. IB, 24 points.

BSc Multimedia with Heritage Management (4 years SW)
UCAS Code: G5KX•Mode: 4 Years Major/Minor

Qualifications/Requirements: GCE, A/AS: 8-12 points. AGNVQ, Merit. ND/C, Merit overall. SQAH, Grades CCCC. SQAV, Individual consideration. IB, 24 points.

BSc Multimedia with Human Geography
UCAS Code: GLNV•Mode: 3 Years Equal Combination

Qualifications/Requirements: GCE, A/AS: 8-12 points. AGNVQ, Merit in Information Technology with 3 additional units. ND/C, Merit overall.

BSc Multimedia with Human Geography
UCAS Code: GLNW•Mode: 4 Years Equal Combination

Qualifications/Requirements: GCE, A/AS: 8-12 points. AGNVQ, Merit in Information Technology with 3 additional units. ND/C, Merit overall.

BSc Multimedia with Human Resource Management
UCAS Code: G566•Mode: 3 Years Single Subject

Qualifications/Requirements: GCE, A/AS: 8-12 points. AGNVQ, Merit in Information Technology with 3 additional units. ND/C, Merit overall. SQAH, Grades CCCC. SQAV, Individual consideration. IB, 26 points.

BSc Multimedia with Human Resource Management
UCAS Code: G571•Mode: 4 Years Single Subject

Qualifications/Requirements: GCE, A/AS: 8-12 points. AGNVQ, Merit in Information Technology with 3 additional units. ND/C, Merit overall. SQAH, Grades CCCC. SQAV, Individual consideration. IB, 24 points.

BSc Multimedia with Information Technology
UCAS Code: G563•Mode: 3 Years Single Subject

Qualifications/Requirements: GCE, A/AS: 8-12 points. AGNVQ, Merit in Information Technology with 3 additional units. ND/C, Merit overall. SQAH, Grades CCCC. SQAV, Individual consideration. IB, 26 points.

BSc Multimedia with Information Technology
UCAS Code: G572•Mode: 4 Years Single Subject

Qualifications/Requirements: GCE, A/AS: 8-12 points. AGNVQ, Merit in Information Technology with 3 additional units. ND/C, Merit overall. SQAH, Grades CCCC. SQAV, Individual consideration. IB, 24 points.

BSc Multimedia with International Business Management
UCAS Code: P4N1•Mode: 3 Years Major/Minor

Qualifications/Requirements: GCE, A/AS: 10 points. AGNVQ, Merit. ND/C, Merit overall and 3 Distinctions. SQAH, Grades CCCC. SQAV, Individual consideration. IB, 26 points.

BSc Multimedia with International Business Management
UCAS Code: P4NC•Mode: 4 Years Major/Minor

Qualifications/Requirements: GCE, A/AS: 10 points. AGNVQ, Merit. ND/C, Merit overall and 3 Distinctions. SQAH, Grades CCCC. SQAV, Individual consideration. IB, 26 points.

BSc Multimedia with International Marketing Management
UCAS Code: P4N5•Mode: 3 Years Major/Minor

Qualifications/Requirements: GCE, A/AS: 12 points. AGNVQ, Merit. ND/C, Merit overall and 3 Distinctions. SQAH, Grades CCCC. SQAV, Individual consideration. IB, 26 points.

BSc Multimedia with International Marketing Management
UCAS Code: P4NM•Mode: 4 Years Major/Minor

Qualifications/Requirements: GCE, A/AS: 12 points. AGNVQ, Merit. ND/C, Merit overall and 3 Distinctions. SQAH, Grades CCCC. SQAV, Individual consideration. IB, 26 points.

BSc Multimedia with Leisure Management
UCAS Code: GNMR•Mode: 4 Years Equal Combination

Qualifications/Requirements: GCE, A/AS: 8-12 points. AGNVQ, Merit in Information Technology with 3 additional units. ND/C, Merit overall. SQAH, Grades CCCC. SQAV, Individual consideration. IB, 24 points.

BSc Multimedia with Leisure Management
UCAS Code: GNN7•Mode: 3 Years Equal Combination

Qualifications/Requirements: GCE, A/AS: 12 points. AGNVQ, Merit in Information Technology with 3 additional units. ND/C, Merit overall and 2 Distinctions. SQAH, Grades CCCC. SQAV, Individual consideration. IB, 26 points.

BSc Multimedia with Marketing Management
UCAS Code: GNMT•Mode: 4 Years Equal Combination

Qualifications/Requirements: GCE, A/AS: 8-12 points. AGNVQ, Merit in Information Technology with 3 additional units. ND/C, Merit overall. SQAH, Grades CCCC. SQAV, Individual consideration. IB, 24 points.

BSc Multimedia with Marketing Management
UCAS Code: NGMM•Mode: 3 Years Equal Combination

Qualifications/Requirements: GCE, A/AS: 8-12 points. AGNVQ, Merit in Information Technology with 3 additional units. ND/C, Merit overall. SQAH, Grades CCCC. SQAV, Individual consideration. IB, 26 points.

BSc Multimedia with Media Communications
UCAS Code: GPMK•Mode: 4 Years Equal Combination

Qualifications/Requirements: GCE, A/AS: 8-12 points. AGNVQ, Merit in Information Technology with 3 additional units. ND/C, Merit overall. SQAH, Grades CCCC. SQAV, Individual consideration. IB, 24 points.

BSc Multimedia with Media Communications
UCAS Code: PGL5•Mode: 3 Years Equal Combination
Qualifications/Requirements: GCE, A/AS: 12 points. AGNVQ, Merit in Information Technology with 3 additional units. ND/C, Merit overall and 2 Distinctions. SQAH, Grades CCCC. SQAV, Individual consideration. IB, 26 points.

BSc Multimedia with Modern Languages (French)
UCAS Code: G5R1•Mode: 3 Years Major/Minor
Qualifications/Requirements: GCE, A/AS: 8-12 points. AGNVQ, Merit in Information Technology with 3 additional units. ND/C, Merit overall. SQAH, Grades CCCC. SQAV, Individual consideration. IB, 26 points.

BSc Multimedia with Modern Languages (French)
UCAS Code: GRMC•Mode: 4 Years Equal Combination
Qualifications/Requirements: GCE, A/AS: 8-12 points. AGNVQ, Merit in Information Technology with 3 additional units. ND/C, Merit overall. SQAH, Grades CCCC. SQAV, Individual consideration. IB, 24 points.

BSc Multimedia with Physical Geography
UCAS Code: GF5W•Mode: 3 Years Equal Combination
Qualifications/Requirements: GCE, A/AS: 8-12 points. AGNVQ, Merit in Information Technology with 3 additional units. ND/C, Merit overall. SQAH, Grades CCCC. SQAV, Individual consideration. IB, 24 points.

BSc Multimedia with Physical Geography
UCAS Code: GFM8•Mode: 4 Years Equal Combination
Qualifications/Requirements: GCE, A/AS: 8-12 points. AGNVQ, Merit in Information Technology with 3 additional units. ND/C, Merit overall. SQAH, Grades CCCC. SQAV, Individual consideration. IB, 24 points.

BSc Multimedia with Psychology
UCAS Code: G5L7•Mode: 3 Years Major/Minor
Qualifications/Requirements: GCE, A/AS: 8-12 points. AGNVQ, Merit in Information Technology with 3 additional units. ND/C, Merit overall. SQAH, Grades CCCC. SQAV, Individual consideration. IB, 26 points.

BSc Multimedia with Psychology
UCAS Code: G5LR•Mode: 4 Years Major/Minor
Qualifications/Requirements: GCE, A/AS: 8-12 points. AGNVQ, Merit in Information Technology with 3 additional units. ND/C, Merit overall. SQAH, Grades CCCC. SQAV, Individual consideration. IB, 24 points.

BSc Multimedia with Sociological Studies
UCAS Code: G5L3•Mode: 3 Years Major/Minor
Qualifications/Requirements: GCE, A/AS: 8-12 points. AGNVQ, Merit in Information Technology with 3 additional units. ND/C, Merit overall. SQAH, Grades CCCC. SQAV, Individual consideration. IB, 26 points.

BSc Multimedia with Sociological Studies
UCAS Code: G5LH•Mode: 4 Years Major/Minor
Qualifications/Requirements: GCE, A/AS: 8-12 points. AGNVQ, Merit in Information Technology with 3 additional units. ND/C, Merit overall. SQAH, Grades CCCC. SQAV, Individual consideration. IB, 24 points.

BSc Multimedia with Tourism Management
UCAS Code: G5PT•Mode: 3 Years Major/Minor
Qualifications/Requirements: GCE, A/AS: 12 points. AGNVQ, Merit in Information Technology with 3 additional units. ND/C, Merit overall and 2 Distinctions. SQAH, Grades CCCC. SQAV, Individual consideration. IB, 26 points.

BSc Multimedia with Tourism Management
UCAS Code: GPMR•Mode: 4 Years Equal Combination
Qualifications/Requirements: GCE, A/AS: 8-12 points. AGNVQ, Merit in Information Technology with 3 additional units. ND/C, Merit overall. SQAH, Grades CCCC. SQAV, Individual consideration. IB, 24 points.

BSc Multimedia with Visual Arts
UCAS Code: G5W1•Mode: 3 Years Major/Minor
Qualifications/Requirements: GCE, A/AS: 10-14 points. AGNVQ, Merit in Art & Design with 3 additional units. ND/C, Merit overall and 2 Distinctions. SQAH, Grades CCCC. SQAV, Individual consideration. IB, 26 points.

BSc Multimedia with Visual Arts
UCAS Code: G5WC•Mode: 4 Years Major/Minor
Qualifications/Requirements: GCE, A/AS: 10-14 points. AGNVQ, Merit in Art & Design with 3 additional units. ND/C, Merit overall and 2 Distinctions. SQAH, Grades CCCC. SQAV, Individual consideration. IB, 26 points.

BSc Multimedia with Women's Studies
UCAS Code: GMMX•Mode: 3 Years Equal Combination
Qualifications/Requirements: GCE, A/AS: 8-12 points. AGNVQ, Merit in Information Technology with 3 additional units. ND/C, Merit overall. SQAH, Grades CCCC. SQAV, Individual consideration. IB, 26 points.

BSc Multimedia with Women's Studies
UCAS Code: GMMY•Mode: 4 Years Equal Combination
Qualifications/Requirements: GCE, A/AS: 8-12 points. AGNVQ, Merit in Information Technology with 3 additional units. ND/C, Merit overall. SQAH, Grades CCCC. SQAV, Individual consideration. IB, 24 points.

Mod Accounting & Financial Management and Multimedia
UCAS Code: NGHM•Mode: 3 Years Equal Combination
Qualifications/Requirements: GCE, A/AS: 8-12 points. AGNVQ, Merit with 3 additional units. ND/C, Merit overall and 2 Distinctions. SQAH, Grades CCCC. SQAV, Individual consideration. IB, 26 points.

Mod Business Management and Multimedia
UCAS Code: NG1N•Mode: 3 Years Equal Combination

Qualifications/Requirements: GCE, A/AS: 12 points. AGNVQ, Merit in Business with 3 additional units. ND/C, Merit overall and 3 Distinctions. SQAH, Grades CCCC. SQAV, Individual consideration. IB, 26 points.

Mod Computing and Media Communications
UCAS Code: GP5L•Mode: 3 Years Equal Combination

Qualifications/Requirements: GCE, A/AS: 8-12 points. AGNVQ, Merit. ND/C, Merit overall and 2 Distinctions. SQAH, Grades CCCC. SQAV, Individual consideration. IB, 24 points.

Mod English Studies and Multimedia
UCAS Code: QG35•Mode: 3 Years Equal Combination

Qualifications/Requirements: GCE, A/AS: 10-12 points. English. AGNVQ, Merit with A level. ND/C, Merit overall and 3 Distinctions. SQAH, Grades CCCC. SQAV, Individual consideration. IB, 26 points.

Mod Film Studies and Multimedia
UCAS Code: WP5K•Mode: 3 Years Equal Combination

Qualifications/Requirements: GCE, A/AS: 8-12 points. AGNVQ, Merit. ND/C, Merit overall. SQAH, Grades CCCC. SQAV, Individual consideration. IB, 24 points.

Mod Financial Services Management and Multimedia
UCAS Code: NG3N•Mode: 3 Years Equal Combination

Qualifications/Requirements: GCE, A/AS: 8-12 points. AGNVQ, Merit with 3 additional units. ND/C, Merit overall and 2 Distinctions. SQAH, Grades CCCC. SQAV, Individual consideration. IB, 26 points.

Mod Garden Design and Multimedia
UCAS Code: WGG5•Mode: 3 Years Equal Combination

Qualifications/Requirements: GCE, A/AS: 8-12 points. AGNVQ, Merit. ND/C, Merit overall. SQAH, Grades CCCC. SQAV, Individual consideration. IB, 24 points.

Mod Heritage Management and Multimedia
UCAS Code: KG95•Mode: 3 Years Equal Combination

Qualifications/Requirements: GCE, A/AS: 8-12 points. AGNVQ, Merit. ND/C, Merit overall. SQAH, Grades CCCC. SQAV, Individual consideration. IB, 24 points.

Mod Human Geography and Multimedia
UCAS Code: GLN8•Mode: 3 Years Equal Combination

Qualifications/Requirements: GCE, A/AS: 10-14 points. AGNVQ, Merit with 3 additional units. ND/C, Merit overall and 2 Distinctions. SQAH, Grades CCCC. SQAV, Individual consideration. IB, 26 points.

Mod Human Resource Management and Multimedia
UCAS Code: G567•Mode: 3 Years Single Subject

Qualifications/Requirements: GCE, A/AS: 8-12 points. AGNVQ, Merit in Business with 3 additional units. ND/C, Merit overall and 2 Distinctions. SQAH, Grades CCCC. SQAV, Individual consideration. IB, 26 points.

Mod Information Technology and Media Communications
UCAS Code: GP5K•Mode: 3 Years Equal Combination

Qualifications/Requirements: GCE, A/AS: 8-12 points. AGNVQ, Merit in Media with 3 additional units. ND/C, Merit overall and 2 Distinctions. SQAH, Grades CCCC. SQAV, Individual consideration. IB, 24 points.

Mod International Business Management and Multimedia
UCAS Code: NP14•Mode: 3 Years Equal Combination

Qualifications/Requirements: GCE, A/AS: 12 points. AGNVQ, Merit. ND/C, Merit overall and 3 Distinctions. SQAH, Grades CCCC. SQAV, Individual consideration. IB, 26 points.

Mod International Marketing Management and Multimedia
UCAS Code: NP54•Mode: 3 Years Equal Combination

Qualifications/Requirements: GCE, A/AS: 12 points. AGNVQ, Merit. ND/C, Merit overall and 3 Distinctions. SQAH, Grades CCCC. SQAV, Individual consideration. IB, 26 points.

Mod Leisure Management and Multimedia
UCAS Code: NGT5•Mode: 3 Years Equal Combination

Qualifications/Requirements: GCE, A/AS: 12-14 points. AGNVQ, Merit in Leisure and Tourism with 3 additional units. ND/C, Merit overall and 2 Distinctions. SQAH, Grades CCCC. SQAV, Individual consideration. IB, 26 points.

Mod Marketing Management and Multimedia
UCAS Code: NGM5•Mode: 3 Years Equal Combination

Qualifications/Requirements: GCE, A/AS: 12 points. AGNVQ, Merit in Business with 3 additional units. ND/C, Merit overall and 2 Distinctions. SQAH, Grades CCCC. SQAV, Individual consideration. IB, 26 points.

Mod Media Communications and Multimedia
UCAS Code: PG45•Mode: 3 Years Equal Combination

Qualifications/Requirements: GCE, A/AS: 12 points. AGNVQ, Merit in Media with 3 additional units. ND/C, Merit overall and 2 Distinctions. SQAH, Grades CCCC. SQAV, Individual consideration. IB, 26 points.

Mod Multimedia and Sociological Studies
UCAS Code: GL53•Mode: 3 Years Equal Combination

Qualifications/Requirements: GCE, A/AS: 12 points. AGNVQ, Merit in Information Technology with 3 additional units. ND/C, Merit overall and 2 Distinctions. SQAH, Grades CCCC. SQAV, Individual consideration. IB, 26 points.

Mod Multimedia and Tourism Management
UCAS Code: GPN7•Mode: 3 Years Equal Combination

Qualifications/Requirements: GCE, A/AS: 12-14 points. AGNVQ, Merit in Information Technology with 3 additional units. ND/C, Merit overall and 2 Distinctions. SQAH, Grades CCCC. SQAV, Individual consideration. IB, 26 points.

Mod Multimedia and Visual Arts
UCAS Code: GW51•Mode: 3 Years Equal Combination

Qualifications/Requirements: GCE, A/AS: 10-14 points. AGNVQ, Merit in Art & Design with 3 additional units. ND/C, Merit overall and 2 Distinctions. SQAH, Grades CCCC. SQAV, Individual consideration. IB, 26 points.

Mod Multimedia and Women's Studies
UCAS Code: GMM9•Mode: 3 Years Equal Combination

Qualifications/Requirements: GCE, A/AS: 8-12 points. AGNVQ, Merit in Information Technology with 3 additional units. ND/C, Merit overall. SQAH, Grades CCCC. SQAV, Individual consideration. IB, 26 points.

HND Multimedia Computing
UCAS Code: 045G•Mode: 2 Years Single Subject

Qualifications/Requirements: GCE, A/AS: 4 points. AGNVQ, Pass. ND/C, National. SQAH, Individual consideration. SQAV, Individual consideration. IB, Individual consideration.

HND Multimedia Computing
UCAS Code: 445G•Mode: 3 Years Single Subject

Qualifications/Requirements: GCE, A/AS: 4 points. AGNVQ, Pass. ND/C, National. SQAH, Individual consideration. SQAV, Individual consideration. IB, Individual consideration.

HND Multimedia Information Technology
UCAS Code: 065G•Mode: 2 Years Single Subject

Qualifications/Requirements: GCE, A/AS: 4 points. AGNVQ, Pass. ND/C, National. SQAH, Individual consideration. SQAV, Individual consideration. IB, Individual consideration.

HND Multimedia Information Technology
UCAS Code: 365G•Mode: 3 Years Single Subject

Qualifications/Requirements: GCE, A/AS: 4 points. AGNVQ, Pass. ND/C, National. SQAH, Individual consideration. SQAV, Individual consideration. IB, Individual consideration.

HND Multimedia Management
UCAS Code: 245G•Mode: 3 Years Single Subject

Qualifications/Requirements: GCE, A/AS: 4 points. AGNVQ, Pass. ND/C, National. SQAH, Individual consideration. SQAV, Individual consideration. IB, Individual consideration.

HND Multimedia Management
UCAS Code: 345G•Mode: 2 Years Single Subject

Qualifications/Requirements: GCE, A/AS: 4 points. AGNVQ, Pass. ND/C, National. SQAH, Individual consideration. SQAV, Individual consideration. IB, Individual consideration.

HND Multimedia Marketing
UCAS Code: 645G•Mode: 2 Years Single Subject

Qualifications/Requirements: GCE, A/AS: 4 points. AGNVQ, Pass. ND/C, National. SQAH, Individual consideration. SQAV, Individual consideration. IB, Individual consideration.

HND Multimedia Marketing
UCAS Code: 745G•Mode: 3 Years Single Subject

Qualifications/Requirements: GCE, A/AS: 4 points. AGNVQ, Pass. ND/C, National. SQAH, Individual consideration. SQAV, Individual consideration. IB, Individual consideration.

HND Multimedia Systems Development
UCAS Code: 145G•Mode: 2 Years Single Subject

Qualifications/Requirements: GCE, A/AS: 4 points. AGNVQ, Pass. ND/C, National. SQAH, Individual consideration. SQAV, Individual consideration. IB, Individual consideration.

HND Multimedia Systems Development
UCAS Code: 545G•Mode: 3 Years Single Subject

Qualifications/Requirements: GCE, A/AS: 4 points. AGNVQ, Pass. ND/C, National. SQAH, Individual consideration. SQAV, Individual consideration. IB, Individual consideration.

CITY UNIVERSITY C60

BEng Media Communication Systems
UCAS Code: G610•Mode: 4 Years Single Subject

Qualifications/Requirements: GCE, A level grades: CCC. AGNVQ, Distinction (in specific programmes) with A/AS. ND/C, 3 Merits and 3 Distinctions. SQAH, Grades BBBBB. SQAV, Individual consideration. IB, 28 points.

BEng Media Communication Systems
UCAS Code: G618•Mode: 3 Years Single Subject

Qualifications/Requirements: GCE, A level grades: CCC. AGNVQ, Distinction (in specific programmes) with A/AS. ND/C, 3 Merits and 3 Distinctions. SQAH, Grades BBBBB. SQAV, Individual consideration. IB, 28 points.

CITY OF BRISTOL COLLEGE C63

HND Multimedia/New Media
UCAS Code: 52GE•Mode: 2 Years Equal Combination

Qualifications/Requirements: An approved subject from restricted list. AGNVQ, Individual consideration. ND/C, Individual consideration. SQAH, Individual consideration. SQAV, Individual consideration. IB, Individual consideration.

HND Multimedia/New Media
UCAS Code: 52GW•Mode: 2 Years Equal Combination

Qualifications/Requirements: An approved subject from restricted list. AGNVQ, Individual consideration. ND/C, Individual consideration. SQAH, Individual consideration. SQAV, Individual consideration. IB, Individual consideration.

CITY COLLEGE MANCHESTER C66

HND Multimedia Design
UCAS Code: 52WE•Mode: 2 Years Equal Combination

Qualifications/Requirements: OTHER. SQAH, Individual consideration. SQAV, Individual consideration. IB, Individual consideration.

HND Multimedia Design
UCAS Code: 52WW•Mode: 2 Years Equal Combination

Qualifications/Requirements: OTHER. SQAH, Individual consideration. SQAV, Individual consideration. IB, Individual consideration.

CORNWALL COLLEGE WITH DUCHY COLLEGE C78

HND Multimedia Design
UCAS Code: 52GW•Mode: 2 Years Equal Combination

Qualifications/Requirements: GCE, A/AS: 4 points. AGNVQ, Pass. ND/C, 4 Merits.

COVENTRY UNIVERSITY C85

BA Multimedia Design and Technology
UCAS Code: GW52•Mode: 3/4 Years Equal Combination

Qualifications/Requirements: GCE, A/AS: 10 points. AGNVQ, Merit. ND/C, 4 Merits. SQAH, Individual consideration. SQAV, Individual consideration. IB, Individual consideration.

BA Multimedia Studies
UCAS Code: P400•Mode: 3/4 Years Single Subject

Qualifications/Requirements: GCE, A/AS: 12 points. AGNVQ, Merit. ND/C, 4 Merits. SQAH, Grades CCCC. SQAV, Individual consideration. IB, Individual consideration.

BSc Multimedia Design and Technology
UCAS Code: GW5F•Mode: 3/4 Years Equal Combination

Qualifications/Requirements: GCE, A/AS: 10 points. AGNVQ, Merit. ND/C, 4 Merits. SQAH, Individual consideration. SQAV, Individual consideration. IB, Individual consideration.

CUMBRIA COLLEGE OF ART AND DESIGN C95

BA Multimedia Design and Digital Animation
UCAS Code: EW52•Mode: 3 Years Equal Combination

Qualifications/Requirements: GCE, A level grades: CD. Foundation Art Course. AGNVQ, Merit (in specific programmes). ND/C, Merit overall and 2 Distinctions. SQAH, Grades BB. SQAV, Individual consideration. IB, Individual consideration.

BA Multimedia Design and Digital Animation
UCAS Code: GW52•Mode: 3 Years Equal Combination

Qualifications/Requirements: GCE, A level grades: CD. Foundation Art Course. AGNVQ, Merit (in specific programmes). ND/C, Merit overall and 2 Distinctions. SQAH, Grades BB. SQAV, Individual consideration. IB, Individual consideration.

DARTINGTON COLLEGE OF ARTS D13

BA Music & Digital Media
UCAS Code: WG35•Mode: 3 Years Equal Combination

Qualifications/Requirements: Please refer to prospectus.

DE MONTFORT UNIVERSITY D26

BA Multimedia Design
UCAS Code: E280•Mode: 3 Years Single Subject

Qualifications/Requirements: GCE, A/AS: 18 points. Foundation Art Course. AGNVQ, Distinction. ND/C, 4 Merits and 4 Distinctions. SQAH, Grades AABB. SQAV, Individual consideration. IB, 32 points.

BA Multimedia Design
UCAS Code: W280•Mode: 3 Years Single Subject

Qualifications/Requirements: GCE, A/AS: 18 points. Foundation Art Course. AGNVQ, Distinction. ND/C, 4 Merits and 4 Distinctions. SQAH, Grades AABB. SQAV, Individual consideration. IB, 32 points.

BA Multimedia Textiles
UCAS Code: EP44•Mode: 3 Years Equal Combination

Qualifications/Requirements: GCE, A/AS: 12 points. AGNVQ, Merit. ND/C, National. SQAH, Grades BBCC. SQAV, National. IB, 26 points.

BA Multimedia Textiles
UCAS Code: JP44•Mode: 3 Years Equal Combination

Qualifications/Requirements: GCE, A/AS: 12 points. AGNVQ, Merit. ND/C, National. SQAH, Grades BBCC. SQAV, National. IB, 26 points.

BSc Multimedia Computing
UCAS Code: G535•Mode: 4 Years Single Subject

Qualifications/Requirements: GCE, A/AS: 12-18 points. AGNVQ, Individual consideration. ND/C, Individual consideration. SQAH, Individual consideration. SQAV, Individual consideration. IB, Individual consideration.

Mod Computing and Media Studies
UCAS Code: GP5L•Mode: 3 Years Equal Combination

Qualifications/Requirements: GCE, A/AS: 12-14 points. AGNVQ, Merit. ND/C, 4 Merits (in specific programmes). SQAH, Grades BBBB including specific subject(s). SQAV, Individual consideration. IB, 30 points.

DONCASTER COLLEGE D52

HND Graphic Design/Multimedia
UCAS Code: 012E•Mode: 2 Years Single Subject

Qualifications/Requirements: Any Art/Design subject. AGNVQ, Pass (in specific programmes). ND/C, National. SQAH, Individual consideration. SQAV, Individual consideration. IB, Individual consideration.

HND Multimedia
UCAS Code: 52GW•Mode: 2 Years Equal Combination

Qualifications/Requirements: GCE, A level grades: E. AGNVQ, Pass. ND/C, National. SQAH, Individual consideration. SQAV, Individual consideration. IB, Individual consideration.

DUDLEY COLLEGE OF TECHNOLOGY D58

HND Multimedia (Computing)
UCAS Code: 045G•Mode: 2 Years Single Subject

Qualifications/Requirements: GCE, A/AS: 2 points. AGNVQ, Individual consideration. ND/C, Individual consideration.

UNIVERSITY OF EAST LONDON E28

BA/BSc Media Studies/Information Technology
UCAS Code: GP5K•Mode: 3 Years Equal Combination

Qualifications/Requirements: GCE, A/AS: 12 points. SQAH, Individual consideration. SQAV, Individual consideration. IB, Individual consideration.

BSc Multimedia and New Technology
UCAS Code: JP9L•Mode: 3 Years Equal Combination

Qualifications/Requirements: GCE, A/AS: 12 points. AGNVQ, Merit. ND/C, Merit overall. SQAH, Individual consideration. SQAV, Individual consideration. IB, Individual consideration.

BSc New Technology and Multimedia
UCAS Code: JP9K•Mode: 3 Years Equal Combination

Qualifications/Requirements: GCE, A/AS: 12 points. AGNVQ, Merit. ND/C, Merit overall. SQAH, Individual consideration. SQAV, Individual consideration. IB, Individual consideration.

FARNBOROUGH COLLEGE OF TECHNOLOGY F66

HND Design Technology (Graphics and Multimedia)
UCAS Code: 042W•Mode: 2 Years Single Subject

Qualifications/Requirements: GCE, A/AS: 6 points. AGNVQ, Pass. ND/C, Individual consideration. SQAH, Individual consideration. SQAV, Individual consideration. IB, Individual consideration.

HND Design Technology (Multimedia, Broadcast Graphics & Animation)
UCAS Code: 52GW•Mode: 2 Years Equal Combination

Qualifications/Requirements: GCE, A/AS: 8 points. AGNVQ, Merit. ND/C, Individual consideration. SQAH, Individual consideration. SQAV, Individual consideration. IB, Individual consideration.

UNIVERSITY OF GLAMORGAN G14

BSc Media Technology
UCAS Code: G500•Mode: 3/4 Years Single Subject

Qualifications/Requirements: GCE, A level grades: CD. AGNVQ, Merit. ND/C, 3 Merits. SQAH, Individual consideration. SQAV, Individual consideration. IB, Individual consideration.

BSc Media Technology
UCAS Code: G502•Mode: 3 Years Single Subject

Qualifications/Requirements: GCE, A level grades: D. AGNVQ, Pass. ND/C, National. SQAH, Individual consideration. SQAV, Individual consideration. IB, Individual consideration.

BSc Multimedia Computing
UCAS Code: G540•Mode: 3/4 Years Single Subject

Qualifications/Requirements: GCE, A/AS: 12 points. AGNVQ, Merit (in specific programmes). ND/C, Individual consideration. SQAH, Individual consideration. SQAV, Individual consideration. IB, Individual consideration.

BSc Multimedia Studies
UCAS Code: PG35•Mode: 3/4 Years Equal Combination

Qualifications/Requirements: GCE, A level grades: BC-BDE. AGNVQ, Merit (in specific programmes). ND/C, 7 Merits (in specific programmes). SQAH, Individual consideration. SQAV, Individual consideration. IB, Individual consideration.

BSc Multimedia Technology
UCAS Code: HP64•Mode: 3/4 Years Equal Combination

Qualifications/Requirements: GCE, A/AS: 12 points. AGNVQ, Merit. ND/C, 3 Merits. SQAH, Individual consideration. SQAV, Individual consideration. IB, Individual consideration.

HND Media Technology
UCAS Code: 005G•Mode: 2/3 Years Single Subject

Qualifications/Requirements: GCE, A level grades: D. AGNVQ, Pass. ND/C, National. SQAH, Individual consideration. SQAV, Individual consideration. IB, Individual consideration.

HND Multimedia
UCAS Code: 53GP•Mode: 2 Years Equal Combination

Qualifications/Requirements: GCE, A/AS: 4 points. AGNVQ, Pass (in specific programmes). ND/C, 3 Merits. SQAH, Individual consideration. SQAV, Individual consideration. IB, Individual consideration.

GLASGOW CALEDONIAN UNIVERSITY G42

BSc Applied Graphics Technology with Multimedia
UCAS Code: GW52•Mode: 2 Years Equal Combination

Qualifications/Requirements: AGNVQ, Not normally sufficient. ND/C, Individual consideration. SQAH, Not normally sufficient. SQAV, Higher National. IB, Individual consideration.

BSc Multimedia Technology
UCAS Code: P421•Mode: 1/2 Years Single Subject

Qualifications/Requirements: AGNVQ, Not normally sufficient. ND/C, Individual consideration. SQAH, Not normally sufficient. SQAV, Higher National. IB, Individual consideration.

UNIVERSITY OF GREENWICH G70

BSc Multimedia Technology
UCAS Code: G503•Mode: 3 Years Single Subject

Qualifications/Requirements: GCE, A/AS: 8-12 points. AGNVQ, Merit (in specific programmes). ND/C, 8 Merits. SQAH, Grades CCC. SQAV, Individual consideration. IB, Individual consideration.

HND Multimedia
UCAS Code: 25WG•Mode: 2 Years Equal Combination

Qualifications/Requirements: Please refer to prospectus.

HALTON COLLEGE H06

HND Multimedia
UCAS Code: 2W5G•Mode: 2 Years Major/Minor

Qualifications/Requirements: Please refer to prospectus.

HERIOT-WATT UNIVERSITY, EDINBURGH H24

BSc Computer Science (Multimedia Systems)
UCAS Code: G540•Mode: 4 Years Single Subject

Qualifications/Requirements: GCE, A level grades: CCD. Maths. AGNVQ, Merit (in specific programmes). ND/C, Merit overall (in specific programmes). SQAH, Grades BBBC including specific subject(s). SQAV, National including specific subject(s). IB, 28 points.

UNIVERSITY OF HERTFORDSHIRE H36

BSc Multimedia Systems
UCAS Code: G5P4•Mode: 4 Years Major/Minor

Qualifications/Requirements: GCE, A/AS: 16 points. AGNVQ, Merit. ND/C, Merit overall. IB, 28 points.

BSc Multimedia Technology
UCAS Code: P410•Mode: 3 Years Single Subject

Qualifications/Requirements: GCE, A/AS: 14 points. ND/C, National.

HND Computing (Multimedia)
UCAS Code: 065G•Mode: 2 Years Single Subject

Qualifications/Requirements: GCE, A/AS: 4 points. AGNVQ, Pass. ND/C, 2 Merits. SQAH, Individual consideration. SQAV, Individual consideration. IB, Individual consideration.

THE UNIVERSITY OF HUDDERSFIELD H60

BA Business Computing with Multimedia
UCAS Code: G504•Mode: 4 Years Single Subject

Qualifications/Requirements: GCE, A/AS: 16 points. AGNVQ, Merit. ND/C, Merit overall. SQAH, Grades BBBB. SQAV, Individual consideration. IB, Individual consideration.

BA Interactive Media
UCAS Code: G5P4•Mode: 4 Years Major/Minor

Qualifications/Requirements: GCE, A/AS: 16 points. AGNVQ, Merit. ND/C, Merit overall. SQAH, Grades BBBB. SQAV, Individual consideration. IB, Individual consideration.

BA Multimedia with Artificial Intelligence
UCAS Code: G5GM•Mode: 4 Years Major/Minor

Qualifications/Requirements: GCE, A/AS: 16 points. AGNVQ, Merit. ND/C, Merit overall. SQAH, Grades BBBB. SQAV, Individual consideration. IB, Individual consideration.

BA Multimedia with Business Computing
UCAS Code: G506•Mode: 4 Years Single Subject

Qualifications/Requirements: GCE, A/AS: 16 points. AGNVQ, Merit. ND/C, Merit overall. SQAH, Grades BBBB. SQAV, Individual consideration. IB, Individual consideration.

BA Multimedia with French
UCAS Code: G5RC•Mode: 4 Years Major/Minor

Qualifications/Requirements: GCE, A/AS: 16 points. AGNVQ, Merit. ND/C, Merit overall. SQAH, Grades BBBB. SQAV, Individual consideration. IB, Individual consideration.

BA Multimedia with German

UCAS Code: G5RF•Mode: 4 Years Major/Minor

Qualifications/Requirements: GCE, A/AS: 16 points. AGNVQ, Merit. ND/C, Merit overall. SQAH, Grades BBBB. SQAV, Individual consideration. IB, Individual consideration.

BA Multimedia with Human-Computer Interaction

UCAS Code: G505•Mode: 4 Years Single Subject

Qualifications/Requirements: GCE, A/AS: 16 points. AGNVQ, Merit. ND/C, Merit overall. SQAH, Grades BBBB. SQAV, Individual consideration. IB, Individual consideration.

BA Multimedia with Software Development

UCAS Code: G5GR•Mode: 4 Years Major/Minor

Qualifications/Requirements: GCE, A/AS: 16 points. AGNVQ, Merit. ND/C, Merit overall. SQAH, Grades BBBB. SQAV, Individual consideration. IB, Individual consideration.

BA Multimedia with Statistics

UCAS Code: G5GK•Mode: 4 Years Major/Minor

Qualifications/Requirements: GCE, A/AS: 16 points. AGNVQ, Merit. ND/C, Merit overall. SQAH, Grades BBBB. SQAV, Individual consideration. IB, Individual consideration.

BA/BSc Multimedia Design

UCAS Code: EW5F•Mode: 3/4 Years Equal Combination

Qualifications/Requirements: GCE, A/AS: 12-14 points. Foundation Art Course. AGNVQ, Merit (in specific programmes). ND/C, Merit overall. SQAH, Grades BBB. SQAV, National including specific subject(s). IB, 26 points.

BA/BSc Multimedia Design

UCAS Code: GW5F•Mode: 3/4 Years Equal Combination

Qualifications/Requirements: GCE, A/AS: 12-14 points. Foundation Art Course. AGNVQ, Merit (in specific programmes). ND/C, Merit overall. SQAH, Grades BBB. SQAV, National including specific subject(s). IB, 26 points.

BSc Multimedia Technology

UCAS Code: H6P4•Mode: 4 Years Major/Minor

Qualifications/Requirements: GCE, A/AS: 14-18 points. Science at A level. AGNVQ, Merit (in specific programmes). ND/C, 5 Merits. SQAH, Grades BBB. SQAV, National including specific subject(s). IB, 28 points.

HND Multimedia

UCAS Code: 4P5G•Mode: 3 Years Major/Minor

Qualifications/Requirements: GCE, A/AS: 6 points. AGNVQ, Pass. ND/C, National. SQAH, Grades CCC. SQAV, Individual consideration. IB, Individual consideration.

LANCASTER UNIVERSITY L14

BSc Computer Science with Multimedia Systems

UCAS Code: G5P4•Mode: 3 Years Major/Minor

Qualifications/Requirements: GCE, A level grades: BCC. AGNVQ, Distinction with 6 additional units or with A/AS. ND/C, Merits and Distinction (in specific programmes). SQAH, Grades BBBBB. SQAV, Individual consideration. IB, 30 points.

LEEDS, TRINITY AND ALL SAINTS COLLEGE L24

BA Digital Media and Culture

UCAS Code: GP54•Mode: 3 Years Equal Combination

Qualifications/Requirements: GCE, A level grades: BCC-CCC. AGNVQ, Merit (in specific programmes). ND/C, Merit overall and 4 Distinctions. SQAH, Grades AABBB. SQAV, Individual consideration. IB, 26 points.

LEEDS METROPOLITAN UNIVERSITY L27

BSc Multimedia Technology

UCAS Code: HP64•Mode: 3/4 Years Equal Combination

Qualifications/Requirements: GCE, A/AS: 12 points. Maths or Science at A level. AGNVQ, Merit (in specific programmes). ND/C, 4 Merits (in specific programmes). SQAH, Grades BBB including specific subject(s). SQAV, Individual consideration. IB, Pass in Diploma including specific subjects.

HND Multimedia Technology

UCAS Code: 46PH•Mode: 2/3 Years Equal Combination

Qualifications/Requirements: GCE, A/AS: 6 points. Maths or Science at A level. AGNVQ, Pass (in specific programmes). ND/C, Merit (in specific programmes). SQAH, Grades BCC. SQAV, Individual consideration. IB, 22 points.

LEEDS COLLEGE OF ART & DESIGN L28

HND Multimedia

UCAS Code: 65GE•Mode: 2 Years Equal Combination

Qualifications/Requirements: Art & Design or Media Studies. AGNVQ, Pass in Art & Design. SQAH, Individual consideration. SQAV, Individual consideration. IB, Individual consideration.

HND Multimedia

UCAS Code: 65GW•Mode: 2 Years Equal Combination

Qualifications/Requirements: Art & Design or Media Studies. AGNVQ, Pass in Art & Design. SQAH, Individual consideration. SQAV, Individual consideration. IB, Individual consideration.

UNIVERSITY OF LINCOLNSHIRE AND HUMBERSIDE L39

BA/BSc Computing and Media Technology

UCAS Code: GP54•Mode: 3 Years Equal Combination

Qualifications/Requirements: GCE, A/AS: 12 points. AGNVQ, Merit. ND/C, 3 Merits and 1 Distinction. SQAH, Grades CCCC. SQAV, Individual consideration. IB, 24 points.

BA/BSc Information Technology and Media Technology

UCAS Code: GP5K•Mode: 3 Years Equal Combination

Qualifications/Requirements: GCE, A/AS: 12 points. AGNVQ, Merit. ND/C, 3 Merits and 1 Distinction. SQAH, Grades CCCC. SQAV, Individual consideration. IB, 24 points.

BSc Computing (Multimedia and Systems Development)

UCAS Code: G503•Mode: 3 Years Single Subject

Qualifications/Requirements: GCE, A/AS: 12 points. AGNVQ, Merit. ND/C, 3 Merits and 1 Distinction. SQAH, Grades CCCC. SQAV, Individual consideration. IB, 24 points.

LIVERPOOL JOHN MOORES UNIVERSITY L51

BA Multimedia Arts

UCAS Code: EG25•Mode: 3 Years Equal Combination

Qualifications/Requirements: Please refer to prospectus.

BSc Multimedia Systems

UCAS Code: G540•Mode: 3 Years Single Subject

Qualifications/Requirements: GCE, A/AS: 12 points. AGNVQ, Distinction. ND/C, Merit overall and 1 Distinction. SQAH, Individual consideration. SQAV, Individual consideration. IB, 26 points.

LONDON GUILDHALL UNIVERSITY L55

BSc Multimedia Systems

UCAS Code: G5W5•Mode: 3 Years Major/Minor

Qualifications/Requirements: GCE, A level grades: CD. AGNVQ, Merit (in specific programmes). ND/C, 4 Merits. SQAH, Individual consideration. SQAV, Individual consideration. IB, 24 points.

Mod Accounting & Multimedia Systems

UCAS Code: GNM4•Mode: 3 Years Equal Combination

Qualifications/Requirements: GCE, A level grades: DD. AGNVQ, Merit (in specific programmes). ND/C, Merit overall. SQAH, Individual consideration. SQAV, Individual consideration. IB, 24 points.

Mod American Studies & Multimedia Systems

UCAS Code: GQM4•Mode: 3 Years Equal Combination

Qualifications/Requirements: GCE, A level grades: CD. AGNVQ, Merit. ND/C, Merit overall and 2 Distinctions. SQAH, Individual consideration. SQAV, Individual consideration. IB, 26 points.

Mod Asia Studies & Multimedia Systems

UCAS Code: GTM5•Mode: 3 Years Equal Combination

Qualifications/Requirements: GCE, A level grades: CD. AGNVQ, Merit. ND/C, Merit overall and 2 Distinctions. SQAH, Individual consideration. SQAV, Individual consideration. IB, 26 points.

Mod Banking & Multimedia Systems

UCAS Code: NG3M•Mode: 3 Years Equal Combination

Qualifications/Requirements: GCE, A level grades: CD. AGNVQ, Merit (in specific programmes). ND/C, Merit overall and 2 Distinctions. SQAH, Individual consideration. SQAV, Individual consideration. IB, 24 points.

Mod Business Economics & Multimedia Systems

UCAS Code: GLMC•Mode: 3 Years Equal Combination

Qualifications/Requirements: GCE, A level grades: DD. AGNVQ, Merit (in specific programmes). ND/C, Merit overall. SQAH, Individual consideration. SQAV, Individual consideration. IB, 24 points.

Mod Communications & Multimedia Systems

UCAS Code: GPM4•Mode: 3 Years Equal Combination

Qualifications/Requirements: GCE, A level grades: CC-CDD. AGNVQ, Merit (in specific programmes). ND/C, Merit overall and 2 Distinctions. SQAH, Individual consideration. SQAV, Individual consideration. IB, 26 points.

Mod Computing & Multimedia Systems

UCAS Code: GG5M•Mode: 3 Years Equal Combination

Qualifications/Requirements: GCE, A level grades: DD. AGNVQ, Merit (in specific programmes). ND/C, Merit overall. SQAH, Individual consideration. SQAV, Individual consideration. IB, 24 points.

Mod Criminology & Multimedia Systems

UCAS Code: GMMH•Mode: 3 Years Equal Combination

Qualifications/Requirements: GCE, A level grades: CD. AGNVQ, Merit. ND/C, Merit overall and 2 Distinctions. SQAH, Individual consideration. SQAV, Individual consideration. IB, 26 points.

Mod Design Studies & Multimedia Systems

UCAS Code: GWM2•Mode: 3 Years Equal Combination

Qualifications/Requirements: GCE, A level grades: CD-DDD. AGNVQ, Merit (in specific programmes). ND/C, Merit overall. SQAH, Individual consideration. SQAV, Individual consideration. IB, 24 points.

Mod Development Studies & Multimedia Systems

UCAS Code: GMM9•Mode: 3 Years Equal Combination

Qualifications/Requirements: GCE, A level grades: DD. AGNVQ, Merit (in specific programmes). ND/C, Merit overall. SQAH, Individual consideration. SQAV, Individual consideration. IB, 24 points.

Mod Economics & Multimedia Systems

UCAS Code: GLM1•Mode: 3 Years Equal Combination

Qualifications/Requirements: GCE, A level grades: DD. AGNVQ, Merit (in specific programmes). ND/C, Merit overall. SQAH, Individual consideration. SQAV, Individual consideration. IB, 24 points.

Mod English Studies & Multimedia Systems

UCAS Code: GQM3•Mode: 3 Years Equal Combination

Qualifications/Requirements: GCE, A level grades: CD-DDD. AGNVQ, Merit (in specific programmes). ND/C, Merit overall and 2 Distinctions. SQAH, Individual consideration. SQAV, Individual consideration. IB, 26 points.

Mod European Studies & Multimedia Systems

UCAS Code: GTM2•Mode: 3 Years Equal Combination

Qualifications/Requirements: GCE, A level grades: DD. AGNVQ, Merit (in specific programmes). ND/C, Merit overall. SQAH, Individual consideration. SQAV, Individual consideration. IB, 24 points.

Mod Financial Services & Multimedia Systems

UCAS Code: GNM3•Mode: 3 Years Equal Combination

Qualifications/Requirements: GCE, A level grades: DD. AGNVQ, Merit (in specific programmes). ND/C, Merit overall. SQAH, Individual consideration. SQAV, Individual consideration. IB, 24 points.

Mod Fine Art & Multimedia Systems

UCAS Code: GWM1•Mode: 3 Years Equal Combination

Qualifications/Requirements: GCE, A level grades: CC-CDD. OTHER. AGNVQ, Merit (in specific programmes). ND/C, Merit overall and 2 Distinctions. SQAH, Individual consideration. SQAV, Individual consideration. IB, 26 points.

Mod French & Multimedia Systems

UCAS Code: GRM1•Mode: 4 Years Equal Combination

Qualifications/Requirements: GCE, A level grades: DD. AGNVQ, Merit (in specific programmes). ND/C, Merit overall. SQAH, Individual consideration. SQAV, Individual consideration. IB, 24 points.

Mod German & Multimedia Systems

UCAS Code: GRM2•Mode: 4 Years Equal Combination

Qualifications/Requirements: GCE, A level grades: DD. AGNVQ, Merit (in specific programmes). ND/C, Merit overall. SQAH, Individual consideration. SQAV, Individual consideration. IB, 24 points.

Mod International Relations & Multimedia Systems

UCAS Code: GMMC•Mode: 3 Years Equal Combination

Qualifications/Requirements: GCE, A level grades: DD. AGNVQ, Merit (in specific programmes). ND/C, Merit overall. SQAH, Individual consideration. SQAV, Individual consideration. IB, 24 points.

Mod Investment & Multimedia Systems

UCAS Code: NGJM•Mode: 3 Years Equal Combination

Qualifications/Requirements: GCE, A level grades: DD. AGNVQ, Merit (in specific programmes). ND/C, Merit overall. SQAH, Individual consideration. SQAV, Individual consideration. IB, 24 points.

Mod Law & Multimedia Systems

UCAS Code: GMM3•Mode: 3 Years Equal Combination

Qualifications/Requirements: GCE, A level grades: CC-CDD. AGNVQ, Merit (in specific programmes). ND/C, Merit overall and 2 Distinctions. SQAH, Individual consideration. SQAV, Individual consideration. IB, 26 points.

Mod Marketing & Multimedia Systems

UCAS Code: GNM5•Mode: 3 Years Equal Combination

Qualifications/Requirements: GCE, A level grades: CD-DDD. AGNVQ, Merit (in specific programmes). ND/C, Merit overall and 2 Distinctions. SQAH, Individual consideration. SQAV, Individual consideration. IB, 26 points.

Mod Mathematics & Multimedia Systems

UCAS Code: GG1M•Mode: 3 Years Equal Combination

Qualifications/Requirements: GCE, A level grades: DD. AGNVQ, Merit (in specific programmes). ND/C, Merit overall. SQAH, Individual consideration. SQAV, Individual consideration. IB, 24 points.

Mod Modern History & Multimedia Systems

UCAS Code: GVM1•Mode: 3 Years Equal Combination

Qualifications/Requirements: GCE, A level grades: DD. AGNVQ, Merit (in specific programmes). ND/C, Merit overall. SQAH, Individual consideration. SQAV, Individual consideration. IB, 24 points.

Mod Multimedia Systems & 3D/Spatial Design

UCAS Code: GWMF•Mode: 3 Years Equal Combination

Qualifications/Requirements: GCE, A level grades: DD. OTHER. AGNVQ, Merit (in specific programmes). ND/C, Merit overall. SQAH, Individual consideration. SQAV, Individual consideration. IB, 24 points.

Mod Multimedia Systems & Philosophy
UCAS Code: GVM7•Mode: 3 Years Equal Combination

Qualifications/Requirements: GCE, A level grades: CD. AGNVQ, Merit. ND/C, Merit overall and 2 Distinctions. SQAH, Individual consideration. SQAV, Individual consideration. IB, 26 points.

Mod Multimedia Systems & Politics
UCAS Code: GMM1•Mode: 3 Years Equal Combination

Qualifications/Requirements: GCE, A level grades: DD. AGNVQ, Merit (in specific programmes). ND/C, Merit overall. SQAH, Individual consideration. SQAV, Individual consideration. IB, 24 points.

Mod Multimedia Systems & Product Development & Manufacture
UCAS Code: GJM4•Mode: 3 Years Equal Combination

Qualifications/Requirements: GCE, A level grades: DD. AGNVQ, Merit (in specific programmes). ND/C, Merit overall. SQAH, Individual consideration. SQAV, Individual consideration. IB, 24 points.

Mod Multimedia Systems & Psychology
UCAS Code: CG8M•Mode: 3 Years Equal Combination

Qualifications/Requirements: GCE, A level grades: CD-DDD. AGNVQ, Merit (in specific programmes). ND/C, Merit overall and 2 Distinctions. SQAH, Individual consideration. SQAV, Individual consideration. IB, 26 points.

Mod Multimedia Systems & Public Policy
UCAS Code: GMMD•Mode: 3 Years Equal Combination

Qualifications/Requirements: GCE, A level grades: CD. AGNVQ, Merit. ND/C, Merit overall and 2 Distinctions. SQAH, Individual consideration. SQAV, Individual consideration. IB, 26 points.

Mod Multimedia Systems & Social Policy & Management
UCAS Code: GLM4•Mode: 3 Years Equal Combination

Qualifications/Requirements: GCE, A level grades: CD-DDD. AGNVQ, Merit (in specific programmes). ND/C, Merit overall. SQAH, Individual consideration. SQAV, Individual consideration. IB, 24 points.

Mod Multimedia Systems & Sociology
UCAS Code: GLM3•Mode: 3 Years Equal Combination

Qualifications/Requirements: GCE, A level grades: CD-DDD. AGNVQ, Merit (in specific programmes). ND/C, Merit overall. SQAH, Individual consideration. SQAV, Individual consideration. IB, 24 points.

Mod Multimedia Systems & Spanish
UCAS Code: GRM4•Mode: 4 Years Equal Combination

Qualifications/Requirements: GCE, A level grades: DD. AGNVQ, Merit (in specific programmes). ND/C, Merit overall. SQAH, Individual consideration. SQAV, Individual consideration. IB, 24 points.

Mod Multimedia Systems & Taxation
UCAS Code: GNMH•Mode: 3 Years Equal Combination

Qualifications/Requirements: GCE, A level grades: DD. AGNVQ, Merit (in specific programmes). ND/C, Merit overall. SQAH, Individual consideration. SQAV, Individual consideration. IB, 24 points.

Mod Multimedia Systems & Textile Furnishing Design
UCAS Code: GJMK•Mode: 3 Years Equal Combination

Qualifications/Requirements: GCE, A level grades: DD. OTHER. AGNVQ, Merit (in specific programmes). ND/C, Merit overall. SQAH, Individual consideration. SQAV, Individual consideration. IB, 24 points.

Mod Multimedia Systems & Transport
UCAS Code: GNM9•Mode: 3 Years Equal Combination

Qualifications/Requirements: GCE, A level grades: CD. AGNVQ, Merit. ND/C, Merit overall and 2 Distinctions. SQAH, Individual consideration. SQAV, Individual consideration. IB, 26 points.

THE MANCHESTER METROPOLITAN UNIVERSITY M40

BSc Applicable Mathematics/Multimedia Technology
UCAS Code: GG16•Mode: 3 Years Equal Combination

Qualifications/Requirements: GCE, A/AS: 16 points. Maths. AGNVQ, Merit (in specific programmes). ND/C, 1 Merit and 3 Distinctions (in specific programmes). SQAH, Grades BBBCC including specific subject(s). SQAV, Individual consideration. IB, 28 points.

BSc Business Economics/Multimedia Technology
UCAS Code: GL6C•Mode: 3 Years Equal Combination

Qualifications/Requirements: GCE, A/AS: 16 points. AGNVQ, Merit (in specific programmes). ND/C, 1 Merit and 3 Distinctions. SQAH, Grades BBBCC. SQAV, Individual consideration. IB, 28 points.

BSc Business Mathematics/Multimedia Technology
UCAS Code: GG1P•Mode: 3 Years Equal Combination

Qualifications/Requirements: GCE, A/AS: 16 points. Maths at A level, Physics or Economics. AGNVQ, Merit (in specific programmes). ND/C, 1 Merit and 3 Distinctions (in specific programmes). SQAH, Grades BBBCC including specific subject(s). SQAV, Individual consideration. IB, 28 points.

BSc Chemistry/Multimedia Technology
UCAS Code: FG16•Mode: 3 Years Equal Combination

Qualifications/Requirements: GCE, A/AS: 12 points. Chemistry. AGNVQ, Merit (in specific programmes). ND/C, 1 Merit and 3 Distinctions (in specific programmes). SQAH, Grades BBBCC including specific subject(s). SQAV, Individual consideration. IB, 28 points.

BSc Economics/Multimedia Technology

UCAS Code: GL61•Mode: 3 Years Equal Combination

Qualifications/Requirements: GCE, A/AS: 16 points. AGNVQ, Merit (in specific programmes). ND/C, 1 Merit and 3 Distinctions. SQAH, Grades BBBCC. SQAV, Individual consideration. IB, 28 points.

BSc Environmental Studies/Multimedia Technology

UCAS Code: FG96•Mode: 3 Years Equal Combination

Qualifications/Requirements: GCE, A/AS: 14 points. AGNVQ, Merit (in specific programmes). ND/C, 1 Merit and 3 Distinctions. SQAH, Grades BBBCC. SQAV, Individual consideration. IB, 28 points.

BSc European Studies/Multimedia Technology

UCAS Code: GT62•Mode: 3 Years Equal Combination

Qualifications/Requirements: GCE, A/AS: 16 points. AGNVQ, Merit (in specific programmes). ND/C, 1 Merit and 3 Distinctions. SQAH, Grades BBBCC. SQAV, Individual consideration. IB, 28 points.

BSc Geography/Multimedia Technology

UCAS Code: GL68•Mode: 3 Years Equal Combination

Qualifications/Requirements: GCE, A/AS: 14 points. AGNVQ, Merit (in specific programmes). ND/C, 1 Merit and 3 Distinctions. SQAH, Grades BBBCC. SQAV, Individual consideration. IB, 28 points.

BSc Materials Science/Multimedia Technology

UCAS Code: FG26•Mode: 3 Years Equal Combination

Qualifications/Requirements: GCE, A/AS: 12 points. Maths, Physics or Chemistry. AGNVQ, Merit (in specific programmes). ND/C, 5 Merits (in specific programmes). SQAH, Grades BCCCC including specific subject(s). SQAV, Individual consideration. IB, 27 points.

BSc Multimedia Computing

UCAS Code: G540•Mode: 4 Years Single Subject

Qualifications/Requirements: GCE, A/AS: 12-18 points. AGNVQ, Distinction (in specific programmes) or Merit (in specific programmes) with 4 additional units or with A/AS. ND/C, Merits and Distinction. SQAH, Grades BBCC. SQAV, Individual consideration. IB, 24 points.

BSc Multimedia Technology and Sociological Studies

UCAS Code: GL63•Mode: 3 Years Equal Combination

Qualifications/Requirements: Please refer to prospectus.

BSc Multimedia Technology/Psychology

UCAS Code: GL67•Mode: 3 Years Equal Combination

Qualifications/Requirements: GCE, A/AS: 18 points. AGNVQ, Distinction (in specific programmes). ND/C, 2 Merits and 4 Distinctions. SQAH, Grades BBBCC including specific subject(s). SQAV, Individual consideration. IB, 29 points.

MID-CHESHIRE COLLEGE M77

HND Multimedia

UCAS Code: 572E•Mode: 2 Years Single Subject

Qualifications/Requirements: GCE, A level grades: A-E. Foundation Art Course or any Art/Design subject. A portfolio of work is required. AGNVQ, Pass in Art & Design. ND/C, National. SQAH, Individual consideration. SQAV, Individual consideration. IB, Individual consideration.

MIDDLESEX UNIVERSITY M80

BSc Multimedia

UCAS Code: G540•Mode: 3 Years Single Subject

Qualifications/Requirements: GCE, A/AS: 12-16 points. AGNVQ, Merit (in specific programmes). ND/C, 5 Merits. SQAH, Individual consideration. SQAV, Individual consideration. IB, 28 points.

NAPIER UNIVERSITY N07

BSc Management and Multimedia Technology

UCAS Code: NG1M•Mode: 3/4 Years Equal Combination

Qualifications/Requirements: GCE, A level grades: DD. ND/C, Individual consideration. SQAH, Grades BBC. SQAV, Individual consideration.

BEng Multimedia Systems

UCAS Code: G540•Mode: 3/4 Years Single Subject

Qualifications/Requirements: GCE, A level grades: CD. ND/C, Individual consideration. SQAH, Grades BBB. SQAV, Individual consideration.

UNIVERSITY COLLEGE NORTHAMPTON N14

Mod Information Systems/Media and Popular Culture

UCAS Code: G5P4•Mode: 3 Years Major/Minor

Qualifications/Requirements: GCE, A/AS: 10-12 points. AGNVQ, Merit. ND/C, 5 Merits. SQAH, Grades CCC. SQAV, Individual consideration. IB, 24 points.

HND Multimedia

UCAS Code: 52GW•Mode: 2 Years Equal Combination

Qualifications/Requirements: GCE, A/AS: 4 points. AGNVQ, Pass. ND/C, National. SQAH, Grades CC. SQAV, Individual consideration. IB, Individual consideration.

NEWCASTLE COLLEGE N23

HND Multimedia
UCAS Code: 52GE•Mode: 2 Years Equal Combination
Qualifications/Requirements: Please refer to prospectus.

HND Multimedia
UCAS Code: 52GW•Mode: 2 Years Equal Combination
Qualifications/Requirements: Please refer to prospectus.

NEW COLLEGE DURHAM N28

HND Multimedia
UCAS Code: 045G•Mode: 2 Years Single Subject
Qualifications/Requirements: GCE, A/AS: 2 points. AGNVQ, Individual consideration. ND/C, National. SQAH, Individual consideration. SQAV, Individual consideration. IB, Individual consideration.

UNIVERSITY OF WALES COLLEGE, NEWPORT N37

BA Multimedia
UCAS Code: EG25•Mode: 3 Years Equal Combination
Qualifications/Requirements: GCE, A/AS: 6 points. OTHER. AGNVQ, Pass (in specific programmes). ND/C, Merits and Distinction. SQAH, Individual consideration. SQAV, Individual consideration. IB, Individual consideration.

BA Multimedia
UCAS Code: WG25•Mode: 3 Years Equal Combination
Qualifications/Requirements: GCE, A/AS: 6 points. OTHER. AGNVQ, Pass (in specific programmes). ND/C, Merits and Distinction. SQAH, Individual consideration. SQAV, Individual consideration. IB, Individual consideration.

NESCOT N49

HND Multimedia
UCAS Code: 45PG•Mode: 2 Years Equal Combination
Qualifications/Requirements: Please refer to prospectus.

THE NORTH EAST WALES INSTITUTE OF HIGHER EDUCATION N56

BA Animation and Multimedia Design
UCAS Code: EH25•Mode: 3 Years Equal Combination
Qualifications/Requirements: OTHER. AGNVQ, Individual consideration. ND/C, Individual consideration. SQAH, Individual consideration. SQAV, Individual consideration. IB, Individual consideration.

BA Animation and Multimedia Design
UCAS Code: WH25•Mode: 3 Years Equal Combination
Qualifications/Requirements: OTHER. AGNVQ, Individual consideration. ND/C, Individual consideration. SQAH, Individual consideration. SQAV, Individual consideration. IB, Individual consideration.

BA Animation with Multimedia Design
UCAS Code: E2HN•Mode: 3 Years Major/Minor
Qualifications/Requirements: Please refer to prospectus.

BA Animation with Multimedia Design
UCAS Code: W2HN•Mode: 3 Years Major/Minor
Qualifications/Requirements: Please refer to prospectus.

BA Ceramics and Multimedia Design
UCAS Code: EH35•Mode: 3 Years Equal Combination
Qualifications/Requirements: Please refer to prospectus.

BA Ceramics and Multimedia Design
UCAS Code: JH35•Mode: 3 Years Equal Combination
Qualifications/Requirements: Please refer to prospectus.

BA Ceramics with Multimedia Design
UCAS Code: E3H5•Mode: 3 Years Major/Minor
Qualifications/Requirements: Please refer to prospectus.

BA Ceramics with Multimedia Design
UCAS Code: J3H5•Mode: 3 Years Major/Minor
Qualifications/Requirements: Please refer to prospectus.

BA Glass and Multimedia Design
UCAS Code: EH3M•Mode: 3 Years Equal Combination
Qualifications/Requirements: Please refer to prospectus.

BA Glass and Multimedia Design
UCAS Code: JH3M•Mode: 3 Years Equal Combination
Qualifications/Requirements: Please refer to prospectus.

BA Glass with Multimedia Design
UCAS Code: E3HM•Mode: 3 Years Major/Minor
Qualifications/Requirements: Please refer to prospectus.

BA Glass with Multimedia Design
UCAS Code: J3HM•Mode: 3 Years Major/Minor
Qualifications/Requirements: Please refer to prospectus.

BA Graphic Design and Multimedia Design
UCAS Code: EH2M•Mode: 3 Years Equal Combination
Qualifications/Requirements: OTHER. AGNVQ, Individual consideration. ND/C, Individual consideration. SQAH, Individual consideration. SQAV, Individual consideration. IB, Individual consideration.

BA Graphic Design and Multimedia Design
UCAS Code: WH2M•Mode: 3 Years Equal Combination
Qualifications/Requirements: OTHER. AGNVQ, Individual consideration. ND/C, Individual consideration. SQAH, Individual consideration. SQAV, Individual consideration. IB, Individual consideration.

BA Graphic Design with Multimedia Design
UCAS Code: E2HM•Mode: 3 Years Major/Minor
Qualifications/Requirements: Please refer to prospectus.

BA Graphic Design with Multimedia Design
UCAS Code: W2HM•Mode: 3 Years Major/Minor
Qualifications/Requirements: Please refer to prospectus.

BA Illustration and Multimedia Design
UCAS Code: EH2N•Mode: 3 Years Equal Combination
Qualifications/Requirements: Please refer to prospectus.

BA Illustration and Multimedia Design
UCAS Code: WH2N•Mode: 3 Years Equal Combination
Qualifications/Requirements: Please refer to prospectus.

BA Illustration with Multimedia Design
UCAS Code: EHF5•Mode: 3 Years Equal Combination
Qualifications/Requirements: Please refer to prospectus.

BA Illustration with Multimedia Design
UCAS Code: WHF5•Mode: 3 Years Equal Combination
Qualifications/Requirements: Please refer to prospectus.

BA Jewellery/Metalwork and Multimedia Design
UCAS Code: EH65•Mode: 3 Years Equal Combination
Qualifications/Requirements: Please refer to prospectus.

BA Jewellery/Metalwork and Multimedia Design
UCAS Code: WH65•Mode: 3 Years Equal Combination
Qualifications/Requirements: Please refer to prospectus.

BA Multimedia
UCAS Code: E1H5•Mode: 3 Years Major/Minor
Qualifications/Requirements: Please refer to prospectus.

BA Multimedia
UCAS Code: W1H5•Mode: 3 Years Major/Minor
Qualifications/Requirements: Please refer to prospectus.

BA Multimedia Design
UCAS Code: E2H5•Mode: 3 Years Major/Minor
Qualifications/Requirements: Please refer to prospectus.

BA Multimedia Design
UCAS Code: W2H5•Mode: 3 Years Major/Minor
Qualifications/Requirements: Please refer to prospectus.

BSc Multimedia Computing
UCAS Code: G564•Mode: 3 Years Single Subject
Qualifications/Requirements: GCE, A/AS: 6-10 points. AGNVQ, Merit (in specific programmes). ND/C, 3 Merits. SQAH, Grades CCC. SQAV, National including specific subject(s). IB, Individual consideration.

HND Multimedia Computing
UCAS Code: 465G•Mode: 2 Years Single Subject
Qualifications/Requirements: Please refer to prospectus.

NORTH EAST WORCESTERSHIRE COLLEGE N58

HND Graphic Design and Multimedia
UCAS Code: 52GE•Mode: 2 Years Equal Combination
Qualifications/Requirements: GCE, A level grades: E. OTHER. AGNVQ, Pass (in specific programmes). ND/C, National.

HND Graphic Design and Multimedia
UCAS Code: 52GW•Mode: 2 Years Equal Combination
Qualifications/Requirements: GCE, A level grades: E. OTHER. AGNVQ, Pass (in specific programmes). ND/C, National.

UNIVERSITY OF NORTH LONDON N63

BSc Multimedia Computing
UCAS Code: G541•Mode: 3/4 Years Single Subject
Qualifications/Requirements: GCE, A/AS: 8 points. Maths, Science, Electronics, Computing or Statistics. AGNVQ, Merit. ND/C, 3 Merits (in specific programmes). SQAH, Grades CCCC including specific subject(s). SQAV, Individual consideration. IB, Individual consideration.

BSc Multimedia Technology and Applications (3 years/4 years SW)
UCAS Code: G540•Mode: 3/4 Years Single Subject
Qualifications/Requirements: GCE, A/AS: 10 points. Maths, Computing, Physics, Statistics or Electronics. AGNVQ, Merit (in specific programmes). ND/C, 4 Merits (in specific programmes). SQAH, Grades CCCC including specific subject(s). SQAV, Individual consideration. IB, Individual consideration.

UNIVERSITY OF NORTHUMBRIA AT NEWCASTLE N77

BA Multimedia Design

UCAS Code: EP25•Mode: 3 Years Equal Combination

Qualifications/Requirements: GCE, A/AS: 14 points. Foundation Art Course. A portfolio of work is required. AGNVQ, Individual consideration. ND/C, National. SQAH, Individual consideration. SQAV, Not normally sufficient. IB, Individual consideration.

BSc Multimedia Computing

UCAS Code: G535•Mode: 4 Years Single Subject

Qualifications/Requirements: GCE, A/AS: 14 points. AGNVQ, Merit. ND/C, 6 Merits. SQAH, Grades BCCC. IB, 26 points.

NORTHUMBERLAND COLLEGE N78

HND TV Production and Multimedia

UCAS Code: 54WE•Mode: 2 Years Equal Combination

Qualifications/Requirements: Art & Design or Media Studies. A portfolio of work is required. AGNVQ, Pass. ND/C, National. SQAH, Individual consideration. SQAV, Individual consideration. IB, Individual consideration.

HND TV Production and Multimedia

UCAS Code: 54WW•Mode: 2 Years Equal Combination

Qualifications/Requirements: Art & Design or Media Studies. A portfolio of work is required. AGNVQ, Pass. ND/C, National. SQAH, Individual consideration. SQAV, Individual consideration. IB, Individual consideration.

THE NOTTINGHAM TRENT UNIVERSITY N91

BSc Multimedia Production

UCAS Code: H685•Mode: 3/4 Years Single Subject

Qualifications/Requirements: GCE, A/AS: 20 points. An approved subject from restricted list. AGNVQ, Individual consideration. ND/C, 4 Distinctions. SQAH, Grades BBBB. SQAV, Individual consideration. IB, Individual consideration.

OXFORD BROOKES UNIVERSITY O66

Mod Accounting and Finance/Multimedia Systems

UCAS Code: GNP4•Mode: 3 Years Equal Combination

Qualifications/Requirements: Please refer to prospectus.

Mod Anthropology/Multimedia Systems

UCAS Code: GLP6•Mode: 3 Years Equal Combination

Qualifications/Requirements: Please refer to prospectus.

Mod Biological Chemistry/Multimedia Systems

UCAS Code: CG7P•Mode: 3 Years Equal Combination

Qualifications/Requirements: Please refer to prospectus.

Mod Biology/Multimedia Systems

UCAS Code: CG1P•Mode: 3 Years Equal Combination

Qualifications/Requirements: Please refer to prospectus.

Mod Business Administration and Management/Multimedia Systems

UCAS Code: GNP1•Mode: 3 Years Equal Combination

Qualifications/Requirements: Please refer to prospectus.

Mod Business Statistics/Multimedia Systems

UCAS Code: GG46•Mode: 3 Years Equal Combination

Qualifications/Requirements: Please refer to prospectus.

Mod Cell Biology/Multimedia Systems

UCAS Code: CGCP•Mode: 3 Years Equal Combination

Qualifications/Requirements: Please refer to prospectus.

Mod Cities and Society/Multimedia Systems

UCAS Code: LGHP•Mode: 3 Years Equal Combination

Qualifications/Requirements: Please refer to prospectus.

Mod Combined Studies/Multimedia Systems

UCAS Code: GYP4•Mode: 3 Years

Qualifications/Requirements: Please refer to prospectus.

Mod Complementary Therapies – Aromatherapy/Multimedia Systems

UCAS Code: GWP8•Mode: 3 Years Equal Combination

Qualifications/Requirements: Please refer to prospectus.

Mod Computer Systems/Multimedia Systems

UCAS Code: G610•Mode: 3 Years Single Subject

Qualifications/Requirements: Please refer to prospectus.

Mod Computing Mathematics/Multimedia Systems

UCAS Code: GGP9•Mode: 3 Years Equal Combination

Qualifications/Requirements: Please refer to prospectus.

Mod Computing Science/Multimedia Systems

UCAS Code: GGPM•Mode: 3 Years Equal Combination

Qualifications/Requirements: Please refer to prospectus.

Mod Computing/Multimedia Systems

UCAS Code: GGP5•Mode: 3 Years Equal Combination

Qualifications/Requirements: Please refer to prospectus.

Mod Ecology/Multimedia Systems

UCAS Code: CG9P•Mode: 3 Years Equal Combination

Qualifications/Requirements: Please refer to prospectus.

Mod Economics/Multimedia Systems
UCAS Code: GLP1•Mode: 3 Years Equal Combination
Qualifications/Requirements: Please refer to prospectus.

Mod Educational Studies/Multimedia Systems
UCAS Code: GX9X•Mode: 3 Years Equal Combination
Qualifications/Requirements: Please refer to prospectus.

Mod Electronics/Multimedia Systems
UCAS Code: GHP6•Mode: 3 Years Equal Combination
Qualifications/Requirements: Please refer to prospectus.

Mod English Studies/Multimedia Systems
UCAS Code: GQP3•Mode: 3 Years Equal Combination
Qualifications/Requirements: Please refer to prospectus.

Mod Environmental Chemistry/Multimedia Systems
UCAS Code: FG1P•Mode: 3 Years Equal Combination
Qualifications/Requirements: Please refer to prospectus.

Mod Environmental Policy/Multimedia Systems
UCAS Code: GKP3•Mode: 3 Years Equal Combination
Qualifications/Requirements: Please refer to prospectus.

Mod Environmental Sciences/Multimedia Systems
UCAS Code: FGXP•Mode: 3 Years Equal Combination
Qualifications/Requirements: Please refer to prospectus.

Mod European Culture and Society/Multimedia Systems
UCAS Code: GTP2•Mode: 3 Years Equal Combination
Qualifications/Requirements: Please refer to prospectus.

Mod Exercise and Health/Multimedia Systems
UCAS Code: BG6P•Mode: 3 Years Equal Combination
Qualifications/Requirements: Please refer to prospectus.

Mod Fine Art/Multimedia Systems
UCAS Code: GWP1•Mode: 3 Years Equal Combination
Qualifications/Requirements: Please refer to prospectus.

Mod Food Science and Nutrition/Multimedia Systems
UCAS Code: DG4P•Mode: 3 Years Equal Combination
Qualifications/Requirements: Please refer to prospectus.

Mod Geography/Multimedia Systems
UCAS Code: GLP8•Mode: 3 Years Equal Combination
Qualifications/Requirements: Please refer to prospectus.

Mod Geology/Multimedia Systems
UCAS Code: FG6P•Mode: 3 Years Equal Combination
Qualifications/Requirements: Please refer to prospectus.

Mod Geotechnics/Multimedia Systems
UCAS Code: GHP2•Mode: 3 Years Equal Combination
Qualifications/Requirements: Please refer to prospectus.

Mod History/Multimedia Systems
UCAS Code: GVP1•Mode: 3 Years Equal Combination
Qualifications/Requirements: Please refer to prospectus.

Mod Hospitality Management Studies/Multimedia Systems
UCAS Code: GNP7•Mode: 3 Years Equal Combination
Qualifications/Requirements: Please refer to prospectus.

Mod Human Biology/Multimedia Systems
UCAS Code: BG1P•Mode: 3 Years Equal Combination
Qualifications/Requirements: Please refer to prospectus.

Mod Information Systems/Multimedia Systems
UCAS Code: GGPN•Mode: 3 Years Equal Combination
Qualifications/Requirements: Please refer to prospectus.

Mod Law/Multimedia Systems
UCAS Code: GMP3•Mode: 3 Years Equal Combination
Qualifications/Requirements: Please refer to prospectus.

Mod Leisure Planning/Multimedia Systems
UCAS Code: GKPH•Mode: 3 Years Equal Combination
Qualifications/Requirements: Please refer to prospectus.

Mod Marketing Management/Multimedia Systems
UCAS Code: GNPN•Mode: 3 Years Equal Combination
Qualifications/Requirements: Please refer to prospectus.

Mod Mathematics/Multimedia Systems
UCAS Code: GWP3•Mode: 3 Years Equal Combination
Qualifications/Requirements: Please refer to prospectus.

Mod Multimedia Systems/Palliative Care
UCAS Code: BGRP•Mode: 3 Years Equal Combination
Qualifications/Requirements: Please refer to prospectus.

Mod Multimedia Systems/Physical Geography
UCAS Code: FGVP•Mode: 3 Years Equal Combination
Qualifications/Requirements: Please refer to prospectus.

Mod Multimedia Systems/Planning Studies
UCAS Code: GKP4•Mode: 3 Years Equal Combination
Qualifications/Requirements: Please refer to prospectus.

Mod Multimedia Systems/Politics
UCAS Code: GMP1•Mode: 3 Years Equal Combination
Qualifications/Requirements: Please refer to prospectus.

Mod Multimedia Systems/Psychology
UCAS Code: CG8P•Mode: 3 Years Equal Combination
Qualifications/Requirements: Please refer to prospectus.

Mod Multimedia Systems/Publishing
UCAS Code: GPP5•Mode: 3 Years Equal Combination
Qualifications/Requirements: Please refer to prospectus.

Mod Multimedia Systems/Rehabilitation
UCAS Code: BGTP•Mode: 3 Years Equal Combination
Qualifications/Requirements: Please refer to prospectus.

Mod Multimedia Systems/Retail Management
UCAS Code: GNP5•Mode: 3 Years Equal Combination
Qualifications/Requirements: Please refer to prospectus.

Mod Multimedia Systems/Sociology
UCAS Code: GLP3•Mode: 3 Years Equal Combination
Qualifications/Requirements: Please refer to prospectus.

Mod Multimedia Systems/Software Engineering
UCAS Code: GGP7•Mode: 3 Years Equal Combination
Qualifications/Requirements: Please refer to prospectus.

Mod Multimedia Systems/Spanish Studies
UCAS Code: GRP4•Mode: 3 Years Equal Combination
Qualifications/Requirements: Please refer to prospectus.

Mod Multimedia Systems/Statistics
UCAS Code: GGP4•Mode: 3 Years Equal Combination
Qualifications/Requirements: Please refer to prospectus.

Mod Multimedia Systems/Telecommunications
UCAS Code: GHPP•Mode: 3 Years Equal Combination
Qualifications/Requirements: Please refer to prospectus.

Mod Multimedia Systems/Tourism
UCAS Code: GPP7•Mode: 3 Years Equal Combination
Qualifications/Requirements: Please refer to prospectus.

Mod Multimedia Systems/Transport and Travel
UCAS Code: GNP9•Mode: 3 Years Equal Combination
Qualifications/Requirements: Please refer to prospectus.

Mod Multimedia Systems/Water Resources
UCAS Code: GHPF•Mode: 3 Years Equal Combination
Qualifications/Requirements: Please refer to prospectus.

OXFORDSHIRE SCHOOL OF ART AND DESIGN O80

HND Multimedia
UCAS Code: 52GE•Mode: 2 Years Equal Combination
Qualifications/Requirements: AGNVQ, Merit in Art & Design. ND/C, National. SQAH, Individual consideration. SQAV, Individual consideration. IB, Individual consideration.

HND Multimedia
UCAS Code: 52GW•Mode: 2 Years Equal Combination
Qualifications/Requirements: AGNVQ, Merit in Art & Design. ND/C, National. SQAH, Individual consideration. SQAV, Individual consideration. IB, Individual consideration.

UNIVERSITY OF PAISLEY P20

BSc Multimedia Systems
UCAS Code: G540•Mode: 3/4 Years Single Subject
Qualifications/Requirements: GCE, A level grades: DE. AGNVQ, Pass (in specific programmes). ND/C, Individual consideration. SQAH, Grades BCC including specific subject(s). SQAV, Higher National. IB, Pass in Diploma.

UNIVERSITY OF PLYMOUTH P60

BSc MediaLab Arts
UCAS Code: GW59•Mode: 4 Years Equal Combination
Qualifications/Requirements: GCE, A/AS: 18 points. An approved subject from restricted list. AGNVQ, Distinction (in specific programmes) with A/AS. ND/C, Individual consideration. SQAH, Grades BBBB. SQAV, Individual consideration. IB, Individual consideration.

BSc Multimedia, Production and Technology
UCAS Code: G610•Mode: 3/4 Years Single Subject
Qualifications/Requirements: GCE, A/AS: 14-18 points. An approved subject from restricted list. AGNVQ, Merit in Engineering in Business in Information Technology in Media. ND/C, 3 Merits. SQAH, Individual consideration. SQAV, Individual consideration. IB, Individual consideration.

HND Sound Engineering and Multimedia Integration
UCAS Code: 086H•Mode: 2 Years Single Subject
Qualifications/Requirements: GCE, A/AS: 4-6 points. AGNVQ, Individual consideration. ND/C, Individual consideration. SQAH, Individual consideration. SQAV, Individual consideration. IB, Individual consideration.

PLYMOUTH COLLEGE OF ART AND DESIGN P65

HND Multimedia

UCAS Code: 082E•Mode: 2 Years Single Subject

Qualifications/Requirements: Foundation Art Course, Art & Design or OTHER. AGNVQ, Pass. ND/C, National. SQAH, Individual consideration. SQAV, Individual consideration. IB, Individual consideration.

HND Multimedia

UCAS Code: 082W•Mode: 2 Years Single Subject

Qualifications/Requirements: OTHER, Art & Design or OTHER. AGNVQ, Pass. ND/C, National. SQAH, Individual consideration. SQAV, Individual consideration. IB, Individual consideration.

READING COLLEGE AND SCHOOL OF ARTS & DESIGN R10

BSc Multimedia Systems and Design

UCAS Code: GW52•Mode: 3 Years Equal Combination

Qualifications/Requirements: Please refer to prospectus.

HND Multimedia Design and Production

UCAS Code: 24WP•Mode: 2 Years Equal Combination

Qualifications/Requirements: Please refer to prospectus.

SAE TECHNOLOGY COLLEGE S05

BA Multimedia Arts

UCAS Code: W280•Mode: 3 Years Single Subject

Qualifications/Requirements: Please refer to prospectus.

SHEFFIELD COLLEGE S22

HND Multimedia Design & Production

UCAS Code: 082E•Mode: 2 Years Single Subject

Qualifications/Requirements: GCE, A/AS: points. AGNVQ, Distinction in Art & Design. ND/C, Individual consideration. SQAH, Individual consideration. SQAV, Individual consideration. IB, Individual consideration.

ST MARTIN'S COLLEGE, LANCASTER: AMBLESIDE: CARLISLE S24

BSc Multimedia Production & Applied Imaging

UCAS Code: HJ65•Mode: 3 Years Equal Combination

Qualifications/Requirements: GCE, A level grades: CD-CEE. Science. AGNVQ, Merit in Science. ND/C, 3 Merits and 2 Distinctions (in specific programmes). SQAH, Grades BCCC including specific subject(s). SQAV, Individual consideration. IB, 28 points.

SOUTHAMPTON INSTITUTE S30

BA Multimedia Design

UCAS Code: E240•Mode: 3 Years Single Subject

Qualifications/Requirements: GCE, A/AS: 10 points. OTHER. AGNVQ, Merit (in specific programmes). ND/C, Merit overall. SQAH, Grades CCCC. SQAV, National. IB, Pass in Diploma.

BA Multimedia Design

UCAS Code: W240•Mode: 3 Years Single Subject

Qualifications/Requirements: GCE, A/AS: 10 points. OTHER. AGNVQ, Merit (in specific programmes). ND/C, Merit overall. SQAH, Grades CCCC. SQAV, National. IB, Pass in Diploma.

SOUTH BANK UNIVERSITY S33

BEng Multimedia Engineering

UCAS Code: GH76•Mode: 3/4 Years Equal Combination

Qualifications/Requirements: GCE, A level grades: DD. Maths and Science or Computing. AGNVQ, Merit. ND/C, 5 Merits. SQAH, Individual consideration. SQAV, Individual consideration. IB, Individual consideration.

Mod Computing and Media Studies

UCAS Code: GP54•Mode: 3 Years Equal Combination

Qualifications/Requirements: GCE, A/AS: 14-18 points. Maths and English at A level. AGNVQ, Distinction. ND/C, 2 Merits and 4 Distinctions. SQAH, Individual consideration. SQAV, Individual consideration. IB, Individual consideration

ST HELENS COLLEGE S51

BA Multimedia Arts

UCAS Code: E560•Mode: 3 Years Single Subject

Qualifications/Requirements: GCE, A/AS: 8-10 points. Art & Design at A level. AGNVQ, Distinction. ND/C, Distinction Overall. SQAH, Individual consideration. SQAV, Individual consideration. IB, Individual consideration.

BA Multimedia Arts

UCAS Code: W560•Mode: 3 Years Single Subject

Qualifications/Requirements: GCE, A/AS: 8-10 points. Art & Design at A level. AGNVQ, Distinction. ND/C, Distinction Overall. SQAH, Individual consideration. SQAV, Individual consideration. IB, Individual consideration.

HND Multimedia (Design)

UCAS Code: 082E•Mode: 2 Years Single Subject

Qualifications/Requirements: GCE, A/AS: 2 points. OTHER. AGNVQ, Merit. ND/C, National. SQAH, Individual consideration. SQAV, Individual consideration. IB, Individual consideration.

HND Multimedia (Design)

UCAS Code: 082W•Mode: 2 Years Single Subject

Qualifications/Requirements: GCE, A/AS: 2 points. OTHER. AGNVQ, Merit. ND/C, National. SQAH, Individual consideration. SQAV, Individual consideration. IB, Individual consideration.

THE COLLEGE OF ST MARK AND ST JOHN S59

BA Information Technology/Media Studies

UCAS Code: G5P4•Mode: 3 Years Major/Minor

Qualifications/Requirements: GCE, A/AS: 4 points. AGNVQ, Merit. ND/C, Merit overall. SQAH, Individual consideration. SQAV, Individual consideration. IB, Pass in Diploma.

BA Media Studies/Information Technology

UCAS Code: P4G5•Mode: 3 Years Major/Minor

Qualifications/Requirements: GCE, A/AS: 16 points. AGNVQ, Merit. ND/C, Merit overall. SQAH, Individual consideration. SQAV, Individual consideration. IB, Individual consideration.

STAFFORDSHIRE UNIVERSITY S72

BA Design: Multimedia Graphics

UCAS Code: E212•Mode: 3 Years Single Subject

Qualifications/Requirements: GCE, A level grades: EE. OTHER. AGNVQ, Merit. ND/C, National. SQAH, Grades BBC. SQAV, Individual consideration. IB, Individual consideration.

BA Design: Multimedia Graphics

UCAS Code: W212•Mode: 3 Years Single Subject

Qualifications/Requirements: GCE, A level grades: EE. OTHER. AGNVQ, Merit. ND/C, National. SQAH, Grades BBC. SQAV, Individual consideration. IB, Individual consideration.

BA Design: Multimedia Graphics (includes Level Zero)

UCAS Code: E227•Mode: 4 Years Single Subject

Qualifications/Requirements: GCE, A level grades: EE. OTHER. AGNVQ, Merit. ND/C, National. SQAH, Grades DDD. SQAV, Individual consideration. IB, Individual consideration.

BA Design: Multimedia Graphics (includes Level Zero)

UCAS Code: W227•Mode: 4 Years Single Subject

Qualifications/Requirements: GCE, A level grades: EE. OTHER. AGNVQ, Merit. ND/C, National. SQAH, Grades DDD. SQAV, Individual consideration. IB, Individual consideration.

BA Design: Multimedia Graphics (includes Semester Zero)

UCAS Code: E247•Mode: 3/4 Years Single Subject

Qualifications/Requirements: GCE, A level grades: EE. OTHER. AGNVQ, Merit. ND/C, National. SQAH, Grades DDD. SQAV, Individual consideration. IB, Individual consideration.

BA Design: Multimedia Graphics (includes Semester Zero)

UCAS Code: W247•Mode: 3/4 Years Single Subject

Qualifications/Requirements: GCE, A level grades: EE. OTHER. AGNVQ, Merit. ND/C, National. SQAH, Grades DDD. SQAV, Individual consideration. IB, Individual consideration.

BA Foundation Interactive Multimedia

UCAS Code: GP5K•Mode: 4/5 Years Equal Combination

Qualifications/Requirements: Please refer to prospectus.

BA Interactive Multimedia

UCAS Code: GP54•Mode: 4 Years Equal Combination

Qualifications/Requirements: GCE, A/AS: 12 points. AGNVQ, Merit. ND/C, Individual consideration. SQAH, Grades CCC. SQAV, Individual consideration. IB, Individual consideration.

BSc Computing and Multimedia

UCAS Code: PG4M•Mode: 3 Years Equal Combination

Qualifications/Requirements: GCE, A/AS: 12 points. AGNVQ, Merit. ND/C, Individual consideration. SQAH, Grades CCC. IB, 27 points.

BSc Foundation Interactive Multimedia

UCAS Code: P408•Mode: 4 Years Single Subject

Qualifications/Requirements: Please refer to prospectus.

BSc Foundation Multimedia Computing

UCAS Code: G548•Mode: 5 Years Single Subject

Qualifications/Requirements: Please refer to prospectus.

BSc Multimedia Computing

UCAS Code: G541•Mode: 4 Years Single Subject

Qualifications/Requirements: GCE, A/AS: 12 points. AGNVQ, Merit. ND/C, Individual consideration. SQAH, Grades CCC. IB, 27 points.

BEng Foundation Multimedia Computing

UCAS Code: G547•Mode: 5 Years Single Subject

Qualifications/Requirements: Please refer to prospectus.

SUFFOLK COLLEGE: AN ACCREDITED COLLEGE OF THE UNIVERSITY OF EAST ANGLIA S81

BA Media Studies and Information Technology

UCAS Code: P4G5•Mode: 3 Years Major/Minor

Qualifications/Requirements: GCE, A level grades: CE. AGNVQ, Pass (in specific programmes). ND/C, National (in specific programmes). SQAH, Individual consideration. SQAV, Individual consideration. IB, Individual consideration.

BA Multimedia and Animation

UCAS Code: EG25•Mode: 3 Years Equal Combination

Qualifications/Requirements: GCE, A level grades: CE. OTHER. AGNVQ, Pass (in specific programmes). ND/C, National (in specific programmes). SQAH, Individual consideration. SQAV, Individual consideration. IB, Individual consideration.

BA Multimedia and Animation

UCAS Code: WG25•Mode: 3 Years Equal Combination

Qualifications/Requirements: GCE, A level grades: CE. OTHER. AGNVQ, Pass (in specific programmes). ND/C, National (in specific programmes). SQAH, Individual consideration. SQAV, Individual consideration. IB, Individual consideration.

BSc Information Technology with Media Studies

UCAS Code: G5P4•Mode: 3 Years Major/Minor

Qualifications/Requirements: GCE, A level grades: EE. AGNVQ, Pass (in specific programmes). ND/C, National (in specific programmes). SQAH, Individual consideration. SQAV, Individual consideration. IB, Individual consideration.

HND Multimedia

UCAS Code: 52GE•Mode: 2 Years Equal Combination

Qualifications/Requirements: GCE, A level grades: E. OTHER. AGNVQ, Pass (in specific programmes). ND/C, National (in specific programmes). SQAH, Individual consideration. SQAV, Individual consideration. IB, Individual consideration.

HND Multimedia

UCAS Code: 52GW•Mode: 2 Years Equal Combination

Qualifications/Requirements: GCE, A level grades: E. OTHER. AGNVQ, Pass (in specific programmes). ND/C, National (in specific programmes). SQAH, Individual consideration. SQAV, Individual consideration. IB, Individual consideration.

UNIVERSITY OF SUNDERLAND S84

BA Interactive Media

UCAS Code: G700•Mode: 3/4 Years Single Subject

Qualifications/Requirements: GCE, A/AS: 10 points. AGNVQ, Merit. ND/C, 4 Merits (in specific programmes). SQAH, Grades CCCC. SQAV, National. IB, 24 points.

BSc Broadcasting and Multimedia Technology

UCAS Code: HP6K•Mode: 3/4 Years Equal Combination

Qualifications/Requirements: GCE, A/AS: 14 points. AGNVQ, Merit. ND/C, 3 Merits (in specific programmes). SQAH, Grades CCC. SQAV, National. IB, 24 points.

BSc Broadcasting and Multimedia Technology (Foundation)

UCAS Code: PP3K•Mode: 3 Years Equal Combination

Qualifications/Requirements: GCE, A/AS: 4 points. AGNVQ, Pass. SQAH, Grades CC.

BSc Computer Studies with Media Studies

UCAS Code: G5P4•Mode: 3 Years Major/Minor

Qualifications/Requirements: GCE, A/AS: 24 points. AGNVQ, Individual consideration. ND/C, Individual consideration. SQAH, Individual consideration. SQAV, Individual consideration. IB, Individual consideration.

BSc Multimedia Technology

UCAS Code: GP5K•Mode: 3/4 Years Equal Combination

Qualifications/Requirements: GCE, A/AS: 12 points. AGNVQ, Merit. ND/C, 4 Merits (in specific programmes). SQAH, Grades CCCC. SQAV, National. IB, 24 points.

Mod Computer Studies and Media Studies

UCAS Code: GP54•Mode: 3/4 Years Equal Combination

Qualifications/Requirements: GCE, A/AS: 24 points. AGNVQ, Distinction (in specific programmes). ND/C, Distinction Overall. SQAH, Grades BBCCC. SQAV, Individual consideration. IB, Individual consideration.

SWANSEA INSTITUTE OF HIGHER EDUCATION S96

BA Multimedia

UCAS Code: EGF5•Mode: 3 Years Equal Combination

Qualifications/Requirements: GCE, A/AS: 12 points. AGNVQ, Pass. ND/C, National. SQAH, Grades CCCC. SQAV, Individual consideration. IB, 24 points.

BA Multimedia

UCAS Code: WGF5•Mode: 3 Years Equal Combination

Qualifications/Requirements: Please refer to prospectus.

BSc Multimedia

UCAS Code: EG45•Mode: 3 Years Equal Combination

Qualifications/Requirements: GCE, A level grades: DD. AGNVQ, Pass. ND/C, National. SQAH, Grades CCCC. SQAV, Individual consideration. IB, 24 points.

BSc Multimedia

UCAS Code: PG45•Mode: 3 Years Equal Combination

Qualifications/Requirements: GCE, A level grades: DD. AGNVQ, Pass. ND/C, National. SQAH, Grades CCCC. SQAV, Individual consideration. IB, 24 points.

BSc Multimedia (Foundation)
UCAS Code: EG4M•Mode: 4 Years Equal Combination

Qualifications/Requirements: Please refer to prospectus.

BSc Multimedia (Foundation)
UCAS Code: PG4M•Mode: 4 Years Equal Combination

Qualifications/Requirements: AGNVQ, Pass.

BEng Computer Systems (Multimedia)
UCAS Code: G500•Mode: 3 Years Single Subject

Qualifications/Requirements: GCE, A/AS: 4 points. AGNVQ, Pass. ND/C, National. SQAH, Grades CCCC. SQAV, Individual consideration. IB, 24 points.

BEng Computer Systems (Multimedia)
UCAS Code: G508•Mode: 4 Years Single Subject

Qualifications/Requirements: Please refer to prospectus.

SWINDON COLLEGE S98

HND Multimedia
UCAS Code: 52GE•Mode: 2 Years Equal Combination

Qualifications/Requirements: Foundation Art Course. A portfolio of work is required. AGNVQ, Merit (in specific programmes). ND/C, National. SQAH, Individual consideration. SQAV, Individual consideration. IB, Individual consideration.

HND Multimedia
UCAS Code: 52GW•Mode: 2 Years Equal Combination

Qualifications/Requirements: Foundation Art Course. A portfolio of work is required. AGNVQ, Merit (in specific programmes). ND/C, National. SQAH, Individual consideration. SQAV, Individual consideration. IB, Individual consideration.

TAMESIDE COLLEGE T10

HND Multimedia
UCAS Code: 042E•Mode: 2 Years Single Subject

Qualifications/Requirements: Please refer to prospectus.

HND Multimedia
UCAS Code: 042W•Mode: 2 Years Single Subject

Qualifications/Requirements: Please refer to prospectus.

UNIVERSITY OF TEESSIDE T20

BA Cultural Studies with Multimedia
UCAS Code: L3GM•Mode: 3 Years Major/Minor

Qualifications/Requirements: GCE, A/AS: 14-16 points. AGNVQ, Merit. ND/C, Individual consideration. SQAH, Individual consideration. SQAV, Individual consideration. IB, Individual consideration.

BA English with Multimedia
UCAS Code: Q3GN•Mode: 3 Years Major/Minor

Qualifications/Requirements: GCE, A/AS: 14-16 points. AGNVQ, Merit. ND/C, Individual consideration. SQAH, Individual consideration. SQAV, Individual consideration. IB, Individual consideration.

BA Graphic Design with Multimedia
UCAS Code: W2GM•Mode: 3 Years Major/Minor

Qualifications/Requirements: Art & Design or Design & Technology. A portfolio of work is required. AGNVQ, Merit (in specific programmes). ND/C, National. SQAH, Grades CCC. SQAV, National. IB, Individual consideration.

BSc Criminology with Multimedia
UCAS Code: MG35•Mode: 3 Years Equal Combination

Qualifications/Requirements: GCE, A/AS: 16 points. AGNVQ, Merit. ND/C, 4 Distinctions. SQAH, Grades CCCC. SQAV, Individual consideration. IB, Individual consideration.

BSc Economics with Multimedia
UCAS Code: L1GM•Mode: 3 Years Major/Minor

Qualifications/Requirements: GCE, A/AS: 10-12 points. Science. AGNVQ, Merit. ND/C, 4 Merits. SQAH, Grades CCC. SQAV, Individual consideration. IB, Individual consideration.

BSc Multimedia
UCAS Code: G540•Mode: 3/4 Years Single Subject

Qualifications/Requirements: GCE, A/AS: 12-16 points. AGNVQ, Merit. ND/C, Individual consideration. SQAH, Grades CCCC. SQAV, Individual consideration. IB, Individual consideration.

BSc Multimedia with Applied Languages
UCAS Code: G5T9•Mode: 3 Years Major/Minor

Qualifications/Requirements: GCE, A/AS: 12-16 points. AGNVQ, Merit. ND/C, Individual consideration. SQAH, Grades CCCC. SQAV, Individual consideration. IB, Individual consideration.

BSc Multimedia with Business Studies
UCAS Code: G5N1•Mode: 3 Years Major/Minor

Qualifications/Requirements: GCE, A/AS: 12-16 points. AGNVQ, Merit. ND/C, Individual consideration. SQAH, Grades CCCC. SQAV, Individual consideration. IB, Individual consideration.

BSc Multimedia with Cultural Studies
UCAS Code: G5L3•Mode: 3 Years Major/Minor

Qualifications/Requirements: GCE, A/AS: 12-16 points. AGNVQ, Merit. ND/C, Individual consideration. SQAH, Grades CCCC. SQAV, Individual consideration. IB, Individual consideration.

BSc Multimedia with Graphic Design

UCAS Code: G5WF•Mode: 3 Years Major/Minor

Qualifications/Requirements: GCE, A/AS: 12-16 points. AGNVQ, Merit. ND/C, Individual consideration. SQAH, Grades CCCC. SQAV, Individual consideration. IB, Individual consideration.

BSc Multimedia with Media Technology and Production

UCAS Code: G5P4•Mode: 3/4 Years Major/Minor

Qualifications/Requirements: GCE, A/AS: 12-16 points. AGNVQ, Merit. ND/C, Individual consideration. SQAH, Grades CCCC. SQAV, Individual consideration. IB, Individual consideration.

BSc Sports Studies with Multimedia

UCAS Code: B6GT•Mode: 3 Years Major/Minor

Qualifications/Requirements: GCE, A/AS: 16 points. AGNVQ, Merit. ND/C, 2 Distinctions. SQAH, Grades CCCC. SQAV, Individual consideration. IB, Individual consideration.

HND Multimedia

UCAS Code: 045G•Mode: 2 Years Single Subject

Qualifications/Requirements: GCE, A/AS: 4 points. AGNVQ, Pass. ND/C, National. SQAH, Grades CC. SQAV, Individual consideration. IB, Individual consideration.

THAMES VALLEY UNIVERSITY T40

BA Digital Arts with Multimedia Computing

UCAS Code: W9GM•Mode: 2/3 Years Major/Minor

Qualifications/Requirements: GCE, A/AS: 8 points. AGNVQ, Merit. ND/C, Merit overall. SQAH, Grades CCCC. IB, 26 points.

BA Media Arts with Multimedia Computing

UCAS Code: W9GN•Mode: 2/3 Years Major/Minor

Qualifications/Requirements: GCE, A/AS: 10 points. AGNVQ, Merit or Distinction. ND/C, Merit overall. SQAH, Grades CCCC. IB, 26 points.

BA Multimedia Computing with Business

UCAS Code: G5N1•Mode: 2/3 Years Major/Minor

Qualifications/Requirements: GCE, A/AS: 8-12 points. AGNVQ, Merit. ND/C, Merit overall. SQAH, Grades CCC. IB, 26 points.

BA Multimedia Computing with English as a Foreign Language

UCAS Code: G5Q1•Mode: 2/3 Years Major/Minor

Qualifications/Requirements: GCE, A/AS: 8-12 points. AGNVQ, Merit. ND/C, Merit overall. SQAH, Grades CCC. IB, 26 points.

BA Multimedia Computing with Human Resource Management

UCAS Code: G5N6•Mode: 3 Years Major/Minor

Qualifications/Requirements: Please refer to prospectus.

BA Multimedia Computing with Marketing

UCAS Code: G5N5•Mode: 3 Years Major/Minor

Qualifications/Requirements: GCE, A/AS: 8-12 points. AGNVQ, Merit. ND/C, Merit overall. SQAH, Grades CCC. IB, 26 points.

BA Multimedia Computing with Radio Broadcasting

UCAS Code: G5H6•Mode: 3 Years Major/Minor

Qualifications/Requirements: GCE, A/AS: 8-12 points. AGNVQ, Merit. ND/C, Merit overall. SQAH, Grades CCC. IB, 26 points.

BA/BSc Multimedia Computing with Information and Knowledge Management

UCAS Code: G5P2•Mode: 3 Years Major/Minor

Qualifications/Requirements: Please refer to prospectus.

BA/BSc Multimedia Computing with Video Production

UCAS Code: G5WM•Mode: 2/3 Years Major/Minor

Qualifications/Requirements: GCE, A/AS: 8 points. Maths. AGNVQ, Merit. ND/C, Merit overall. SQAH, Grades CCCC. IB, 26 points.

BSc Information Systems with Multimedia Computing

UCAS Code: G523•Mode: 2/3 Years Single Subject

Qualifications/Requirements: GCE, A/AS: 8 points. AGNVQ, Merit. ND/C, Merit overall. SQAH, Grades CCCC. IB, 26 points.

BSc Multimedia Computing with Advertising

UCAS Code: G5P3•Mode: 2/3 Years Major/Minor

Qualifications/Requirements: GCE, A/AS: 8 points. Maths. AGNVQ, Merit. ND/C, Merit overall. SQAH, Grades CCCC. IB, 26 points.

BSc Multimedia Computing with Digital Arts

UCAS Code: G5W9•Mode: 2/3 Years Major/Minor

Qualifications/Requirements: GCE, A/AS: 8 points. Maths. AGNVQ, Merit. ND/C, Merit overall. SQAH, Grades CCCC. IB, 26 points.

BSc Multimedia Computing with Information Systems

UCAS Code: G5GM•Mode: 2/3 Years Major/Minor

Qualifications/Requirements: GCE, A/AS: 8 points. Maths. AGNVQ, Merit. ND/C, Merit overall. SQAH, Grades CCCC. IB, 26 points.

BSc Multimedia Computing with Media Studies

UCAS Code: G5WX•Mode: 2/3 Years Major/Minor

Qualifications/Requirements: GCE, A/AS: 8 points. Maths. AGNVQ, Merit. ND/C, Merit overall. SQAH, Grades CCCC. IB, 26 points.

BSc Multimedia Computing with Sound and Music Recording

UCAS Code: G5WH•Mode: 2/3 Years Major/Minor

Qualifications/Requirements: GCE, A/AS: 8 points. Maths. AGNVQ, Merit. ND/C, Merit overall. SQAH, Grades CCCC. IB, 26 points.

Mod Advertising with Multimedia Computing

UCAS Code: P3G5•Mode: 2/3 Years Major/Minor

Qualifications/Requirements: GCE, A/AS: 8 points. AGNVQ, Merit. ND/C, Merit overall. SQAH, Grades CCCC. IB, 26 points.

Mod Journalism with Multimedia Computing

UCAS Code: P6G5•Mode: 2/3 Years Major/Minor

Qualifications/Requirements: GCE, A/AS: 8 points. AGNVQ, Merit. ND/C, Merit overall. SQAH, Grades CCCC. IB, 26 points.

Mod Multimedia Computing with Art and Design History

UCAS Code: G5VK•Mode: 2/3 Years Major/Minor

Qualifications/Requirements: GCE, A/AS: 8 points. Maths. AGNVQ, Merit. ND/C, Merit overall. SQAH, Grades CCCC. IB, 26 points.

Mod Multimedia Computing with Digital Animation

UCAS Code: G5W2•Mode: 2/3 Years Major/Minor

Qualifications/Requirements: GCE, A/AS: 8 points. Maths. AGNVQ, Merit. ND/C, Merit overall. SQAH, Grades CCCC. IB, 26 points.

Mod Multimedia Computing with Music

UCAS Code: G5W3•Mode: 2/3 Years Major/Minor

Qualifications/Requirements: GCE, A/AS: 8 points. Maths. AGNVQ, Merit. ND/C, Merit overall. SQAH, Grades CCCC. IB, 26 points.

Mod Multimedia Computing with Photography

UCAS Code: G5W5•Mode: 2/3 Years Major/Minor

Qualifications/Requirements: GCE, A/AS: 8 points. Maths. AGNVQ, Merit. ND/C, Merit overall. SQAH, Grades CCCC. IB, 26 points.

UNIVERSITY OF ULSTER U20

BSc Interactive Multimedia Design

UCAS Code: G532•Mode: 4 Years Single Subject

Qualifications/Requirements: GCE, A level grades: BBC. AGNVQ, Distinction with 6 additional units or with A level. ND/C, Merit overall and 4 Distinctions. SQAH, Grades ABBB. SQAV, Individual consideration. IB, 32 points.

WARWICKSHIRE COLLEGE, ROYAL LEAMINGTON SPA AND MORETON MORRELL W25

HND Design (Multimedia)

UCAS Code: 065E•Mode: 2 Years Single Subject

Qualifications/Requirements: AGNVQ, Individual consideration. ND/C, Individual consideration. SQAH, Individual consideration. IB, Individual consideration.

HNC Design (Multimedia)

UCAS Code: 165E•Mode: 1 Years Single Subject

Qualifications/Requirements: AGNVQ, Individual consideration. ND/C, Individual consideration. SQAH, Individual consideration. IB, Individual consideration.

WEST HERTS COLLEGE, WATFORD W40

HND Multimedia

UCAS Code: 24PE•Mode: 2 Years Equal Combination

Qualifications/Requirements: GCE, A/AS: 2 points.

UNIVERSITY OF WESTMINSTER W50

BSc Multimedia Computing

UCAS Code: GP54•Mode: 3 Years Equal Combination

Qualifications/Requirements: GCE, A/AS: 12 points. AGNVQ, Distinction. ND/C, 5 Merits. SQAH, Individual consideration. SQAV, Individual consideration. IB, Individual consideration.

HND Multimedia Computing

UCAS Code: 045G•Mode: 2 Years Single Subject

Qualifications/Requirements: Please refer to prospectus.

WESTMINSTER COLLEGE W52

HND Design (Multimedia)

UCAS Code: 002W•Mode: 2 Years Single Subject

Qualifications/Requirements: GCE, A/AS: 4 points. OTHER. AGNVQ, Merit (in specific programmes). ND/C, Merit overall (in specific programmes). SQAH, Individual consideration. SQAV, Individual consideration. IB, Individual consideration.

WIGAN AND LEIGH COLLEGE W67

HND Multimedia Computing

UCAS Code: 45PG•Mode: 2 Years Equal Combination

Qualifications/Requirements: Please refer to prospectus.

UNIVERSITY OF WOLVERHAMPTON W75

BA Design for Multimedia

UCAS Code: E201•Mode: 4 Years Single Subject

Qualifications/Requirements: GCE, A/AS: 12 points. Art, Art & Design or Design & Technology. A portfolio of work is required. AGNVQ, Merit. ND/C, 4 Merits. SQAH, Grades BBBB. IB, 24 points.

BA Design for Multimedia

UCAS Code: W201•Mode: 4 Years Single Subject

Qualifications/Requirements: GCE, A/AS: 12 points. Art, Art & Design or Design & Technology. A portfolio of work is required. AGNVQ, Merit. ND/C, 4 Merits. SQAH, Grades BBBB. IB, 24 points.

BA Interactive Multimedia Communication

UCAS Code: E270•Mode: 4 Years Single Subject

Qualifications/Requirements: GCE, A/AS: 18 points. AGNVQ, Distinction. ND/C, 4 Merits. SQAH, Grades AAAA. SQAV, Individual consideration. IB, 24 points.

BA Interactive Multimedia Communication

UCAS Code: P300•Mode: 4 Years Single Subject

Qualifications/Requirements: GCE, A/AS: 18 points. AGNVQ, Distinction. ND/C, 4 Merits. SQAH, Grades AAAA. SQAV, Individual consideration. IB, 24 points.

BA Interactive Multimedia Communication (4 years SW)

UCAS Code: W270•Mode: 4 Years Single Subject

Qualifications/Requirements: GCE, A/AS: 18 points. AGNVQ, Distinction. ND/C, 4 Merits. SQAH, Grades AAAA. SQAV, Individual consideration. IB, 24 points.

BSc Computer Science (Multimedia Technology) (4 years SW)

UCAS Code: G560•Mode: 4 Years Single Subject

Qualifications/Requirements: GCE, A/AS: 10 points. AGNVQ, Merit. ND/C, 4 Merits. SQAH, Grades CCCC. SQAV, Individual consideration. IB, 24 points.

WORCESTER COLLEGE OF TECHNOLOGY W81

HND Multimedia

UCAS Code: 012E•Mode: 2 Years Single Subject

Qualifications/Requirements: GCE, A/AS: 2 points. AGNVQ, Merit. ND/C, National.

YORKSHIRE COAST COLLEGE OF FURTHER AND HIGHER EDUCATION Y80

HND Multimedia (Design)

UCAS Code: 25WE•Mode: 2 Years Equal Combination

Qualifications/Requirements: Please refer to prospectus.

HND Multimedia (Design)

UCAS Code: 25WG•Mode: 2 Years Equal Combination

Qualifications/Requirements: Please refer to prospectus.

SOFTWARE ENGINEERING

UNIVERSITY OF ABERTAY DUNDEE A30

HND Computing: Software Development

UCAS Code: 005G • Mode: 2 Years Single Subject

Qualifications/Requirements: GCE, A level grades: D. AGNVQ, Individual consideration. ND/C, Individual consideration. SQAH, Grades CC. SQAV, Individual consideration. IB, Individual consideration.

THE UNIVERSITY OF WALES, ABERYSTWYTH A40

BEng Software Engineering (including integrated industrial & professional training)

UCAS Code: G700 • Mode: 4 Years Single Subject

Qualifications/Requirements: GCE, A/AS: 20 points. AGNVQ, Merit. ND/C, 3 Merits and 2 Distinctions. SQAH, Grades BBBCC. SQAV, Individual consideration. IB, 30 points.

MEng Software Engineering (including integrated industrial & professional training)

UCAS Code: G701 • Mode: 5 Years Single Subject

Qualifications/Requirements: GCE, A/AS: 24 points. AGNVQ, Not normally sufficient. ND/C, Not normally sufficient. SQAH, Not normally sufficient. SQAV, Not normally sufficient. IB, Not normally sufficient.

ANGLIA POLYTECHNIC UNIVERSITY A60

BSc Computer Software Development

UCAS Code: G700 • Mode: 3 Years Single Subject

Qualifications/Requirements: GCE, A/AS: 14 points. Science. AGNVQ, Merit (in specific programmes). ND/C, 6 Merits. SQAH, Grades BBCC. SQAV, Individual consideration. IB, Pass in Diploma.

BSc Software Engineering

UCAS Code: G701 • Mode: 3 Years Single Subject

Qualifications/Requirements: GCE, A/AS: 14 points. Science at A level. AGNVQ, Merit (in specific programmes). ND/C, 6 Merits. SQAH, Grades BBCC. SQAV, Individual consideration. IB, Pass in Diploma.

UNIVERSITY OF BATH B16

BSc Computer Software Theory

UCAS Code: G700 • Mode: 3 Years Single Subject

Qualifications/Requirements: GCE, A/AS: 22-24 points. Maths at A level. AGNVQ, Individual consideration. ND/C, Individual consideration. SQAH, CSYS required. SQAV, Individual consideration. IB, 32 points.

BSc Computer Software Theory (4 year SW)

UCAS Code: G701 • Mode: 4 Years Single Subject

Qualifications/Requirements: GCE, A/AS: 22-24 points. Maths at A level. AGNVQ, Individual consideration. ND/C, Individual consideration. SQAH, CSYS required. SQAV, Individual consideration. IB, 32 points.

HND Computing (Business Information Technology)

UCAS Code: 027G • Mode: 2 Years Single Subject

Qualifications/Requirements: GCE, A/AS: 2-6 points. AGNVQ, Pass. ND/C, National. SQAH, Individual consideration. SQAV, Individual consideration. IB, Individual consideration.

HND Computing (Software Engineering)

UCAS Code: 007G • Mode: 2 Years Single Subject

Qualifications/Requirements: Please refer to prospectus.

HNC Computing (Business Information Technology)

UCAS Code: 127G • Mode: 1 Years Single Subject

Qualifications/Requirements: Please refer to prospectus.

BELL COLLEGE OF TECHNOLOGY B26

HND Business Information Systems

UCAS Code: 027G • Mode: 2 Years Single Subject

Qualifications/Requirements: GCE, A level grades - DD-D. An approved subject from restricted list. AGNVQ, Pass (in specific programmes). ND/C, National (in specific programmes). SQAH, Grades CC including specific subject(s). SQAV, National including specific subject(s). IB, Individual consideration

HND Computing (Software Development)

UCAS Code: 007G • Mode: 2 Years Single Subject

Qualifications/Requirements: GCE, A level grades: EE. Maths, English or Computing. AGNVQ, Individual consideration. ND/C, Individual consideration. SQAH, Grades CC including specific subject(s). SQAV, Individual consideration. IB, Individual consideration.

THE UNIVERSITY OF BIRMINGHAM B32

BSc Computer Science/Software Engineering

UCAS Code: GG57 • Mode: 3 Years Equal Combination

Qualifications/Requirements: GCE, A level grades: BBB. Science at A level. AGNVQ, Individual consideration. ND/C, Individual consideration. SQAH, Individual consideration. SQAV, Individual consideration. IB, 32 points.

BSc Computer Science/Software Engineering with Business Studies
UCAS Code: G5N1 • Mode: 3 Years Major/Minor

Qualifications/Requirements: GCE, A level grades: BBB. Science at A level. AGNVQ, Individual consideration. ND/C, Individual consideration. SQAH, Individual consideration. SQAV, Individual consideration. IB, 32 points.

MEng Electronic and Software Engineering (e)
UCAS Code: GH76 • Mode: 3/4 Years Equal Combination

Qualifications/Requirements: GCE, A level grades: BBB-BCC. Maths at A level and Physics. AGNVQ, Individual consideration. ND/C, 2 Merits and 4 Distinctions (in specific programmes). SQAH, Individual consideration. SQAV, Individual consideration. IB, 32 points.

BLACKPOOL AND THE FYLDE COLLEGE B41

HND Computing (Business Information Technology)
UCAS Code: 027G • Mode: 2 Years Single Subject

Qualifications/Requirements: GCE, A/AS: 2 points. AGNVQ, Pass. ND/C, National. SQAH, Individual consideration. SQAV, Individual consideration. IB, Individual consideration.

HND Computing (Software Engineering)
UCAS Code: 007G • Mode: 2 Years Single Subject

Qualifications/Requirements: GCE, A/AS: 2 points. AGNVQ, Pass. ND/C, National. SQAH, Individual consideration. SQAV, Individual consideration. IB, Individual consideration.

BOLTON INSTITUTE OF HIGHER EDUCATION B44

BA/BSc Leisure Computing Technology and Simulation/Virtual Environment
UCAS Code: G740 • Mode: 3 Years Single Subject

Qualifications/Requirements: Please refer to prospectus.

BSc Leisure Computing Technology
UCAS Code: G710 • Mode: 3 Years Single Subject

Qualifications/Requirements: GCE, A/AS: 10 points. AGNVQ, Merit. ND/C, 3 Merits. SQAH, Individual consideration. SQAV, Individual consideration. IB, Individual consideration.

BSc Software Development
UCAS Code: G700 • Mode: 3 Years Single Subject

Qualifications/Requirements: GCE, A level grades: CD. AGNVQ, Merit. ND/C, Merit overall. SQAH, Grades BBCC. SQAV, Individual consideration. IB, 24 points.

Mod Business Information Systems and Community Studies
UCAS Code: GL75 • Mode: 3 Years Equal Combination

Qualifications/Requirements: GCE, A level grades - CD. AGNVQ, Merit. ND/C, Merit overall. SQAH, Grades BBCC. SQAV, Individual consideration. IB, 24 points.

Mod Business Studies and Leisure Computing Technology
UCAS Code: GN71 • Mode: 3 Years Equal Combination

Qualifications/Requirements: GCE, A/AS: 10 points. AGNVQ, Merit. ND/C, 3 Merits. SQAH, Individual consideration. SQAV, Individual consideration. IB, Individual consideration.

Mod Civil Engineering and Simulation/Virtual Environment
UCAS Code: HG27 • Mode: 3 Years Equal Combination

Qualifications/Requirements: GCE, A/AS: 10 points. Maths or Science. AGNVQ, Merit (in specific programmes). ND/C, 3 Merits. SQAH, Individual consideration. SQAV, Individual consideration. IB, Individual consideration.

Mod Leisure Computing Technical and Manufacturing Systems Design
UCAS Code: GH77 • Mode: 3 Years Equal Combination

Qualifications/Requirements: GCE, A/AS: 10 points. AGNVQ, Merit (in specific programmes). ND/C, 3 Merits. SQAH, Individual consideration. SQAV, Individual consideration. IB, Individual consideration.

Mod Leisure Computing Technology
UCAS Code: G711 • Mode: 3/4 Years Single Subject

Qualifications/Requirements: GCE, A/AS: 4 points. AGNVQ, Pass. ND/C, National. SQAH, Individual consideration. SQAV, Individual consideration. IB, Individual consideration.

Mod Leisure Computing Technology and Marketing
UCAS Code: GN75 • Mode: 3 Years Equal Combination

Qualifications/Requirements: GCE, A/AS: 10 points. AGNVQ, Merit. ND/C, 3 Merits. SQAH, Individual consideration. SQAV, Individual consideration. IB, Individual consideration.

Mod Leisure Computing Technology and Mathematics
UCAS Code: GG71 • Mode: 3 Years Equal Combination

Qualifications/Requirements: GCE, A/AS: 10 points. Maths at A level. AGNVQ, Merit. ND/C, 3 Merits. SQAH, Individual consideration. SQAV, Individual consideration. IB, Individual consideration.

Mod Leisure Computing Technology and Music Technology

UCAS Code: GW7H • Mode: 3 Years Equal Combination

Qualifications/Requirements: GCE, A/AS: 10 points. AGNVQ, Merit. ND/C, 3 Merits. SQAH, Individual consideration. SQAV, Individual consideration. IB, Individual consideration.

Mod Leisure Computing Technology and Product Design

UCAS Code: GH7T • Mode: 3 Years Equal Combination

Qualifications/Requirements: GCE, A/AS: 10 points. Maths, Science, OTHER, Design & Technology or Business Studies. AGNVQ, Merit (in specific programmes). ND/C, 3 Merits. SQAH, Individual consideration. SQAV, Individual consideration. IB, Individual consideration.

Mod Leisure Computing Technology and Statistics

UCAS Code: GG74 • Mode: 3 Years Equal Combination

Qualifications/Requirements: GCE, A/AS: 10 points. AGNVQ, Merit. ND/C, 3 Merits. SQAH, Individual consideration. SQAV, Individual consideration. IB, Individual consideration.

Mod Manufacturing Systems Design and Simulation/Virtual Environment

UCAS Code: HG77 • Mode: 3 Years Equal Combination

Qualifications/Requirements: GCE, A/AS: 10 points. Maths or Science. AGNVQ, Merit (in specific programmes). ND/C, 3 Merits. SQAH, Individual consideration. SQAV, Individual consideration. IB, Individual consideration.

BOURNEMOUTH UNIVERSITY B50

BSc Software Engineering Management

UCAS Code: G700 • Mode: 4 Years Single Subject

Qualifications/Requirements: GCE, A/AS: 16-18 points. AGNVQ, Distinction (in specific programmes). ND/C, Merit overall and Distinction. SQAH, Grades CCCC. SQAV, Individual consideration. IB, 30 points.

THE UNIVERSITY OF BRADFORD B56

BEng Software Engineering

UCAS Code: G701 • Mode: 3 Years Single Subject

Qualifications/Requirements: GCE, A/AS: 18 points. AGNVQ, Distinction (in specific programmes) with 4 additional units or with A/AS. ND/C, 3 Merits and 2 Distinctions. SQAH, Individual consideration. SQAV, Individual consideration. IB, Individual consideration.

BEng Software Engineering

UCAS Code: G702 • Mode: 4 Years Single Subject

Qualifications/Requirements: GCE, A/AS: 18 points. AGNVQ, Distinction (in specific programmes) with 4 additional units or with A/AS. ND/C, 3 Merits and 2 Distinctions (in specific programmes). SQAH, Individual consideration. SQAV, Individual consideration. IB, Individual consideration.

MEng Software Engineering

UCAS Code: G700 • Mode: 4 Years Single Subject

Qualifications/Requirements: GCE, A/AS: 24 points. AGNVQ, Distinction in Science with 4 additional units or with A/AS. ND/C, 3 Merits and 2 Distinctions. SQAH, Individual consideration. SQAV, Individual consideration. IB, Individual consideration.

BRADFORD COLLEGE B60

BA Business Systems Management

UCAS Code: N1G7 • Mode: 3 Years Major/Minor

Qualifications/Requirements: GCE, A/AS: 8 points. AGNVQ, Merit. ND/C, Merits and Distinction. SQAH, Individual consideration. SQAV, Individual consideration. IB, Pass in Diploma.

HND Business Information Technology

UCAS Code: 027G • Mode: 2 Years Single Subject

Qualifications/Requirements: GCE, A/AS: 4 points. AGNVQ, Merit. ND/C, 3 Merits. SQAH, Individual consideration. SQAV, National. IB, Pass in Diploma.

HND Software Engineering

UCAS Code: 007G • Mode: 2 Years Single Subject

Qualifications/Requirements: GCE, A/AS: 2 points. AGNVQ, Pass. ND/C, National (in specific programmes). SQAH, Individual consideration. SQAV, Individual consideration.

UNIVERSITY OF BRIGHTON B72

BSc Software Engineering

UCAS Code: G700 • Mode: 4 Years Single Subject

Qualifications/Requirements: GCE, A/AS: 18 points. AGNVQ, Merit (in specific programmes) with 6 additional units or with A level. ND/C, Distinction Overall (in specific programmes). SQAH, Grades BBBC including specific subject(s). SQAV, Individual consideration. IB, 27 points.

BEng Electronic Systems with Software

UCAS Code: H6G7 • Mode: 3/4 Years Major/Minor

Qualifications/Requirements: GCE, A/AS: 10-18 points. Maths and Physics, Science or Design & Technology. AGNVQ, Merit (in specific programmes) with A level or. ND/C, 5 Merits (in specific programmes). SQAH, Grades BBBBC including specific subject(s). SQAV, Individual consideration. IB, Pass in Diploma including specific subjects.

HND Computing (Software Engineering)

UCAS Code: 007G • Mode: 2 Years Single Subject

Qualifications/Requirements: GCE, A/AS: 4 points. AGNVQ, Merit. ND/C, Merit overall. SQAH, Grades CC. SQAV, Individual consideration. IB, Pass in Diploma.

UNIVERSITY OF THE WEST OF ENGLAND, BRISTOL B80

BSc Software Engineering

UCAS Code: G700 • Mode: 4 Years Single Subject

Qualifications/Requirements: GCE, A/AS: 16 points. Computing, Maths, Science or Electronics. AGNVQ, Merit (in specific programmes). ND/C, Merit overall. SQAH, Grades BBB including specific subject(s). SQAV, National including specific subject(s). IB, 24 points.

BRUNEL UNIVERSITY B84

Mod Leisure Management/Computer Studies

UCAS Code: N7G5 • Mode: 3 Years Major/Minor

Qualifications/Requirements: GCE, A/AS: 18 points. AGNVQ, Distinction. ND/C, Merit overall (in specific programmes). SQAH, Grades BCCC including specific subject(s). SQAV, Individual consideration. IB, 26 points.

BUCKINGHAMSHIRE CHILTERNS UNIVERSITY COLLEGE B94

BSc Business Information Technology

UCAS Code: G720 • Mode: 3 Years Single Subject

Qualifications/Requirements: GCE, A/AS: 8-10 points. AGNVQ, Merit. ND/C, Merit overall. SQAH, Grades CCCC. SQAV, Individual consideration. IB, Individual consideration.

BSc Business Information Technology (1/2 years conversion to Degree)

UCAS Code: G721 • Mode: 1/2 Years Single Subject

Qualifications/Requirements: ND/C, Higher National. SQAV, Individual consideration.

BSc Business Information Technology with Image Processing

UCAS Code: G7W2 • Mode: 3 Years Major/Minor

Qualifications/Requirements: GCE, A/AS: 8-10 points. AGNVQ, Merit. ND/C, Merit overall. SQAH, Grades CCCC. SQAV, Individual consideration. IB, Individual consideration.

BSc Business Information Technology with Marketing

UCAS Code: G7N5 • Mode: 3 Years Major/Minor

Qualifications/Requirements: GCE, A/AS: 8-10 points. AGNVQ, Merit. ND/C, Merit overall. SQAH, Grades CCCC. SQAV, Individual consideration. IB, Individual consideration.

BSc Business Information Technology with Multimedia

UCAS Code: G7P4 • Mode: 3 Years Major/Minor

Qualifications/Requirements: GCE, A/AS: 8-10 points. AGNVQ, Merit. ND/C, Merit overall. SQAH, Grades CCCC. SQAV, Individual consideration. IB, Individual consideration.

HND Business Information Technology

UCAS Code: 027G • Mode: 2 Years Single Subject

Qualifications/Requirements: GCE, A/AS: 4-6 points. AGNVQ, Pass. ND/C, National. SQAH, Grades CCC. SQAV, Individual consideration. IB, Individual consideration.

CANTERBURY CHRIST CHURCH UNIVERSITY COLLEGE C10

HND Software Development

UCAS Code: 107G • Mode: 2 Years Single Subject

Qualifications/Requirements: GCE, A level grades: D. AGNVQ, Pass. ND/C, National. SQAH, Individual consideration. SQAV, Individual consideration. IB, 24 points.

UNIVERSITY OF WALES INSTITUTE, CARDIFF C20

BA Business Studies with Information Systems Management

UCAS Code: N1G7 • Mode: 3 Years Major/Minor

Qualifications/Requirements: GCE, A/AS: 12-14 points. AGNVQ, Distinction. ND/C, 4 Merits and 2 Distinctions. SQAH, Grades CCCC. SQAV, Individual consideration. IB, Individual consideration.

HND Business Information Technology

UCAS Code: 027G • Mode: 2 Years Single Subject

Qualifications/Requirements: GCE, A/AS: 2 points. AGNVQ, Merit (in specific programmes). ND/C, 3 Merits. SQAH, Grades CC. SQAV, Individual consideration. IB, Individual consideration.

CARMARTHENSHIRE COLLEGE C22

HND Business Information Technology

UCAS Code: 027G • Mode: 2 Years Single Subject

Qualifications/Requirements: AGNVQ, Individual consideration. ND/C, Individual consideration. SQAH, Individual consideration. SQAV, Individual consideration. IB, Individual consideration.

HND Software Engineering

UCAS Code: 006G • Mode: 2 Years Single Subject

Qualifications/Requirements: AGNVQ, Individual consideration. ND/C, Individual consideration. SQAH, Individual consideration. SQAV, Individual consideration. IB, Individual consideration.

UNIVERSITY OF CENTRAL ENGLAND IN BIRMINGHAM C25

BSc Business Information Technology

UCAS Code: G720 • Mode: 3 Years Single Subject

Qualifications/Requirements: GCE, A/AS: 12 points. AGNVQ, Merit. ND/C, Merit overall. SQAH, Individual consideration. SQAV, Individual consideration. IB, Individual consideration.

BSc Software Engineering

UCAS Code: G700 • Mode: 3 Years Single Subject

Qualifications/Requirements: GCE, A/AS: 12 points. AGNVQ, Merit. ND/C, Merit overall. SQAH, Individual consideration. SQAV, Individual consideration. IB, Individual consideration.

BEng Software Design for Engineering Systems

UCAS Code: G701 • Mode: 3/4 Years Single Subject

Qualifications/Requirements: GCE, A/AS: 16 points. Maths at A level. AGNVQ, Merit in Engineering with 1 additional unit. ND/C, 5 Merits (in specific programmes). SQAH, Grades BBCCC including specific subject(s). SQAV, Individual consideration. IB, 28 points.

BEng Software Design for Engineering Systems Foundation Year

UCAS Code: G709 • Mode: 4/5 Years Single Subject

Qualifications/Requirements: GCE, A/AS: 4 points. AGNVQ, Pass. ND/C, 1 Merit. SQAH, Grades CC. SQAV, Individual consideration. IB, 24 points.

HND Business Information Technology

UCAS Code: 027G • Mode: 2 Years Single Subject

Qualifications/Requirements: GCE, A/AS: 4 points. AGNVQ, Pass. ND/C, 2 Merits. SQAH, Individual consideration. SQAV, Individual consideration. IB, Individual consideration.

HND Engineering (Software Development)

UCAS Code: 007G • Mode: 2/3 Years Single Subject

Qualifications/Requirements: GCE, A/AS: 4 points. Maths or Physics at A level. AGNVQ, Pass in Engineering with 1 additional unit. ND/C, 1 Merit (in specific programmes). SQAH, Grades CC including specific subject(s). SQAV, Individual consideration. IB, 24 points.

HND Software Engineering

UCAS Code: 107G • Mode: 2 Years Single Subject

Qualifications/Requirements: GCE, A/AS: 4 points. AGNVQ, Pass. ND/C, 2 Merits. SQAH, Individual consideration. SQAV, Individual consideration. IB, Individual consideration.

BSc Software Engineering

UCAS Code: G700 • Mode: 3/4 Years Single Subject

Qualifications/Requirements: GCE, A/AS: 14 points. AGNVQ, Distinction (in specific programmes) with 6 additional units or with A/AS. ND/C, Merit overall. SQAH, Grades BBCC. SQAV, Individual consideration. IB, 26 points.

HND Business with Business Information Technology

UCAS Code: 75GN • Mode: 2 Years Equal Combination

Qualifications/Requirements: GCE, A/AS: 8 points. AGNVQ, Merit. ND/C, National. SQAH, Grades CCC. SQAV, Individual consideration. IB, 24 points.

HND Computing (Business Computing)

UCAS Code: 017G • Mode: 2 Years Single Subject

Qualifications/Requirements: GCE, A/AS: 4 points. AGNVQ, Merit. ND/C, 2 Merits. SQAH, Grades CC. SQAV, Individual consideration. IB, 24 points.

HND Computing (Software Engineering)

UCAS Code: 007G • Mode: 2 Years Single Subject

Qualifications/Requirements: GCE, A level grades: EE. Maths or Science. AGNVQ, Merit. ND/C, National. SQAH, Grades CC. SQAV, Individual consideration. IB, 24 points.

CHELTENHAM & GLOUCESTER COLLEGE OF HIGHER EDUCATION C50

BSc Business Computer Systems with International Business Management

UCAS Code: G7N1 • Mode: 3 Years Major/Minor

Qualifications/Requirements: GCE, A/AS: 10 points. AGNVQ, Merit. ND/C, Merit overall and 3 Distinctions. SQAH, Grades CCCC. SQAV, Individual consideration. IB, 26 points.

BSc Business Computer Systems with International Marketing Management

UCAS Code: G7N5 • Mode: 3 Years Major/Minor

Qualifications/Requirements: GCE, A/AS: 12 points. AGNVQ, Merit. ND/C, Merit overall and 3 Distinctions. SQAH, Grades CCCC. SQAV, Individual consideration. IB, 26 points.

BSc Business Information Technology and Hospitality Management (Catering)

UCAS Code: NG75 • Mode: 3 Years Equal Combination

Qualifications/Requirements: GCE, A/AS: 8 points. AGNVQ, Merit. ND/C, Merit overall. SQAH, Grades CCCC. SQAV, Individual consideration. IB, 24 points.

BSc Business Information Technology with International Business Management

UCAS Code: G7NC • Mode: 3 Years Major/Minor

Qualifications/Requirements: GCE, A/AS: 10 points. AGNVQ, Merit. ND/C, Merit overall and 3 Distinctions. SQAH, Grades CCCC. SQAV, Individual consideration. IB, 26 points.

BSc Business Information Technology with International Marketing Management

UCAS Code: G7NM • Mode: 3 Years Major/Minor

Qualifications/Requirements: GCE, A/AS: 12 points. AGNVQ, Merit. ND/C, Merit overall and 3 Distinctions. SQAH, Grades CCCC. SQAV, Individual consideration. IB, 26 points.

Mod Business Computer Systems and International Business Management

UCAS Code: GN71 • Mode: 3 Years Equal Combination

Qualifications/Requirements: GCE, A/AS: 10 points. AGNVQ, Merit. ND/C, Merit overall and 3 Distinctions. SQAH, Grades CCCC. SQAV, Individual consideration. IB, 26 points.

Mod Business Computer Systems and International Marketing Management

UCAS Code: GN75 • Mode: 3 Years Equal Combination

Qualifications/Requirements: GCE, A/AS: 10 points. AGNVQ, Merit. ND/C, Merit overall and 3 Distinctions. SQAH, Grades CCCC. SQAV, Individual consideration. IB, 26 points.

Mod Business Information Technology and Hospitality Management (Hotel)

UCAS Code: GNM7 • Mode: 3 Years Equal Combination

Qualifications/Requirements: GCE, A/AS: 10-14 points. AGNVQ, Merit. ND/C, Merit overall. SQAH, Grades CCCC. SQAV, Individual consideration. IB, 24 points.

Mod Business Information Technology and International Business Management

UCAS Code: GN7C • Mode: 3 Years Equal Combination

Qualifications/Requirements: GCE, A/AS: 10 points. AGNVQ, Merit. ND/C, Merit overall and 3 Distinctions. SQAH, Grades CCCC. SQAV, Individual consideration. IB, 26 points.

Mod Business Information Technology and International Marketing Management

UCAS Code: GN7M • Mode: 3 Years Equal Combination

Qualifications/Requirements: GCE, A/AS: 12 points. AGNVQ, Merit. ND/C, Merit overall and 3 Distinctions. SQAH, Grades CCCC. SQAV, Individual consideration. IB, 26 points.

Mod Computing and International Business Management

UCAS Code: GN7D • Mode: 3 Years Equal Combination

Qualifications/Requirements: GCE, A/AS: 10 points. AGNVQ, Merit. ND/C, Merit overall and 3 Distinctions. SQAH, Grades CCCC. SQAV, Individual consideration. IB, 26 points.

Mod Computing and International Marketing Management

UCAS Code: GN7N • Mode: 3 Years Equal Combination

Qualifications/Requirements: GCE, A/AS: 10 points. AGNVQ, Merit. ND/C, Merit overall and 3 Distinctions. SQAH, Grades CCCC. SQAV, Individual consideration. IB, 26 points.

CHESTER: A COLLEGE OF THE UNIVERSITY OF LIVERPOOL C55

HND Computing (Business Information Technology)

UCAS Code: 127G • Mode: 2 Years Single Subject

Qualifications/Requirements: GCE, A/AS: 6 points. AGNVQ, Pass (in specific programmes). ND/C, National. SQAH, Grades DDD. SQAV, National. IB, 24 points.

HNC Computing (Business Information Technology)

UCAS Code: 027G • Mode: 1 Years Single Subject

Qualifications/Requirements: SQAH, Grades DDD. IB, 24 points.

CITY UNIVERSITY C60

BEng Software Engineering (4 years SW)

UCAS Code: G700 • Mode: 4 Years Single Subject

Qualifications/Requirements: GCE, A/AS: 22 points. Maths, Physics, Computing or Science. AGNVQ, Distinction (in specific programmes) with A level. ND/C, Merit overall and 3 Distinctions. SQAH, Grades ABBBB. SQAV, Individual consideration. IB, 28 points.

CITY OF BRISTOL COLLEGE C63

HND Computing (Computing, Software Engineering or Business IT)

UCAS Code: 65GG • Mode: 2 Years Equal Combination

Qualifications/Requirements: An approved subject from restricted list. AGNVQ, Individual consideration. ND/C, Individual consideration. SQAH, Individual consideration. SQAV, Individual consideration. IB, Individual consideration.

HND Business Information Technology

UCAS Code: 027G • Mode: 2 Years Single Subject

Qualifications/Requirements: GCE, A/AS: 6 points. Computing or Italian. AGNVQ, Pass. ND/C, National.

COVENTRY UNIVERSITY C85

BSc Software Engineering

UCAS Code: G700 • Mode: 3/4 Years Single Subject

Qualifications/Requirements: GCE, A/AS: 12 points. AGNVQ, Merit. ND/C, 4 Merits. SQAH, Grades CCCC. SQAV, Individual consideration. IB, Individual consideration.

CRANFIELD UNIVERSITY C90

BEng Software Engineering

UCAS Code: G700 • Mode: 3/4 Years Single Subject

Qualifications/Requirements: GCE, A level grades: BC-CCC. AGNVQ, Distinction (in specific programmes). ND/C, 4 Merits. SQAH, Individual consideration. SQAV, Individual consideration. IB, Individual consideration.

DE MONTFORT UNIVERSITY D26

BSc Software Engineering

UCAS Code: G700 • Mode: 4 Years Single Subject

Qualifications/Requirements: GCE, A/AS: 16 points. AGNVQ, Individual consideration. ND/C, 2 Merits and 2 Distinctions. SQAH, Grades BBBB. SQAV, Individual consideration. IB, 28 points.

UNIVERSITY OF DERBY D39

BSc Business Information Systems

UCAS Code: G720 • Mode: 4 Years Single Subject

Qualifications/Requirements: GCE, A/AS: 12-14 points. AGNVQ, Merit (in specific programmes). ND/C, Merit overall (in specific programmes). SQAH, Grades BCCC. SQAV, Individual consideration. IB, 26 points.

BSc Computer Studies (Software Engineering)

UCAS Code: G5G7 • Mode: 4 Years Major/Minor

Qualifications/Requirements: GCE, A/AS: 12-14 points. AGNVQ, Merit (in specific programmes). ND/C, Merit overall (in specific programmes). SQAH, Grades BCCC. SQAV, Individual consideration. IB, 26 points.

HND Business Information Systems

UCAS Code: 027G • Mode: 2 Years Single Subject

Qualifications/Requirements: GCE, A/AS: 4 points. AGNVQ, Pass (in specific programmes). ND/C, Pass (in specific programmes). SQAH, Grades CCD including specific subject(s). SQAV, Individual consideration. IB, Pass in Diploma including specific subjects.

DUDLEY COLLEGE OF TECHNOLOGY D58

HND Business Information Technology

UCAS Code: 007G • Mode: 2 Years Single Subject

Qualifications/Requirements: GCE, A/AS: 2 points. AGNVQ, Individual consideration. ND/C, Individual consideration.

HND Software Engineering

UCAS Code: 107G • Mode: 2 Years Single Subject

Qualifications/Requirements: GCE, A/AS: 2 points. AGNVQ, Individual consideration. ND/C, Individual consideration.

THE UNIVERSITY OF DURHAM D86

BSc Software Engineering

UCAS Code: G700 • Mode: 3 Years Single Subject

Qualifications/Requirements: GCE, A/AS: 22 points. Two Sciences at A level. AGNVQ, Individual consideration. ND/C, Individual consideration. SQAH, Grades AABBB including specific subject(s). SQAV, Individual consideration. IB, 30 points.

UNIVERSITY OF EAST ANGLIA E14

BSc Information Systems and Software Engineering

UCAS Code: G515 • Mode: 3 Years Single Subject

Qualifications/Requirements: GCE, A level grades: BBC-BCC. AGNVQ, Distinction with A level. ND/C, Merit overall and 3 Distinctions. SQAH, Grades BBBB. IB, 28 points.

UNIVERSITY OF EAST LONDON E28

BSc Software Engineering

UCAS Code: G700 • Mode: 4 Years Single Subject

Qualifications/Requirements: GCE, A/AS: 10 points. AGNVQ, Merit (in specific programmes). ND/C, 6 Merits. SQAH, Individual consideration. SQAV, Individual consideration. IB, Individual consideration.

BEng Electronics and Software Engineering
UCAS Code: GH76 • Mode: 4 Years Equal Combination

Qualifications/Requirements: GCE, A level grades: CCC. Maths and Physics at A level. ND/C, Merit overall (in specific programmes). SQAH, Grades BBBB including specific subject(s). SQAV, National including specific subject(s). IB, Pass in Diploma including specific subjects.

BEng Software Engineering
UCAS Code: G700 • Mode: 4 Years Single Subject

Qualifications/Requirements: GCE, A level grades: CCC. Maths at A level. AGNVQ, Pass. ND/C, Merit overall (in specific programmes). SQAH, Grades BBBB including specific subject(s). SQAV, National including specific subject(s). IB, Pass in Diploma including specific subjects.

THE UNIVERSITY OF ESSEX E70

BSc Computer Science (Software Engineering)
UCAS Code: G700 • Mode: 3 Years Single Subject

Qualifications/Requirements: GCE, A/AS: 20 points. AGNVQ, Distinction. ND/C, Merit overall and 2 Distinctions. SQAH, Grades BBBB. SQAV, Individual consideration. IB, 28 points.

BEng Computers & Networks
UCAS Code: H6G7 • Mode: 3/4 Years Major/Minor

Qualifications/Requirements: GCE, A level grades: CCC. Physics and Maths at A level. Entry point may vary according to qualifications held. AGNVQ, Distinction. ND/C, Merit overall (in specific programmes). SQAH, Grades BBBC. SQAV, Individual consideration. IB, 28 points.

FARNBOROUGH COLLEGE OF TECHNOLOGY F66

HND Software Engineering
UCAS Code: 007G • Mode: 2 Years Single Subject

Qualifications/Requirements: GCE, A level grades: E. AGNVQ, Pass. ND/C, National. SQAH, Individual consideration. SQAV, Individual consideration. IB, Individual consideration.

UNIVERSITY OF GLAMORGAN G14

BSc Financial Information Technology
UCAS Code: G7N3 • Mode: 4 Years Major/Minor

Qualifications/Requirements: GCE, A level grades: CD. AGNVQ, Merit. ND/C, 3 Merits. SQAH, Individual consideration. SQAV, Individual consideration. IB, Individual consideration.

BSc Software Engineering
UCAS Code: G700 • Mode: 3/4 Years Single Subject

Qualifications/Requirements: GCE, A/AS: 12 points. AGNVQ, Merit (in specific programmes). ND/C, Individual consideration. SQAH, Individual consideration. SQAV, Individual consideration. IB, Individual consideration.

HND Software Engineering
UCAS Code: 007G • Mode: 2 Years Single Subject

Qualifications/Requirements: GCE, A/AS: 4 points. AGNVQ, Pass (in specific programmes). ND/C, 3 Merits. SQAH, Individual consideration. SQAV, Individual consideration. IB, Individual consideration.

UNIVERSITY OF GLASGOW G28

BSc Electronic and Software Engineering
UCAS Code: GH76 • Mode: 4 Years Equal Combination

Qualifications/Requirements: GCE, A level grades: BBC-CCC. Maths and Science. AGNVQ, Merit. ND/C, National. SQAH, Grades BBBB including specific subject(s). SQAV, National. IB, 24 points.

BSc Software Engineering
UCAS Code: G530 • Mode: 4 Years Single Subject

Qualifications/Requirements: GCE, A level grades: BBC-CCC. Two Sciences. AGNVQ, Merit. ND/C, National. SQAH, Grades BBBB including specific subject(s). SQAV, National. IB, 24 points.

BEng Electronic and Software Engineering
UCAS Code: GH7P • Mode: 4 Years Equal Combination

Qualifications/Requirements: GCE, A level grades: BCC. Maths and Physics. AGNVQ, Merit in Engineering. ND/C, Merit overall. SQAH, Grades BBBBC including specific subject(s). SQAV, National including specific subject(s). IB, 24 points.

UNIVERSITY OF GREENWICH G70

BSc Computer Networking
UCAS Code: G720 • Mode: 3/4 Years Single Subject

Qualifications/Requirements: GCE, A/AS: 14 points. Science, Computing or Maths. AGNVQ, Merit (in specific programmes). ND/C, 4 Merits. SQAH, Grades CCCC. SQAV, Individual consideration. IB, 24 points.

BSc Computer Networking (Extended)
UCAS Code: G721 • Mode: 4/5 Years Single Subject

Qualifications/Requirements: Please refer to prospectus.

BEng Computer Systems with Software Engineering

UCAS Code: G6G7 • Mode: 3/4 Years Major/Minor

Qualifications/Requirements: GCE, A/AS: 14 points. Science, Computing or Maths. AGNVQ, Merit (in specific programmes). ND/C, 4 Merits. SQAH, Grades CCCC. SQAV, Individual consideration. IB, 24 points.

BEng Computer Systems with Software Engineering (Extended)

UCAS Code: G6GR • Mode: 4/5 Years Major/Minor

Qualifications/Requirements: GCE, A/AS: 4 points. AGNVQ, Individual consideration. ND/C, Individual consideration. SQAH, Individual consideration. SQAV, Individual consideration. IB, Individual consideration.

HND Business Information Systems

UCAS Code: 017G • Mode: 2 Years Single Subject

Qualifications/Requirements: GCE, A/AS: 2 points. AGNVQ, Pass (in specific programmes). ND/C, 2 Merits. SQAH, Grades C. SQAV, Individual consideration. IB, Individual consideration.

HND Engineering (Software Engineering)

UCAS Code: 007G • Mode: 2 Years Single Subject

Qualifications/Requirements: Please refer to prospectus.

HEREFORDSHIRE COLLEGE OF TECHNOLOGY H16

HND Computing (Business Information Technology)

UCAS Code: 006G • Mode: 2 Years Single Subject

Qualifications/Requirements: GCE, A/AS: 2 points. AGNVQ, Pass. ND/C, Individual consideration. SQAH, Grades CC. SQAV, Individual consideration.

HND Computing (Software Engineering)

UCAS Code: 007G • Mode: 2 Years Single Subject

Qualifications/Requirements: GCE, A/AS: 2 points. AGNVQ, Pass. ND/C, Individual consideration. SQAH, Grades CC. SQAV, Individual consideration.

HERIOT-WATT UNIVERSITY, EDINBURGH H24

BSc Computer Science (Software Engineering)

UCAS Code: G700 • Mode: 4 Years Single Subject

Qualifications/Requirements: GCE, A level grades: CCD. Maths. AGNVQ, Merit (in specific programmes). ND/C, Merit overall (in specific programmes). SQAH, Grades BBBC including specific subject(s). SQAV, National including specific subject(s). IB, 28 points.

MEng Software Engineering

UCAS Code: G701 • Mode: 5 Years Single Subject

Qualifications/Requirements: GCE, A level grades: CDD. Maths and Physics. AGNVQ, Merit (in specific programmes). ND/C, (in specific programmes). SQAH, Grades BBBC including specific subject(s). SQAV, National including specific subject(s). IB, 28 points.

MEng Software Engineering

UCAS Code: G702 • Mode: 4 Years Single Subject

Qualifications/Requirements: GCE, A level grades: BCC. Maths and Physics. AGNVQ, Not normally sufficient. ND/C, Higher National. SQAH, CSYS required. SQAV, Higher National. IB, Individual consideration.

UNIVERSITY OF HERTFORDSHIRE H36

BSc Business Information Systems

UCAS Code: G710 • Mode: 3/4 Years Single Subject

Qualifications/Requirements: GCE, A/AS: 18 points. AGNVQ, Merit or Distinction in Information Technology with A/AS. ND/C, Distinction Overall. SQAH, Grades BBBB. IB, 28 points.

BSc Software Engineering

UCAS Code: G5G7 • Mode: 4 Years Major/Minor

Qualifications/Requirements: GCE, A/AS: 16 points. AGNVQ, Merit. ND/C, Merit overall. SQAH, Individual consideration. SQAV, Individual consideration. IB, 28 points.

BSc Software Systems for the Arts and Media

UCAS Code: G701 • Mode: 3 Years Single Subject

Qualifications/Requirements: GCE, A/AS: 10 points. AGNVQ, Pass. ND/C, Merit overall. SQAH, Individual consideration. SQAV, Individual consideration. IB, Individual consideration.

BSc Software Systems for the Arts and Media (Extended)

UCAS Code: G708 • Mode: 4 Years Single Subject

Qualifications/Requirements: Please refer to prospectus.

HND Business Decision Analysis

UCAS Code: 017G • Mode: 2 Years Single Subject

Qualifications/Requirements: GCE, A/AS: 4 points. AGNVQ, Pass or Merit (in specific programmes) with A/AS. ND/C, National. SQAH, Grades CCCC. SQAV, Individual consideration. IB, 24 points.

HND Computing (Software Engineering)

UCAS Code: 007G • Mode: 2 Years Single Subject

Qualifications/Requirements: GCE, A/AS: 4 points. AGNVQ, Pass. ND/C, 2 Merits. SQAH, Individual consideration. SQAV, Individual consideration. IB, Individual consideration.

THE UNIVERSITY OF HUDDERSFIELD H60

BA Business Computing with Software Development
UCAS Code: G5G7 • Mode: 4 Years Major/Minor

Qualifications/Requirements: GCE, A/AS: 16 points. AGNVQ, Merit. ND/C, Merit overall. SQAH, Grades BBBB. SQAV, Individual consideration. IB, Individual consideration.

BSc Computing (Software Development)
UCAS Code: G700 • Mode: 4 Years Single Subject

Qualifications/Requirements: GCE, A/AS: 16 points. AGNVQ, Merit. ND/C, Merit overall. SQAH, Grades BBBB. SQAV, Individual consideration. IB, Individual consideration.

BSc Software Development with Artificial Intelligence
UCAS Code: G768 • Mode: 4 Years Single Subject

Qualifications/Requirements: GCE, A/AS: 16 points. AGNVQ, Merit. ND/C, Merit overall. SQAH, Grades BBBB. SQAV, Individual consideration. IB, Individual consideration.

BSc Software Development with Business
UCAS Code: G7N1 • Mode: 4 Years Major/Minor

Qualifications/Requirements: GCE, A/AS: 16 points. AGNVQ, Merit. ND/C, Merit overall. SQAH, Grades BBBB. SQAV, Individual consideration. IB, Individual consideration.

BSc Software Development with Business Computing
UCAS Code: G7G5 • Mode: 4 Years Major/Minor

Qualifications/Requirements: GCE, A/AS: 16 points. AGNVQ, Merit. ND/C, Merit overall. SQAH, Grades BBBB. SQAV, Individual consideration. IB, Individual consideration.

BSc Software Development with French
UCAS Code: G7R1 • Mode: 4 Years Major/Minor

Qualifications/Requirements: GCE, A/AS: 16 points. AGNVQ, Merit. ND/C, Merit overall. SQAH, Grades BBBB. SQAV, Individual consideration. IB, Individual consideration.

BSc Software Development with German
UCAS Code: G7R2 • Mode: 4 Years Major/Minor

Qualifications/Requirements: GCE, A/AS: 16 points. AGNVQ, Merit. ND/C, Merit overall. SQAH, Grades BBBB. SQAV, Individual consideration. IB, Individual consideration.

BSc Software Development with Human-Computer Interaction
UCAS Code: G7GM • Mode: 4 Years Major/Minor

Qualifications/Requirements: GCE, A/AS: 16 points. AGNVQ, Merit. ND/C, Merit overall. SQAH, Grades BBBB. SQAV, Individual consideration. IB, Individual consideration.

BSc Software Development with Management
UCAS Code: G7NC • Mode: 4 Years Major/Minor

Qualifications/Requirements: GCE, A/AS: 16 points. AGNVQ, Merit. ND/C, Merit overall. SQAH, Grades BBBB. SQAV, Individual consideration. IB, Individual consideration.

BSc Software Development with Mathematics
UCAS Code: G7G1 • Mode: 4 Years Major/Minor

Qualifications/Requirements: GCE, A/AS: 16 points. AGNVQ, Merit. ND/C, Merit overall. SQAH, Grades BBBB. SQAV, Individual consideration. IB, Individual consideration.

BSc Software Development with Multimedia
UCAS Code: G7GN • Mode: 4 Years Major/Minor

Qualifications/Requirements: GCE, A/AS: 16 points. AGNVQ, Merit. ND/C, Merit overall. SQAH, Grades BBBB. SQAV, Individual consideration. IB, Individual consideration.

BSc Software Development with Operational Research
UCAS Code: G7N2 • Mode: 4 Years Major/Minor

Qualifications/Requirements: GCE, A/AS: 16 points. AGNVQ, Merit. ND/C, Merit overall. SQAH, Grades BBBB. SQAV, Individual consideration. IB, Individual consideration.

BSC Software Development with Statistics
UCAS Code: G7G4 • Mode: 4 Years Major/Minor

Qualifications/Requirements: GCE, A/AS: 16 points. AGNVQ, Merit. ND/C, Merit overall. SQAH, Grades BBBB. SQAV, Individual consideration. IB, Individual consideration.

MEng Software Engineering
UCAS Code: G701 • Mode: 5 Years Single Subject

Qualifications/Requirements: GCE, A/AS: 24 points. AGNVQ, Distinction. ND/C, Merit overall and 2 Distinctions. SQAH, Individual consideration. SQAV, Individual consideration. IB, Individual consideration.

HND Software Engineering

UCAS Code: 007G • Mode: 3 Years Single Subject

Qualifications/Requirements: GCE, A/AS: 6 points. AGNVQ, Pass. ND/C, National. SQAH, Grades CCC. SQAV, Individual consideration. IB, Individual consideration.

THE UNIVERSITY OF HULL H72

BSc Software Engineering

UCAS Code: G700 • Mode: 3 Years Single Subject

Qualifications/Requirements: GCE, A/AS: 18 points. Maths at A level. AGNVQ, Merit (in specific programmes) with A level. ND/C, Merit overall (in specific programmes). SQAH, Grades BBBCC. SQAV, Individual consideration. IB, 26 points.

BSc Software Engineering

UCAS Code: G701 • Mode: 4 Years Single Subject

Qualifications/Requirements: GCE, A level grades: CD. AGNVQ, Pass (in specific programmes) with A level. ND/C, National. SQAH, Grades CCCCD. SQAV, Individual consideration. IB, 24 points.

BSc Software Engineering with Industrial Experience

UCAS Code: G702 • Mode: 4 Years Single Subject

Qualifications/Requirements: GCE, A/AS: 18 points. Maths at A level. AGNVQ, Merit (in specific programmes) with A level. ND/C, Merit overall (in specific programmes). SQAH, Grades BBBCC. SQAV, Individual consideration. IB, 26 points.

BSc Software Engineering with Study Abroad

UCAS Code: G703 • Mode: 4 Years Single Subject

Qualifications/Requirements: GCE, A/AS: 18 points. Maths at A level. AGNVQ, Merit (in specific programmes) with A level. ND/C, Merit overall (in specific programmes). SQAH, Grades BBBCC. SQAV, Individual consideration. IB, 26 points.

IMPERIAL COLLEGE OF SCIENCE, TECHNOLOGY AND MEDICINE (UNIVERSITY OF LONDON) I50

MEng Computing (Software Engineering) (h)

UCAS Code: G700 • Mode: 4 Years Single Subject

Qualifications/Requirements: GCE, A level grades: AAB. Maths at A level. AGNVQ, Pass. ND/C, Individual consideration. SQAH, Individual consideration. SQAV, Individual consideration. IB, Individual consideration.

THE UNIVERSITY OF KENT AT CANTERBURY K24

BSc Software Engineering

UCAS Code: G700 • Mode: 3 Years Single Subject

Qualifications/Requirements: AGNVQ, Individual consideration. ND/C, Individual consideration. SQAH, Individual consideration. SQAV, Individual consideration. IB, Individual consideration.

KINGSTON UNIVERSITY K84

BSc Software Engineering

UCAS Code: G700 • Mode: 4 Years Single Subject

Qualifications/Requirements: GCE, A/AS: 18 points. AGNVQ, Merit (in specific programmes) with 3 additional units or with A/AS. ND/C, Individual consideration. SQAH, Individual consideration. SQAV, Individual consideration. IB, 26 points.

LANCASTER UNIVERSITY L14

BSc Computer Science with Software Engineering

UCAS Code: G5G7 • Mode: 3 Years Major/Minor

Qualifications/Requirements: GCE, A level grades: BCC. AGNVQ, Distinction with 6 additional units or with A/AS. ND/C, Individual consideration. SQAH, Grades BBBBB. SQAV, Individual consideration. IB, 30 points.

LEEDS METROPOLITAN UNIVERSITY L27

BSc Business Computing

UCAS Code: G710 • Mode: 3/4 Years Single Subject

Qualifications/Requirements: GCE, A/AS: 12-14 points. AGNVQ, Distinction (in specific programmes) or Merit (in specific programmes). ND/C, 3 Merits and 1 Distinction. SQAH, Grades BBBC. SQAV, Individual consideration. IB, 26 points.

HND Business Computing

UCAS Code: 017G • Mode: 2/3 Years Single Subject

Qualifications/Requirements: GCE, A/AS: 6-8 points. AGNVQ, Pass (in specific programmes) or Merit (in specific programmes). ND/C, 2 to 3 Merits. SQAH, Grades BCC. SQAV, Individual consideration. IB, Individual consideration.

UNIVERSITY OF LEICESTER L34

BEng Electronic & Software Engineering

UCAS Code: HG67 • Mode: 3 Years Equal Combination

Qualifications/Requirements: GCE, A/AS: 18 points. Maths at A level and Physics. AGNVQ, Merit in Engineering with 6 additional units with A/AS. ND/C, 2 Merits and 3 Distinctions (in specific programmes). SQAH, Grades BBBBC including specific subject(s). SQAV, Individual consideration. IB, 26 points.

BEng Electronic and Software Engineering with Industry

UCAS Code: HG6T • Mode: 4 Years Equal Combination

Qualifications/Requirements: GCE, A/AS: 18 points. Maths at A level and Physics. AGNVQ, Merit in Engineering with 6 additional units with A/AS. ND/C, 2 Merits and 3 Distinctions (in specific programmes). SQAH, Grades BBBBC including specific subject(s). SQAV, Individual consideration. IB, 26 points.

MEng Electronic & Software Engineering

UCAS Code: HGP7 • Mode: 4 Years Equal Combination

Qualifications/Requirements: GCE, A/AS: 22-24 points. Maths at A level and Physics. AGNVQ, Distinction in Engineering with 6 additional units with A/AS. ND/C, 5 Distinctions (in specific programmes). SQAH, Grades ABBBB including specific subject(s). SQAV, Individual consideration. IB, 30 points.

MEng Electronic & Software Engineering with Industry

UCAS Code: HGQ7 • Mode: 5 Years Equal Combination

Qualifications/Requirements: GCE, A/AS: 22-24 points. Maths at A level and Physics. AGNVQ, Distinction in Engineering with 6 additional units with A/AS. ND/C, 5 Distinctions (in specific programmes). SQAH, Grades ABBBB including specific subject(s). SQAV, Individual consideration. IB, 30 points.

LIVERPOOL JOHN MOORES UNIVERSITY L51

BSc Software Engineering

UCAS Code: G700 • Mode: 4 Years Single Subject

Qualifications/Requirements: GCE, A/AS: 12 points. AGNVQ, Distinction (in specific programmes). ND/C, Merit overall and 1 Distinction. SQAH, Individual consideration. SQAV, Individual consideration. IB, 26 points.

LLANDRILLO COLLEGE, NORTH WALES L53

HND Business Information Technology

UCAS Code: 027G • Mode: 2 Years Single Subject

Qualifications/Requirements: Please refer to prospectus.

HND Software Engineering

UCAS Code: 007G • Mode: 2 Years Single Subject

Qualifications/Requirements: Please refer to prospectus.

LONDON GUILDHALL UNIVERSITY L55

BSc Business Information Technology

UCAS Code: G720 • Mode: 3 Years Single Subject

Qualifications/Requirements: Please refer to prospectus.

Mod Business Economics & Business Information Technology

UCAS Code: GL7C • Mode: 3 Years Equal Combination

Qualifications/Requirements: GCE, A level grades - DD. AGNVQ, Merit (in specific programmes). ND/C, Merit overall. SQAH, Individual consideration. SQAV, Individual consideration. IB, 24 points.

Mod Business Information Technology & 3D/Spatial Design

UCAS Code: GW7F • Mode: 3 Years Equal Combination

Qualifications/Requirements: GCE, A level grades - DD. OTHER. AGNVQ, Merit (in specific programmes). ND/C, Merit overall. SQAH, Individual consideration. SQAV, Individual consideration. IB, 24 points.

Mod Business Information Technology & Communications

UCAS Code: GP74 • Mode: 3 Years Equal Combination

Qualifications/Requirements: GCE, A level grades - CC-CDD. AGNVQ, Distinction (in specific programmes). ND/C, Merit overall and 4 Distinctions. SQAH, Individual consideration. SQAV, Individual consideration. IB, 26 points.

Mod Business Information Technology & Criminology

UCAS Code: GM7H • Mode: 3 Years Equal Combination

Qualifications/Requirements: GCE, A level grades - CD. AGNVQ, Merit. ND/C, Merit overall and 2 Distinctions. SQAH, Individual consideration. SQAV, Individual consideration. IB, 26 points.

Mod Business Information Technology & Design Studies

UCAS Code: GW72 • Mode: 3 Years Equal Combination

Qualifications/Requirements: GCE, A level grades - CD-DDD. AGNVQ, Merit (in specific programmes). ND/C, Merit overall and 2 Distinctions. SQAH, Individual consideration. SQAV, Individual consideration. IB, 24 points.

Mod Business Information Technology & Development Studies

UCAS Code: GM79 • Mode: 3 Years Equal Combination

Qualifications/Requirements: GCE, A level grades - DD. AGNVQ, Merit (in specific programmes). ND/C, Merit overall. SQAH, Individual consideration. SQAV, Individual consideration. IB, 24 points.

Mod Business Information Technology & Economics

UCAS Code: GL71 • Mode: 3 Years Equal Combination

Qualifications/Requirements: GCE, A level grades - DD. AGNVQ, Merit (in specific programmes). ND/C, Merit overall. SQAH, Individual consideration. SQAV, Individual consideration. IB, 24 points.

Mod Business Information Technology & English Studies

UCAS Code: GQ73 • Mode: 3 Years Equal Combination

Qualifications/Requirements: GCE, A level grades - CD-DDD. AGNVQ, Merit (in specific programmes). ND/C, Merit overall and 2 Distinctions. SQAH, Individual consideration. SQAV, Individual consideration. IB, 24 points.

Mod Business Information Technology & European Studies

UCAS Code: GT72 • Mode: 3 Years Equal Combination

Qualifications/Requirements: GCE, A level grades - DD. AGNVQ, Merit (in specific programmes). ND/C, Merit overall. SQAH, Individual consideration. SQAV, Individual consideration. IB, 24 points.

Mod Business Information Technology & Financial Services

UCAS Code: GN73 • Mode: 3 Years Equal Combination

Qualifications/Requirements: GCE, A level grades - DD. AGNVQ, Merit (in specific programmes). ND/C, Merit overall. SQAH, Individual consideration. SQAV, Individual consideration. IB, 24 points.

Mod Business Information Technology & Fine Art

UCAS Code: GW71 • Mode: 3 Years Equal Combination

Qualifications/Requirements: GCE, A level grades - CC-CDD. OTHER. AGNVQ, Merit (in specific programmes). ND/C, Merit overall and 2 Distinctions. SQAH, Individual consideration. SQAV, Individual consideration. IB, 26 points.

Mod Business Information Technology & French

UCAS Code: GR71 • Mode: 4 Years Equal Combination

Qualifications/Requirements: GCE, A level grades - DD. AGNVQ, Merit (in specific programmes). ND/C, Merit overall. SQAH, Individual consideration. SQAV, Individual consideration. IB, 24 points.

Mod Business Information Technology & German

UCAS Code: GR72 • Mode: 4 Years Equal Combination

Qualifications/Requirements: GCE, A level grades - DD. AGNVQ, Merit (in specific programmes). ND/C, Merit overall. SQAH, Individual consideration. SQAV, Individual consideration. IB, 24 points.

Mod Business Information Technology & Insurance

UCAS Code: GN7J • Mode: 3 Years Equal Combination

Qualifications/Requirements: GCE, A level grades - DD. AGNVQ, Merit (in specific programmes). ND/C, Merit overall. SQAH, Individual consideration. SQAV, Individual consideration. IB, 24 points.

Mod Business Information Technology & International Relations

UCAS Code: GM7C • Mode: 3 Years Equal Combination

Qualifications/Requirements: GCE, A level grades - DD. AGNVQ, Merit (in specific programmes). ND/C, Merit overall. SQAH, Individual consideration. SQAV, Individual consideration. IB, 24 points.

Mod Business Information Technology & Investment

UCAS Code: NGJ7 • Mode: 3 Years Equal Combination

Qualifications/Requirements: GCE, A level grades - DD. AGNVQ, Merit (in specific programmes). ND/C, Merit overall. SQAH, Individual consideration. SQAV, Individual consideration. IB, 24 points.

Mod Business Information Technology & Law

UCAS Code: GM73 • Mode: 3 Years Equal Combination

Qualifications/Requirements: GCE, A level grades - CC-CDD. AGNVQ, Merit (in specific programmes). ND/C, Merit overall and 2 Distinctions. SQAH, Individual consideration. SQAV, Individual consideration. IB, 26 points.

Mod Business Information Technology & Marketing

UCAS Code: GN75 • Mode: 3 Years Equal Combination

Qualifications/Requirements: GCE, A level grades - CD-DDD. AGNVQ, Merit (in specific programmes). ND/C, Merit overall and 2 Distinctions. SQAH, Individual consideration. SQAV, Individual consideration. IB, 26 points.

Mod Business Information Technology & Modern History

UCAS Code: GV71 • Mode: 3 Years Equal Combination

Qualifications/Requirements: GCE, A level grades - DD. AGNVQ, Merit (in specific programmes). ND/C, Merit overall. SQAH, Individual consideration. SQAV, Individual consideration. IB, 24 points.

Mod Business Information Technology & Multimedia Systems

UCAS Code: GG7M • Mode: 3 Years Equal Combination

Qualifications/Requirements: GCE, A level grades - DD. AGNVQ, Merit (in specific programmes). ND/C, Merit overall. SQAH, Individual consideration. SQAV, Individual consideration. IB, 24 points.

Mod Business Information Technology & Philosophy

UCAS Code: GV77 • Mode: 3 Years Equal Combination

Qualifications/Requirements: GCE, A level grades - CD. AGNVQ, Merit. ND/C, Merit overall and 2 Distinctions. SQAH, Individual consideration. SQAV, Individual consideration. IB, 26 points.

Mod Business Information Technology & Politics

UCAS Code: GM71 • Mode: 3 Years Equal Combination

Qualifications/Requirements: GCE, A level grades - DD. AGNVQ, Merit (in specific programmes). ND/C, Merit overall. SQAH, Individual consideration. SQAV, Individual consideration. IB, 24 points.

Mod Business Information Technology & Product Development & Manufacture

UCAS Code: GJ74 • Mode: 3 Years Equal Combination

Qualifications/Requirements: GCE, A level grades - DD. AGNVQ, Merit (in specific programmes). ND/C, Merit overall. SQAH, Individual consideration. SQAV, Individual consideration. IB, 24 points.

Mod Business Information Technology & Psychology

UCAS Code: CG87 • Mode: 3 Years Equal Combination

Qualifications/Requirements: GCE, A level grades - CD-DDD. AGNVQ, Merit (in specific programmes). ND/C, Merit overall and 2 Distinctions. SQAH, Individual consideration. SQAV, Individual consideration. IB, 26 points.

Mod Business Information Technology & Public Policy

UCAS Code: GM7D • Mode: 3 Years Equal Combination

Qualifications/Requirements: GCE, A level grades - CD. AGNVQ, Merit. ND/C, Merit overall and 2 Distinctions. SQAH, Individual consideration. SQAV, Individual consideration. IB, 26 points.

Mod Business Information Technology & Social Policy & Management

UCAS Code: GL74 • Mode: 3 Years Equal Combination

Qualifications/Requirements: GCE, A level grades - CD-DDD. AGNVQ, Merit (in specific programmes). ND/C, Merit overall. SQAH, Individual consideration. SQAV, Individual consideration. IB, 24 points.

Mod Business Information Technology & Sociology

UCAS Code: GL73 • Mode: 3 Years Equal Combination

Qualifications/Requirements: GCE, A level grades - CD-DDD. AGNVQ, Merit (in specific programmes). ND/C, Merit overall. SQAH, Individual consideration. SQAV, Individual consideration. IB, 24 points.

Mod Business Information Technology & Spanish

UCAS Code: GR74 • Mode: 4 Years Equal Combination

Qualifications/Requirements: GCE, A level grades - DD. AGNVQ, Merit (in specific programmes). ND/C, Merit overall. SQAH, Individual consideration. SQAV, Individual consideration. IB, 24 points.

Mod Business Information Technology & Taxation

UCAS Code: GN7H • Mode: 3 Years Equal Combination

Qualifications/Requirements: GCE, A level grades - DD. AGNVQ, Merit (in specific programmes). ND/C, Merit overall. SQAH, Individual consideration. SQAV, Individual consideration. IB, 24 points.

Mod Business Information Technology & Textile Furnishing Design

UCAS Code: GJ7K • Mode: 3 Years Equal Combination

Qualifications/Requirements: GCE, A level grades - DD. OTHER. AGNVQ, Merit (in specific programmes). ND/C, Merit overall. SQAH, Individual consideration. SQAV, Individual consideration. IB, 24 points.

Mod Business Information Technology & Transport

UCAS Code: GN79 • Mode: 3 Years Equal Combination

Qualifications/Requirements: GCE, A level grades - CD. AGNVQ, Merit. ND/C, Merit overall and 2 Distinctions. SQAH, Individual consideration. SQAV, Individual consideration. IB, 26 points.

LOUGHBOROUGH UNIVERSITY L79

BEng Electronics and Software Engineering

UCAS Code: GH56 • Mode: 3 Years Equal Combination

Qualifications/Requirements: Please refer to prospectus.

BEng Electronics and Software Engineering

UCAS Code: GH5P • Mode: 4 Years Equal Combination

Qualifications/Requirements: Please refer to prospectus.

MEng Electronics and Software Engineering

UCAS Code: HG65 • Mode: 4 Years Equal Combination

Qualifications/Requirements: Please refer to prospectus.

MEng Electronics and Software Engineering

UCAS Code: HG6M • Mode: 5 Years Equal Combination

Qualifications/Requirements: Please refer to prospectus.

UNIVERSITY OF LUTON L93

BSc Software Engineering

UCAS Code: G700 • Mode: 3 Years Single Subject

Qualifications/Requirements: GCE, A/AS: 14 points. AGNVQ, Distinction or Merit with 6 additional units or with A/AS. ND/C, Individual consideration. SQAH, Grades BBCC. SQAV, Individual consideration. IB, 32 points.

THE UNIVERSITY OF MANCHESTER M20

BSc Software Engineering

UCAS Code: G700 • Mode: 3 Years Single Subject

Qualifications/Requirements: GCE, A level grades: BBB. Maths at A level. AGNVQ, Distinction with A level. ND/C, Individual consideration. SQAH, Grades BBBBB including specific subject(s). SQAV, Individual consideration. IB, 30 points.

BSc Software Engineering with Industrial Experience

UCAS Code: G701 • Mode: 3 Years Single Subject

Qualifications/Requirements: GCE, A level grades: BBB. Maths at A level. AGNVQ, Distinction with A level. ND/C, Individual consideration. SQAH, Grades BBBBB including specific subject(s). SQAV, Individual consideration. IB, 30 points.

THE UNIVERSITY OF MANCHESTER INSTITUTE OF SCIENCE AND TECHNOLOGY (UMIST) M25

BEng Software Engineering

UCAS Code: G700 • Mode: 3 Years Single Subject

Qualifications/Requirements: GCE, A level grades: BBC. Maths or Physics. AGNVQ, Distinction (in specific programmes). ND/C, Individual consideration. SQAH, Grades ABBBB. SQAV, Individual consideration. IB, 30 points.

BEng Software Engineering with Industrial Experience

UCAS Code: G702 • Mode: 4 Years Single Subject

Qualifications/Requirements: GCE, A level grades: BBC. Maths or Physics. AGNVQ, Distinction (in specific programmes). ND/C, Individual consideration. SQAH, Grades ABBBB. SQAV, Individual consideration. IB, 30 points.

MEng Software Engineering

UCAS Code: G701 • Mode: 4 Years Single Subject

Qualifications/Requirements: GCE, A level grades: BBC. Maths or Physics. AGNVQ, Distinction (in specific programmes). ND/C, Individual consideration. SQAH, Grades ABBBB. SQAV, Individual consideration. IB, 30 points.

THE MANCHESTER METROPOLITAN UNIVERSITY M40

BSc Business Information Technology Management (HND top-up)

UCAS Code: G7N1 • Mode: 1 Years Major/Minor

Qualifications/Requirements: AGNVQ, Not normally sufficient. ND/C, Higher National. SQAH, Not normally sufficient. SQAV, Higher National. IB, Not normally sufficient.

BSc Software Engineering

UCAS Code: G700 • Mode: 4 Years Single Subject

Qualifications/Requirements: GCE, A/AS: 12-18 points. AGNVQ, Distinction (in specific programmes) or Merit (in specific programmes) with 4 additional units or with A/AS. ND/C, Merits and Distinction. SQAH, Grades BBCC. SQAV, National including specific subject(s). IB, 24 points.

BSc Software Engineering (Foundation)

UCAS Code: G708 • Mode: 5 Years Single Subject

Qualifications/Requirements: GCE, A level grades: E. Maths or Physics. AGNVQ, Pass (in specific programmes). ND/C, 2 Merits (in specific programmes). SQAH, Individual consideration. SQAV, Individual consideration. IB, Individual consideration.

HND Software Engineering

UCAS Code: 007G • Mode: 2 Years Single Subject

Qualifications/Requirements: GCE, A/AS: 6-10 points. AGNVQ, Merit (in specific programmes). ND/C, Merit overall. SQAH, Grades CCC. SQAV, Individual consideration. IB, Pass in Diploma.

MID-CHESHIRE COLLEGE M77

HND Business Information Technology

UCAS Code: 027G • Mode: 2 Years Single Subject

Qualifications/Requirements: GCE, A level grades - E. AGNVQ, Individual consideration. ND/C, National. SQAH, Individual consideration. SQAV, Individual consideration. IB, Individual consideration.

NAPIER UNIVERSITY N07

BA Business Information Technology

UCAS Code: G720 • Mode: 3/4 Years Single Subject

Qualifications/Requirements: GCE, A level grades - C-CC. AGNVQ, Distinction. ND/C, Individual consideration. SQAH, Grades BBCC including specific subject(s). SQAV, Individual consideration. IB, Individual consideration.

BEng Software Engineering

UCAS Code: G700 • Mode: 3/4 Years Single Subject

Qualifications/Requirements: GCE, A level grades: CD. AGNVQ, Individual consideration. ND/C, Individual consideration. SQAH, Grades BBB. SQAV, Individual consideration. IB, Individual consideration.

HND Software Engineering

UCAS Code: 007G • Mode: 2 Years Single Subject

Qualifications/Requirements: GCE, A level grades: D. AGNVQ, Individual consideration. ND/C, Individual consideration. SQAH, Grades CC. SQAV, Individual consideration. IB, Individual consideration.

UNIVERSITY OF NEWCASTLE UPON TYNE N21

BSc Software Engineering

UCAS Code: G700 • Mode: 3 Years Single Subject

Qualifications/Requirements: GCE, A/AS: 20 points. AGNVQ, Individual consideration. ND/C, Individual consideration. SQAH, Individual consideration. SQAV, Individual consideration. IB, Individual consideration.

BEng Microelectronics and Software Engineering

UCAS Code: GH76 • Mode: 3 Years Equal Combination

Qualifications/Requirements: GCE, A/AS: 20 points. Maths at A level and Science. AGNVQ, Individual consideration. ND/C, 7 Merits (in specific programmes). SQAH, CSYS required. SQAV, Higher National including specific subject(s). IB, 28 points.

MEng Microelectronics and Software Engineering

UCAS Code: GH7Q • Mode: 4 Years Equal Combination

Qualifications/Requirements: GCE, A/AS: 26 points. Maths at A level and Science. AGNVQ, Individual consideration. ND/C, 3 Merits and 3 Distinctions (in specific programmes). SQAH, CSYS required. SQAV, Higher National including specific subject(s). IB, 32 points.

NEWCASTLE COLLEGE N23

HND Computing (Business Information Technology)

UCAS Code: 027G • Mode: 2 Years Single Subject

Qualifications/Requirements: Please refer to prospectus.

HND Computing (Software Engineering)

UCAS Code: 007G • Mode: 2 Years Single Subject

Qualifications/Requirements: Please refer to prospectus.

NEW COLLEGE DURHAM N28

HND Computing (Business Information Technology)

UCAS Code: 027G • Mode: 2 Years Single Subject

Qualifications/Requirements: GCE, A/AS: 2 points. AGNVQ, Individual consideration. ND/C, National. SQAH, Individual consideration. SQAV, Individual consideration. IB, Individual consideration.

UNIVERSITY OF WALES COLLEGE, NEWPORT N37

BSc Business Information Technology

UCAS Code: GN71 • Mode: 3 Years Equal Combination

Qualifications/Requirements: GCE, A/AS: 8-10 points. AGNVQ, Distinction (in specific programmes). ND/C, 3 Merits. SQAH, Individual consideration. SQAV, Individual consideration. IB, Individual consideration.

HND Business Information Technology

UCAS Code: 17NG • Mode: 2 Years Equal Combination

Qualifications/Requirements: GCE, A/AS: 2 points. AGNVQ, Pass (in specific programmes). ND/C, National. SQAH, Individual consideration. SQAV, Individual consideration. IB, Individual consideration.

NORTHBROOK COLLEGE SUSSEX N41

HND Computing (Information Technology Management)

UCAS Code: 027G • Mode: 2 Years Single Subject

Qualifications/Requirements: ND/C, National. SQAH, Individual consideration. SQAV, Individual consideration. IB, Individual consideration.

NESCOT N49

HND Computer Automation and Networking

UCAS Code: 027G • Mode: 2 Years Single Subject

Qualifications/Requirements: GCE, A level grades: D. Science. AGNVQ, Pass. ND/C, National. SQAH, Individual consideration. SQAV, National including specific subject(s). IB, Pass in Diploma.

UNIVERSITY OF NORTH LONDON N63

BSc Computer Software Development

UCAS Code: G610 • Mode: 3/4 Years Single Subject

Qualifications/Requirements: GCE, A/AS: 8 points. Two Sciences. AGNVQ, Merit (in specific programmes). ND/C, 3 Merits (in specific programmes). SQAH, Grades CCCC including specific subject(s). SQAV, Individual consideration. IB, Individual consideration.

NORWICH: CITY COLLEGE N82

HND Software Engineering

UCAS Code: 007G • Mode: 2 Years Single Subject

Qualifications/Requirements: GCE, A level grades: EE. AGNVQ, Merit. ND/C, National. SQAH, Grades CC. SQAV, Individual consideration. IB, Individual consideration.

OXFORD BROOKES UNIVERSITY O66

BA/BSc Software Engineering/Spanish Studies

UCAS Code: GR74 • Mode: 3 Years Equal Combination

Qualifications/Requirements: Please refer to prospectus.

BSc Software Engineering (FT or SW)
UCAS Code: G700 • Mode: 4 Years Single Subject

Qualifications/Requirements: GCE, A level grades: DDD-CD. AGNVQ, Merit. ND/C, Individual consideration. SQAH, Individual consideration. SQAV, Individual consideration. IB, Individual consideration.

Mod Accounting and Finance/Software Engineering
UCAS Code: GN74 • Mode: 3 Years Equal Combination

Qualifications/Requirements: GCE, A level grades: CDD-BCC. AGNVQ, Merit or Distinction with 3 additional units. ND/C, Individual consideration. SQAH, Individual consideration. SQAV, Individual consideration. IB, Individual consideration.

Mod Anthropology/Software Engineering
UCAS Code: GL76 • Mode: 3 Years Equal Combination

Qualifications/Requirements: GCE, A level grades: CDD-BCC. AGNVQ, Merit with A/AS. ND/C, Individual consideration. SQAH, Individual consideration. SQAV, Individual consideration. IB, Individual consideration.

Mod Biological Chemistry/Software Engineering
UCAS Code: CG77 • Mode: 3 Years Equal Combination

Qualifications/Requirements: GCE, A level grades: DDD-BCC. Science. AGNVQ, Merit in Science or Merit with A/AS. ND/C, Individual consideration. SQAH, Individual consideration. SQAV, Individual consideration. IB, Individual consideration.

Mod Biology/Software Engineering
UCAS Code: CG17 • Mode: 3 Years Equal Combination

Qualifications/Requirements: GCE, A level grades: DD-BC. Science. AGNVQ, Merit in Science. ND/C, Individual consideration. SQAH, Individual consideration. SQAV, Individual consideration. IB, Individual consideration.

Mod Business Statistics/Software Engineering
UCAS Code: GG47 • Mode: 3 Years Equal Combination

Qualifications/Requirements: Please refer to prospectus.

Mod Cell Biology/Software Engineering
UCAS Code: CGC7 • Mode: 3 Years Equal Combination

Qualifications/Requirements: GCE, A level grades: DDD-BCC. Science. AGNVQ, Merit in Science or Merit with A/AS. ND/C, Individual consideration. SQAH, Individual consideration. SQAV, Individual consideration. IB, Individual consideration.

Mod Cities and Society/Software Engineering
UCAS Code: GL7H • Mode: 3 Years Equal Combination

Qualifications/Requirements: GCE, A level grades: DD-CCC. AGNVQ, Merit. ND/C, Individual consideration. SQAH, Individual consideration. SQAV, Individual consideration. IB, Individual consideration.

Mod Combined Studies/Software Engineering
UCAS Code: GY74 • Mode: 3 Years

Qualifications/Requirements: AGNVQ, Not normally sufficient. ND/C, Individual consideration. SQAH, Not normally sufficient. SQAV, Individual consideration. IB, Not normally sufficient.

Mod Complementary Therapies – Aromatherapy/Software Engineering
UCAS Code: GW78 • Mode: 3 Years Equal Combination

Qualifications/Requirements: Please refer to prospectus.

Mod Computer Systems/Software Engineering
UCAS Code: GG67 • Mode: 3 Years Equal Combination

Qualifications/Requirements: GCE, A level grades: CDD-BC. AGNVQ, Merit. ND/C, Individual consideration. SQAH, Individual consideration. SQAV, Individual consideration. IB, Individual consideration.

Mod Computing Mathematics/Software Engineering
UCAS Code: GG79 • Mode: 3 Years Equal Combination

Qualifications/Requirements: GCE, A level grades: CD-BC. AGNVQ, Merit. ND/C, Individual consideration. SQAH, Individual consideration. SQAV, Individual consideration. IB, Individual consideration.

Mod Computing Science/Software Engineering
UCAS Code: GG57 • Mode: 3 Years Equal Combination

Qualifications/Requirements: GCE, A level grades: DD-CCC. AGNVQ, Merit. ND/C, Individual consideration. SQAH, Individual consideration. SQAV, Individual consideration. IB, Individual consideration.

Mod Computing/Software Engineering
UCAS Code: GG75 • Mode: 3 Years Equal Combination

Qualifications/Requirements: GCE, A level grades: CDD-BCC. AGNVQ, Merit. ND/C, Individual consideration. SQAH, Individual consideration. SQAV, Individual consideration. IB, Individual consideration.

Mod Ecology/Software Engineering
UCAS Code: CG97 • Mode: 3 Years Equal Combination

Qualifications/Requirements: GCE, A level grades: CD-BC. Science. AGNVQ, Merit in Science. ND/C, Individual consideration. SQAH, Individual consideration. SQAV, Individual consideration. IB, Individual consideration.

Mod Economics/Software Engineering
UCAS Code: GL71 • Mode: 3 Years Equal Combination

Qualifications/Requirements: GCE, A level grades: CDD-BB. AGNVQ, Merit or Merit with 3 additional units. ND/C, Individual consideration. SQAH, Individual consideration. SQAV, Individual consideration. IB, Individual consideration.

Mod Educational Studies/Software Engineering

UCAS Code: GX79 • Mode: 3 Years Equal Combination

Qualifications/Requirements: GCE, A level grades: CC-BC. AGNVQ, Merit or Merit with 3 additional units. ND/C, Individual consideration. SQAH, Individual consideration. SQAV, Individual consideration. IB, Individual consideration.

Mod Electronics/Software Engineering

UCAS Code: GH76 • Mode: 3 Years Equal Combination

Qualifications/Requirements: GCE, A level grades: CC-BC. Science or Maths at A level. AGNVQ, Merit (in specific programmes). ND/C, Individual consideration. SQAH, Individual consideration. SQAV, Individual consideration. IB, Individual consideration.

Mod English Studies/Software Engineering

UCAS Code: GQ73 • Mode: 3 Years Equal Combination

Qualifications/Requirements: GCE, A level grades: CD-AB. AGNVQ, Merit with A/AS. ND/C, Individual consideration. SQAH, Individual consideration. SQAV, Individual consideration. IB, Individual consideration.

Mod Environmental Chemistry/Software Engineering

UCAS Code: GF71 • Mode: 3 Years Equal Combination

Qualifications/Requirements: GCE, A level grades - DD-CCC. Science. AGNVQ, Merit in Science. ND/C, Individual consideration. SQAH, Individual consideration. SQAV, Individual consideration. IB, Individual consideration.

Mod Environmental Design and Conservation/Software Engineering

UCAS Code: FG97 • Mode: 3 Years Equal Combination

Qualifications/Requirements: GCE, A level grades - DD-CCC. AGNVQ, Merit. ND/C, Individual consideration. SQAH, Individual consideration. SQAV, Individual consideration. IB, Individual consideration.

Mod Environmental Policy/Software Engineering

UCAS Code: KG37 • Mode: 3 Years Equal Combination

Qualifications/Requirements: GCE, A level grades - DD-CCC. AGNVQ, Merit. ND/C, Individual consideration. SQAH, Individual consideration. SQAV, Individual consideration. IB, Individual consideration.

Mod Environmental Sciences/Software Engineering

UCAS Code: FGX7 • Mode: 3 Years Equal Combination

Qualifications/Requirements: GCE, A level grades - CD-BC. Science. AGNVQ, Merit or Distinction in Science. ND/C, Individual consideration. SQAH, Individual consideration. SQAV, Individual consideration. IB, Individual consideration.

Mod European Culture and Society/Software Engineering

UCAS Code: TGG7 • Mode: 3 Years Equal Combination

Qualifications/Requirements: GCE, A level grades: CD-CCC. AGNVQ, Merit with A/AS or Merit with 3 additional units. ND/C, Individual consideration. SQAH, Individual consideration. SQAV, Individual consideration. IB, Individual consideration.

Mod Exercise and Health/Software Engineering

UCAS Code: GB76 • Mode: 3 Years Equal Combination

Qualifications/Requirements: GCE, A level grades - DD-BC. Science. AGNVQ, Merit in Science. ND/C, Individual consideration. SQAH, Individual consideration. SQAV, Individual consideration. IB, Individual consideration.

Mod Fine Art/Software Engineering

UCAS Code: GW71 • Mode: 3 Years Equal Combination

Qualifications/Requirements: GCE, A level grades: CDD-BC. Art. A portfolio of work is required. AGNVQ, Merit in Art & Design with A/AS. ND/C, Individual consideration. SQAH, Individual consideration. SQAV, Individual consideration. IB, Individual consideration.

Mod French Studies/Software Engineering

UCAS Code: RG17 • Mode: 3 Years Equal Combination

Qualifications/Requirements: Please refer to prospectus.

Mod Geographic Information Science/Software Engineering

UCAS Code: GF78 • Mode: 3 Years Equal Combination

Qualifications/Requirements: Please refer to prospectus.

Mod Geography/Software Engineering

UCAS Code: GL78 • Mode: 3 Years Equal Combination

Qualifications/Requirements: GCE, A level grades - CDD-BB. AGNVQ, Merit. ND/C, Individual consideration. SQAH, Individual consideration. SQAV, Individual consideration. IB, Individual consideration.

Mod Geology/Software Engineering

UCAS Code: FG67 • Mode: 3 Years Equal Combination

Qualifications/Requirements: GCE, A level grades - DD-BC. Maths or Science. AGNVQ, Pass in Science or Merit. ND/C, Individual consideration. SQAH, Individual consideration. SQAV, Individual consideration. IB, Individual consideration.

Mod Geotechnics/Software Engineering

UCAS Code: GH72 • Mode: 3 Years Equal Combination

Qualifications/Requirements: GCE, A level grades: DD-BC. Science, Maths, Design & Technology or Electronics. AGNVQ, Merit (in specific programmes). ND/C, Individual consideration. SQAH, Individual consideration. SQAV, Individual consideration. IB, Individual consideration.

Mod German Studies/Software Engineering

UCAS Code: GR7G • Mode: 4 Years Equal Combination

Qualifications/Requirements: GCE, A level grades: CDD-BC. German. AGNVQ, Merit with A level. ND/C, Individual consideration. SQAH, Individual consideration. SQAV, Individual consideration. IB, Individual consideration.

Mod History of Art/Software Engineering

UCAS Code: GV74 • Mode: 3 Years Equal Combination

Qualifications/Requirements: GCE, A level grades: CDD-BCC. AGNVQ, Merit with A/AS. ND/C, Individual consideration. SQAH, Individual consideration. SQAV, Individual consideration. IB, Individual consideration.

Mod History/Software Engineering

UCAS Code: GV71 • Mode: 3 Years Equal Combination

Qualifications/Requirements: GCE, A level grades: CDD-BB. AGNVQ, Merit with A/AS. ND/C, Individual consideration. SQAH, Individual consideration. SQAV, Individual consideration. IB, Individual consideration.

Mod Hospitality Management Studies/Software Engineering

UCAS Code: GN77 • Mode: 3 Years Equal Combination

Qualifications/Requirements: GCE, A level grades: BB-BC. AGNVQ, Merit or Merit with 3 additional units. ND/C, Individual consideration. SQAH, Individual consideration. SQAV, Individual consideration. IB, Individual consideration.

Mod Human Biology/Software Engineering

UCAS Code: BG17 • Mode: 3 Years Equal Combination

Qualifications/Requirements: GCE, A level grades: DD-BCC. Science. AGNVQ, Merit in Science. ND/C, Individual consideration. SQAH, Individual consideration. SQAV, Individual consideration. IB, Individual consideration.

Mod Information Systems/Software Engineering

UCAS Code: GGM7 • Mode: 3 Years Equal Combination

Qualifications/Requirements: GCE, A level grades: CDD-BC. AGNVQ, Merit. ND/C, Individual consideration. SQAH, Individual consideration. SQAV, Individual consideration. IB, Individual consideration.

Mod Law/Software Engineering

UCAS Code: GM73 • Mode: 3 Years Equal Combination

Qualifications/Requirements: GCE, A level grades: CDD-BBB. AGNVQ, Merit or Distinction with 3 additional units. ND/C, Individual consideration. SQAH, Individual consideration. SQAV, Individual consideration. IB, Individual consideration.

Mod Leisure Planning/Software Engineering

UCAS Code: KGH7 • Mode: 3 Years Equal Combination

Qualifications/Requirements: GCE, A level grades: DDD-BCC. AGNVQ, Merit with A/AS. ND/C, Individual consideration. SQAH, Individual consideration. SQAV, Individual consideration. IB, Individual consideration.

Mod Marketing Management/Software Engineering

UCAS Code: GN7N • Mode: 3 Years Equal Combination

Qualifications/Requirements: GCE, A level grades: CDD-BCC. AGNVQ, Merit or Distinction with 3 additional units. ND/C, Individual consideration. SQAH, Individual consideration. SQAV, Individual consideration. IB, Individual consideration.

Mod Mathematics/Software Engineering

UCAS Code: GG71 • Mode: 3 Years Equal Combination

Qualifications/Requirements: GCE, A level grades: DD-BC. Maths. AGNVQ, Merit with A level. ND/C, Individual consideration. SQAH, Individual consideration. SQAV, Individual consideration. IB, Individual consideration.

Mod Music/Software Engineering

UCAS Code: GW73 • Mode: 3 Years Equal Combination

Qualifications/Requirements: GCE, A level grades: DD-BC. Music. AGNVQ, Merit. ND/C, Individual consideration. SQAH, Individual consideration. SQAV, Individual consideration. IB, Individual consideration.

Mod Palliative Care/Software Engineering

UCAS Code: BGR7 • Mode: 3 Years Equal Combination

Qualifications/Requirements: Please refer to prospectus.

Mod Physical Geography/Software Engineering

UCAS Code: FGV7 • Mode: 3 Years Equal Combination

Qualifications/Requirements: GCE, A level grades - DDD-BCC. AGNVQ, Merit with A/AS. ND/C, Individual consideration. SQAH, Individual consideration. SQAV, Individual consideration. IB, Individual consideration.

Mod Planning Studies/Software Engineering

UCAS Code: GK74 • Mode: 3 Years Equal Combination

Qualifications/Requirements: GCE, A level grades: CD-CC. AGNVQ, Merit. ND/C, Individual consideration. SQAH, Individual consideration. SQAV, Individual consideration. IB, Individual consideration.

Mod Politics/Software Engineering

UCAS Code: GM71 • Mode: 3 Years Equal Combination

Qualifications/Requirements: GCE, A level grades: CD-AB. AGNVQ, Merit with A/AS. ND/C, Individual consideration. SQAH, Individual consideration. SQAV, Individual consideration. IB, Individual consideration.

Mod Psychology/Software Engineering

UCAS Code: CG87 • Mode: 3 Years Equal Combination

Qualifications/Requirements: GCE, A level grades - CC-BC. AGNVQ, Merit with A/AS. ND/C, Individual consideration. SQAH, Individual consideration. SQAV, Individual consideration. IB, Individual consideration.

Mod Publishing/Software Engineering

UCAS Code: GP75 • Mode: 3 Years Equal Combination

Qualifications/Requirements: GCE, A level grades: CDD-CCC. AGNVQ, Merit or Merit (in specific programmes) with 3 additional units. ND/C, Individual consideration. SQAH, Individual consideration. SQAV, Individual consideration. IB, Individual consideration.

Mod Rehabilitation/Software Engineering

UCAS Code: BGT7 • Mode: 3 Years Equal Combination

Qualifications/Requirements: Please refer to prospectus.

Mod Retail Management/Software Engineering

UCAS Code: GN75 • Mode: 3 Years Equal Combination

Qualifications/Requirements: GCE, A level grades: CC-BBC. AGNVQ, Merit in Business with 4 additional units. ND/C, Individual consideration. SQAH, Individual consideration. SQAV, Individual consideration. IB, Individual consideration.

Mod Sociology/Software Engineering

UCAS Code: GL73 • Mode: 3 Years Equal Combination

Qualifications/Requirements: GCE, A level grades: CDD-BCC. AGNVQ, Merit with A/AS. ND/C, Individual consideration. SQAH, Individual consideration. SQAV, Individual consideration. IB, Individual consideration.

Mod Software Engineering/Statistics

UCAS Code: GG74 • Mode: 3 Years Equal Combination

Qualifications/Requirements: GCE, A level grades: DD-BC. AGNVQ, Merit. ND/C, Individual consideration. SQAH, Individual consideration. SQAV, Individual consideration. IB, Individual consideration.

Mod Software Engineering/ Telecommunications

UCAS Code: GH7P • Mode: 3 Years Equal Combination

Qualifications/Requirements: GCE, A level grades: DD-CCC. Science or Maths. AGNVQ, Merit (in specific programmes). ND/C, Individual consideration. SQAH, Individual consideration. SQAV, Individual consideration. IB, Individual consideration.

Mod Software Engineering/Tourism

UCAS Code: GP77 • Mode: 3 Years Equal Combination

Qualifications/Requirements: GCE, A level grades: CDD-BC. AGNVQ, Merit or Merit with 3 additional units. ND/C, Individual consideration. SQAH, Individual consideration. SQAV, Individual consideration. IB, Individual consideration.

Mod Software Engineering/Transport and Travel

UCAS Code: GN79 • Mode: 3 Years Equal Combination

Qualifications/Requirements: GCE, A level grades: DDD-BC. AGNVQ, Merit. ND/C, Individual consideration. SQAH, Individual consideration. SQAV, Individual consideration. IB, Individual consideration.

Mod Software Engineering/Water Resources

UCAS Code: GH7F • Mode: 3 Years Equal Combination

Qualifications/Requirements: GCE, A level grades: DD-CCC. AGNVQ, Merit with A/AS. ND/C, Individual consideration. SQAH, Individual consideration. SQAV, Individual consideration. IB, Individual consideration.

UNIVERSITY OF PAISLEY P20

BA Business Information Technology and Accounting

UCAS Code: GN74 • Mode: 3/4/5 Years Equal Combination

Qualifications/Requirements: GCE, A level grades - CD. AGNVQ, Pass in Information Technology. ND/C, Individual consideration. SQAH, Grades BCCC including specific subject(s). SQAV, Higher National. IB, Pass in Diploma.

BA Business Information Technology and Human Resource Management

UCAS Code: GN71 • Mode: 3/4/5 Years Equal Combination

Qualifications/Requirements: GCE, A level grades - CD. AGNVQ, Pass in Information Technology. ND/C, Individual consideration. SQAH, Grades BCCC including specific subject(s). SQAV, Higher National. IB, Pass in Diploma.

BA Business Information Technology and Languages

UCAS Code: GT72 • Mode: 3/4/5 Years Equal Combination

Qualifications/Requirements: GCE, A level grades - CD. AGNVQ, Pass in Information Technology. ND/C, Individual consideration. SQAH, Grades BCCC including specific subject(s). SQAV, Higher National. IB, Pass in Diploma.

BA Business Information Technology and Marketing

UCAS Code: GN75 • Mode: 3/4/5 Years Equal Combination

Qualifications/Requirements: GCE, A level grades - CD. AGNVQ, Pass in Information Technology. ND/C, Individual consideration. SQAH, Grades BCCC including specific subject(s). SQAV, Higher National. IB, Pass in Diploma.

BSc Business Information Technology

UCAS Code: G720 • Mode: 3/4/5 Years Single Subject

Qualifications/Requirements: GCE, A level grades - CD. AGNVQ, Pass in Information Technology. ND/C, Individual consideration. SQAH, Grades BBCC including specific subject(s). SQAV, Higher National. IB, Pass in Diploma.

BSc Business Information Technology and Multimedia

UCAS Code: G7G5 • Mode: 3/4/5 Years Major/Minor

Qualifications/Requirements: GCE, A level grades - CD. AGNVQ, Pass in Information Technology. ND/C, Individual consideration. SQAH, Grades BCCC including specific subject(s). SQAV, Higher National. IB, Pass in Diploma.

BSc Software Engineering

UCAS Code: G700 • Mode: 3/4/5 Years Single Subject

Qualifications/Requirements: GCE, A level grades: CC. Maths. AGNVQ, Pass in Information Technology. ND/C, Individual consideration. SQAH, Grades BCCC including specific subject(s). SQAV, National including specific subject(s). IB, Pass in Diploma.

BSc Software Engineering with French/German/Spanish

UCAS Code: G7TF • Mode: 3/4/5 Years Major/Minor

Qualifications/Requirements: GCE, A level grades: CC. Maths. AGNVQ, Pass in Information Technology. ND/C, Individual consideration. SQAH, Grades BCCC including specific subject(s). SQAV, Higher National. IB, Pass in Diploma.

UNIVERSITY OF PORTSMOUTH P80

BSc Software Engineering

UCAS Code: G700 • Mode: 4 Years Single Subject

Qualifications/Requirements: GCE, A/AS: 20 points. AGNVQ, Distinction. ND/C, Merits and 2 Distinctions. SQAH, Grades BBBB. SQAV, Individual consideration. IB, 30 points.

HND Software Engineering

UCAS Code: 007G • Mode: 2 Years Single Subject

Qualifications/Requirements: GCE, A/AS: 6 points. AGNVQ, Merit. ND/C, 3 Merits. SQAH, Grades CCC. SQAV, Individual consideration. IB, Pass in Diploma.

THE QUEEN'S UNIVERSITY OF BELFAST Q75

BEng Electronic and Software Engineering (including professional experience) (4 years SW)

UCAS Code: HG67 • Mode: 4 Years Equal Combination

Qualifications/Requirements: GCE, A level grades: BCC. Maths and Physics. AGNVQ, Individual consideration. ND/C, Individual consideration. SQAH, Grades BBBC. SQAV, Individual consideration. IB, 29 points.

RAVENSBOURNE COLLEGE OF DESIGN AND COMMUNICATION R06

BSc Communications and Technology

UCAS Code: G720 • Mode: 2 Years Single Subject

Qualifications/Requirements: Maths, Physics, Computing or Electronics. AGNVQ, Pass in Media. ND/C, National. SQAH, Individual consideration. SQAV, Individual consideration. IB, Individual consideration.

ROEHAMPTON INSTITUTE LONDON R48

BA/BSc Business Computing and Early Childhood Studies

UCAS Code: XG97 • Mode: 3 Years Equal Combination

Qualifications/Requirements: Please refer to prospectus.

BSc Business Computing

UCAS Code: G710 • Mode: 3 Years Single Subject

Qualifications/Requirements: GCE, A/AS: 12-16 points. AGNVQ, Merit (in specific programmes). ND/C, 2 Merits and 2 Distinctions. SQAH, Grades BCC. SQAV, National including specific subject(s). IB, 26 points.

Mod Business Computing & English as a Foreign Language

UCAS Code: GQ7J • Mode: 3 Years Equal Combination

Qualifications/Requirements: GCE, A/AS: 12-16 points. An approved subject from restricted list. AGNVQ, Merit (in specific programmes). ND/C, 4 Merits (in specific programmes). SQAH, Grades BCC. SQAV, National including specific subject(s). IB, 26 points.

Mod Business Computing and Business Studies

UCAS Code: GN71 • Mode: 3 Years Equal Combination

Qualifications/Requirements: GCE, A/AS: 12-16 points. AGNVQ, Merit (in specific programmes). ND/C, 3 Distinctions. SQAH, Grades BCC. SQAV, National including specific subject(s). IB, 26 points.

Mod Business Computing and Childhood Studies

UCAS Code: GX7Y • Mode: 3 Years Equal Combination

Qualifications/Requirements: GCE, A/AS: 12-16 points. AGNVQ, Merit. ND/C, 3 Distinctions. SQAH, Grades BCC. SQAV, National including specific subject(s). IB, 26 points.

Mod Business Computing and Cultural Studies

UCAS Code: GV79 • Mode: 3 Years Equal Combination

Qualifications/Requirements: GCE, A/AS: 12-16 points. AGNVQ, Merit. ND/C, 3 Distinctions. SQAH, Grades BCC. SQAV, National including specific subject(s). IB, 26 points.

Mod Business Computing and Dance Studies

UCAS Code: GW74 • Mode: 3 Years Equal Combination

Qualifications/Requirements: GCE, A level grades - CC. AGNVQ, Merit (in specific programmes) with A level. ND/C, 3 Distinctions (in specific programmes). SQAH, Grades BCC. SQAV, National including specific subject(s). IB, 30 points.

Mod Business Computing and Drama & Theatre Studies

UCAS Code: GW7L • Mode: 3 Years Equal Combination

Qualifications/Requirements: GCE, A/AS: 16-20 points. English or Theatre Studies at A level. AGNVQ, Merit (in specific programmes) with A level. ND/C, 2 Merits and 2 Distinctions (in specific programmes). SQAH, Grades BBC. SQAV, National including specific subject(s). IB, 30 points.

Mod Business Computing and Education

UCAS Code: GX79 • Mode: 3 Years Equal Combination

Qualifications/Requirements: GCE, A/AS: 12-16 points. AGNVQ, Merit (in specific programmes). ND/C, 3 Distinctions. SQAH, Grades BCC. SQAV, National including specific subject(s). IB, 26 points.

Mod Business Computing and English Language & Linguistics

UCAS Code: GQ7H • Mode: 3 Years Equal Combination

Qualifications/Requirements: GCE, A/AS: 12-16 points. AGNVQ, Merit (in specific programmes) with A level. ND/C, 2 Merits and 2 Distinctions. SQAH, Grades BCC. SQAV, National including specific subject(s). IB, 30 points.

Mod Business Computing and English Literature

UCAS Code: GQ73 • Mode: 3 Years Equal Combination

Qualifications/Requirements: GCE, A/AS: 12-16 points. English at A level. AGNVQ, Merit with A level. ND/C, 2 Merits and 2 Distinctions. SQAH, Grades BCC. SQAV, National including specific subject(s). IB, 30 points.

Mod Business Computing and Film & Television Studies

UCAS Code: GP74 • Mode: 3 Years Equal Combination

Qualifications/Requirements: GCE, A/AS: 16-20 points. An approved subject from restricted list. AGNVQ, Distinction (in specific programmes) with A level. ND/C, 3 Distinctions. SQAH, Grades BBC. SQAV, National including specific subject(s). IB, 30 points.

Mod Business Computing and French

UCAS Code: GR71 • Mode: 4 Years Equal Combination

Qualifications/Requirements: GCE, A/AS: 12-16 points. French at A level. AGNVQ, Merit with A level. ND/C, 3 Distinctions (in specific programmes). SQAH, Grades BCC. SQAV, National including specific subject(s). IB, 26 points.

Mod Business Computing and French Studies

UCAS Code: GR7D • Mode: 3 Years Equal Combination

Qualifications/Requirements: GCE, A/AS: 12-16 points. French at A level. AGNVQ, Merit with A level. ND/C, 3 Distinctions (in specific programmes). SQAH, Grades BCC. SQAV, National including specific subject(s). IB, 26 points.

Mod Business Computing and Health Studies

UCAS Code: GB79 • Mode: 3 Years Equal Combination

Qualifications/Requirements: GCE, A/AS: 12-16 points. AGNVQ, Merit (in specific programmes). ND/C, 3 Distinctions. SQAH, Grades BCC. SQAV, National including specific subject(s). IB, 26 points.

Mod Business Computing and History

UCAS Code: GV71 • Mode: 3 Years Equal Combination

Qualifications/Requirements: GCE, A/AS: 12-16 points. History at A level. AGNVQ, Merit with A level. ND/C, 3 Distinctions. SQAH, Grades BCC. SQAV, National including specific subject(s). IB, 26 points.

Mod Business Computing and Human & Social Biology

UCAS Code: GC7C • Mode: 3 Years Equal Combination

Qualifications/Requirements: GCE, A/AS: 12-16 points. AGNVQ, Merit (in specific programmes). ND/C, 3 Distinctions. SQAH, Grades BCC. SQAV, National including specific subject(s). IB, 26 points.

Mod Business Computing and Leisure Management

UCAS Code: GN77 • Mode: 3 Years Equal Combination

Qualifications/Requirements: GCE, A/AS: 12-16 points. AGNVQ, Merit (in specific programmes). ND/C, 4 Merits (in specific programmes). SQAH, Grades BCC. SQAV, National including specific subject(s). IB, 26 points.

Mod Business Computing and Music
UCAS Code: GW73 • Mode: 3 Years Equal Combination

Qualifications/Requirements: GCE, A/AS: 12-16 points. Music at A level. AGNVQ, Merit with A level. ND/C, 3 Distinctions. SQAH, Grades BCC. SQAV, National including specific subject(s). IB, 26 points.

Mod Business Computing and Psychology
UCAS Code: GL77 • Mode: 3 Years Equal Combination

Qualifications/Requirements: GCE, A/AS: 12-16 points. AGNVQ, Merit with A level. ND/C, 3 Distinctions. SQAH, Grades BBC. SQAV, National including specific subject(s). IB, 30 points.

Mod Business Computing and Social Anthropology
UCAS Code: GC79 • Mode: 3 Years Equal Combination

Qualifications/Requirements: GCE, A/AS: 12-16 points. AGNVQ, Merit (in specific programmes). ND/C, 3 Distinctions. SQAH, Grades BCC. SQAV, National including specific subject(s). IB, 26 points.

Mod Business Computing and Social Policy & Administration
UCAS Code: GL74 • Mode: 3 Years Equal Combination

Qualifications/Requirements: GCE, A/AS: 12-16 points. AGNVQ, Merit (in specific programmes). ND/C, 3 Distinctions. SQAH, Grades BCC. SQAV, National including specific subject(s). IB, 26 points.

Mod Business Computing and Sociology
UCAS Code: GL73 • Mode: 3 Years Equal Combination

Qualifications/Requirements: GCE, A/AS: 12-16 points. AGNVQ, Merit (in specific programmes). ND/C, 3 Distinctions. SQAH, Grades BCC. SQAV, National including specific subject(s). IB, 26 points.

Mod Business Computing and Spanish
UCAS Code: GR74 • Mode: 3 Years Equal Combination

Qualifications/Requirements: GCE, A/AS: 12-16 points. AGNVQ, Merit with A level. ND/C, 3 Distinctions. SQAH, Grades BCC. SQAV, National including specific subject(s). IB, 26 points.

Mod Business Computing and Sport Studies
UCAS Code: GB76 • Mode: 3 Years Equal Combination

Qualifications/Requirements: GCE, A/AS: 12-16 points. Science. AGNVQ, Merit (in specific programmes). ND/C, 2 Merits and 2 Distinctions. SQAH, Grades BCC. SQAV, National including specific subject(s). IB, 28 points.

Mod Business Computing and Theology & Religious Studies
UCAS Code: GV78 • Mode: 3 Years Equal Combination

Qualifications/Requirements: GCE, A/AS: 12-16 points. AGNVQ, Merit (in specific programmes). ND/C, 3 Distinctions. SQAH, Grades BCC. SQAV, National including specific subject(s). IB, 26 points.

Mod Business Computing and Women's Studies
UCAS Code: GM79 • Mode: 3 Years Equal Combination

Qualifications/Requirements: GCE, A/AS: 12-16 points. AGNVQ, Merit (in specific programmes). ND/C, 3 Distinctions. SQAH, Grades BCC. SQAV, National including specific subject(s). IB, 26 points.

THE UNIVERSITY OF SALFORD S03

HND Software
UCAS Code: 107G • Mode: 2 Years Single Subject

Qualifications/Requirements: GCE, A/AS: 8 points. AGNVQ, Merit (in specific programmes). ND/C, National. SQAH, Grades CC. SQAV, Individual consideration. IB, Individual consideration.

SANDWELL COLLEGE S08

HND Animation and Graphics
UCAS Code: 67HG • Mode: 2 Years Equal Combination

Qualifications/Requirements: GCE, A level grades: DD-EE. Art, Art & Design, Computing, Design & Technology or Maths. AGNVQ, Pass (in specific programmes). ND/C, National. SQAH, Individual consideration. SQAV, Individual consideration. IB, Individual consideration.

HND Software Engineering
UCAS Code: 007G • Mode: 2 Years Single Subject

Qualifications/Requirements: GCE, A level grades: DD-EE. Computing, Design & Technology, Maths or Physics. AGNVQ, Pass. ND/C, National. SQAH, Individual consideration. SQAV, Individual consideration. IB, Individual consideration.

THE UNIVERSITY OF SHEFFIELD S18

MEng Software Engineering
UCAS Code: G700 • Mode: 3/4 Years Single Subject

Qualifications/Requirements: GCE, A/AS: 24 points. Maths at A level. AGNVQ, Distinction with A level. ND/C, 3 Merits and 3 Distinctions. SQAH, Grades AABB including specific subject(s). SQAV, Individual consideration. IB, 32 points.

MEng Software Engineering (including Foundation Year)
UCAS Code: G701 • Mode: 4/5 Years Single Subject

Qualifications/Requirements: GCE, A level grades: BBB-BCC. AGNVQ, Merit or Distinction with A level. ND/C, 3 Merits and 3 Distinctions. SQAH, Grades ABBB. IB, 30 points.

MEng Software Engineering and Law

UCAS Code: GM73 • Mode: 4 Years Equal Combination

Qualifications/Requirements: GCE, A level grades: AAB. AGNVQ, Not normally sufficient. ND/C, 2 Merits and 4 Distinctions (in specific programmes). SQAH, Grades AAAA including specific subject(s). SQAV, Individual consideration. IB, 35 points.

SHEFFIELD HALLAM UNIVERSITY S21

BSc Business Network Engineering

UCAS Code: G710 • Mode: 3 Years Single Subject

Qualifications/Requirements: GCE, A/AS: 14 points. Science. AGNVQ, Distinction (in specific programmes). ND/C, 5 Merits. SQAH, Individual consideration. SQAV, Individual consideration. IB, Individual consideration.

BSc Computing (Software Engineering)

UCAS Code: G700 • Mode: 4 Years Single Subject

Qualifications/Requirements: GCE, A/AS: 12 points. AGNVQ, Merit. ND/C, 6 Merits. SQAH, Individual consideration. SQAV, Individual consideration. IB, Individual consideration.

HND Business Network Engineering

UCAS Code: 017G • Mode: 2 Years Single Subject

Qualifications/Requirements: GCE, A level grades - E. Maths or Science. AGNVQ, Pass. ND/C, National.

HND Software Engineering

UCAS Code: 007G • Mode: 3 Years Single Subject

Qualifications/Requirements: GCE, A/AS: 6 points. AGNVQ, Pass. ND/C, 2 Merits. SQAH, Individual consideration. SQAV, Individual consideration. IB, Individual consideration.

UNIVERSITY OF SOUTHAMPTON S27

MEng Software Engineering

UCAS Code: G700 • Mode: 4 Years Single Subject

Qualifications/Requirements: GCE, A/AS: 26 points. Maths at A level. AGNVQ, Distinction with A level. ND/C, Individual consideration. SQAH, CSYS required. SQAV, Individual consideration. IB, 32 points.

SOUTHAMPTON INSTITUTE S30

BSc Software Engineering

UCAS Code: G700 • Mode: 3 Years Single Subject

Qualifications/Requirements: GCE, A/AS: 12 points. AGNVQ, Merit (in specific programmes). ND/C, Merit overall. SQAH, Grades CCCC. SQAV, National. IB, Pass in Diploma.

BSc Software Engineering (with foundation)

UCAS Code: G708 • Mode: 4 Years Single Subject

Qualifications/Requirements: GCE, A/AS: 2-4 points. AGNVQ, Pass (in specific programmes). ND/C, National. SQAH, Grades CCCC. SQAV, National. IB, Pass in Diploma.

HND Business Information Technology

UCAS Code: 027G • Mode: 2 Years Single Subject

Qualifications/Requirements: GCE, A/AS: 2 points. AGNVQ, Pass (in specific programmes). ND/C, National. SQAH, Grades CCCC. SQAV, National. IB, Pass in Diploma.

SOUTH BANK UNIVERSITY S33

BA/BSc Business Information Technology and Music

UCAS Code: WG37 • Mode: 3 Years Equal Combination

Qualifications/Requirements: GCE, A/AS: 12-16 points. Maths and Music. AGNVQ, Merit. ND/C, Merit overall. SQAH, Individual consideration. SQAV, Individual consideration. IB, Individual consideration.

BSc Business Information Technology

UCAS Code: G720 • Mode: 3/4 Years Single Subject

Qualifications/Requirements: GCE, A level grades - CC. Computing, Business Studies at A level, Maths or Science. AGNVQ, Merit in Business or in Information Technology. ND/C, 5 Merits. SQAH, Individual consideration. SQAV, Individual consideration. IB, Individual consideration.

BSc Business Information Technology

UCAS Code: G725 • Mode: 1/2 Years Single Subject

Qualifications/Requirements: ND/C, Higher National. SQAH, Individual consideration. SQAV, Individual consideration. IB, Individual consideration.

BEng Software Engineering for Real Time Systems

UCAS Code: HG67 • Mode: 3/4 Years Equal Combination

Qualifications/Requirements: GCE, A level grades: DD. Maths and Science at A level. AGNVQ, Merit. ND/C, Merits and Distinction. SQAH, Individual consideration. SQAV, Individual consideration. IB, Individual consideration.

Mod Business Information Technology & Health Studies

UCAS Code: GL74 • Mode: 3 Years Equal Combination

Qualifications/Requirements: GCE, A/AS: 12-16 points. Maths and Science at A level. AGNVQ, Merit. ND/C, 4 Merits and 2 Distinctions. SQAH, Individual consideration. SQAV, Individual consideration. IB, Individual consideration.

Mod Business Information Technology and Accounting

UCAS Code: GN74 • Mode: 3 Years Equal Combination

Qualifications/Requirements: GCE, A/AS: 12-16 points. Maths at A level. AGNVQ, Merit. ND/C, 4 Merits and 2 Distinctions. SQAH, Individual consideration. SQAV, Individual consideration. IB, Individual consideration.

Mod Business Information Technology and Crime Studies

UCAS Code: GM7H • Mode: 3 Years Equal Combination

Qualifications/Requirements: GCE, A/AS: 12-18 points. AGNVQ, Merit. ND/C, Merit overall. SQAH, Individual consideration. SQAV, Individual consideration. IB, 28 points.

Mod Business Information Technology and French

UCAS Code: GR71 • Mode: 3 Years Equal Combination

Qualifications/Requirements: GCE, A/AS: 12-16 points. Maths and French at A level. AGNVQ, Merit. ND/C, 4 Merits and 2 Distinctions. SQAH, Individual consideration. SQAV, Individual consideration. IB, Individual consideration.

Mod Business Information Technology and German

UCAS Code: GR7F • Mode: 3 Years Equal Combination

Qualifications/Requirements: GCE, A/AS: 14-18 points. Maths and German at A level. AGNVQ, Merit. ND/C, 2 Merits and 4 Distinctions. SQAH, Individual consideration. SQAV, Individual consideration. IB, Individual consideration.

Mod Business Information Technology and German ab initio

UCAS Code: GR72 • Mode: 3 Years Equal Combination

Qualifications/Requirements: GCE, A/AS: 12-16 points. Maths at A level. AGNVQ, Merit. ND/C, 4 Merits and 2 Distinctions. SQAH, Individual consideration. SQAV, Individual consideration. IB, Individual consideration.

Mod Business Information Technology and Human Geography

UCAS Code: GL78 • Mode: 3 Years Equal Combination

Qualifications/Requirements: GCE, A/AS: 12-16 points. Maths and Geography at A level. AGNVQ, Merit. ND/C, 4 Merits and 2 Distinctions. SQAH, Individual consideration. SQAV, Individual consideration. IB, Individual consideration.

Mod Business Information Technology and Human Resource Management

UCAS Code: GN76 • Mode: 3 Years Equal Combination

Qualifications/Requirements: GCE, A/AS: 12 points. Maths at A level. AGNVQ, Merit. ND/C, 4 Merits and 2 Distinctions. SQAH, Individual consideration. SQAV, Individual consideration. IB, Individual consideration.

Mod Business Information Technology and Law

UCAS Code: GM73 • Mode: 3 Years Equal Combination

Qualifications/Requirements: GCE, A/AS: 14-18 points. Maths at A level. AGNVQ, Merit. ND/C, 2 Merits and 4 Distinctions. SQAH, Individual consideration. SQAV, Individual consideration. IB, Individual consideration.

Mod Business Information Technology and Management

UCAS Code: GN71 • Mode: 3 Years Equal Combination

Qualifications/Requirements: GCE, A/AS: 12-16 points. Maths at A level. AGNVQ, Merit. ND/C, 4 Merits and 2 Distinctions. SQAH, Individual consideration. SQAV, Individual consideration. IB, Individual consideration.

Mod Business Information Technology and Marketing

UCAS Code: GN75 • Mode: 3 Years Equal Combination

Qualifications/Requirements: GCE, A/AS: 12-16 points. Maths at A level. AGNVQ, Merit. ND/C, 2 Merits and 4 Distinctions. SQAH, Individual consideration. SQAV, Individual consideration. IB, Individual consideration.

Mod Business Information Technology and Media Studies

UCAS Code: GP74 • Mode: 3 Years Equal Combination

Qualifications/Requirements: GCE, A/AS: 14-18 points. Maths and English at A level. AGNVQ, Distinction. ND/C, 4 Merits and 2 Distinctions. SQAH, Individual consideration. SQAV, Individual consideration. IB, Individual consideration.

Mod Business Information Technology and Politics

UCAS Code: GM71 • Mode: 3 Years Equal Combination

Qualifications/Requirements: GCE, A/AS: 12-16 points. Maths at A level. AGNVQ, Merit. ND/C, 4 Merits and 2 Distinctions. SQAH, Individual consideration. SQAV, Individual consideration. IB, Individual consideration.

Mod Business Information Technology and Product Design

UCAS Code: GH77 • Mode: 3 Years Equal Combination

Qualifications/Requirements: GCE, A/AS: 12-16 points. Maths and Art & Design at A level. AGNVQ, Merit. ND/C, 4 Merits and 2 Distinctions. SQAH, Individual consideration. SQAV, Individual consideration. IB, Individual consideration.

Mod Business Information Technology and Psychology

UCAS Code: CG87 • Mode: 3 Years Equal Combination

Qualifications/Requirements: GCE, A/AS: 14-18 points. Science at A level. AGNVQ, Merit. ND/C, 2 Merits and 4 Distinctions. SQAH, Individual consideration. SQAV, Individual consideration. IB, Individual consideration.

Mod Business Information Technology and Spanish

UCAS Code: GR7K • Mode: 3 Years Equal Combination

Qualifications/Requirements: GCE, A/AS: 14-18 points. Spanish and Maths at A level. AGNVQ, Merit. ND/C, 2 Merits and 4 Distinctions. SQAH, Individual consideration. SQAV, Individual consideration. IB, Individual consideration.

Mod Business Information Technology and Spanish - ab initio

UCAS Code: GR74 • Mode: 3 Years Equal Combination

Qualifications/Requirements: GCE, A/AS: 12-16 points. Maths at A level. AGNVQ, Merit. ND/C, 4 Merits and 2 Distinctions. SQAH, Individual consideration. SQAV, Individual consideration. IB, Individual consideration.

Mod Business Information Technology and Tourism

UCAS Code: GP77 • Mode: 3 Years Equal Combination

Qualifications/Requirements: GCE, A/AS: 12-16 points. Maths and a modern foreign language at A level. AGNVQ, Merit. ND/C, 4 Merits and 2 Distinctions. SQAH, Individual consideration. SQAV, Individual consideration. IB, Individual consideration.

Mod Business Information Technology and World Theatre

UCAS Code: GW74 • Mode: 3 Years Equal Combination

Qualifications/Requirements: GCE, A/AS: 14-18 points. Maths at A level. AGNVQ, Merit. ND/C, 2 Merits and 4 Distinctions. SQAH, Individual consideration. SQAV, Individual consideration. IB, Individual consideration.

HND Business Information Technology

UCAS Code: 027G • Mode: 2 Years Single Subject

Qualifications/Requirements: GCE, A level grades - D. Business Studies, Computing or Science. AGNVQ, Pass in Business or in Information Technology. ND/C, Individual consideration. SQAH, Individual consideration. SQAV, Individual consideration. IB, Individual consideration.

ST HELENS COLLEGE S51

HND Software Engineering

UCAS Code: 007G • Mode: 2 Years Single Subject

Qualifications/Requirements: GCE, A/AS: 2-6 points. Maths, Physics, Electronics or Computing at A level. AGNVQ, Pass (in specific programmes). ND/C, National. SQAH, Individual consideration. SQAV, Individual consideration. IB, Individual consideration.

STAFFORDSHIRE UNIVERSITY S72

BSc Business Computing

UCAS Code: G711 • Mode: 4 Years Single Subject

Qualifications/Requirements: GCE, A/AS: 12 points. AGNVQ, Merit. ND/C, Individual consideration. SQAH, Grades CCC. SQAV, Individual consideration. IB, 27 points.

BSc Foundation Software Engineering

UCAS Code: G708 • Mode: 5 Years Single Subject

Qualifications/Requirements: Please refer to prospectus.

BSc Software Engineering

UCAS Code: G700 • Mode: 4 Years Single Subject

Qualifications/Requirements: GCE, A/AS: 12 points. AGNVQ, Merit. ND/C, Individual consideration. SQAH, Grades CCC. SQAV, Individual consideration. IB, 27 points.

BEng Foundation Software Engineering

UCAS Code: G709 • Mode: 5 Years Single Subject

Qualifications/Requirements: Please refer to prospectus.

MEng Software Engineering

UCAS Code: G701 • Mode: 5 Years Single Subject

Qualifications/Requirements: GCE, A/AS: 12 points. AGNVQ, Merit. ND/C, Individual consideration. SQAH, Grades CCC. SQAV, Individual consideration. IB, 27 points.

HND Software Engineering

UCAS Code: 007G • Mode: 2 Years Single Subject

Qualifications/Requirements: GCE, A/AS: 4 points. AGNVQ, Pass. ND/C, National. SQAH, Grades CC. SQAV, Individual consideration. IB, 24 points.

THE UNIVERSITY OF STIRLING S75

BA Business Computing

UCAS Code: G710 • Mode: 4 Years Single Subject

Qualifications/Requirements: GCE, A level grades - CCD. AGNVQ, Merit (in specific programmes) with 2 additional units or with A/AS. ND/C, Merit overall and 2 Distinctions. SQAH, Grades BBCC. SQAV, Higher National. IB, 28 points.

BSc Software Engineering

UCAS Code: G700 • Mode: 4 Years Single Subject

Qualifications/Requirements: GCE, A level grades: CCD. AGNVQ, Merit (in specific programmes) with 2 additional units or with A/AS. ND/C, Merit overall. SQAH, Grades BBCC. SQAV, Higher National. IB, 28 points.

STOCKPORT COLLEGE OF FURTHER & HIGHER EDUCATION S76

HND Computing (Business Information Technology)

UCAS Code: 005G • Mode: 2 Years Single Subject

Qualifications/Requirements: English and Maths. AGNVQ, Pass in Information Technology. ND/C, National. SQAH, Individual consideration. SQAV, Individual consideration. IB, Individual consideration.

HND Computing (Software Engineering)

UCAS Code: 65GG • Mode: 2 Years Equal Combination

Qualifications/Requirements: Please refer to prospectus.

THE UNIVERSITY OF STRATHCLYDE S78

BSc Software Engineering

UCAS Code: G700 • Mode: 4 Years Single Subject

Qualifications/Requirements: GCE, A level grades: BCC. Maths at A level. ND/C, Individual consideration. SQAH, Grades BBBC including specific subject(s). SQAV, Individual consideration. IB, 28 points.

SUFFOLK COLLEGE: AN ACCREDITED COLLEGE OF THE UNIVERSITY OF EAST ANGLIA S81

BSc Software Engineering

UCAS Code: G700 • Mode: 3 Years Single Subject

Qualifications/Requirements: GCE, A level grades: EE. AGNVQ, Pass (in specific programmes). ND/C, National (in specific programmes). SQAH, Individual consideration. SQAV, Individual consideration. IB, Individual consideration.

HND Software Engineering

UCAS Code: 007G • Mode: 2 Years Single Subject

Qualifications/Requirements: GCE, A level grades: E. AGNVQ, Pass (in specific programmes). ND/C, National (in specific programmes). SQAH, Individual consideration. SQAV, Individual consideration. IB, Individual consideration.

UNIVERSITY OF SUNDERLAND S84

BA Business Information Technology

UCAS Code: G720 • Mode: 3 Years Single Subject

Qualifications/Requirements: Please refer to prospectus.

SWANSEA INSTITUTE OF HIGHER EDUCATION S96

BSc Business Information Technology

UCAS Code: G710 • Mode: 3 Years Single Subject

Qualifications/Requirements: GCE, A/AS: 4-20 points. An approved subject from restricted list. AGNVQ, Merit. ND/C, 2 Merits. SQAH, Grades CCCC. SQAV, Individual consideration. IB, 24 points.

BSc Software Engineering

UCAS Code: G700 • Mode: 3 Years Single Subject

Qualifications/Requirements: GCE, A level grades: EE. AGNVQ, Merit. ND/C, 2 Merits. SQAH, Grades CCCC. SQAV, Individual consideration. IB, 24 points.

HND Computer Software Engineering

UCAS Code: 007G • Mode: 2 Years Single Subject

Qualifications/Requirements: Please refer to prospectus.

SWINDON COLLEGE S98

HND Business Information Technology

UCAS Code: 027G • Mode: 2 Years Single Subject

Qualifications/Requirements: GCE, A/AS: 4-12 points. AGNVQ, Merit (in specific programmes). ND/C, Individual consideration. SQAH, Individual consideration. SQAV, Individual consideration. IB, Individual consideration.

UNIVERSITY OF TEESSIDE T20

BSc Business Computing

UCAS Code: G710 • Mode: 4 Years Single Subject

Qualifications/Requirements: GCE, A/AS: 12-16 points. AGNVQ, Merit. ND/C, Individual consideration. SQAH, Grades CCCC. SQAV, Individual consideration. IB, Individual consideration.

BSc Software Engineering

UCAS Code: G700 • Mode: 4 Years Single Subject

Qualifications/Requirements: GCE, A/AS: 12-16 points. AGNVQ, Merit. ND/C, Individual consideration. SQAH, Grades CCCC. SQAV, Individual consideration. IB, Individual consideration.

HND Software Engineering

UCAS Code: 007G • Mode: 2 Years Single Subject

Qualifications/Requirements: GCE, A/AS: 4 points. AGNVQ, Pass. ND/C, National. SQAH, Grades CC. SQAV, Individual consideration. IB, Individual consideration.

THAMES VALLEY UNIVERSITY T40

BA/BSc Business Studies with Information Systems

UCAS Code: N1G5 • Mode: 2/3/4 Years Major/Minor

Qualifications/Requirements: GCE, A/AS: 8 points. AGNVQ, Merit. ND/C, Merit overall. SQAH, Grades CCCC. IB, 26 points.

BSc Software Design

UCAS Code: G701 • Mode: 3 Years Single Subject

Qualifications/Requirements: Please refer to prospectus.

Mod Information and Knowledge Management with Software Design

UCAS Code: P2G7 • Mode: 2/3 Years Major/Minor

Qualifications/Requirements: GCE, A/AS: 8 points. AGNVQ, Merit. ND/C, Merit overall. SQAH, Grades CCCC. IB, 26 points.

Mod Information Systems with Software Design

UCAS Code: G5G7 • Mode: 2/3 Years Major/Minor

Qualifications/Requirements: GCE, A/AS: 8 points. AGNVQ, Merit. ND/C, Merit overall. SQAH, Grades CCCC. IB, 26 points.

UNIVERSITY OF ULSTER U20

MEng Software Engineering (4/5 years SW including Diploma in Industrial Studies)

UCAS Code: G700 • Mode: 4/5 Years Single Subject

Qualifications/Requirements: GCE, A level grades: BCC. AGNVQ, Distinction in Information Technology with 6 additional units or with A level. ND/C, Merit overall and 4 Distinctions. SQAH, Grades BBBC. SQAV, Individual consideration. IB, 29 points.

WARWICKSHIRE COLLEGE, ROYAL LEAMINGTON SPA AND MORETON MORRELL W25

HND Computing (Business Information Technology)

UCAS Code: 027G • Mode: 2 Years Single Subject

Qualifications/Requirements: GCE, A/AS: 1 points. AGNVQ, Individual consideration. ND/C, Individual consideration. SQAH, Individual consideration. SQAV, Individual consideration. IB, Individual consideration.

HNC Computing (Business Information Technology)

UCAS Code: 127G • Mode: 1 Years Single Subject

Qualifications/Requirements: GCE, A/AS: 1 points. AGNVQ, Individual consideration. ND/C, Individual consideration. SQAH, Individual consideration. SQAV, Individual consideration. IB, Individual consideration.

UNIVERSITY OF WESTMINSTER W50

BSc Business Information Technology

UCAS Code: G710 • Mode: 4 Years Single Subject

Qualifications/Requirements: GCE, A level grades - BBC. AGNVQ, Distinction. ND/C, Merit overall and 3 Distinctions. SQAH, Grades BBBB. SQAV, Individual consideration. IB, 28 points.

BSc Software Engineering

UCAS Code: G700 • Mode: 3 Years Single Subject

Qualifications/Requirements: GCE, A/AS: 12 points. AGNVQ, Distinction (in specific programmes). ND/C, 4 Merits. SQAH, Individual consideration. SQAV, Individual consideration. IB, Individual consideration.

BSc Software Engineering with International Foundation

UCAS Code: G7Q3 • Mode: 4 Years Major/Minor

Qualifications/Requirements: GCE, A level grades: D. AGNVQ, Pass. ND/C, National.

Mod Business Computing

UCAS Code: G711 • Mode: 3 Years Single Subject

Qualifications/Requirements: GCE, A level grades - CC. AGNVQ, Distinction. ND/C, Merit overall. SQAH, Grades BBCC. IB, 26 points.

WESTMINSTER COLLEGE W52

HND Business Information Technology

UCAS Code: 027G • Mode: 2 Years Single Subject

Qualifications/Requirements: GCE, A/AS: 8 points. An approved subject from restricted list. AGNVQ, Individual consideration. ND/C, Individual consideration. SQAH, Individual consideration. SQAV, Individual consideration. IB, 24 points.

WEST THAMES COLLEGE W65

HND Business Information Technology

UCAS Code: 027G • Mode: 2 Years Single Subject

Qualifications/Requirements: GCE, A/AS: 1 points. AGNVQ, Pass. ND/C, National. SQAH, Individual consideration. SQAV, Individual consideration. IB, Individual consideration.

HNC Business Information Technology

UCAS Code: 127G • Mode: 1 Years Single Subject

Qualifications/Requirements: GCE, A/AS: 1 points. AGNVQ, Pass. ND/C, National. SQAH, Individual consideration. SQAV, Individual consideration. IB, Individual consideration.

WIGAN AND LEIGH COLLEGE W67

HND Business Information Technology

UCAS Code: 027G • Mode: 2 Years Single Subject

Qualifications/Requirements: GCE, A level grades - DD. AGNVQ, Pass. ND/C, National.

HND Software Engineering

UCAS Code: 007G • Mode: 2 Years Single Subject

Qualifications/Requirements: GCE, A level grades: DD. AGNVQ, Pass. ND/C, National.

UNIVERSITY OF WOLVERHAMPTON W75

BSc Business Information Technology

UCAS Code: G720 • Mode: 1/2 Years Single Subject

Qualifications/Requirements: Please refer to prospectus.

BSc Computer Science (Software Engineering)

UCAS Code: G700 • Mode: 4 Years Single Subject

Qualifications/Requirements: GCE, A/AS: 18 points. AGNVQ, Merit. ND/C, 4 Merits. SQAH, Grades CCCC. SQAV, Individual consideration. IB, 24 points.

WRITTLE COLLEGE W85

BSc Business Information Systems

UCAS Code: G720 • Mode: 3 Years Single Subject

Qualifications/Requirements: GCE, A/AS: 10 points. An approved subject from restricted list. AGNVQ, Merit. ND/C, Merit overall. SQAH, Individual consideration. SQAV, Individual consideration. IB, Individual consideration.

THE UNIVERSITY OF YORK Y50

MEng Computer Systems and Software Engineering

UCAS Code: GG67 • Mode: 4 Years Equal Combination

Qualifications/Requirements: GCE, A/AS: 26-28 points. Maths and any Physical Science or Biology at A level. AGNVQ, Distinction (in specific programmes) with A level. ND/C, Higher National (in specific programmes). SQAH, CSYS required. SQAV, Higher National including specific subject(s). IB, 34 points.

YORK COLLEGE OF FURTHER AND HIGHER EDUCATION Y70

HND Computing (Business Information Technology)

UCAS Code: 027G • Mode: 2 Years Single Subject

Qualifications/Requirements: Please refer to prospectus.

HND Computing (Software Engineering)

UCAS Code: 007G • Mode: 2 Years Single Subject

Qualifications/Requirements: Please refer to prospectus.

OTHER INFORMATICS SCIENCES

THE UNIVERSITY OF ABERDEEN A20

MA Computing/Statistics

UCAS Code: GG4M • Mode: 4 Years Equal Combination

Qualifications/Requirements: GCE, A level grades: CCC. Maths. AGNVQ, Merit (in specific programmes). ND/C, Individual consideration. SQAH, Grades BBBB including specific subject(s). SQAV, Individual consideration. IB, 30 points.

UNIVERSITY OF ABERTAY DUNDEE A30

BSc Bioinformatics

UCAS Code: GC51 • Mode: 4/5 Years Equal Combination

Qualifications/Requirements: Please refer to prospectus.

THE UNIVERSITY OF WALES, ABERYSTWYTH A40

BSc Computer Science/Mathematics

UCAS Code: GG15 • Mode: 3 Years Equal Combination

Qualifications/Requirements: GCE, A/AS: 20 points. Maths at A level. AGNVQ, Merit with A level. ND/C, Not normally sufficient. SQAH, Grades BBBCC including specific subject(s). SQAV, Individual consideration. IB, 30 points.

BSc Computer Science/Statistics

UCAS Code: GG45 • Mode: 3 Years Equal Combination

Qualifications/Requirements: GCE, A/AS: 20 points. Maths at A level. AGNVQ, Merit with A level. ND/C, Not normally sufficient. SQAH, Grades BBBCC including specific subject(s). SQAV, Individual consideration. IB, 30 points.

ANGLIA POLYTECHNIC UNIVERSITY A60

BA/BSc Graphic Design and Internet Technology

UCAS Code: GW62 • Mode: 3 Years Equal Combination

Qualifications/Requirements: GCE, A/AS: 14 points. Art and Maths or Physics at A level. AGNVQ, Merit in Art & Design. ND/C, 6 Merits. SQAH, Grades BBBC. SQAV, Individual consideration. IB, Pass in Diploma.

BA/BSc Graphic Design and Real Time Computer Systems

UCAS Code: WG2M • Mode: 3 Years Equal Combination

Qualifications/Requirements: GCE, A/AS: 14 points. Art at A level. AGNVQ, Merit in Art & Design. ND/C, 6 Merits. SQAH, Grades BBBC. SQAV, Individual consideration. IB, Pass in Diploma.

BSc Communications Network

UCAS Code: GH5P • Mode: 3 Years Equal Combination

Qualifications/Requirements: GCE, A/AS: 14 points. AGNVQ, Merit. ND/C, 6 Merits. SQAH, Grades BBBC. SQAV, Individual consideration. IB, Pass in Diploma.

BSc Communications Technology

UCAS Code: GH5Q • Mode: 3 Years Equal Combination

Qualifications/Requirements: GCE, A/AS: 14 points. AGNVQ, Merit. ND/C, 6 Merits. SQAH, Grades BBBC. SQAV, Individual consideration. IB, Pass in Diploma.

BSc Computer Science and Internet Technology

UCAS Code: GG56 • Mode: 3 Years Equal Combination

Qualifications/Requirements: GCE, A/AS: 12 points. AGNVQ, Merit in Science. ND/C, 4 Merits. SQAH, Grades BBCC. SQAV, National. IB, Pass in Diploma.

BSc Imaging Science and Real Time Computer Systems

UCAS Code: BG8M • Mode: 3 Years Equal Combination

Qualifications/Requirements: GCE, A/AS: 12 points. AGNVQ, Merit in Science. ND/C, 4 Merits. SQAH, Grades BCCC. SQAV, National. IB, Pass in Diploma including specific subjects.

BSc Internet Technology and Maths with Statistics

UCAS Code: GG16 • Mode: 3 Years Equal Combination

Qualifications/Requirements: GCE, A/AS: 12 points. AGNVQ, Merit (in specific programmes). ND/C, 4 Merits. SQAH, Grades BCCC. SQAV, National. IB, Individual consideration.

BSc Printmaking and Real Time Computer Systems

UCAS Code: GW51 • Mode: 3 Years Equal Combination

Qualifications/Requirements: GCE, A/AS: 14 points. Art at A level. AGNVQ, Merit in Art & Design. ND/C, 6 Merits. SQAH, Grades BBBC. SQAV, Individual consideration. IB, Pass in Diploma.

BSc Surveying and Information Technology

UCAS Code: GK5G • Mode: 3 Years Equal Combination

Qualifications/Requirements: GCE, A/AS: 14 points. AGNVQ, Merit. ND/C, 6 Merits. SQAH, Grades BBCC. SQAV, National. IB, Pass in Diploma.

ASTON UNIVERSITY A80

BSc Computer Science/Mathematics

UCAS Code: GG15 • Mode: 3/4 Years Equal Combination

Qualifications/Requirements: GCE, A/AS: 20 points. Maths. AGNVQ, Distinction (in specific programmes) with A level. ND/C, Not normally sufficient. SQAH, Grades ABBBB including specific subject(s). SQAV, Individual consideration. IB, 31 points.

BSc Computer Science/Mathematics (Year Zero)

UCAS Code: GGCN • Mode: 4/5 Years Equal Combination

Qualifications/Requirements: Please refer to prospectus.

BSc Geographical Information Systems/ Mathematics

UCAS Code: FG81 • Mode: 3/4 Years Equal Combination

Qualifications/Requirements: GCE, A/AS: 20 points. Maths at A level. AGNVQ, Distinction (in specific programmes) with A level. SQAH, Grades BBBBB including specific subject(s). IB, 30 points.

BSc Geographical Information Systems/ Mathematics (Year Zero)

UCAS Code: FGVC • Mode: 4/5 Years Equal Combination

Qualifications/Requirements: Please refer to prospectus.

UNIVERSITY OF WALES, BANGOR B06

BSc Computer Systems with Business Studies

UCAS Code: H6N1 • Mode: 3 Years Major/Minor

Qualifications/Requirements: GCE, A/AS: 18 points. AGNVQ, Merit (in specific programmes) with 6 additional units or with A level. ND/C, 3 Merits. SQAH, Grades BBCC. SQAV, Individual consideration. IB, 26 points.

BSc Computer Systems with French

UCAS Code: H6R1 • Mode: 4 Years Major/Minor

Qualifications/Requirements: GCE, A level grades: CCD. French at A level. AGNVQ, Merit (in specific programmes) with A level. ND/C, 3 Merits (in specific programmes). SQAH, Grades BBCC including specific subject(s). SQAV, Not normally sufficient. IB, 26 points.

BSc Computer Systems with German

UCAS Code: H6R2 • Mode: 4 Years Major/Minor

Qualifications/Requirements: GCE, A level grades: CCD. German at A level. AGNVQ, Merit (in specific programmes) with A level. ND/C, 3 Merits (in specific programmes). SQAH, Grades BBCC. SQAV, Individual consideration. IB, 26 points.

BSc Computer Systems with Psychology

UCAS Code: H6C8 • Mode: 3 Years Major/Minor

Qualifications/Requirements: GCE, A/AS: 18 points. AGNVQ, Merit (in specific programmes) with 6 additional units or with A level. ND/C, 3 Merits. SQAH, Grades BBCC. SQAV, Individual consideration. IB, 26 points.

UNIVERSITY OF BATH B16

BEng Networks & Information Engineering

UCAS Code: GG56 • Mode: 3 Years Equal Combination

Qualifications/Requirements: GCE, A/AS: 20 points. Maths and Physics or Electronics at A level. AGNVQ, Individual consideration. ND/C, Individual consideration. SQAH, CSYS required. SQAV, Individual consideration. IB, 30 points.

BEng Networks & Information Engineering

UCAS Code: GG5P • Mode: 4 Years Equal Combination

Qualifications/Requirements: GCE, A/AS: 20 points. Maths and Physics or Electronics at A level. AGNVQ, Individual consideration. ND/C, Individual consideration. SQAH, CSYS required. SQAV, Individual consideration. IB, 30 points.

MEng Networks & Information Engineering

UCAS Code: GG6M • Mode: 4 Years Equal Combination

Qualifications/Requirements: GCE, A/AS: 24 points. Maths and Physics or Electronics at A level. AGNVQ, Individual consideration. ND/C, Individual consideration. SQAH, CSYS required. SQAV, Individual consideration. IB, 32 points.

MEng Networks & Information Engineering

UCAS Code: GG6N • Mode: 5 Years Equal Combination

Qualifications/Requirements: GCE, A/AS: 24 points. Maths and Physics or Electronics at A level. AGNVQ, Individual consideration. ND/C, Individual consideration. SQAH, CSYS required. SQAV, Individual consideration. IB, 32 points.

BATH SPA UNIVERSITY COLLEGE B20

Mod Modular Degree Programme (Options)

UCAS Code: Y400 • Mode: 3 Years Single Subject

Qualifications/Requirements: GCE, A level grades – EE. AGNVQ, Pass. ND/C, National. SQAH, Individual consideration. SQAV, Individual consideration. IB, Pass in Diploma.

DipHE Modular Programme
UCAS Code: Y460 • Mode: 2 Years Single Subject
Qualifications/Requirements: GCE, A level grades – EE. Geography at A level. AGNVQ, Pass. ND/C, National. SQAH, Individual consideration. SQAV, Individual consideration. IB, Pass in Diploma.

THE UNIVERSITY OF BIRMINGHAM B32

BA Archaeology & Ancient History/ Computer Studies
UCAS Code: GV56 • Mode: 3 Years Equal Combination
Qualifications/Requirements: GCE, A level grades – BBB. AGNVQ, Distinction with A/AS. ND/C, Individual consideration. SQAH, Grades ABBBB. SQAV, Individual consideration. IB, 32 points.

BOLTON INSTITUTE OF HIGHER EDUCATION B44

Mod Accountancy and Computing
UCAS Code: GN54 • Mode: 3 Years Equal Combination
Qualifications/Requirements: GCE, A level grades: CD. AGNVQ, Merit. ND/C, Merit overall. SQAH, Grades BBCC. SQAV, Individual consideration. IB, 24 points.

Mod Architectural Technology and Simulation/Virtual Environment
UCAS Code: KG27 • Mode: 3 Years Equal Combination
Qualifications/Requirements: GCE, A/AS: 10 points. AGNVQ, Merit. ND/C, 3 Merits. SQAH, Individual consideration. SQAV, Individual consideration. IB, Individual consideration.

Mod Art & Design History and Business Information Systems
UCAS Code: GV54 • Mode: 3 Years Equal Combination
Qualifications/Requirements: GCE, A level grades – CD. AGNVQ, Merit. ND/C, Merit overall. SQAH, Grades BBCC. SQAV, Individual consideration. IB, 24 points.

Mod Computing and Mathematics
UCAS Code: GG15 • Mode: 3 Years Equal Combination
Qualifications/Requirements: GCE, A level grades: CD. Maths at A level. AGNVQ, Individual consideration. ND/C, Individual consideration. SQAH, Grades BBCC. SQAV, Individual consideration. IB, 24 points.

Mod Construction and Simulation/Virtual Environment
UCAS Code: KGG7 • Mode: 3 Years Equal Combination
Qualifications/Requirements: GCE, A/AS: 10 points. AGNVQ, Merit. ND/C, 3 Merits. SQAH, Individual consideration. SQAV, Individual consideration. IB, Individual consideration.

Mod Design and Business Information Systems
UCAS Code: GW52 • Mode: 3 Years Equal Combination
Qualifications/Requirements: GCE, A level grades: CD. AGNVQ, Merit. ND/C, Merit overall. SQAH, Grades BBCC. SQAV, Individual consideration. IB, 24 points.

Mod Design and Computing
UCAS Code: GW5F • Mode: 3 Years Equal Combination
Qualifications/Requirements: GCE, A level grades: CD. AGNVQ, Merit. ND/C, Merit overall. SQAH, Grades BBCC. SQAV, Individual consideration. IB, 24 points.

Mod Design and Simulation/Virtual Environment
UCAS Code: GW72 • Mode: 3 Years Equal Combination
Qualifications/Requirements: GCE, A/AS: 10 points. AGNVQ, Merit. ND/C, Merit overall. SQAH, Individual consideration. SQAV, Individual consideration. IB, Individual consideration.

Mod Enterprise Development and Leisure Computing Technology
UCAS Code: GN7D • Mode: 3 Years Equal Combination
Qualifications/Requirements: GCE, A level grades: CD. AGNVQ, Merit. ND/C, Merit overall. SQAH, Grades BBCC. SQAV, Individual consideration. IB, 24 points.

Mod Law and Leisure Computing Technology
UCAS Code: GM73 • Mode: 3 Years Equal Combination
Qualifications/Requirements: GCE, A level grades: CD. AGNVQ, Merit. ND/C, Merit overall. SQAH, Individual consideration. SQAV, Individual consideration. IB, Individual consideration.

Mod Manufacturing Systems Design and Simulation/Virtual Environment
UCAS Code: HGRT • Mode: 4 Years Equal Combination
Qualifications/Requirements: GCE, A/AS: 4 points. Maths or Science. AGNVQ, Pass (in specific programmes). ND/C, National. SQAH, Individual consideration. SQAV, Individual consideration. IB, Individual consideration.

Mod Product Design and Simulation/ Virtual Environment
UCAS Code: HGT7 • Mode: 3 Years Equal Combination
Qualifications/Requirements: GCE, A/AS: 10 points. Maths, Science, Art & Design or Design & Technology. AGNVQ, Merit (in specific programmes). ND/C, 3 Merits. SQAH, Individual consideration. SQAV, Individual consideration. IB, Individual consideration.

Mod Product Design and Simulation/ Virtual Environment

UCAS Code: HGTT • Mode: 4 Years Equal Combination

Qualifications/Requirements: GCE, A/AS: 4 points. Maths, Science, Art & Design or Design & Technology. AGNVQ, Pass (in specific programmes). ND/C, National. SQAH, Individual consideration. SQAV, Individual consideration. IB, Individual consideration.

Mod Product Design/Simulation & Virtual Environment

UCAS Code: HG7R • Mode: 4/5 Years Equal Combination

Qualifications/Requirements: GCE, A/AS: 4 points. Maths, Science, Art & Design or Design & Technology. AGNVQ, Pass. ND/C, National. SQAH, Individual consideration. SQAV, Individual consideration. IB, Individual consideration.

Mod Simulation/Virtual Environment and Technology Management

UCAS Code: GN79 • Mode: 3 Years Equal Combination

Qualifications/Requirements: GCE, A/AS: 10 points. Maths, Science or Business Studies. AGNVQ, Merit (in specific programmes). ND/C, 3 Merits. SQAH, Individual consideration. SQAV, Individual consideration. IB, Individual consideration.

Mod Simulation/Virtual Environment and Technology Management

UCAS Code: GN7X • Mode: 4 Years Equal Combination

Qualifications/Requirements: GCE, A/AS: 4 points. Maths, Science or Business Studies. AGNVQ, Pass (in specific programmes). ND/C, National. SQAH, Individual consideration. SQAV, Individual consideration. IB, Individual consideration.

Mod Simulation/Virtual Environment and Textiles

UCAS Code: GJ74 • Mode: 3 Years Equal Combination

Qualifications/Requirements: GCE, A/AS: 10 points. Maths or Science. AGNVQ, Merit (in specific programmes). ND/C, 3 Merits. SQAH, Individual consideration. SQAV, Individual consideration. IB, Individual consideration.

Mod Simulation/Virtual Environment and Transport Studies

UCAS Code: GJ79 • Mode: 3 Years Equal Combination

Qualifications/Requirements: GCE, A/AS: 10 points. Maths or Science. AGNVQ, Merit (in specific programmes). ND/C, 3 Merits. SQAH, Individual consideration. SQAV, Individual consideration. IB, Individual consideration.

Mod Visual Arts and Business Information Systems

UCAS Code: GW51 • Mode: 3 Years Equal Combination

Qualifications/Requirements: GCE, A level grades – CD. AGNVQ, Merit. ND/C, Merit overall. SQAH, Grades BBCC. SQAV, Individual consideration. IB, 24 points.

Mod Visual Arts and Computing

UCAS Code: GW5C • Mode: 3 Years Equal Combination

Qualifications/Requirements: GCE, A level grades: CD. AGNVQ, Merit. ND/C, Merit overall. SQAH, Grades BBCC. SQAV, Individual consideration. IB, 24 points.

BOURNEMOUTH UNIVERSITY B50

BA Computer Visualisation and Animation

UCAS Code: W270 • Mode: 3 Years Single Subject

Qualifications/Requirements: GCE, A/AS: 24 points. Art or Maths at A level. AGNVQ, Distinction (in specific programmes). ND/C, Merit overall and 5 Distinctions. SQAH, Grades ABBBB. SQAV, Individual consideration. IB, 34 points.

BSc Applied Psychology and Computing

UCAS Code: C878 • Mode: 3 Years Single Subject

Qualifications/Requirements: GCE, A/AS: 14-16 points. Science at A level. AGNVQ, Merit (in specific programmes). ND/C, Merit overall and 3 Distinctions. SQAH, Grades CCCCCC. SQAV, Individual consideration. IB, 30 points.

BSc Applied Psychology and Computing

UCAS Code: C879 • Mode: 4 Years Single Subject

Qualifications/Requirements: GCE, A/AS: 12-14 points. AGNVQ, Merit (in specific programmes). ND/C, Merit overall and Distinction. SQAH, Grades CCCCCC. SQAV, Individual consideration. IB, Individual consideration.

BSc Business Communication Systems

UCAS Code: G523 • Mode: 4 Years Single Subject

Qualifications/Requirements: GCE, A/AS: 10-12 points. AGNVQ, Merit. ND/C, 3 Merits. SQAH, Individual consideration. SQAV, Individual consideration. IB, 28 points.

BEng Computer and Communication Technology

UCAS Code: H621 • Mode: 4/5 Years Single Subject

Qualifications/Requirements: GCE, A level grades: DE. AGNVQ, Merit (in specific programmes). ND/C, Individual consideration. SQAH, Individual consideration. SQAV, Individual consideration. IB, Individual consideration.

THE UNIVERSITY OF BRADFORD B56

BSc Computing and Information Systems

UCAS Code: G520 • Mode: 3 Years Single Subject

Qualifications/Requirements: GCE, A/AS: 18 points. AGNVQ, Distinction (in specific programmes) with 4 additional units or with A/AS. ND/C, 3 Merits and 2 Distinctions. SQAH, Individual consideration. SQAV, Individual consideration. IB, Individual consideration.

BSc Computing and Information Systems

UCAS Code: G521 • Mode: 4 Years Single Subject

Qualifications/Requirements: GCE, A/AS: 18 points. AGNVQ, Distinction (in specific programmes) with 4 additional units or with A/AS. ND/C, 3 Merits and 2 Distinctions. SQAH, Individual consideration. SQAV, Individual consideration. IB, Individual consideration.

BSc Networks Information Management

UCAS Code: G530 • Mode: 3 Years Single Subject

Qualifications/Requirements: GCE, A/AS: 18 points. AGNVQ, Distinction (in specific programmes) with 4 additional units or with A/AS. ND/C, 3 Merits and 2 Distinctions. SQAH, Individual consideration. SQAV, Individual consideration. IB, Individual consideration.

BSc Networks Information Management

UCAS Code: G531 • Mode: 4 Years Single Subject

Qualifications/Requirements: GCE, A/AS: 18 points. AGNVQ, Distinction (in specific programmes) with 4 additional units or with A/AS. ND/C, 3 Merits and 2 Distinctions. SQAH, Individual consideration. SQAV, Individual consideration. IB, Individual consideration.

UNIVERSITY OF BRIGHTON B72

BA Computing and Information Systems (4 years SW)

UCAS Code: G560 • Mode: 4 Years Single Subject

Qualifications/Requirements: GCE, A/AS: 18 points. AGNVQ, Merit with 6 additional units or with A/AS. ND/C, Distinction Overall. SQAH, Grades BBBC. SQAV, Individual consideration. IB, 27 points.

BA Upper Primary/Lower Secondary (7-14) Information Technology with QTS

UCAS Code: XG65 • Mode: 4 Years Equal Combination

Qualifications/Requirements: GCE, A/AS: 12 points. Italian. AGNVQ, Merit (in specific programmes). ND/C, Merit overall (in specific programmes). SQAH, Individual consideration. SQAV, Individual consideration. IB, Individual consideration.

BSc Mathematics for Computing

UCAS Code: G170 • Mode: 3/4 Years Single Subject

Qualifications/Requirements: GCE, A/AS: 14 points. Maths at A level. AGNVQ, Not normally sufficient. ND/C, Not normally sufficient. SQAH, Grades BBCC including specific subject(s). SQAV, Not normally sufficient. IB, Pass in Diploma including specific subjects.

UNIVERSITY OF THE WEST OF ENGLAND, BRISTOL B80

BA English and Information Systems

UCAS Code: QG37 • Mode: 3 Years Equal Combination

Qualifications/Requirements: GCE, A/AS: 18-20 points. English Literature. AGNVQ, Merit (in specific programmes) with A/AS. ND/C, 4 Merits and 2 Distinctions (in specific programmes). SQAH, Grades BBB including specific subject(s). SQAV, Individual consideration. IB, 28 points.

BA Geography and Information Systems

UCAS Code: LG87 • Mode: 3 Years Equal Combination

Qualifications/Requirements: GCE, A/AS: 18 points. Geography. AGNVQ, Merit with A/AS or Distinction. ND/C, 5 Merits and 1 Distinction. SQAH, Grades BBBC. SQAV, Individual consideration. IB, 26 points.

BA Law and Information Systems

UCAS Code: MG35 • Mode: 3 Years Equal Combination

Qualifications/Requirements: GCE, A/AS: 18-20 points. AGNVQ, Merit with A/AS Distinction. ND/C, 5 Merits and 1 Distinction. SQAH, Grades BBBC. SQAV, Individual consideration. IB, 26 points.

BA Marketing and Information Systems

UCAS Code: NG55 • Mode: 3 Years Equal Combination

Qualifications/Requirements: GCE, A/AS: 16 points. AGNVQ, Merit. ND/C, 6 Merits. SQAH, Grades BBB. SQAV, Individual consideration. IB, 24 points.

BSc Chemistry and Information Technology in Science

UCAS Code: FG15 • Mode: 3/4 Years Equal Combination

Qualifications/Requirements: GCE, A/AS: 10 points. Chemistry. AGNVQ, Merit. ND/C, 3 Merits. SQAH, Grades BB including specific subject(s). SQAV, National including specific subject(s). IB, 24 points.

BSc Computing and Information Systems

UCAS Code: G501 • Mode: 4 Years Single Subject

Qualifications/Requirements: GCE, A/AS: 16 points. AGNVQ, Merit (in specific programmes). ND/C, Merit overall. SQAH, Grades BBB. SQAV, Individual consideration. IB, 24 points.

BSc Computing and Mathematics

UCAS Code: GG15 • Mode: 3/4 Years Equal Combination

Qualifications/Requirements: GCE, A/AS: 16 points. Maths. AGNVQ, Merit (in specific programmes) with A level. ND/C, 4 Merits (in specific programmes). SQAH, Grades BBB including specific subject(s). SQAV, National including specific subject(s). IB, 24 points.

BSc Computing and Statistics
UCAS Code: G4GM • Mode: 3/4 Years Major/Minor

Qualifications/Requirements: GCE, A/AS: 16 points. Maths or Statistics. AGNVQ, Merit (in specific programmes) with A level. ND/C, 4 Merits (in specific programmes). SQAH, Grades BBB including specific subject(s). SQAV, National including specific subject(s). IB, 24 points.

BSc Environmental Science and Information Technology in Science
UCAS Code: FG95 • Mode: 3/4 Years Equal Combination

Qualifications/Requirements: GCE, A/AS: 6 points. Science. AGNVQ, Pass in Science. ND/C, National (in specific programmes). SQAH, Grades CCC including specific subject(s). SQAV, Individual consideration. IB, 24 points.

BSc Health Science and Information Systems
UCAS Code: BG97 • Mode: 3 Years Equal Combination

Qualifications/Requirements: GCE, A/AS: 16 points. AGNVQ, Merit (in specific programmes). ND/C, 6 Merits (in specific programmes). SQAH, Grades BBB including specific subject(s). SQAV, Individual consideration. IB, 24 points.

BSc Psychology and Information Technology in Science
UCAS Code: CG85 • Mode: 3/4 Years Equal Combination

Qualifications/Requirements: GCE, A/AS: 10 points. Science. AGNVQ, Merit. ND/C, 3 Merits (in specific programmes). SQAH, Grades BB including specific subject(s). SQAV, Individual consideration. IB, 24 points.

BSc Statistics and Information Systems
UCAS Code: GG45 • Mode: 3/4 Years Equal Combination

Qualifications/Requirements: GCE, A/AS: 16 points. Maths or Statistics. AGNVQ, Merit with A level. ND/C, Merit overall. SQAH, Grades BBB including specific subject(s). SQAV, National including specific subject(s). IB, 24 points.

BRUNEL UNIVERSITY B84

BSc Health Information Science
UCAS Code: L4G5 • Mode: 3 Years Major/Minor

Qualifications/Requirements: GCE, A/AS: 12-20 points. Biology. AGNVQ, Merit in Health & Social Care. ND/C, Individual consideration. SQAH, Individual consideration. SQAV, Individual consideration. IB, Individual consideration.

BUCKINGHAMSHIRE CHILTERNS UNIVERSITY COLLEGE B94

BA E-Commerce Business
UCAS Code: GN51 • Mode: 3 Years Equal Combination

Qualifications/Requirements: GCE, A/AS: 8-12 points. AGNVQ, Merit. ND/C, Merit overall. SQAH, Grades CCCC. SQAV, Individual consideration. IB, 27 points.

CANTERBURY CHRIST CHURCH UNIVERSITY COLLEGE C10

Mod Information Technology with Early Childhood Studies
UCAS Code: G5X9 • Mode: 3 Years Major/Minor

Qualifications/Requirements: GCE, A level grades: CC. AGNVQ, Merit. ND/C, Merit overall. SQAH, Individual consideration. SQAV, Individual consideration. IB, 24 points.

Mod IT with Science (Natural) (with foundation)
UCAS Code: G5YD • Mode: 4 Years

Qualifications/Requirements: Please refer to prospectus.

Mix Early Childhood Studies and Information Technology
UCAS Code: XG95 • Mode: 3 Years Equal Combination

Qualifications/Requirements: GCE, A level grades: CC. AGNVQ, Merit. ND/C, Merit overall. SQAH, Individual consideration. SQAV, Individual consideration. IB, 24 points.

Mix English and Information Technology
UCAS Code: QG35 • Mode: 3 Years Equal Combination

Qualifications/Requirements: GCE, A level grades: CC. English at A level. AGNVQ, Merit with A level. ND/C, Merit overall. SQAH, Individual consideration. SQAV, Individual consideration. IB, 24 points.

Mix Geography and Information Technology
UCAS Code: GL58 • Mode: 3 Years Equal Combination

Qualifications/Requirements: GCE, A level grades – DD. AGNVQ, Merit. ND/C, Merit overall. SQAH, Individual consideration. SQAV, Individual consideration. IB, 24 points.

Mix History and Information Technology
UCAS Code: VG15 • Mode: 3 Years Equal Combination

Qualifications/Requirements: GCE, A level grades: DD. History at A level. AGNVQ, Merit with A level. ND/C, Merit overall. SQAH, Individual consideration. SQAV, Individual consideration. IB, 24 points.

Mix Information Technology and Mathematics

UCAS Code: GG15 • Mode: 3 Years Equal Combination

Qualifications/Requirements: GCE, A level grades: DD. Maths at A level. AGNVQ, Individual consideration. ND/C, Individual consideration. SQAH, Individual consideration. SQAV, Individual consideration. IB, 24 points.

Mix Information Technology and Statistics & Operational Research

UCAS Code: GG45 • Mode: 3 Years Equal Combination

Qualifications/Requirements: GCE, A level grades: DD. Maths at A level. AGNVQ, Individual consideration. ND/C, Individual consideration. SQAH, Individual consideration. SQAV, Individual consideration. IB, 24 points.

CARDIFF UNIVERSITY C15

BSc Computing and Mathematics

UCAS Code: GGD5 • Mode: 3 Years Equal Combination

Qualifications/Requirements: GCE, A/AS: 22 points. Maths at A level. AGNVQ, Individual consideration. ND/C, 3 Merits and 3 Distinctions (in specific programmes). SQAH, Individual consideration. SQAV, Individual consideration. IB, Individual consideration.

BSc Preliminary Year

UCAS Code: Y121 • Mode: 4 Years Single Subject

Qualifications/Requirements: AGNVQ, Pass. ND/C, 3 Merits. SQAH, Individual consideration. SQAV, Individual consideration. IB, Individual consideration.

UNIVERSITY OF CENTRAL ENGLAND IN BIRMINGHAM C25

BSc Industrial Information Technology

UCAS Code: G560 • Mode: 3/4 Years Single Subject

Qualifications/Requirements: GCE, A/AS: 16 points. AGNVQ, Merit. ND/C, 5 Merits. SQAH, Grades BBCCC. SQAV, Individual consideration. IB, 28 points.

BEng Industrial Information Technology Foundation Year

UCAS Code: G568 • Mode: 4/5 Years Single Subject

Qualifications/Requirements: GCE, A/AS: 4 points. AGNVQ, Pass. ND/C, 1 Merit. SQAH, Grades CC. SQAV, Individual consideration. IB, 24 points.

UNIVERSITY OF CENTRAL LANCASHIRE C30

BA Economics and Business Information Systems

UCAS Code: LG15 • Mode: 3 Years Equal Combination

Qualifications/Requirements: GCE, A/AS: 14 points. AGNVQ, Distinction (in specific programmes). ND/C, Merit overall. SQAH, Grades BBBC. SQAV, Individual consideration. IB, 28 points.

BA Management and Business Information Systems

UCAS Code: NG1M • Mode: 3 Years Equal Combination

Qualifications/Requirements: GCE, A/AS: 14 points. AGNVQ, Distinction in Business with 6 additional units or with A/AS. ND/C, Merit overall. SQAH, Grades BBBC. SQAV, Individual consideration. IB, 28 points.

BSc Management Information Systems

UCAS Code: G521 • Mode: 1 Years Single Subject

Qualifications/Requirements: ND/C, Higher National.

BSc Mathematics and Business Information Systems

UCAS Code: GG1M • Mode: 3 Years Equal Combination

Qualifications/Requirements: GCE, A/AS: 12 points. Maths. AGNVQ, Merit (in specific programmes) with 6 additional units or with A level. ND/C, Merit overall. SQAH, Grades BBB including specific subject(s). SQAV, Individual consideration. IB, 26 points.

BSc Statistics and Business Information Systems

UCAS Code: GG45 • Mode: 3 Years Equal Combination

Qualifications/Requirements: GCE, A/AS: 12 points. AGNVQ, Merit (in specific programmes) with 6 additional units or with A level. ND/C, Merit overall. SQAH, Grades BBB. SQAV, Individual consideration. IB, 24 points.

CHELTENHAM & GLOUCESTER COLLEGE OF HIGHER EDUCATION C50

BSc Business Information Technology with Business Management

UCAS Code: G5N1 • Mode: 3 Years Major/Minor

Qualifications/Requirements: GCE, A/AS: 8-12 points. AGNVQ, Merit. ND/C, Merit overall. SQAH, Grades CCCC. SQAV, Individual consideration. IB, 24 points.

BSc Computing and Information Technology

UCAS Code: G527 • Mode: 3 Years Single Subject

Qualifications/Requirements: GCE, A/AS: 8 points. AGNVQ, Merit. ND/C, Merit overall. SQAH, Grades CCCC. SQAV, Individual consideration. IB, 24 points.

BSc Computing with Information Technology

UCAS Code: G526 • Mode: 3 Years Single Subject

Qualifications/Requirements: GCE, A/AS: 8 points. AGNVQ, Merit. ND/C, Merit overall. SQAH, Grades CCCC. SQAV, Individual consideration. IB, 24 points.

BSc Environmental Science and Information Technology

UCAS Code: FGX7 • Mode: 3 Years Equal Combination

Qualifications/Requirements: GCE, A/AS: 8-12 points. AGNVQ, Merit. ND/C, Merit overall. SQAH, Grades CCCC. SQAV, Individual consideration. IB, 24 points.

BSc Geography and Information Technology

UCAS Code: GFNV • Mode: 3 Years Equal Combination

Qualifications/Requirements: GCE, A/AS: 12 points. AGNVQ, Merit. ND/C, Merit overall. SQAH, Grades CCCC. SQAV, Individual consideration. IB, 24 points.

BSc Geology and Information Technology

UCAS Code: FG65 • Mode: 3 Years Equal Combination

Qualifications/Requirements: GCE, A/AS: 8 points. AGNVQ, Merit with 3 additional units. ND/C, Merit overall. SQAH, Grades CCCC. SQAV, Individual consideration. IB, 26 points.

Mod English Studies and Multimedia

UCAS Code: QG35 • Mode: 3 Years Equal Combination

Qualifications/Requirements: GCE, A/AS: 10-12 points. English. AGNVQ, Merit with A level. ND/C, Merit overall and 3 Distinctions. SQAH, Grades CCCC. SQAV, Individual consideration. IB, 26 points.

Mod Environmental Policy and Information Technology

UCAS Code: FG95 • Mode: 3 Years Equal Combination

Qualifications/Requirements: GCE, A/AS: 8-12 points. AGNVQ, Merit with 3 additional units. ND/C, Merit overall. SQAH, Grades CCCC. SQAV, Individual consideration. IB, 24 points.

Mod Human Geography and Information Technology

UCAS Code: GL58 • Mode: 3/4 Years Equal Combination

Qualifications/Requirements: GCE, A/AS: 8-12 points. AGNVQ, Merit with 3 additional units. ND/C, Merit overall. SQAH, Grades CCCC. SQAV, Individual consideration. IB, 24 points.

CHESTER: A COLLEGE OF THE UNIVERSITY OF LIVERPOOL C55

BSc Mathematics, Statistics and Computing

UCAS Code: G900 • Mode: 3 Years Single Subject

Qualifications/Requirements: GCE, A/AS: 12 points. Maths at A level. AGNVQ, Not normally sufficient. ND/C, Not normally sufficient. SQAH, Grades CCCC. SQAV, Not normally sufficient. IB, 24 points.

UNIVERSITY COLLEGE CHICHESTER C58

BA Information and Communications Technology with Education (QTS)

UCAS Code: G5X5 • Mode: 4 Years Major/Minor

Qualifications/Requirements: GCE, A/AS: 12 points. AGNVQ, Merit (in specific programmes). ND/C, Merits and Distinction. SQAH, Individual consideration. SQAV, Individual consideration. IB, Individual consideration.

COLCHESTER INSTITUTE C75

HND Computing and Information Technology

UCAS Code: 005G • Mode: 2 Years Single Subject

Qualifications/Requirements: GCE, A/AS: 8 points. Italian. AGNVQ, Merit. ND/C, Merit overall. SQAH, Individual consideration. SQAV, Individual consideration. IB, Individual consideration.

COVENTRY UNIVERSITY C85

BA Extended Business Information Technology (Overseas)

UCAS Code: GN5C • Mode: 4 Years Equal Combination

Qualifications/Requirements: GCE, A/AS: 2 points. AGNVQ, Pass. ND/C, National. SQAH, Individual consideration. SQAV, Individual consideration. IB, Pass in Diploma.

BSc Geographical Information Systems

UCAS Code: G562 • Mode: 3/4 Years Single Subject

Qualifications/Requirements: GCE, A/AS: 12 points. AGNVQ, Merit. ND/C, 4 Merits. SQAH, Individual consideration. SQAV, Individual consideration. IB, Individual consideration.

BSc Network Computing

UCAS Code: G530 • Mode: 3/4 Years Single Subject

Qualifications/Requirements: GCE, A/AS: 12 points. AGNVQ, Merit. ND/C, 4 Merits. SQAH, Grades CCCC. SQAV, Individual consideration. IB, Individual consideration.

BSc Statistics and Computing

UCAS Code: GG45 • Mode: 3/4 Years Equal Combination

Qualifications/Requirements: GCE, A/AS: 12-16 points. Maths, Computing or Statistics at A level. AGNVQ, Merit. ND/C, 3 Merits (in specific programmes). SQAH, Individual consideration. SQAV, Individual consideration. IB, Individual consideration.

DE MONTFORT UNIVERSITY D26

BSc Computer and Information Systems

UCAS Code: G520 • Mode: 3/4 Years Single Subject

Qualifications/Requirements: GCE, A/AS: 12-18 points. AGNVQ, Distinction. ND/C, 2 Merits and 2 Distinctions. SQAH, Grades BBBC including specific subject(s). SQAV, Individual consideration. IB, 30 points.

BSc Computer-Based Mathematics

UCAS Code: G170 • Mode: 3 Years Single Subject

Qualifications/Requirements: GCE, A/AS: 12 points. Maths. AGNVQ, Merit (in specific programmes). ND/C, 3 Merits and 1 Distinction. SQAH, Individual consideration. SQAV, Individual consideration. IB, Individual consideration.

BSc Computing and Mathematics

UCAS Code: GG15 • Mode: 3 Years Equal Combination

Qualifications/Requirements: GCE, A/AS: 8-14 points. AGNVQ, Merit. ND/C, 4 Merits (in specific programmes). SQAH, Grades BBBC including specific subject(s). SQAV, Individual consideration. IB, 28 points.

BSc Computing and Medical Statistics

UCAS Code: GG4N • Mode: 3 Years Equal Combination

Qualifications/Requirements: GCE, A/AS: 8-14 points. AGNVQ, Merit. ND/C, 4 Merits (in specific programmes). SQAH, Grades BBCC including specific subject(s). SQAV, Individual consideration. IB, 28 points.

BSc Computing and Statistics

UCAS Code: GGK5 • Mode: 3 Years Equal Combination

Qualifications/Requirements: GCE, A/AS: 8-14 points. AGNVQ, Merit. ND/C, 4 Merits (in specific programmes). SQAH, Grades BBCC including specific subject(s). SQAV, Individual consideration. IB, 28 points.

BSc Economics and Information Systems

UCAS Code: GL51 • Mode: 3 Years Equal Combination

Qualifications/Requirements: GCE, A/AS: 12 points. AGNVQ, Merit (in specific programmes). ND/C, 3 Merits and 3 Distinctions. SQAH, Individual consideration. SQAV, Individual consideration. IB, Individual consideration.

BSc Mathematics and Technology

UCAS Code: GJ15 • Mode: 4 Years Equal Combination

Qualifications/Requirements: GCE, A/AS: 12 points. Maths. AGNVQ, Merit (in specific programmes). ND/C, 3 Merits and 1 Distinction. SQAH, Individual consideration. SQAV, Individual consideration. IB, Individual consideration.

BSc Medical and Health Statistics

UCAS Code: GG54 • Mode: 3/4 Years Equal Combination

Qualifications/Requirements: GCE, A/AS: 10-14 points. AGNVQ, Merit. ND/C, Merit overall. SQAH, Individual consideration. SQAV, Individual consideration. IB, Individual consideration.

BSc Statistics and Technology

UCAS Code: GJ45 • Mode: 4 Years Equal Combination

Qualifications/Requirements: Please refer to prospectus.

Mod Accounting (Vocational) and Computing

UCAS Code: NG4M • Mode: 3 Years Equal Combination

Qualifications/Requirements: Please refer to prospectus.

UNIVERSITY OF DERBY D39

BSc Applicable Mathematics and Computing

UCAS Code: GG15 • Mode: 4 Years Equal Combination

Qualifications/Requirements: GCE, A/AS: 10 points. Maths at A level. AGNVQ, Merit with A level. ND/C, 3 Merits (in specific programmes). SQAH, Grades CCCD including specific subject(s). SQAV, Individual consideration. IB, Pass in Diploma including specific subjects.

BSc Digital Entertainment

UCAS Code: GG57 • Mode: 3 Years Equal Combination

Qualifications/Requirements: GCE, A/AS: 12-14 points. AGNVQ, Merit (in specific programmes). ND/C, Merit overall (in specific programmes). SQAH, Grades CCCC. SQAV, Individual consideration. IB, 26 points.

BSc Health Care Information Management

UCAS Code: G580 • Mode: 3 Years Single Subject

Qualifications/Requirements: GCE, A/AS: 12 points. AGNVQ, Merit (in specific programmes). ND/C, 4 Merits and 1 Distinction. SQAH, Grades CCCD. SQAV, Individual consideration. IB, 25 points.

BSc Mathematics, Statistics and Computing

UCAS Code: G900 • Mode: 4 Years Single Subject

Qualifications/Requirements: GCE, A/AS: 10 points. Maths at A level. AGNVQ, Merit with A level. ND/C, 3 Merits (in specific programmes). SQAH, Grades CCCD including specific subject(s). SQAV, Individual consideration. IB, Pass in Diploma including specific subjects.

BSc Operational Research and Systems Analysis

UCAS Code: G434 • Mode: 4 Years Single Subject

Qualifications/Requirements: GCE, A/AS: 10 points. Maths at A level. AGNVQ, Merit with A level. ND/C, 3 Merits (in specific programmes). SQAH, Grades CCCD including specific subject(s). SQAV, Individual consideration. IB, Pass in Diploma including specific subjects.

BSc Virtual Product Design Technology

UCAS Code: GH77 • Mode: 3/4 Years Equal Combination

Qualifications/Requirements: GCE, A/AS: 12-14 points. AGNVQ, Distinction (in specific programmes). ND/C, 8 Merits. SQAH, Grades BBBC. SQAV, Individual consideration. IB, 28 points.

UNIVERSITY OF DUNDEE D65

BSc Accountancy and Applied Computing

UCAS Code: GN54 • Mode: 4 Years Equal Combination

Qualifications/Requirements: GCE, A/AS: 14 points. Science at A level. AGNVQ, Merit in Science. ND/C, 5 Merits (in specific programmes). SQAH, Grades BBBC including specific subject(s). SQAV, National including specific subject(s). IB, 25 points.

BSc Applied Computing and Digital Microelectronics

UCAS Code: GHM6 • Mode: 4 Years Equal Combination

Qualifications/Requirements: GCE, A/AS: 14 points. Maths and Science at A level. AGNVQ, Merit in Science with A level. ND/C, 5 Merits (in specific programmes). SQAH, Grades BBBC including specific subject(s). SQAV, National including specific subject(s). IB, 25 points.

BSc Mathematics and Applied Computing

UCAS Code: GG15 • Mode: 4 Years Equal Combination

Qualifications/Requirements: GCE, A/AS: 14 points. Maths. AGNVQ, Merit in Science with A level. ND/C, 5 Merits (in specific programmes). SQAH, Grades BBBC including specific subject(s). SQAV, National including specific subject(s). IB, 25 points.

BSc Psychology and Applied Computing

UCAS Code: LG75 • Mode: 4 Years Equal Combination

Qualifications/Requirements: GCE, A/AS: 14 points. Science at A level. AGNVQ, Merit in Science. ND/C, 5 Merits (in specific programmes). SQAH, Grades BBBC including specific subject(s). SQAV, National including specific subject(s). IB, 25 points.

UNIVERSITY OF EAST ANGLIA E14

BSc Decision Support Systems

UCAS Code: G506 • Mode: 3 Years Single Subject

Qualifications/Requirements: GCE, A level grades: BBC-BCC. AGNVQ, Distinction with A level. ND/C, Merit overall and 3 Distinctions. SQAH, Grades BBBB. IB, 28 points.

UNIVERSITY OF EAST LONDON E28

BA Education & Community Studies/ Information Technology

UCAS Code: GX59 • Mode: 3 Years Equal Combination

Qualifications/Requirements: GCE, A/AS: 12 points. AGNVQ, Merit (in specific programmes). ND/C, Merit overall. SQAH, Individual consideration. SQAV, Individual consideration. IB, Individual consideration.

BA European Studies/Information Technology

UCAS Code: T2G5 • Mode: 3 Years Major/Minor

Qualifications/Requirements: GCE, A/AS: 12 points. AGNVQ, Merit. ND/C, Merit overall. SQAH, Individual consideration. SQAV, Individual consideration. IB, Individual consideration.

BA History of Art, Design & Film/ Information Technology

UCAS Code: GV54 • Mode: 3 Years Equal Combination

Qualifications/Requirements: GCE, A/AS: 12 points. AGNVQ, Merit (in specific programmes). ND/C, Merit overall. SQAH, Individual consideration. SQAV, Individual consideration. IB, Individual consideration.

BA History/Information Technology

UCAS Code: GV51 • Mode: 3 Years Equal Combination

Qualifications/Requirements: GCE, A/AS: 12 points. AGNVQ, Merit (in specific programmes). ND/C, Merit overall. SQAH, Individual consideration. SQAV, Individual consideration. IB, Individual consideration.

BA Law/Information Technology

UCAS Code: GM53 • Mode: 3 Years Equal Combination

Qualifications/Requirements: GCE, A/AS: 12 points. AGNVQ, Distinction. ND/C, Merit overall. SQAH, Individual consideration. SQAV, Individual consideration. IB, Individual consideration.

BA Printed Textiles and Surface Decoration/Information Technology

UCAS Code: J4G5 • Mode: 3 Years Major/Minor

Qualifications/Requirements: GCE, A/AS: 12 points. AGNVQ, Merit. ND/C, Merit overall. SQAH, Individual consideration. SQAV, Individual consideration. IB, Individual consideration.

BA Social Sciences/Information Technology

UCAS Code: GL5H • Mode: 3 Years Equal Combination

Qualifications/Requirements: GCE, A/AS: 12 points. AGNVQ, Merit in Business. ND/C, Merit overall. SQAH, Individual consideration. SQAV, Individual consideration. IB, Individual consideration.

BA Third World Development/ Information Technology
UCAS Code: GM5Y • Mode: 3 Years Equal Combination

Qualifications/Requirements: GCE, A/AS: 12 points. AGNVQ, Merit. ND/C, Merit overall. SQAH, Individual consideration. SQAV, Individual consideration. IB, Individual consideration.

BSc Distributed Information Systems
UCAS Code: G521 • Mode: 4 Years Single Subject

Qualifications/Requirements: GCE, A/AS: 10 points. AGNVQ, Merit in Information Technology. ND/C, 6 Merits. SQAH, Individual consideration. SQAV, Individual consideration. IB, 24 points.

BSc Environmental Sciences/Information Technology
UCAS Code: GF59 • Mode: 3 Years Equal Combination

Qualifications/Requirements: GCE, A/AS: 12 points. AGNVQ, Merit (in specific programmes). ND/C, Merit overall. SQAH, Individual consideration. SQAV, Individual consideration. IB, Individual consideration.

BSc Geographical Information Systems
UCAS Code: G5F8 • Mode: 3 Years Major/Minor

Qualifications/Requirements: GCE, A/AS: 12 points. AGNVQ, Merit. ND/C, 3 Merits. SQAH, Individual consideration. SQAV, Individual consideration. IB, Individual consideration.

BSc Property and Planning Informatics
UCAS Code: K4G5 • Mode: 3 Years Major/Minor

Qualifications/Requirements: GCE, A/AS: 12 points. AGNVQ, Merit. ND/C, 3 Merits. SQAH, Individual consideration. SQAV, Individual consideration. IB, Individual consideration.

BSc Spatial Business Informatics
UCAS Code: G523 • Mode: 3 Years Single Subject

Qualifications/Requirements: GCE, A/AS: 12 points. AGNVQ, Merit. ND/C, 3 Merits. SQAH, Individual consideration. SQAV, Individual consideration. IB, Individual consideration.

Mod Three-Subject Degree
UCAS Code: Y600 • Mode: 3 Years Single Subject

Qualifications/Requirements: GCE, A/AS: 12 points. Science. AGNVQ, Merit in Science. ND/C, Merit (in specific programmes). SQAH, Individual consideration. SQAV, Individual consideration. IB, Individual consideration.

EDGE HILL COLLEGE OF HIGHER EDUCATION E42

BSc Geography and Information Systems
UCAS Code: GL58 • Mode: 3 Years Equal Combination

Qualifications/Requirements: GCE, A level grades – CD. Geography and Environmental Science at A level. AGNVQ, Pass with A level. ND/C, Individual consideration. SQAH, Grades BBCC including specific subject(s). SQAV, Individual consideration. IB, Pass in Diploma.

BSc Information Systems and Mathematics
UCAS Code: GG15 • Mode: 3 Years Equal Combination

Qualifications/Requirements: GCE, A level grades: CD. Maths at A level. AGNVQ, Merit or Pass with A/AS. ND/C, 3 Merits and 3 Distinctions. SQAH, Grades BBCC. SQAV, Individual consideration. IB, Pass in Diploma.

BSc Marketing and Information Systems
UCAS Code: NG15 • Mode: 3 Years Equal Combination

Qualifications/Requirements: GCE, A level grades: CC. AGNVQ, Merit or Pass with A/AS. ND/C, Individual consideration. SQAH, Grades BBCC. SQAV, Individual consideration. IB, Pass in Diploma.

BSc Sports Studies and Information Systems
UCAS Code: BG65 • Mode: 3 Years Equal Combination

Qualifications/Requirements: GCE, A level grades – BC. AGNVQ, Distinction or Merit with A/AS. ND/C, 2 Merits and 4 Distinctions. SQAH, Grades BBCC. SQAV, Individual consideration. IB, Pass in Diploma.

THE UNIVERSITY OF EDINBURGH E56

BSc Computer Science and Mathematics
UCAS Code: GG15 • Mode: 4 Years Equal Combination

Qualifications/Requirements: GCE, A level grades: BBC. Maths at A level. AGNVQ, Pass. ND/C, Merit overall (in specific programmes). SQAH, Grades ABBC including specific subject(s). SQAV, National including specific subject(s). IB, Pass in Diploma including specific subjects.

UNIVERSITY OF GLASGOW G28

BSc Computing Science/Statistics
UCAS Code: GG45 • Mode: 4 Years Equal Combination

Qualifications/Requirements: GCE, A level grades: BBC-CCC. Two Sciences. AGNVQ, Merit. ND/C, National. SQAH, Grades BBBB including specific subject(s). SQAV, National. IB, 24 points.

GOLDSMITHS COLLEGE (UNIVERSITY OF LONDON) G56

BSc Computing and Information Systems

UCAS Code: G520 • Mode: 3/4 Years Single Subject

Qualifications/Requirements: GCE, A level grades: CCC. AGNVQ, Merit. ND/C, Merit overall. SQAH, Grades BBBBC. SQAV, National. IB, Pass in Diploma including specific subjects.

BSc Statistics, Computer Science and Applicable Maths

UCAS Code: GG45 • Mode: 3/4 Years Equal Combination

Qualifications/Requirements: GCE, A level grades: DD. Maths at A level. AGNVQ, Merit. ND/C, Merit overall. SQAH, Grades BCCCC. SQAV, National. IB, Pass in Diploma.

UNIVERSITY OF GREENWICH G70

BA Law & Information Systems

UCAS Code: GM53 • Mode: 3 Years Equal Combination

Qualifications/Requirements: Please refer to prospectus.

BSc Geographical Information Systems

UCAS Code: FG8M • Mode: 4/5 Years Equal Combination

Qualifications/Requirements: GCE, A/AS: 4 points. AGNVQ, Individual consideration. ND/C, Individual consideration. SQAH, Individual consideration. SQAV, Individual consideration. IB, Individual consideration.

BSc Geographical Information Systems and Remote Sensing

UCAS Code: FGV5 • Mode: 3 Years Equal Combination

Qualifications/Requirements: Please refer to prospectus.

BSc Geographical Information Systems with Remote Sensing

UCAS Code: FG8N • Mode: 3 Years Equal Combination

Qualifications/Requirements: Please refer to prospectus.

BSc Mathematics, Statistics and Computing (FT/SW)

UCAS Code: G900 • Mode: 3/4 Years Single Subject

Qualifications/Requirements: GCE, A level grades: CE. Maths at A level. AGNVQ, Individual consideration. ND/C, 3 Merits (in specific programmes). SQAH, Grades CCC including specific subject(s). SQAV, Individual consideration. IB, Individual consideration.

HND Computing and Information Systems

UCAS Code: 025G • Mode: 2 Years Single Subject

Qualifications/Requirements: GCE, A/AS: 2 points. AGNVQ, Pass (in specific programmes). ND/C, 2 Merits. SQAH, Grades C. SQAV, Individual consideration. IB, Individual consideration.

HND Mathematics, Statistics and Computing

UCAS Code: 009G • Mode: 2 Years Single Subject

Qualifications/Requirements: GCE, A/AS: 2 points. Maths at A level. AGNVQ, Individual consideration. ND/C, 1 Merit (in specific programmes). SQAH, Grades C including specific subject(s). SQAV, Individual consideration. IB, Individual consideration.

UNIVERSITY OF HERTFORDSHIRE H36

BA Literature with Humanities Computing

UCAS Code: Q3GM • Mode: 3 Years Major/Minor

Qualifications/Requirements: GCE, A/AS: 14 points. AGNVQ, Individual consideration. ND/C, Merits and Distinction. SQAH, Grades CCCCC. SQAV, Individual consideration. IB, 28 points.

BA Marketing/Information Systems

UCAS Code: NG55 • Mode: 3/4 Years Equal Combination

Qualifications/Requirements: GCE, A/AS: 18 points. AGNVQ, Merit or Distinction in Business with A/AS. ND/C, Distinction Overall. SQAH, Grades BBBB. SQAV, Individual consideration. IB, 28 points.

BSc Environmental Studies/Mathematics

UCAS Code: F9G1 • Mode: 3/4 Years Major/Minor

Qualifications/Requirements: GCE, A/AS: 14 points. Maths. AGNVQ, Merit (in specific programmes) with A level. ND/C, Merit overall (in specific programmes). SQAH, Grades BCCC including specific subject(s). SQAV, Individual consideration. IB, 24 points.

BSc Mathematics/Computing

UCAS Code: G1G5 • Mode: 3/4 Years Major/Minor

Qualifications/Requirements: GCE, A/AS: 14 points. Maths. AGNVQ, Merit (in specific programmes) with A level. ND/C, Merit overall (in specific programmes). SQAH, Grades BCCC including specific subject(s). SQAV, Individual consideration. IB, 24 points.

BSc Statistics/Computing

UCAS Code: G4G5 • Mode: 3/4 Years Major/Minor

Qualifications/Requirements: GCE, A/AS: 14 points. AGNVQ, Merit (in specific programmes). ND/C, Merit overall. SQAH, Grades BCCC. SQAV, Individual consideration. IB, 24 points.

HND Digital Modelling and Animation

UCAS Code: 52GE • Mode: 2 Years Equal Combination

Qualifications/Requirements: GCE, A/AS: 4 points. Foundation Art Course. A portfolio of work is required. AGNVQ, Individual consideration. ND/C, Not normally sufficient. SQAH, Individual consideration. SQAV, Individual consideration. IB, Individual consideration.

HND Digital Modelling and Animation

UCAS Code: 52GW • Mode: 2 Years Equal Combination

Qualifications/Requirements: GCE, A/AS: 4 points. Foundation Art Course. A portfolio of work is required. AGNVQ, Individual consideration. ND/C, Not normally sufficient. SQAH, Individual consideration. SQAV, Individual consideration. IB, Individual consideration.

THE UNIVERSITY OF HUDDERSFIELD H60

BA/BSc Virtual Reality Design

UCAS Code: EW5G • Mode: 3/4 Years Equal Combination

Qualifications/Requirements: GCE, A/AS: 12-14 points. Foundation Art Course. AGNVQ, Merit (in specific programmes). ND/C, Merit overall. SQAH, Grades BBB. SQAV, National including specific subject(s). IB, 26 points.

BA/BSc Virtual Reality Design

UCAS Code: GW5G • Mode: 3/4 Years Equal Combination

Qualifications/Requirements: GCE, A/AS: 12-14 points. Foundation Art Course. AGNVQ, Merit (in specific programmes). ND/C, Merit overall. SQAH, Grades BBB. SQAV, National including specific subject(s). IB, 26 points.

BSc Scientific Computing

UCAS Code: GG15 • Mode: 3/4 Years Equal Combination

Qualifications/Requirements: GCE, A/AS: 16 points. Maths at A level. AGNVQ, Merit with A level. ND/C, Merit overall. SQAH, Grades BBBB. SQAV, Individual consideration. IB, Individual consideration.

BSc Virtual Reality Systems

UCAS Code: GG57 • Mode: 3/4 Years Equal Combination

Qualifications/Requirements: GCE, A/AS: 14-18 points. Science at A level. AGNVQ, Merit (in specific programmes). ND/C, 5 Merits. SQAH, Grades BBB. SQAV, National including specific subject(s). IB, 28 points.

THE UNIVERSITY OF HULL H72

BSc Computer Science with Information Engineering

UCAS Code: G560 • Mode: 3 Years Single Subject

Qualifications/Requirements: GCE, A/AS: 18 points. AGNVQ, Merit (in specific programmes) with A level. ND/C, Merit overall. SQAH, Grades BBBCC. SQAV, Individual consideration. IB, 26 points.

BSc Computer Science with Information Engineering

UCAS Code: G568 • Mode: 4 Years Single Subject

Qualifications/Requirements: GCE, A level grades: CD. AGNVQ, Merit (in specific programmes) with A level. ND/C, National. SQAH, Grades CCCCD. SQAV, Individual consideration. IB, 24 points.

KEELE UNIVERSITY K12

BSc Applied Environmental Science and Computer Science

UCAS Code: FGY5 • Mode: 3 Years Equal Combination

Qualifications/Requirements: GCE, A level grades – BCC-CCD. Science. AGNVQ, Merit (in specific programmes) with A level. ND/C, Individual consideration. SQAH, CSYS required. SQAV, Individual consideration. IB, 26 points.

BSc Computer Science and Mathematics

UCAS Code: GG15 • Mode: 3 Years Equal Combination

Qualifications/Requirements: GCE, A level grades: BCC-CCD. Maths at A level. AGNVQ, Merit (in specific programmes) with A level. ND/C, Individual consideration. SQAH, CSYS required. SQAV, Individual consideration. IB, 26 points.

BSc Computer Science and Statistics

UCAS Code: GG45 • Mode: 3 Years Equal Combination

Qualifications/Requirements: GCE, A level grades: BCC-CCD. Maths at A level. AGNVQ, Merit (in specific programmes) with A level. ND/C, Individual consideration. SQAH, CSYS required. SQAV, Individual consideration. IB, 26 points.

Mod Computer Science and International History

UCAS Code: GV5C • Mode: 3 Years Equal Combination

Qualifications/Requirements: GCE, A level grades: BCC-CCC. Maths or Science. AGNVQ, Distinction (in specific programmes) with A level. ND/C, Individual consideration. SQAH, CSYS required. SQAV, Individual consideration. IB, 28 points.

THE UNIVERSITY OF KENT AT CANTERBURY K24

BA English & American Literature/ Computing

UCAS Code: QG35 • Mode: 3 Years Equal Combination

Qualifications/Requirements: GCE, A/AS: 22 points. English at A level. AGNVQ, Individual consideration. ND/C, 2 Merits and 4 Distinctions. SQAH, Individual consideration. SQAV, Individual consideration. IB, 30 points.

BA English, American & Post-Colonial Literature/Computing

UCAS Code: GQ5J • Mode: 3 Years Equal Combination

Qualifications/Requirements: GCE, A/AS: 22 points. English at A level. AGNVQ, Individual consideration. ND/C, 2 Merits and 4 Distinctions. SQAH, Individual consideration. SQAV, Individual consideration. IB, 30 points.

BA Film Studies/Computing

UCAS Code: WG55 • Mode: 3 Years Equal Combination

Qualifications/Requirements: GCE, A/AS: 22 points. AGNVQ, Individual consideration. ND/C, 2 Merits and 4 Distinctions. SQAH, Individual consideration. SQAV, Individual consideration. IB, 30 points.

BSc Statistics and Computer Science

UCAS Code: GG45 • Mode: 3/4 Years Equal Combination

Qualifications/Requirements: GCE, A/AS: 20 points. Maths at A level. AGNVQ, Merit with A level. ND/C, Individual consideration. SQAH, Individual consideration. SQAV, Individual consideration. IB, 28 points.

KINGSTON UNIVERSITY K84

BA Mathematics & Information Technology (7-11 years)

UCAS Code: XG69 • Mode: 3 Years Equal Combination

Qualifications/Requirements: GCE, A/AS: 12 points. AGNVQ, Individual consideration. ND/C, Individual consideration. SQAH, Individual consideration. SQAV, Individual consideration. IB, Individual consideration.

BA Mathematics and Information Technology (3-8 years)

UCAS Code: XG29 • Mode: 3 Years Equal Combination

Qualifications/Requirements: GCE, A/AS: 12 points. AGNVQ, Individual consideration. ND/C, Individual consideration. SQAH, Individual consideration. SQAV, Individual consideration. IB, Individual consideration.

BSc Geographical Information Systems

UCAS Code: GL58 • Mode: 3 Years Equal Combination

Qualifications/Requirements: GCE, A/AS: 10-12 points. Geography or Italian. AGNVQ, Individual consideration. ND/C, (in specific programmes). SQAH, Grades BCCC. SQAV, Individual consideration. IB, Individual consideration.

BSc Geographical Information Systems (Foundation)

UCAS Code: LG85 • Mode: 4 Years Equal Combination

Qualifications/Requirements: GCE, A/AS: 12 points. AGNVQ, Individual consideration. ND/C, Individual consideration. SQAH, Individual consideration. SQAV, Individual consideration. IB, Individual consideration.

BSc Mathematics & Computing

UCAS Code: GG15 • Mode: 3 Years Equal Combination

Qualifications/Requirements: GCE, A/AS: 12-14 points. Maths. AGNVQ, Individual consideration. ND/C, 3 Merits (in specific programmes). SQAH, Grades CCC. SQAV, Individual consideration. IB, Individual consideration.

HND Geographical Information Systems

UCAS Code: 85LG • Mode: 2 Years Equal Combination

Qualifications/Requirements: GCE, A/AS: 6 points. Geography or Italian. AGNVQ, Individual consideration. ND/C, (in specific programmes). SQAH, Grades CC. SQAV, Individual consideration. IB, Individual consideration.

THE UNIVERSITY OF WALES, LAMPETER L07

BA Medieval Studies and Information Technology

UCAS Code: VG1M • Mode: 3 Years Equal Combination

Qualifications/Requirements: GCE, A/AS: 16 points. AGNVQ, Individual consideration. ND/C, Individual consideration. SQAH, Individual consideration. SQAV, Individual consideration. IB, Individual consideration.

BA Modern Historical Studies and Information Technology

UCAS Code: VG1N • Mode: 3 Years Equal Combination

Qualifications/Requirements: GCE, A/AS: 16 points. History. AGNVQ, Individual consideration. ND/C, Individual consideration. SQAH, Individual consideration. SQAV, Individual consideration. IB, Individual consideration.

BA Philosophical Studies and Information Technology

UCAS Code: GV57 • Mode: 3 Years Equal Combination

Qualifications/Requirements: GCE, A/AS: 16 points. AGNVQ, Individual consideration. ND/C, Individual consideration. SQAH, Individual consideration. SQAV, Individual consideration. IB, Individual consideration.

BA Religious Studies and Information Technology

UCAS Code: GV58 • Mode: 3 Years Equal Combination

Qualifications/Requirements: GCE, A/AS: 14 points. AGNVQ, Individual consideration. ND/C, Individual consideration. SQAH, Individual consideration. SQAV, Individual consideration. IB, Individual consideration.

BA Theology and Information Technology

UCAS Code: GV5V • Mode: 3 Years Equal Combination

Qualifications/Requirements: GCE, A/AS: 14 points. AGNVQ, Individual consideration. ND/C, Individual consideration. SQAH, Individual consideration. SQAV, Individual consideration. IB, Individual consideration.

BA Victorian Studies and Information Technology

UCAS Code: VG15 • Mode: 3 Years Equal Combination

Qualifications/Requirements: GCE, A/AS: 14 points. AGNVQ, Individual consideration. ND/C, Individual consideration. SQAH, Individual consideration. SQAV, Individual consideration. IB, Individual consideration.

Mod Diploma in Information Technology/ Management

UCAS Code: GP52 • Mode: 2/3 Years Equal Combination

Qualifications/Requirements: GCE, A/AS: 14 points. AGNVQ, Individual consideration. ND/C, Individual consideration. SQAH, Individual consideration. SQAV, Individual consideration. IB, Individual consideration.

LANCASTER UNIVERSITY L14

BSc Computer Science and Mathematics

UCAS Code: GG15 • Mode: 3 Years Equal Combination

Qualifications/Requirements: GCE, A/AS: 20 points. Maths at A level. AGNVQ, Distinction with 6 additional units or with A/AS. ND/C, Individual consideration. SQAH, Grades BBBBB including specific subject(s). SQAV, Individual consideration. IB, 30 points.

UNIVERSITY OF LEEDS L23

BSc Computer Science-Mathematics

UCAS Code: GG15 • Mode: 3/4 Years Equal Combination

Qualifications/Requirements: GCE, A level grades: BBC. Maths at A level. AGNVQ, Individual consideration. ND/C, 1 Merit and 5 Distinctions (in specific programmes). SQAH, Individual consideration. SQAV, Individual consideration. IB, 30 points.

BSc Computer Science-Statistics

UCAS Code: GG45 • Mode: 3/4 Years Equal Combination

Qualifications/Requirements: GCE, A level grades: BBC. Maths at A level. AGNVQ, Individual consideration. ND/C, 1 Merit and 5 Distinctions (in specific programmes). SQAH, Individual consideration. SQAV, Individual consideration. IB, 30 points.

LEEDS METROPOLITAN UNIVERSITY L27

BA Secondary Information Technology

UCAS Code: XG75 • Mode: 2 Years Equal Combination

Qualifications/Requirements: AGNVQ, Not normally sufficient. ND/C, Higher National. SQAH, Not normally sufficient. SQAV, Higher National. IB, Not normally sufficient.

UNIVERSITY OF LINCOLNSHIRE AND HUMBERSIDE L39

BA/BSc Business and Information Technology

UCAS Code: G720 • Mode: 3 Years Single Subject

Qualifications/Requirements: GCE, A/AS: 12 points. AGNVQ, Merit. ND/C, 3 Merits and 1 Distinction. SQAH, Grades CCCC. SQAV, Individual consideration. IB, 24 points.

BA/BSc Computing and Information Technology

UCAS Code: G523 • Mode: 3 Years Single Subject

Qualifications/Requirements: GCE, A/AS: 12 points. AGNVQ, Merit. ND/C, 3 Merits and 1 Distinction. SQAH, Grades CCCC. SQAV, Individual consideration. IB, 24 points.

BA/BSc Economics and Information Systems

UCAS Code: GL51 • Mode: 3 Years Equal Combination

Qualifications/Requirements: GCE, A/AS: 16 points. AGNVQ, Distinction. ND/C, 1 Merit and 3 Distinctions. SQAH, Grades BBCCC. SQAV, Individual consideration. IB, 24 points.

BA/BSc European Studies and Information Technology

UCAS Code: GT5F • Mode: 3 Years Equal Combination

Qualifications/Requirements: GCE, A/AS: 12 points. AGNVQ, Merit. ND/C, 3 Merits and 1 Distinction. SQAH, Grades CCCC. SQAV, Individual consideration. IB, 24 points.

BA/BSc Health Studies and Information Systems

UCAS Code: GL54 • Mode: 3 Years Equal Combination

Qualifications/Requirements: GCE, A/AS: 16 points. AGNVQ, Distinction. ND/C, 1 Merit and 3 Distinctions. SQAH, Grades BBCCC. SQAV, Individual consideration. IB, 24 points.

BA/BSc Human Resource Management and Information Technology

UCAS Code: GN5P • Mode: 3 Years Equal Combination

Qualifications/Requirements: GCE, A/AS: 12 points. AGNVQ, Merit. ND/C, 3 Merits and 1 Distinction. SQAH, Grades CCCC. SQAV, Individual consideration. IB, 24 points.

BA/BSc Management and Information Systems

UCAS Code: NG1N • Mode: 3 Years Equal Combination

Qualifications/Requirements: GCE, A/AS: 14 points. AGNVQ, Merit. ND/C, 2 Merits and 2 Distinctions. SQAH, Grades BCCC. SQAV, Individual consideration. IB, 24 points.

Mod Applied Development Studies and Information Systems

UCAS Code: MG95 • Mode: 3 Years Equal Combination

Qualifications/Requirements: AGNVQ, Individual consideration. ND/C, Individual consideration. SQAH, Individual consideration. SQAV, Individual consideration. IB, Individual consideration.

Mod Applied Social Science and Computing
UCAS Code: LGH5 • Mode: 3 Years Equal Combination

Qualifications/Requirements: AGNVQ, Individual consideration. ND/C, Individual consideration. SQAH, Individual consideration. SQAV, Individual consideration. IB, Individual consideration.

Mod Architectural Technology and Interactive Design
UCAS Code: KG2N • Mode: 3 Years Equal Combination

Qualifications/Requirements: AGNVQ, Individual consideration. ND/C, Individual consideration. SQAH, Individual consideration. SQAV, Individual consideration. IB, Individual consideration.

Mod English and Information Systems
UCAS Code: GQM3 • Mode: 3 Years Equal Combination

Qualifications/Requirements: GCE, A/AS: 16 points. AGNVQ, Distinction. ND/C, 1 Merit and 3 Distinctions. SQAH, Grades BBCCC. SQAV, Individual consideration. IB, 24 points.

Mod Environmental Biology and Information Systems
UCAS Code: CG95 • Mode: 3 Years Equal Combination

Qualifications/Requirements: AGNVQ, Individual consideration. ND/C, Individual consideration. SQAH, Individual consideration. SQAV, Individual consideration. IB, Individual consideration.

Mod Environmental Studies and Information Systems
UCAS Code: FGYM • Mode: 3 Years Equal Combination

Qualifications/Requirements: AGNVQ, Individual consideration. ND/C, Individual consideration. SQAH, Individual consideration. SQAV, Individual consideration. IB, Individual consideration.

Mod European Studies and Interactive Design
UCAS Code: TG2N • Mode: 3 Years Equal Combination

Qualifications/Requirements: AGNVQ, Individual consideration. ND/C, Individual consideration. SQAH, Individual consideration. SQAV, Individual consideration. IB, Individual consideration.

Mod Finance and Information Systems
UCAS Code: GNMJ • Mode: 3 Years Equal Combination

Qualifications/Requirements: AGNVQ, Individual consideration. ND/C, Individual consideration. SQAH, Individual consideration. SQAV, Individual consideration. IB, Individual consideration.

Mod Fine Art and Information Technology
UCAS Code: WG1M • Mode: 3 Years Equal Combination

Qualifications/Requirements: AGNVQ, Individual consideration. ND/C, Individual consideration. SQAH, Individual consideration. SQAV, Individual consideration. IB, Individual consideration.

Mod Fine Art and Interactive Design
UCAS Code: WG1N • Mode: 3 Years Equal Combination

Qualifications/Requirements: AGNVQ, Individual consideration. ND/C, Individual consideration. SQAH, Individual consideration. SQAV, Individual consideration. IB, Individual consideration.

Mod Food Studies and Information Systems
UCAS Code: DGKM • Mode: 3 Years Equal Combination

Qualifications/Requirements: AGNVQ, Individual consideration. ND/C, Individual consideration. SQAH, Individual consideration. SQAV, Individual consideration. IB, Individual consideration.

Mod Forensic Science and Information Systems
UCAS Code: BG1M • Mode: 3 Years Equal Combination

Qualifications/Requirements: AGNVQ, Individual consideration. ND/C, Individual consideration. SQAH, Individual consideration. SQAV, Individual consideration. IB, Individual consideration.

Mod Graphic Design and Information Technology
UCAS Code: WG2M • Mode: 3 Years Equal Combination

Qualifications/Requirements: AGNVQ, Individual consideration. ND/C, Individual consideration. SQAH, Individual consideration. SQAV, Individual consideration. IB, Individual consideration.

Mod Graphic Design and Interactive Design
UCAS Code: WG2N • Mode: 3 Years Equal Combination

Qualifications/Requirements: AGNVQ, Individual consideration. ND/C, Individual consideration. SQAH, Individual consideration. SQAV, Individual consideration. IB, Individual consideration.

Mod History and Information Systems
UCAS Code: GVM1 • Mode: 3 Years Equal Combination

Qualifications/Requirements: GCE, A/AS: 16 points. AGNVQ, Distinction. ND/C, 1 Merit and 3 Distinctions. SQAH, Grades BBCCC. SQAV, Individual consideration. IB, 24 points.

Mod Illustration and Information Technology
UCAS Code: WGGM • Mode: 3 Years Equal Combination

Qualifications/Requirements: AGNVQ, Individual consideration. ND/C, Individual consideration. SQAH, Individual consideration. SQAV, Individual consideration. IB, Individual consideration.

Mod Illustration and Interactive Design
UCAS Code: WGGN • Mode: 3 Years Equal Combination

Qualifications/Requirements: AGNVQ, Individual consideration. ND/C, Individual consideration. SQAH, Individual consideration. SQAV, Individual consideration. IB, Individual consideration.

LIVERPOOL HOPE L46

BA Identity Studies and Information Technology
UCAS Code: LG35 • Mode: 3 Years Equal Combination

Qualifications/Requirements: Please refer to prospectus.

BA Music/Information Technology
UCAS Code: GW53 • Mode: 3 Years Equal Combination

Qualifications/Requirements: GCE, A/AS: 12 points. Music. AGNVQ, Merit in Management or Pass with A level. GCSEs or equivalent needed. ND/C, 8 Merits. SQAH, Individual consideration. SQAV, Individual consideration. IB, Individual consideration.

BA Sociology/Information Technology
UCAS Code: GL53 • Mode: 3 Years Equal Combination

Qualifications/Requirements: GCE, A/AS: 12 points. AGNVQ, Merit. ND/C, 8 Merits. SQAH, Individual consideration. SQAV, Individual consideration. IB, Individual consideration.

BSc Health and Information Technology
UCAS Code: BG95 • Mode: 3 Years Equal Combination

Qualifications/Requirements: Please refer to prospectus.

BSc Mathematics/Information Technology
UCAS Code: GG51 • Mode: 3 Years Equal Combination

Qualifications/Requirements: GCE, A/AS: 10 points. Maths. AGNVQ, Pass with A level. ND/C, 6 Merits. SQAH, Individual consideration. SQAV, Individual consideration. IB, Individual consideration.

BSc Psychology/Information Technology
UCAS Code: GC58 • Mode: 3 Years Equal Combination

Qualifications/Requirements: GCE, A/AS: 10 points. AGNVQ, Merit. ND/C, 6 Merits. SQAH, Individual consideration. SQAV, Individual consideration. IB, Individual consideration.

LIVERPOOL JOHN MOORES UNIVERSITY L51

BSc Applied Statistics and Computing
UCAS Code: G440 • Mode: 4 Years Single Subject

Qualifications/Requirements: GCE, A/AS: 10 points. Maths at A level. AGNVQ, Distinction (in specific programmes) with A level. ND/C, 3 Merits (in specific programmes). SQAH, Individual consideration. SQAV, Individual consideration. IB, 26 points.

BSc Mathematics, Statistics and Computing
UCAS Code: G920 • Mode: 4 Years Single Subject

Qualifications/Requirements: GCE, A/AS: 10 points. Maths. AGNVQ, Distinction (in specific programmes) with A level. ND/C, 3 Merits (in specific programmes). SQAH, Individual consideration. SQAV, Individual consideration. IB, 26 points.

LLANDRILLO COLLEGE, NORTH WALES L53

HND Personal Computer Technology
UCAS Code: 065G • Mode: 2 Years Single Subject

Qualifications/Requirements: Please refer to prospectus.

LONDON GUILDHALL UNIVERSITY L55

BSc Computing and Information Systems
UCAS Code: G520 • Mode: 3 Years Single Subject

Qualifications/Requirements: GCE, A level grades: CD. AGNVQ, Merit (in specific programmes). ND/C, 4 Merits. SQAH, Individual consideration. SQAV, Individual consideration. IB, 24 points.

Mod American Studies & Computing
UCAS Code: GQ54 • Mode: 3 Years Equal Combination

Qualifications/Requirements: GCE, A level grades: CD. AGNVQ, Merit. ND/C, Merit overall and 2 Distinctions. SQAH, Individual consideration. SQAV, Individual consideration. IB, 26 points.

Mod Banking & Business Information Technology
UCAS Code: NG37 • Mode: 3 Years Equal Combination

Qualifications/Requirements: GCE, A level grades – DD. AGNVQ, Merit (in specific programmes). ND/C, Merit overall. SQAH, Individual consideration. SQAV, Individual consideration. IB, 24 points.

Mod Business & Business Information Technology
UCAS Code: GN71 • Mode: 3 Years Equal Combination

Qualifications/Requirements: GCE, A level grades – CD-DDD. AGNVQ, Merit (in specific programmes). ND/C, Merit overall and 4 Distinctions. SQAH, Individual consideration. SQAV, Individual consideration. IB, 26 points.

Mod Business & Multimedia Systems
UCAS Code: GNM1 • Mode: 3 Years Equal Combination

Qualifications/Requirements: GCE, A level grades – CD-DDD. AGNVQ, Merit (in specific programmes). ND/C, Merit overall and 2 Distinctions. SQAH, Individual consideration. SQAV, Individual consideration. IB, 26 points.

Mod Computing & Mathematics
UCAS Code: GG15 • Mode: 3 Years Equal Combination

Qualifications/Requirements: GCE, A level grades: DD. AGNVQ, Merit (in specific programmes). ND/C, Merit overall. SQAH, Individual consideration. SQAV, Individual consideration. IB, 24 points.

Mod Criminology & Multimedia Systems

UCAS Code: GMMH • Mode: 3 Years Equal Combination

Qualifications/Requirements: GCE, A level grades: CD. AGNVQ, Merit. ND/C, Merit overall and 2 Distinctions. SQAH, Individual consideration. SQAV, Individual consideration. IB, 26 points.

Mod Design Studies & Multimedia Systems

UCAS Code: GWM2 • Mode: 3 Years Equal Combination

Qualifications/Requirements: GCE, A level grades: CD-DDD. AGNVQ, Merit (in specific programmes). ND/C, Merit overall. SQAH, Individual consideration. SQAV, Individual consideration. IB, 24 points.

Mod Development Studies & Multimedia Systems

UCAS Code: GMM9 • Mode: 3 Years Equal Combination

Qualifications/Requirements: GCE, A level grades: DD. AGNVQ, Merit (in specific programmes). ND/C, Merit overall. SQAH, Individual consideration. SQAV, Individual consideration. IB, 24 points.

Mod English Studies & Multimedia Systems

UCAS Code: GQM3 • Mode: 3 Years Equal Combination

Qualifications/Requirements: GCE, A level grades: CD-DDD. AGNVQ, Merit (in specific programmes). ND/C, Merit overall and 2 Distinctions. SQAH, Individual consideration. SQAV, Individual consideration. IB, 26 points.

Mod Mathematics & Multimedia Systems

UCAS Code: GG1M • Mode: 3 Years Equal Combination

Qualifications/Requirements: GCE, A level grades: DD. AGNVQ, Merit (in specific programmes). ND/C, Merit overall. SQAH, Individual consideration. SQAV, Individual consideration. IB, 24 points.

Mod Modular Programme

UCAS Code: Y400 • Mode: 3 Years Single Subject

Qualifications/Requirements: GCE, A level grades – CC-DD. AGNVQ, Merit (in specific programmes). ND/C, Merit overall. SQAH, Individual consideration. SQAV, Individual consideration. IB, 24 points.

Mod Modular Programme

UCAS Code: Y420 • Mode: 3 Years

Qualifications/Requirements: GCE, A level grades – EE. AGNVQ, Pass. ND/C, Merit overall. SQAH, Individual consideration. SQAV, Individual consideration. IB, 24 points.

LOUGHBOROUGH UNIVERSITY L79

BA Library and Information Management

UCAS Code: P200 • Mode: 3 Years Single Subject

Qualifications/Requirements: GCE, A/AS: 18 points. AGNVQ, Distinction with 6 additional units or with A level. ND/C, 3 Merits and 2 Distinctions. SQAH, Individual consideration. SQAV, Individual consideration. IB, 28 points.

BA Library and Information Management (4 years SW)

UCAS Code: P201 • Mode: 4 Years Single Subject

Qualifications/Requirements: GCE, A/AS: 18 points. AGNVQ, Distinction with 6 additional units or with A level. ND/C, 3 Merits and 2 Distinctions. SQAH, Individual consideration. SQAV, Individual consideration. IB, 28 points.

UNIVERSITY OF LUTON L93

Mod Modular Credit Scheme

UCAS Code: Y400 • Mode: 3 Years Single Subject

Qualifications/Requirements: GCE, A/AS: 12-16 points. AGNVQ, Distinction or Merit with 6 additional units or with A/AS. ND/C, Merits and Distinction. SQAH, Grades BBCC. SQAV, Individual consideration. IB, 32 points.

HND Geographical Information Systems

UCAS Code: 58GF • Mode: 2 Years Equal Combination

Qualifications/Requirements: GCE, A/AS: 6 points. AGNVQ, Merit or Pass with 6 additional units or with A/AS. ND/C, Merit. SQAH, Grades CCCC. SQAV, Individual consideration. IB, 26 points.

THE UNIVERSITY OF MANCHESTER M20

BSc Computer Science and Mathematics

UCAS Code: GG15 • Mode: 3 Years Equal Combination

Qualifications/Requirements: GCE, A level grades: ABC. Maths at A level. AGNVQ, Not normally sufficient. ND/C, Not normally sufficient. SQAH, Grades ABBBB. SQAV, Not normally sufficient. IB, 30 points.

BSc Computing and Information Systems with Industrial Experience

UCAS Code: G507 • Mode: 4 Years Single Subject

Qualifications/Requirements: GCE, A level grades: BBB. Maths at A level. AGNVQ, Distinction in Science with A level. ND/C, Individual consideration. SQAH, Grades BBBBB including specific subject(s). SQAV, Not normally sufficient. IB, 30 points.

BSc Computing and Information Systems

UCAS Code: G506 • Mode: 3 Years Single Subject

Qualifications/Requirements: GCE, A level grades: BBB. Maths at A level. AGNVQ, Distinction in Science with A level. ND/C, Individual consideration. SQAH, Grades BBBBB including specific subject(s). SQAV, Not normally sufficient. IB, 30 points.

THE UNIVERSITY OF MANCHESTER INSTITUTE OF SCIENCE AND TECHNOLOGY (UMIST) M25

BSc Management and Information Technology

UCAS Code: GN51 • Mode: 3 Years Equal Combination

Qualifications/Requirements: GCE, A level grades: BBB. AGNVQ, Distinction (in specific programmes) with A/AS. ND/C, 3 Merits and 4 Distinctions. SQAH, CSYS required. SQAV, Individual consideration. IB, 32 points.

THE MANCHESTER METROPOLITAN UNIVERSITY M40

BA/BSc Information and Communication

UCAS Code: G563 • Mode: 3 Years Single Subject

Qualifications/Requirements: GCE, A/AS: 10-12 points. AGNVQ, Individual consideration. ND/C, Individual consideration. SQAH, Individual consideration. SQAV, Individual consideration. IB, Individual consideration.

BSc Applicable Mathematics/Computing Science

UCAS Code: GG15 • Mode: 3 Years Equal Combination

Qualifications/Requirements: GCE, A/AS: 16 points. Maths. AGNVQ, Merit (in specific programmes). ND/C, 1 Merit and 3 Distinctions (in specific programmes). SQAH, Grades BBBCC including specific subject(s). SQAV, Individual consideration. IB, 28 points.

BSc Applicable Mathematics/Multimedia Technology

UCAS Code: GG16 • Mode: 3 Years Equal Combination

Qualifications/Requirements: GCE, A/AS: 16 points. Maths. AGNVQ, Merit (in specific programmes). ND/C, 1 Merit and 3 Distinctions (in specific programmes). SQAH, Grades BBBCC including specific subject(s). SQAV, Individual consideration. IB, 28 points.

BSc Business Mathematics/Multimedia Technology

UCAS Code: GG1P • Mode: 3 Years Equal Combination

Qualifications/Requirements: GCE, A/AS: 16 points. Maths at A level, Physics or Economics. AGNVQ, Merit (in specific programmes). ND/C, 1 Merit and 3 Distinctions (in specific programmes). SQAH, Grades BBBCC including specific subject(s). SQAV, Individual consideration. IB, 28 points.

BSc Economics/Information Systems

UCAS Code: GL5C • Mode: 3 Years Equal Combination

Qualifications/Requirements: GCE, A/AS: 16 points. AGNVQ, Merit (in specific programmes). ND/C, 1 Merit and 3 Distinctions. SQAH, Grades BBBCC. SQAV, Individual consideration. IB, 28 points.

BSc Environmental Studies/Information Systems

UCAS Code: GF59 • Mode: 3 Years Equal Combination

Qualifications/Requirements: GCE, A/AS: 16 points. AGNVQ, Merit (in specific programmes). ND/C, 1 Merit and 3 Distinctions. SQAH, Grades BBBCC. SQAV, Individual consideration. IB, 28 points.

BSc Environmental Studies/Multimedia Technology

UCAS Code: FG96 • Mode: 3 Years Equal Combination

Qualifications/Requirements: GCE, A/AS: 14 points. AGNVQ, Merit (in specific programmes). ND/C, 1 Merit and 3 Distinctions. SQAH, Grades BBBCC. SQAV, Individual consideration. IB, 28 points.

BSc European Studies/Information Systems

UCAS Code: GT5F • Mode: 3 Years Equal Combination

Qualifications/Requirements: GCE, A/AS: 16 points. AGNVQ, Merit (in specific programmes). ND/C, 1 Merit and 3 Distinctions. SQAH, Grades BBBCC. SQAV, Individual consideration. IB, 28 points.

BSc Geography/Information Systems

UCAS Code: GL5V • Mode: 3 Years Equal Combination

Qualifications/Requirements: GCE, A/AS: 16 points. AGNVQ, Merit (in specific programmes). ND/C, 1 Merit and 3 Distinctions. SQAH, Grades BBBCC. SQAV, Individual consideration. IB, 28 points.

BSc Geography/Multimedia Technology

UCAS Code: GL68 • Mode: 3 Years Equal Combination

Qualifications/Requirements: GCE, A/AS: 14 points. AGNVQ, Merit (in specific programmes). ND/C, 1 Merit and 3 Distinctions. SQAH, Grades BBBCC. SQAV, Individual consideration. IB, 28 points.

MIDDLESEX UNIVERSITY M80

HND Networking and Computer Systems

UCAS Code: 65GG • Mode: 2 Years Equal Combination

Qualifications/Requirements: GCE, A level grades: E. AGNVQ, Pass (in specific programmes). ND/C, 3 Merits. SQAH, Individual consideration. SQAV, Individual consideration. IB, 24 points.

NAPIER UNIVERSITY N07

BA Librarianship and Information Studies
UCAS Code: P100 • Mode: 3/4 Years Single Subject

Qualifications/Requirements: GCE, A level grades: CD. English. AGNVQ, Individual consideration. ND/C, Individual consideration. SQAH, Grades BCC. SQAV, Individual consideration. IB, Individual consideration.

BSc Management and Information Technology
UCAS Code: NG15 • Mode: 3/4 Years Equal Combination

Qualifications/Requirements: GCE, A level grades: DD. ND/C, Individual consideration. SQAH, Grades BBC. SQAV, Individual consideration.

BSc Network Computing (HND top-up)
UCAS Code: G530 • Mode: 1 Years Single Subject

Qualifications/Requirements: Please refer to prospectus.

UNIVERSITY COLLEGE NORTHAMPTON N14

BA Education/Information Systems
UCAS Code: X9G5 • Mode: 3 Years Major/Minor

Qualifications/Requirements: GCE, A/AS: 10-12 points. AGNVQ, Merit. ND/C, 5 Merits. SQAH, Grades BCC. SQAV, Individual consideration. IB, 24 points.

BA Law/Information Systems
UCAS Code: M3G5 • Mode: 3 Years Major/Minor

Qualifications/Requirements: GCE, A/AS: 10 points. AGNVQ, Merit. ND/C, 3 Merits and 2 Distinctions. SQAH, Grades BCC. SQAV, Individual consideration. IB, 24 points.

Mod Drama/Information Systems
UCAS Code: W4G5 • Mode: 3 Years Major/Minor

Qualifications/Requirements: GCE, A/AS: 12 points. AGNVQ, Merit. ND/C, 5 Merits and 1 Distinction. SQAH, Grades BCC. SQAV, Individual consideration. IB, 24 points.

Mod Ecology/Information Systems
UCAS Code: C9G5 • Mode: 3 Years Major/Minor

Qualifications/Requirements: GCE, A/AS: 8 points. AGNVQ, Merit. ND/C, 5 Merits. SQAH, Grades CCC. SQAV, Individual consideration. IB, 24 points.

Mod Energy Management/Information Systems
UCAS Code: J9G5 • Mode: 3 Years Major/Minor

Qualifications/Requirements: GCE, A/AS: 8 points. AGNVQ, Merit. ND/C, 5 Merits. SQAH, Grades CCC. SQAV, Individual consideration. IB, 24 points.

Mod Environmental Chemistry/Information Systems
UCAS Code: F1G5 • Mode: 3 Years Major/Minor

Qualifications/Requirements: GCE, A/AS: 8 points. AGNVQ, Merit. ND/C, 5 Merits. SQAH, Grades CCC. SQAV, Individual consideration. IB, 24 points.

Mod Geography/Information Systems
UCAS Code: F8G5 • Mode: 3 Years Major/Minor

Qualifications/Requirements: GCE, A/AS: 10-12 points. Geography. AGNVQ, Merit. ND/C, 5 Merits. SQAH, Grades CCC. SQAV, Individual consideration. IB, 24 points.

Mod Human Biological Studies/Information Systems
UCAS Code: B1G5 • Mode: 3 Years Major/Minor

Qualifications/Requirements: GCE, A/AS: 8-10 points. Science. AGNVQ, Merit. ND/C, 5 Merits. SQAH, Grades CCC. SQAV, Individual consideration. IB, 24 points.

Mod Industrial Archaeology/Information Systems
UCAS Code: V6G5 • Mode: 3 Years Major/Minor

Qualifications/Requirements: GCE, A/AS: 10 points. AGNVQ, Merit. ND/C, 5 Merits. SQAH, Grades CCC. SQAV, Individual consideration. IB, 24 points.

Mod Industrial Enterprise/Information Systems
UCAS Code: H1G5 • Mode: 3 Years Major/Minor

Qualifications/Requirements: GCE, A/AS: 8 points. AGNVQ, Merit. ND/C, 5 Merits. SQAH, Grades CCC. SQAV, Individual consideration. IB, 24 points.

Mod Information Systems/Education
UCAS Code: G5X9 • Mode: 3 Years Major/Minor

Qualifications/Requirements: GCE, A/AS: 10-12 points. AGNVQ, Merit. ND/C, 5 Merits. SQAH, Grades CCC. SQAV, Individual consideration. IB, 24 points.

Mod Management Science/Information Systems
UCAS Code: G4G5 • Mode: 3 Years Major/Minor

Qualifications/Requirements: GCE, A/AS: 8 points. AGNVQ, Merit. ND/C, 5 Merits. SQAH, Grades CCC. SQAV, Individual consideration. IB, 24 points.

Mod Mathematics/Information Systems
UCAS Code: G1G5 • Mode: 3 Years Major/Minor

Qualifications/Requirements: GCE, A/AS: 8-10 points. Maths. AGNVQ, Individual consideration. ND/C, Individual consideration. SQAH, Grades CCC. SQAV, Individual consideration. IB, 24 points.

Mod Media & Popular Culture/ Information Systems

UCAS Code: P4G5 • Mode: 3 Years Major/Minor

Qualifications/Requirements: GCE, A/AS: 14 points. AGNVQ, Merit. ND/C, Merit. SQAH, Grades BBC. SQAV, Individual consideration. IB, 24 points.

Mod Sociology/Information Systems

UCAS Code: L3G5 • Mode: 3 Years Major/Minor

Qualifications/Requirements: GCE, A/AS: 10-12 points. AGNVQ, Merit. ND/C, 5 Merits. SQAH, Grades CCC. SQAV, Individual consideration. IB, 24 points.

Mod Sport Studies/Information Systems

UCAS Code: B6G5 • Mode: 3 Years Major/Minor

Qualifications/Requirements: GCE, A/AS: 14 points. AGNVQ, Merit. ND/C, Merits and 2 Distinctions. SQAH, Grades BBB. IB, 24 points.

Mod Wastes Management/Information Systems

UCAS Code: F9GM • Mode: 3 Years Major/Minor

Qualifications/Requirements: GCE, A/AS: 8 points. AGNVQ, Merit. ND/C, 5 Merits. SQAH, Grades CCC. SQAV, Individual consideration. IB, 24 points.

UNIVERSITY OF NEWCASTLE UPON TYNE N21

BSc Computing Science and Mathematics (3 or 4 years)

UCAS Code: GG15 • Mode: 3 Years Equal Combination

Qualifications/Requirements: GCE, A/AS: 20 points. Maths at A level. AGNVQ, Pass. ND/C, Individual consideration. SQAH, Grades AAAB including specific subject(s). SQAV, Individual consideration. IB, 28 points.

BSc Computing Science and Statistics (3 or 4 years)

UCAS Code: GG45 • Mode: 3 Years Equal Combination

Qualifications/Requirements: GCE, A/AS: 20 points. Maths at A level. AGNVQ, Pass. ND/C, Individual consideration. SQAH, Grades AAAB including specific subject(s). SQAV, Individual consideration. IB, 28 points.

UNIVERSITY OF WALES COLLEGE, NEWPORT N37

BA Sports Studies and Information Technology

UCAS Code: BG65 • Mode: 3 Years Equal Combination

Qualifications/Requirements: GCE, A/AS: 10 points. AGNVQ, Distinction (in specific programmes). ND/C, Merits and Distinction. SQAH, Individual consideration. SQAV, Individual consideration. IB, Individual consideration.

BSc Industrial Information Technology

UCAS Code: G5H6 • Mode: 3 Years Major/Minor

Qualifications/Requirements: GCE, A/AS: 8-10 points. AGNVQ, Distinction (in specific programmes). ND/C, National. SQAH, Individual consideration. SQAV, Individual consideration. IB, Individual consideration.

NORTHBROOK COLLEGE SUSSEX N41

BSc Computing and Information Systems for Business

UCAS Code: GG56 • Mode: 3 Years Equal Combination

Qualifications/Requirements: Please refer to prospectus.

THE NORTH EAST WALES INSTITUTE OF HIGHER EDUCATION N56

BA Social Science and Information Technology

UCAS Code: LG35 • Mode: 3 Years Equal Combination

Qualifications/Requirements: GCE, A/AS: 6-12 points. AGNVQ, Merit (in specific programmes). ND/C, 3 to 4 Merits. SQAH, Individual consideration. SQAV, Individual consideration. IB, Individual consideration.

BSc Environmental Studies and Information Technology

UCAS Code: FG95 • Mode: 3 Years Equal Combination

Qualifications/Requirements: GCE, A/AS: 4-10 points. AGNVQ, Merit (in specific programmes). ND/C, 3 to 4 Merits. SQAH, Individual consideration. SQAV, Individual consideration. IB, Individual consideration.

BSc Geography and Information Technology

UCAS Code: FG8M • Mode: 3 Years Equal Combination

Qualifications/Requirements: GCE, A/AS: 4-10 points. AGNVQ, Merit (in specific programmes). ND/C, 3 to 4 Merits. SQAH, Individual consideration. SQAV, Individual consideration. IB, Individual consideration.

BSc Sports Science and Information Technology

UCAS Code: BG65 • Mode: 3 Years Equal Combination

Qualifications/Requirements: GCE, A/AS: 4-10 points. AGNVQ, Merit (in specific programmes). ND/C, 3 to 4 Merits. SQAH, Individual consideration. SQAV, Individual consideration. IB, Individual consideration.

NORTH EAST WORCESTERSHIRE COLLEGE N58

HND Graphic Design and Multimedia

UCAS Code: 52GE • Mode: 2 Years Equal Combination

Qualifications/Requirements: GCE, A level grades: E. OTHER. AGNVQ, Pass (in specific programmes). ND/C, National.

HND Graphic Design and Multimedia

UCAS Code: 52GW • Mode: 2 Years Equal Combination

Qualifications/Requirements: GCE, A level grades: E. OTHER. AGNVQ, Pass (in specific programmes). ND/C, National.

UNIVERSITY OF NORTH LONDON N63

HND Computing and Mathematical Sciences

UCAS Code: 009G • Mode: 2 Years Single Subject

Qualifications/Requirements: GCE, A/AS: 4 points. Science. AGNVQ, Pass. ND/C, National (in specific programmes). SQAH, Grades DDD including specific subject(s). SQAV, Individual consideration. IB, Pass in Diploma.

UNIVERSITY OF NORTHUMBRIA AT NEWCASTLE N77

BSc Applied Business Computing

UCAS Code: G710 • Mode: 1 Years Single Subject

Qualifications/Requirements: AGNVQ, Not normally sufficient. ND/C, Higher National. SQAH, Not normally sufficient. IB, Not normally sufficient.

BSc Information and Communication Management

UCAS Code: G560 • Mode: 3 Years Single Subject

Qualifications/Requirements: GCE, A/AS: 12 points. AGNVQ, Merit. ND/C, 3 Merits and 1 Distinction. SQAH, Grades CCCC. SQAV, Individual consideration. IB, 24 points.

THE UNIVERSITY OF NOTTINGHAM N84

BSc Digital Business

UCAS Code: GN5D • Mode: 3 Years Equal Combination

Qualifications/Requirements: GCE, A level grades – ABB-BBB. Maths at A level. ND/C, Individual consideration. SQAH, Individual consideration. SQAV, Individual consideration. IB, Individual consideration.

THE NOTTINGHAM TRENT UNIVERSITY N91

BSc Environmental Conservation & Management & Computing

UCAS Code: FG95 • Mode: 3 Years Equal Combination

Qualifications/Requirements: GCE, A/AS: 12 points. Science. AGNVQ, Individual consideration. ND/C, Individual consideration. SQAH, Grades C. SQAV, Individual consideration. IB, Pass in Diploma.

BSc Environmental Systems and Monitoring and IT for Sciences

UCAS Code: FGXM • Mode: 3 Years Equal Combination

Qualifications/Requirements: GCE, A/AS: 12 points. Science. AGNVQ, Individual consideration. ND/C, Individual consideration. SQAH, Grades C. SQAV, Individual consideration. IB, Pass in Diploma.

BSc Environmental Systems and Monitoring and Mathematics

UCAS Code: FGX1 • Mode: 3 Years Equal Combination

Qualifications/Requirements: GCE, A/AS: 12 points. Maths at A level. AGNVQ, Individual consideration. ND/C, Individual consideration. SQAH, Grades C. SQAV, Individual consideration. IB, Pass in Diploma.

BSc Sport & Exercise Science and Computing

UCAS Code: BG65 • Mode: 3 Years Equal Combination

Qualifications/Requirements: GCE, A/AS: 16 points. Biology and Physical Education or Sports Studies. AGNVQ, Individual consideration. ND/C, Individual consideration. SQAH, Grades B. SQAV, Individual consideration. IB, Pass in Diploma.

BSc Sport & Exercise Science and Information Technology for Science

UCAS Code: BG6M • Mode: 3 Years Equal Combination

Qualifications/Requirements: GCE, A/AS: 16 points. Biology at A level and Physical Education or Sports Studies. AGNVQ, Individual consideration. ND/C, Individual consideration. SQAH, Grades B. SQAV, Individual consideration. IB, Pass in Diploma.

OXFORD BROOKES UNIVERSITY O66

BSc Computer Technology

UCAS Code: H610 • Mode: 3/4 Years Single Subject

Qualifications/Requirements: GCE, A level grades: CCC. Science or Maths. AGNVQ, Merit (in specific programmes). ND/C, Individual consideration. SQAH, Individual consideration. SQAV, Individual consideration. IB, Individual consideration.

Mod Business Administration and Management/Information Systems

UCAS Code: GNM1 • Mode: 3 Years Equal Combination

Qualifications/Requirements: GCE, A level grades – CDD-BBC. AGNVQ, Merit in Business or Merit in Business with 4 additional units. ND/C, Individual consideration. SQAH, Individual consideration. SQAV, Individual consideration. IB, Individual consideration.

Mod Business Administration and Management/Intelligent Systems

UCAS Code: GN81 • Mode: 3 Years Equal Combination

Qualifications/Requirements: GCE, A level grades – CD-BBC. AGNVQ, Merit or Merit in Business with 4 additional units. ND/C, Individual consideration. SQAH, Individual consideration. SQAV, Individual consideration. IB, Individual consideration.

Mod Business Administration & Management/Computing

UCAS Code: GN51 • Mode: 3 Years Equal Combination

Qualifications/Requirements: GCE, A level grades – CCC-BBC. AGNVQ, Merit or Merit in Business with 4 additional units. ND/C, Individual consideration. SQAH, Individual consideration. SQAV, Individual consideration. IB, Individual consideration.

Mod Business Administration and Management/Geographic Information Science

UCAS Code: FN8C • Mode: 3 Years Equal Combination

Qualifications/Requirements: Please refer to prospectus.

Mod Business Administration and Management/Multimedia Systems

UCAS Code: GNP1 • Mode: 3 Years Equal Combination

Qualifications/Requirements: Please refer to prospectus.

Mod Business Administration and Management/Software Engineering

UCAS Code: GN71 • Mode: 3 Years Equal Combination

Qualifications/Requirements: GCE, A level grades – CCC-BBC. AGNVQ, Merit or Merit in Business with 4 additional units. ND/C, Individual consideration. SQAH, Individual consideration. SQAV, Individual consideration. IB, Individual consideration.

Mod Business Statistics/Computer Systems

UCAS Code: GG4P • Mode: 3 Years Equal Combination

Qualifications/Requirements: Please refer to prospectus.

Mod Business Statistics/Computing

UCAS Code: GG4M • Mode: 3 Years Equal Combination

Qualifications/Requirements: Please refer to prospectus.

Mod Business Statistics/Computing Science

UCAS Code: GG4N • Mode: 3 Years Equal Combination

Qualifications/Requirements: Please refer to prospectus.

Mod Business Statistics/Geographic Information Science

UCAS Code: FG84 • Mode: 3 Years Equal Combination

Qualifications/Requirements: Please refer to prospectus.

Mod Business Statistics/Multimedia Systems

UCAS Code: GG46 • Mode: 3 Years Equal Combination

Qualifications/Requirements: Please refer to prospectus.

Mod Cities and Society/Computing

UCAS Code: GLMH • Mode: 3 Years Equal Combination

Qualifications/Requirements: GCE, A level grades: DD-CCC. AGNVQ, Merit. ND/C, Individual consideration. SQAH, Individual consideration. SQAV, Individual consideration. IB, Individual consideration.

Mod Combined Studies/Computing

UCAS Code: GY54 • Mode: 3 Years

Qualifications/Requirements: AGNVQ, Not normally sufficient. ND/C, Individual consideration. SQAH, Not normally sufficient. SQAV, Individual consideration. IB, Not normally sufficient.

Mod Combined Studies/Computing Science

UCAS Code: YG45 • Mode: 3 Years

Qualifications/Requirements: AGNVQ, Not normally sufficient. ND/C, Individual consideration. SQAH, Not normally sufficient. SQAV, Individual consideration. IB, Not normally sufficient.

Mod Complementary Therapies – Aromatherapy/Computing

UCAS Code: GW58 • Mode: 3 Years Equal Combination

Qualifications/Requirements: Please refer to prospectus.

Mod Complementary Therapies – Aromatherapy/Computing Science

UCAS Code: GW5V • Mode: 3 Years Equal Combination

Qualifications/Requirements: Please refer to prospectus.

Mod Complementary Therapies – Aromatherapy/Information Systems

UCAS Code: GW5W • Mode: 3 Years Equal Combination

Qualifications/Requirements: Please refer to prospectus.

Mod Computing Mathematics/ Information Systems

UCAS Code: GGM9 • Mode: 3 Years Equal Combination

Qualifications/Requirements: GCE, A level grades: CD-BC. AGNVQ, Merit. ND/C, Individual consideration. SQAH, Individual consideration. SQAV, Individual consideration. IB, Individual consideration.

Mod Computing Science/Educational Studies

UCAS Code: XG95 • Mode: 3 Years Equal Combination

Qualifications/Requirements: GCE, A level grades: DD-BCC. AGNVQ, Merit with 3 additional units. ND/C, Individual consideration. SQAH, Individual consideration. SQAV, Individual consideration. IB, Individual consideration.

Mod Computing Science/Information Systems

UCAS Code: G512 • Mode: 3 Years Single Subject

Qualifications/Requirements: GCE, A level grades: DD-CCC. AGNVQ, Merit. ND/C, Individual consideration. SQAH, Individual consideration. SQAV, Individual consideration. IB, Individual consideration.

Mod Computing/Information Systems

UCAS Code: G510 • Mode: 3 Years Single Subject

Qualifications/Requirements: GCE, A level grades: CDD-BC. AGNVQ, Merit. ND/C, Individual consideration. SQAH, Individual consideration. SQAV, Individual consideration. IB, Individual consideration.

Mod Computing/Mathematics

UCAS Code: GG15 • Mode: 2/3 Years Equal Combination

Qualifications/Requirements: GCE, A level grades: DD-BC. Maths. AGNVQ, Merit with A level. ND/C, Individual consideration. SQAH, Individual consideration. SQAV, Individual consideration. IB, Individual consideration.

Mod Computing/Statistics

UCAS Code: GG45 • Mode: 2/3 Years Equal Combination

Qualifications/Requirements: GCE, A level grades: DD-BC. AGNVQ, Merit. ND/C, Individual consideration. SQAH, Individual consideration. SQAV, Individual consideration. IB, Individual consideration.

Mod Ecology/Information Systems

UCAS Code: CG9M • Mode: 3 Years Equal Combination

Qualifications/Requirements: GCE, A level grades: CD-BC. AGNVQ, Merit in Science. ND/C, Individual consideration. SQAH, Individual consideration. SQAV, Individual consideration. IB, Individual consideration.

Mod Ecology/Multimedia Systems

UCAS Code: CG9P • Mode: 3 Years Equal Combination

Qualifications/Requirements: Please refer to prospectus.

Mod Economics/Information Systems

UCAS Code: GLM1 • Mode: 3 Years Equal Combination

Qualifications/Requirements: GCE, A level grades: CDD-BB. AGNVQ, Merit or Merit with 3 additional units. ND/C, Individual consideration. SQAH, Individual consideration. SQAV, Individual consideration. IB, Individual consideration.

Mod Educational Studies/Information Systems

UCAS Code: GXM9 • Mode: 3 Years Equal Combination

Qualifications/Requirements: GCE, A level grades: CDD-BB. AGNVQ, Merit or Merit with 3 additional units. ND/C, Individual consideration. SQAH, Individual consideration. SQAV, Individual consideration. IB, Individual consideration.

Mod Educational Studies/Intelligent Systems

UCAS Code: GX89 • Mode: 3 Years Equal Combination

Qualifications/Requirements: GCE, A level grades: CD-CC. AGNVQ, Merit or Merit with 3 additional units. ND/C, Individual consideration. SQAH, Individual consideration. SQAV, Individual consideration. IB, Individual consideration.

Mod Educational Studies/Multimedia Systems

UCAS Code: GX9X • Mode: 3 Years Equal Combination

Qualifications/Requirements: Please refer to prospectus.

Mod Educational Studies/Software Engineering

UCAS Code: GX79 • Mode: 3 Years Equal Combination

Qualifications/Requirements: GCE, A level grades: CC-BC. AGNVQ, Merit or Merit with 3 additional units. ND/C, Individual consideration. SQAH, Individual consideration. SQAV, Individual consideration. IB, Individual consideration.

Mod English Studies/Information Systems

UCAS Code: GQM3 • Mode: 3 Years Equal Combination

Qualifications/Requirements: GCE, A level grades: CDD-BCC. AGNVQ, Merit with A/AS. ND/C, Individual consideration. SQAH, Individual consideration. SQAV, Individual consideration. IB, Individual consideration.

Mod English Studies/Intelligent Systems

UCAS Code: GQ83 • Mode: 3 Years Equal Combination

Qualifications/Requirements: GCE, A level grades: CD-AB. AGNVQ, Merit with A/AS. ND/C, Individual consideration. SQAH, Individual consideration. SQAV, Individual consideration. IB, Individual consideration.

Mod English Studies/Multimedia Systems

UCAS Code: GQP3 • Mode: 3 Years Equal Combination

Qualifications/Requirements: Please refer to prospectus.

Mod English Studies/Statistics

UCAS Code: GQ43 • Mode: 3 Years Equal Combination

Qualifications/Requirements: GCE, A level grades: DD-AB. AGNVQ, Merit with A/AS. ND/C, Individual consideration. SQAH, Individual consideration. SQAV, Individual consideration. IB, Individual consideration.

Mod Environmental Design and Conservation/Information Systems

UCAS Code: FGXN • Mode: 3 Years Equal Combination

Qualifications/Requirements: GCE, A level grades – DD-CCC. AGNVQ, Merit. ND/C, Individual consideration. SQAH, Individual consideration. SQAV, Individual consideration. IB, Individual consideration.

Mod Environmental Design and Conservation/Intelligent Systems

UCAS Code: FG98 • Mode: 3 Years Equal Combination

Qualifications/Requirements: GCE, A level grades – DD-CCC. AGNVQ, Merit. ND/C, Individual consideration. SQAH, Individual consideration. SQAV, Individual consideration. IB, Individual consideration.

Mod Environmental Chemistry/ Information Systems

UCAS Code: GF5C • Mode: 3 Years Equal Combination

Qualifications/Requirements: GCE, A level grades – DD-BC. Science. AGNVQ, Merit in Science. ND/C, Individual consideration. SQAH, Individual consideration. SQAV, Individual consideration. IB, Individual consideration.

Mod Environmental Chemistry/ Intelligent Systems

UCAS Code: GF81 • Mode: 3 Years Equal Combination

Qualifications/Requirements: GCE, A level grades – DD-CCC. Science. AGNVQ, Merit in Science. ND/C, Individual consideration. SQAH, Individual consideration. SQAV, Individual consideration. IB, Individual consideration.

Mod Environmental Chemistry/ Multimedia Systems

UCAS Code: FG1P • Mode: 3 Years Equal Combination

Qualifications/Requirements: Please refer to prospectus.

Mod Environmental Design and Conservation/Multimedia Systems

UCAS Code: FG9P • Mode: 3 Years Equal Combination

Qualifications/Requirements: Please refer to prospectus.

Mod Environmental Policy/Information Systems

UCAS Code: KG3M • Mode: 3 Years Equal Combination

Qualifications/Requirements: GCE, A level grades – DD-CCC. AGNVQ, Merit. ND/C, Individual consideration. SQAH, Individual consideration. SQAV, Individual consideration. IB, Individual consideration.

Mod Environmental Policy/Intelligent Systems

UCAS Code: KG38 • Mode: 3 Years Equal Combination

Qualifications/Requirements: GCE, A level grades – DD-CCC. Science. AGNVQ, Merit in Science. ND/C, Individual consideration. SQAH, Individual consideration. SQAV, Individual consideration. IB, Individual consideration.

Mod Environmental Sciences/Information Systems

UCAS Code: FGXM • Mode: 3 Years Equal Combination

Qualifications/Requirements: GCE, A level grades – CD-BC. Science. AGNVQ, Merit or Distinction in Science. ND/C, Individual consideration. SQAH, Individual consideration. SQAV, Individual consideration. IB, Individual consideration.

Mod Environmental Sciences/Intelligent Systems

UCAS Code: FGX8 • Mode: 3 Years Equal Combination

Qualifications/Requirements: GCE, A level grades – CD. Science. AGNVQ, Merit or Distinction in Science. ND/C, Individual consideration. SQAH, Individual consideration. SQAV, Individual consideration. IB, Individual consideration.

Mod European Culture and Society/ Information Systems

UCAS Code: TGGM • Mode: 3 Years Equal Combination

Qualifications/Requirements: GCE, A level grades: CD-CCC. AGNVQ, Merit with A/AS or Merit with 3 additional units. ND/C, Individual consideration. SQAH, Individual consideration. SQAV, Individual consideration. IB, Individual consideration.

Mod European Culture and Society/ Intelligent Systems

UCAS Code: TGG8 • Mode: 3 Years Equal Combination

Qualifications/Requirements: GCE, A level grades: CD-CCC. AGNVQ, Merit with A/AS or Merit with 3 additional units. ND/C, Individual consideration. SQAH, Individual consideration. SQAV, Individual consideration. IB, Individual consideration.

Mod European Culture and Society/ Multimedia Systems

UCAS Code: GTP2 • Mode: 3 Years Equal Combination

Qualifications/Requirements: Please refer to prospectus.

Mod Exercise and Health/Information Systems

UCAS Code: GBM6 • Mode: 3 Years Equal Combination

Qualifications/Requirements: GCE, A level grades – DD-BCD. Science. AGNVQ, Merit in Science. ND/C, Individual consideration. SQAH, Individual consideration. SQAV, Individual consideration. IB, Individual consideration.

Mod Fine Art/Information Systems

UCAS Code: GWM1 • Mode: 3 Years Equal Combination

Qualifications/Requirements: GCE, A level grades: CDD-BC. Art. A portfolio of work is required. AGNVQ, Merit in Art & Design with A/AS. ND/C, Individual consideration. SQAH, Individual consideration. SQAV, Individual consideration. IB, Individual consideration.

Mod Food Science and Nutrition/ Information Systems

UCAS Code: DG4M • Mode: 3 Years Equal Combination

Qualifications/Requirements: GCE, A level grades – DD-BC. Science. AGNVQ, Merit in Science. ND/C, Individual consideration. SQAH, Individual consideration. SQAV, Individual consideration. IB, Individual consideration.

Mod Food Science and Nutrition/ Intelligent Systems

UCAS Code: DG48 • Mode: 3 Years Equal Combination

Qualifications/Requirements: GCE, A level grades – DD-CD. Science. AGNVQ, Merit in Science. ND/C, Individual consideration. SQAH, Individual consideration. SQAV, Individual consideration. IB, Individual consideration.

Mod Food Science and Nutrition/ Multimedia Systems

UCAS Code: DG4P • Mode: 3 Years Equal Combination

Qualifications/Requirements: Please refer to prospectus.

Mod Geographic Information Science/ Information Systems

UCAS Code: GFMW • Mode: 3 Years Equal Combination

Qualifications/Requirements: Please refer to prospectus.

Mod Geographic Information Science/ Mathematics

UCAS Code: GF18 • Mode: 3 Years Equal Combination

Qualifications/Requirements: Please refer to prospectus.

Mod Geographic Information Science/ Multimedia Systems

UCAS Code: GFP8 • Mode: 3 Years Equal Combination

Qualifications/Requirements: Please refer to prospectus.

Mod Geographic Information Science/ Statistics

UCAS Code: FG8K • Mode: 3 Years Equal Combination

Qualifications/Requirements: Please refer to prospectus.

Mod Geography/Information Systems

UCAS Code: GLM8 • Mode: 3 Years Equal Combination

Qualifications/Requirements: GCE, A level grades – BC-BB. AGNVQ, Merit. ND/C, Individual consideration. SQAH, Individual consideration. SQAV, Individual consideration. IB, Individual consideration.

Mod Geology/Information Systems

UCAS Code: FG6N • Mode: 3 Years Equal Combination

Qualifications/Requirements: GCE, A level grades – DD-BC. Maths or Science. AGNVQ, Pass in Science or Merit. ND/C, Individual consideration. SQAH, Individual consideration. SQAV, Individual consideration. IB, Individual consideration.

Mod Geotechnics/Information Systems

UCAS Code: GHN2 • Mode: 3 Years Equal Combination

Qualifications/Requirements: GCE, A level grades: DD-BCD. Science, Maths, Design & Technology or Electronics. AGNVQ, Merit (in specific programmes). ND/C, Individual consideration. SQAH, Individual consideration. SQAV, Individual consideration. IB, Individual consideration.

Mod History of Art/Information Systems

UCAS Code: GVN4 • Mode: 3 Years Equal Combination

Qualifications/Requirements: GCE, A level grades: CDD-BCC. AGNVQ, Merit with A/AS. ND/C, Individual consideration. SQAH, Individual consideration. SQAV, Individual consideration. IB, Individual consideration.

Mod History of Art/Multimedia Systems

UCAS Code: GVP4 • Mode: 3 Years Equal Combination

Qualifications/Requirements: Please refer to prospectus.

Mod History/Information Systems

UCAS Code: GVN1 • Mode: 3 Years Equal Combination

Qualifications/Requirements: GCE, A level grades: BC-BB. AGNVQ, Merit. ND/C, Individual consideration. SQAH, Individual consideration. SQAV, Individual consideration. IB, Individual consideration.

Mod History/Multimedia Systems

UCAS Code: GVP1 • Mode: 3 Years Equal Combination

Qualifications/Requirements: Please refer to prospectus.

Mod Hospitality Management Studies/ Information Systems

UCAS Code: GNM7 • Mode: 3 Years Equal Combination

Qualifications/Requirements: GCE, A level grades: CC-BC. AGNVQ, Merit or Merit with 3 additional units. ND/C, Individual consideration. SQAH, Individual consideration. SQAV, Individual consideration. IB, Individual consideration.

Mod Hospitality Management Studies/ Multimedia Systems

UCAS Code: GNP7 • Mode: 3 Years Equal Combination

Qualifications/Requirements: Please refer to prospectus.

Mod Human Biology/Information Systems

UCAS Code: BG1M • Mode: 3 Years Equal Combination

Qualifications/Requirements: GCE, A level grades: DD-BC. Science. AGNVQ, Merit in Science. ND/C, Individual consideration. SQAH, Individual consideration. SQAV, Individual consideration. IB, Individual consideration.

Mod Mapping and Cartography/ Information Systems

UCAS Code: FG8M • Mode: 3 Years Equal Combination

Qualifications/Requirements: GCE, A level grades: DDD-BC. AGNVQ, Merit. ND/C, Individual consideration. SQAH, Individual consideration. SQAV, Individual consideration. IB, Individual consideration.

UNIVERSITY OF PLYMOUTH P60

BSc Computing and Informatics

UCAS Code: G501 • Mode: 4 Years Single Subject

Qualifications/Requirements: GCE, A/AS: 16 points. An approved subject from restricted list. AGNVQ, Distinction (in specific programmes) with A/AS. ND/C, Distinction and Merit (in specific programmes). SQAH, Grades BBBC. SQAV, Individual consideration. IB, Individual consideration.

UNIVERSITY OF PORTSMOUTH P80

BSc Computing and Statistics

UCAS Code: GG45 • Mode: 3 Years Equal Combination

Qualifications/Requirements: GCE, A/AS: 20 points. Maths at A level. AGNVQ, Distinction. ND/C, Merits and 2 Distinctions (in specific programmes). SQAH, Grades BBBB. SQAV, Individual consideration. IB, Individual consideration.

BSc Entertainment Technology

UCAS Code: GH56 • Mode: 3/4 Years Equal Combination

Qualifications/Requirements: GCE, A/AS: 20 points. AGNVQ, Distinction. ND/C, Merits and 2 Distinctions. SQAH, Grades BBBB. SQAV, Individual consideration. IB, 30 points.

HND Computer Animation

UCAS Code: 072E • Mode: 2 Years Single Subject

Qualifications/Requirements: OTHER. AGNVQ, Individual consideration. ND/C, Individual consideration. SQAH, Individual consideration. SQAV, Individual consideration. IB, Individual consideration.

HND Computer Animation

UCAS Code: 072W • Mode: 2 Years Single Subject

Qualifications/Requirements: OTHER. AGNVQ, Individual consideration. ND/C, Individual consideration. SQAH, Individual consideration. SQAV, Individual consideration. IB, Individual consideration.

QUEEN MARY AND WESTFIELD COLLEGE (UNIVERSITY OF LONDON) Q50

BSc Computer Science and Mathematics

UCAS Code: GG15 • Mode: 3 Years Equal Combination

Qualifications/Requirements: GCE, A level grades: BCC. Maths at A level. AGNVQ, Not normally sufficient. ND/C, Not normally sufficient. SQAH, Grades BBBBB including specific subject(s). IB, 28 points.

THE QUEEN'S UNIVERSITY OF BELFAST Q75

BSc Management and Information Systems

UCAS Code: NG15 • Mode: 3 Years Equal Combination

Qualifications/Requirements: Please refer to prospectus.

THE ROBERT GORDON UNIVERSITY R36

BSc Computing and Information

UCAS Code: G520 • Mode: 3/4 Years Single Subject

Qualifications/Requirements: GCE, A level grades: CE. AGNVQ, Individual consideration. ND/C, National. SQAH, Grades BBC including specific subject(s). SQAV, Individual consideration. IB, Individual consideration.

HND Computing: Information Technology

UCAS Code: 005G • Mode: 2 Years Single Subject

Qualifications/Requirements: GCE, A level grades: EE. AGNVQ, Individual consideration. ND/C, National. SQAH, Grades BC including specific subject(s). SQAV, Individual consideration. IB, Individual consideration.

ROEHAMPTON INSTITUTE LONDON R48

Mod Applied Consumer Studies and Business Computing

UCAS Code: GN79 • Mode: 3 Years Equal Combination

Qualifications/Requirements: GCE, A/AS: 12-16 points. AGNVQ, Merit (in specific programmes). ND/C, 3 Distinctions. SQAH, Grades CCC. SQAV, National including specific subject(s). IB, 26 points.

Mod Biological Anthropology and Business Computing

UCAS Code: GC7Y • Mode: 3 Years Equal Combination

Qualifications/Requirements: GCE, A/AS: 12-16 points. Biology. AGNVQ, Merit (in specific programmes) with A/AS. ND/C, 3 Distinctions (in specific programmes). SQAH, Grades CCC. SQAV, National including specific subject(s). IB, 26 points.

Mod Biology and Business Computing

UCAS Code: CG17 • Mode: 3 Years Equal Combination

Qualifications/Requirements: GCE, A/AS: 12-16 points. Biology at A level. AGNVQ, Merit (in specific programmes). ND/C, 3 Distinctions. SQAH, Grades BCC. SQAV, National including specific subject(s). IB, 26 points.

Mod Business Computing and Health & Social Care

UCAS Code: GLT4 • Mode: 3 Years Equal Combination

Qualifications/Requirements: GCE, A/AS: 12-16 points. AGNVQ, Merit. ND/C, 3 Distinctions. SQAH, Grades BCC. SQAV, National including specific subject(s). IB, 26 points.

ROYAL HOLLOWAY, UNIVERSITY OF LONDON R72

BSc Foundation Year

UCAS Code: Y100 • Mode: 4 Years Single Subject

Qualifications/Requirements: GCE, A level grades – CCC. AGNVQ, Individual consideration. ND/C, Individual consideration. SQAH, Individual consideration. IB, Individual consideration.

BSc Management and Information Systems
UCAS Code: NG15 • Mode: 3 Years Equal Combination
Qualifications/Requirements: GCE, A level grades: BBB. Maths. AGNVQ, Distinction with A/AS. ND/C, Individual consideration. SQAH, Individual consideration. IB, 30 points.

Mix Foundation Programme
UCAS Code: Y408 • Mode: 4 Years Single Subject
Qualifications/Requirements: Please refer to prospectus.

THE UNIVERSITY OF SALFORD S03

BA Art and Creative Technology
UCAS Code: E1G5 • Mode: 3 Years Major/Minor
Qualifications/Requirements: AGNVQ, Individual consideration. ND/C, Individual consideration. SQAH, Individual consideration. SQAV, Individual consideration. IB, Individual consideration.

BA Art and Creative Technology
UCAS Code: W1G5 • Mode: 3 Years Major/Minor
Qualifications/Requirements: AGNVQ, Individual consideration. ND/C, Individual consideration. SQAH, Individual consideration. SQAV, Individual consideration. IB, Individual consideration.

BSc Computer Science and Information Systems
UCAS Code: G505 • Mode: 4/5 Years Single Subject
Qualifications/Requirements: GCE, A/AS: 8 points. AGNVQ, Merit (in specific programmes). ND/C, National. SQAH, Individual consideration. SQAV, Individual consideration. IB, Individual consideration.

BSc Computer Science and Information Systems
UCAS Code: G506 • Mode: 3/4 Years Single Subject
Qualifications/Requirements: GCE, A/AS: 18 points. AGNVQ, Merit or Distinction. ND/C, Merits and Distinction. SQAH, Individual consideration. SQAV, Individual consideration. IB, Individual consideration.

BSc Economics and Information Technology
UCAS Code: LG15 • Mode: 3/4 Years Equal Combination
Qualifications/Requirements: GCE, A level grades: BCC-CCD. Economics or Italian at A level. AGNVQ, Individual consideration. ND/C, Individual consideration. SQAH, Individual consideration. SQAV, Individual consideration. IB, Individual consideration.

BSc Geography and Information Technology
UCAS Code: GF58 • Mode: 3/4 Years Equal Combination
Qualifications/Requirements: GCE, A level grades – BCC-CCD. Geography or Italian at A level. AGNVQ, Individual consideration. ND/C, Individual consideration. SQAH, Individual consideration. SQAV, Individual consideration. IB, Individual consideration.

BSc Management Science and Information Systems
UCAS Code: NG15 • Mode: 3/4 Years Equal Combination
Qualifications/Requirements: Please refer to prospectus.

BSc Management Science and Information Systems with Studies in North America
UCAS Code: NG1M • Mode: 3/4 Years Equal Combination
Qualifications/Requirements: Please refer to prospectus.

BSc Physics and Information Technology
UCAS Code: FG35 • Mode: 3/4 Years Equal Combination
Qualifications/Requirements: GCE, A level grades – BCC-CDD. Physics or Italian at A level. AGNVQ, Individual consideration. ND/C, Individual consideration. SQAH, Individual consideration. SQAV, Individual consideration. IB, Individual consideration.

BSc Physiology and Information Technology
UCAS Code: CG95 • Mode: 3/4 Years Equal Combination
Qualifications/Requirements: GCE, A level grades – BCC-CCD. Italian. AGNVQ, Individual consideration. ND/C, Individual consideration. SQAH, Individual consideration. SQAV, Individual consideration. IB, Individual consideration.

UNIVERSITY COLLEGE SCARBOROUGH S10

BA Information & Communications Technology (with QTS)
UCAS Code: XG5C • Mode: 3 Years Equal Combination
Qualifications/Requirements: GCE, A/AS: 10 points. ND/C, 3 Merits. SQAH, Grades CCC. IB, 27 points.

SHEFFIELD HALLAM UNIVERSITY S21

BA Primary Education (Information and Communications Technology)
UCAS Code: XG55 • Mode: 3 Years Equal Combination
Qualifications/Requirements: Please refer to prospectus.

BSc European Computing
UCAS Code: G7R1 • Mode: 4 Years Major/Minor

Qualifications/Requirements: GCE, A/AS: 12 points. AGNVQ, Merit. ND/C, 6 Merits. SQAH, Individual consideration. SQAV, Individual consideration. IB, Individual consideration.

BSc Mathematics and Technology
UCAS Code: GJ19 • Mode: 3/4 Years Equal Combination

Qualifications/Requirements: GCE, A/AS: 14 points. Maths, Statistics or Physics at A level. AGNVQ, Pass (in specific programmes). ND/C, National (in specific programmes). SQAH, Individual consideration. SQAV, Individual consideration. IB, Individual consideration.

ST MARTIN'S COLLEGE, LANCASTER: AMBLESIDE: CARLISLE S24

BA Information & Communications Technology/Education (2+2, 5-11 years)
UCAS Code: X5G5 • Mode: 4 Years Major/Minor

Qualifications/Requirements: GCE, A level grades: CD-CEE. Italian at A level. AGNVQ, Merit in Information Technology. ND/C, Not normally sufficient. SQAH, Grades BCCC including specific subject(s). SQAV, Individual consideration. IB, 28 points.

BA Information & Communications Technology/Education (2+2, 7-14 years)
UCAS Code: X6GM • Mode: 4 Years Major/Minor

Qualifications/Requirements: GCE, A level grades: CD-CEE. Italian at A level. AGNVQ, Merit in Information Technology. ND/C, Not normally sufficient. SQAH, Grades BCCC including specific subject(s). SQAV, Individual consideration. IB, 28 points.

BA Information & Communications Technology/Education (nlp)
UCAS Code: X2G5 • Mode: 4 Years Major/Minor

Qualifications/Requirements: GCE, A level grades: CD-CEE. Italian at A level. AGNVQ, Merit in Information Technology. ND/C, Not normally sufficient. SQAH, Grades BCCC including specific subject(s). SQAV, Individual consideration. IB, 28 points.

BA Information & Communications Technology/Education (upr)
UCAS Code: X4G5 • Mode: 4 Years Major/Minor

Qualifications/Requirements: GCE, A level grades: CD-CEE. Italian at A level. AGNVQ, Merit in Information Technology. ND/C, Not normally sufficient. SQAH, Grades BCCC including specific subject(s). SQAV, Individual consideration. IB, 28 points.

BA Technology/Education
UCAS Code: X6GN • Mode: 4 Years Major/Minor

Qualifications/Requirements: GCE, A level grades: CD-CEE. Italian at A level. AGNVQ, Merit in Information Technology. ND/C, Not normally sufficient. SQAH, Grades BBCC including specific subject(s). SQAV, Individual consideration. IB, 28 points.

UNIVERSITY OF SOUTHAMPTON S27

BSc Mathematics and Information Technology
UCAS Code: GG15 • Mode: 3 Years Equal Combination

Qualifications/Requirements: GCE, A level grades: BBC. Maths at A level. AGNVQ, Individual consideration. ND/C, Individual consideration. SQAH, Grades ABBBB. SQAV, Individual consideration. IB, 30 points.

SOUTH BANK UNIVERSITY S33

BA Business Administration (Managing Business Information Processing)
UCAS Code: NG15 • Mode: 3 Years Equal Combination

Qualifications/Requirements: GCE, A level grades – CC. AGNVQ, Distinction. ND/C, 3 Distinctions. SQAH, Individual consideration. SQAV, Individual consideration. IB, Individual consideration.

BA Business Studies (Managing Business Information Processing)
UCAS Code: N1G5 • Mode: 4 Years Major/Minor

Qualifications/Requirements: GCE, A level grades – CC. AGNVQ, Distinction. ND/C, 3 Distinctions. SQAH, Individual consideration. SQAV, Individual consideration. IB, Individual consideration.

Mod English Studies and Business Information Technology
UCAS Code: GQ73 • Mode: 3 Years Equal Combination

Qualifications/Requirements: GCE, A/AS: 14-18 points. Maths and English at A level. AGNVQ, Merit with A level. ND/C, Not normally sufficient. SQAH, Individual consideration. SQAV, Individual consideration. IB, Individual consideration.

UNIVERSITY OF ST ANDREWS S36

BSc Computer Science/Mathematics
UCAS Code: GG15 • Mode: 4 Years Equal Combination

Qualifications/Requirements: GCE, A level grades: BCC. Maths at A level. AGNVQ, Individual consideration. ND/C, Individual consideration. SQAH, Grades BBBC including specific subject(s). SQAV, Individual consideration. IB, 28 points.

BSc Computer Science/Statistics
UCAS Code: GG45 • Mode: 3/4 Years Equal Combination

Qualifications/Requirements: GCE, A level grades: BCC. Maths at A level. AGNVQ, Individual consideration. ND/C, Individual consideration. SQAH, Grades BBBC including specific subject(s). SQAV, Individual consideration. IB, 28 points.

THE COLLEGE OF ST MARK AND ST JOHN S59

BA English (Literary Studies)/Information Technology
UCAS Code: Q3G5 • Mode: 3 Years Major/Minor

Qualifications/Requirements: GCE, A/AS: 14-16 points. English Literature. AGNVQ, Merit. ND/C, Individual consideration. SQAH, Individual consideration. SQAV, Individual consideration. IB, Individual consideration.

BA English Language Studies/ Information Technology
UCAS Code: Q1G5 • Mode: 3 Years Major/Minor

Qualifications/Requirements: GCE, A/AS: 12 points. AGNVQ, Merit. ND/C, Merit overall. SQAH, Individual consideration. SQAV, Individual consideration. IB, Individual consideration.

BA Geography/Information Technology
UCAS Code: L8G5 • Mode: 3 Years Major/Minor

Qualifications/Requirements: GCE, A/AS: 8-10 points. Geography. AGNVQ, Merit. ND/C, Merit overall. SQAH, Individual consideration. SQAV, Individual consideration. IB, Individual consideration.

BA Public Relations/Information Technology
UCAS Code: P3GN • Mode: 3 Years Major/Minor

Qualifications/Requirements: GCE, A/AS: 16 points. AGNVQ, Merit. ND/C, Merit overall. SQAH, Individual consideration. SQAV, Individual consideration. IB, Individual consideration.

BA Sports Science & Coaching/ Information Technology
UCAS Code: B6G5 • Mode: 3 Years Major/Minor

Qualifications/Requirements: GCE, A/AS: 12-14 points. AGNVQ, Merit. ND/C, Merit overall. SQAH, Individual consideration. SQAV, Individual consideration. IB, Individual consideration.

BEd Information Technology/Primary Education
UCAS Code: X5G5 • Mode: 4 Years Major/Minor

Qualifications/Requirements: Please refer to prospectus.

STAFFORDSHIRE UNIVERSITY S72

BA Business and Data Analysis
UCAS Code: NG27 • Mode: 3 Years Equal Combination

Qualifications/Requirements: Please refer to prospectus.

BSc Business Communication Technology
UCAS Code: G725 • Mode: 3 Years Single Subject

Qualifications/Requirements: GCE, A/AS: 12 points. AGNVQ, Merit. ND/C, 3 Merits. SQAH, Grades CCC. SQAV, Individual consideration. IB, 24 points.

BSc Computing with Applicable Mathematics
UCAS Code: G151 • Mode: 4 Years Single Subject

Qualifications/Requirements: GCE, A/AS: 12 points. AGNVQ, Merit. ND/C, Individual consideration. SQAH, Grades CCC. IB, 27 points.

BSc Foundation Biology and Computing
UCAS Code: CG1M • Mode: 4 Years Equal Combination

Qualifications/Requirements: GCE, A/AS: 4 points. AGNVQ, Pass. ND/C, National. SQAH, Grades CCC. SQAV, Individual consideration. IB, 24 points.

BSc Foundation Business Computing
UCAS Code: G712 • Mode: 5 Years Single Subject

Qualifications/Requirements: Please refer to prospectus.

BSc Foundation Business Information Technology
UCAS Code: G561 • Mode: 5 Years Single Subject

Qualifications/Requirements: Please refer to prospectus.

BSc Foundation for Business, Computing, Engineering, and Health
UCAS Code: Y400 • Mode: 1 Years

Qualifications/Requirements: GCE, A/AS: 4 points. AGNVQ, Pass. ND/C, National. SQAH, Grades DDD. IB, 24 points.

BSc Management and Business Computing
UCAS Code: N111 • Mode: 3 Years Single Subject

Qualifications/Requirements: GCE, A/AS: 12 points. AGNVQ, Merit (in specific programmes). ND/C, Merit overall. SQAH, Grades BBC. SQAV, Individual consideration. IB, 24 points.

BSc Simulation and Virtual Reality
UCAS Code: GG5R • Mode: 3 Years Equal Combination

Qualifications/Requirements: GCE, A/AS: 12 points. AGNVQ, Merit. ND/C, 3 Merits. SQAH, Grades CCC. SQAV, Individual consideration. IB, 24 points.

BSc Sport Sciences and Information Systems

UCAS Code: BG65 • Mode: 3 Years Equal Combination

Qualifications/Requirements: GCE, A/AS: 14 points. Science at A level. AGNVQ, Distinction. ND/C, 3 Merits and 3 Distinctions. SQAH, Grades BBC. SQAV, Individual consideration. IB, 28 points.

BSc Technology Management

UCAS Code: GNM1 • Mode: 4 Years Equal Combination

Qualifications/Requirements: GCE, A/AS: 12 points. AGNVQ, Merit. ND/C, Individual consideration. SQAH, Grades CCC. SQAV, Individual consideration. IB, 27 points.

BEng Foundation Computer Graphics, Imaging and Visualisation

UCAS Code: GW5G • Mode: 5 Years Equal Combination

Qualifications/Requirements: GCE, A/AS: 12 points. AGNVQ, Merit. ND/C, Individual consideration. SQAH, Grades CCC. IB, 27 points.

BEng Foundation Computer Graphics, Imaging and Visualisation

UCAS Code: GWM2 • Mode: 5 Years Equal Combination

Qualifications/Requirements: Please refer to prospectus.

Mod Computing/Information Systems

UCAS Code: G529 • Mode: 3 Years Single Subject

Qualifications/Requirements: GCE, A/AS: 12 points. AGNVQ, Merit. ND/C, Individual consideration. SQAH, Grades CCC. SQAV, Individual consideration. IB, 27 points.

Mod Environmental Studies/Information Systems

UCAS Code: GF59 • Mode: 3 Years Equal Combination

Qualifications/Requirements: GCE, A/AS: 12 points. AGNVQ, Merit. ND/C, Individual consideration. SQAH, Grades BBC. SQAV, Individual consideration. IB, 24 points.

Mod Geography/Information Systems

UCAS Code: GL5V • Mode: 3 Years Equal Combination

Qualifications/Requirements: GCE, A level grades – CC. AGNVQ, Merit. ND/C, 3 Merits and 1 Distinction. SQAH, Grades BBC. SQAV, Individual consideration. IB, 24 points.

Mod History of Art and Design/ Information Systems

UCAS Code: GV54 • Mode: 3 Years Equal Combination

Qualifications/Requirements: GCE, A/AS: 12 points. AGNVQ, Merit. ND/C, Individual consideration. SQAH, Grades BBB. SQAV, Individual consideration. IB, 27 points.

THE UNIVERSITY OF STIRLING S75

BSc Computing Science/Mathematics/ Education

UCAS Code: GX97 • Mode: 4/5 Years Equal Combination

Qualifications/Requirements: GCE, A level grades: CCD. Maths at A level. AGNVQ, Merit (in specific programmes) with 2 additional units or with A level. ND/C, Merit overall. SQAH, Grades BBCC including specific subject(s). SQAV, Higher National. IB, 28 points.

BSc Education/Computing Science

UCAS Code: GX57 • Mode: 4/5 Years Equal Combination

Qualifications/Requirements: GCE, A level grades: CCD. AGNVQ, Merit (in specific programmes) with 6 additional units or with A/AS. ND/C, Merit overall. SQAH, Grades BBCC including specific subject(s). SQAV, Higher National. IB, 28 points.

BSc Environmental Science/Computing Science

UCAS Code: FG95 • Mode: 4 Years Equal Combination

Qualifications/Requirements: GCE, A level grades – CCC. A level: Science, Electronics, Geography, Geology or Maths. Specific GCSEs required. AGNVQ, Merit in Science with 3 additional units or with A/AS. ND/C, Merit overall and 1 Distinction. SQAH, Grades BBBC including specific subject(s). SQAV, Higher National.

STRANMILLIS UNIVERSITY COLLEGE: A COLLEGE OF THE QUEEN'S UNIVERSITY OF BELFAST S79

BEd Information Technology with Education

UCAS Code: XG55 • Mode: 3 Years Equal Combination

Qualifications/Requirements: Please refer to prospectus.

SUFFOLK COLLEGE: AN ACCREDITED COLLEGE OF THE UNIVERSITY OF EAST ANGLIA S81

BA Early Childhood Studies and Information Technology

UCAS Code: XG95 • Mode: 3 Years Equal Combination

Qualifications/Requirements: GCE, A level grades: DE. AGNVQ, Merit. ND/C, Merit overall. SQAH, Individual consideration. SQAV, Individual consideration. IB, Individual consideration.

BA Fine & Applied Arts and Information Technology

UCAS Code: EG15 • Mode: 3 Years Equal Combination

Qualifications/Requirements: GCE, A level grades: CE. OTHER. AGNVQ, Pass (in specific programmes). ND/C, National (in specific programmes). SQAH, Individual consideration. SQAV, Individual consideration. IB, Individual consideration.

BA Fine & Applied Arts and Information Technology

UCAS Code: WG15 • Mode: 3 Years Equal Combination

Qualifications/Requirements: GCE, A level grades: CE. OTHER. AGNVQ, Pass (in specific programmes). ND/C, National (in specific programmes). SQAH, Individual consideration. SQAV, Individual consideration. IB, Individual consideration.

BA Graphic Design and Information Technology

UCAS Code: EG2N • Mode: 3 Years Equal Combination

Qualifications/Requirements: GCE, A level grades: CE. OTHER. AGNVQ, Pass (in specific programmes). ND/C, National (in specific programmes). SQAH, Individual consideration. SQAV, Individual consideration. IB, Individual consideration.

BA Graphic Design and Information Technology

UCAS Code: WG2N • Mode: 3 Years Equal Combination

Qualifications/Requirements: GCE, A level grades: CE. OTHER. AGNVQ, Pass (in specific programmes). ND/C, National (in specific programmes). SQAH, Individual consideration. SQAV, Individual consideration. IB, Individual consideration.

BA History and Information Technology

UCAS Code: VG15 • Mode: 3 Years Equal Combination

Qualifications/Requirements: GCE, A level grades: CE. AGNVQ, Pass (in specific programmes). ND/C, National (in specific programmes). SQAH, Individual consideration. SQAV, Individual consideration. IB, Individual consideration.

BA Illustration and Information Technology

UCAS Code: EG2M • Mode: 3 Years Equal Combination

Qualifications/Requirements: GCE, A level grades: CE. OTHER. AGNVQ, Pass (in specific programmes). ND/C, National (in specific programmes). SQAH, Individual consideration. SQAV, Individual consideration. IB, Individual consideration.

BA Illustration and Information Technology

UCAS Code: WG2M • Mode: 3 Years Equal Combination

Qualifications/Requirements: GCE, A level grades: CE. OTHER. AGNVQ, Pass (in specific programmes). ND/C, National (in specific programmes). SQAH, Individual consideration. SQAV, Individual consideration. IB, Individual consideration.

BA Literary Studies and Information Technology

UCAS Code: QG25 • Mode: 3 Years Equal Combination

Qualifications/Requirements: GCE, A level grades: CE. English. AGNVQ, Pass (in specific programmes). ND/C, National (in specific programmes). SQAH, Individual consideration. SQAV, Individual consideration. IB, Individual consideration.

BA Model Design and Information Technology

UCAS Code: GW5F • Mode: 3 Years Equal Combination

Qualifications/Requirements: GCE, A level grades: CE. OTHER. AGNVQ, Pass (in specific programmes). ND/C, National (in specific programmes). SQAH, Individual consideration. SQAV, Individual consideration. IB, Individual consideration.

BA Spatial Design and Information Technology

UCAS Code: GW5G • Mode: 3 Years Equal Combination

Qualifications/Requirements: GCE, A level grades: CE. OTHER. AGNVQ, Pass (in specific programmes). ND/C, National (in specific programmes). SQAH, Individual consideration. SQAV, Individual consideration. IB, Individual consideration.

BSc Animal Science & Conservation and Information Technology

UCAS Code: DGF5 • Mode: 3 Years Equal Combination

Qualifications/Requirements: GCE, A level grades: EE. Science. AGNVQ, Merit in Science or in Environmental Studies. ND/C, 3 Merits (in specific programmes). SQAH, Individual consideration. SQAV, Individual consideration. IB, Individual consideration.

BSc Environmental Studies and Information Technology

UCAS Code: FG95 • Mode: 3 Years Equal Combination

Qualifications/Requirements: GCE, A level grades – EE. Science or Geography. AGNVQ, Pass in Science or in Environmental Studies. ND/C, 3 Merits (in specific programmes). SQAH, Individual consideration. SQAV, Individual consideration. IB, Individual consideration.

BSc Food Production and Information Technology

UCAS Code: DG45 • Mode: 3 Years Equal Combination

Qualifications/Requirements: GCE, A level grades – EE. Science. AGNVQ, Merit in Environmental Studies or in Science. ND/C, 3 Merits (in specific programmes). SQAH, Individual consideration. SQAV, Individual consideration. IB, Individual consideration.

BSc Food Production with Information Technology

UCAS Code: D4G5 • Mode: 3 Years Major/Minor

Qualifications/Requirements: GCE, A level grades – EE. Science. AGNVQ, Merit in Environmental Studies or in Science. ND/C, 3 Merits (in specific programmes). SQAH, Individual consideration. SQAV, Individual consideration. IB, Individual consideration.

BSc Information Technology with Early Childhood Studies

UCAS Code: G5XX • Mode: 3 Years Major/Minor

Qualifications/Requirements: GCE, A level grades: EE. AGNVQ, Pass (in specific programmes). ND/C, National (in specific programmes). SQAH, Individual consideration. SQAV, Individual consideration. IB, Individual consideration.

UNIVERSITY OF SUNDERLAND S84

BA Information Technology Education (11-18 years)

UCAS Code: XG75 • Mode: 2 Years Equal Combination

Qualifications/Requirements: AGNVQ, Not normally sufficient. ND/C, Higher National (in specific programmes). SQAH, Not normally sufficient. SQAV, Higher National including specific subject(s). IB, Not normally sufficient.

BA Information Technology Education (11-18 years)

UCAS Code: XG7M • Mode: 3 Years Equal Combination

Qualifications/Requirements: GCE, A/AS: 12 points. Computing or Italian. AGNVQ, Merit (in specific programmes) with A level. ND/C, Higher National (in specific programmes). SQAH, Grades CCCCC including specific subject(s). SQAV, National including specific subject(s). IB, 24 points.

BA Information Technology Education (Key Stage 2/3) (11-18 years)

UCAS Code: XG6M • Mode: 3 Years Equal Combination

Qualifications/Requirements: GCE, A/AS: 12 points. Computing or Italian. AGNVQ, Merit (in specific programmes) with A level. ND/C, Higher National (in specific programmes). SQAH, Grades CCCCC including specific subject(s). SQAV, National including specific subject(s). IB, 24 points.

BSc Computer Studies with Studies in Education & Training

UCAS Code: G5X9 • Mode: 3 Years Major/Minor

Qualifications/Requirements: Please refer to prospectus.

THE UNIVERSITY OF SURREY S87

BSc Computing & Information Technology (h)

UCAS Code: G560 • Mode: 3 Years Single Subject

Qualifications/Requirements: GCE, A level grades: BBB-BCC. AGNVQ, Distinction (in specific programmes) with A/AS. ND/C, Merit overall and 4 Distinctions. SQAH, Individual consideration. SQAV, Individual consideration. IB, Individual consideration.

BSc Computing & Information Technology (h)

UCAS Code: G561 • Mode: 4 Years Single Subject

Qualifications/Requirements: GCE, A level grades: BBB-BCC. AGNVQ, Distinction (in specific programmes) with A/AS. ND/C, Merit overall and 4 Distinctions. SQAH, Individual consideration. SQAV, Individual consideration. IB, Individual consideration.

BSc Mathematics and Computing Science with a European language (k)

UCAS Code: GGC5 • Mode: 3 Years Equal Combination

Qualifications/Requirements: GCE, A/AS: 18-20 points. Maths and a modern foreign language at A level. AGNVQ, Individual consideration. ND/C, Individual consideration. SQAH, CSYS required. SQAV, Individual consideration. IB, Individual consideration.

BSc Mathematics and Computing Science (i)

UCAS Code: GG1M • Mode: 3 Years Equal Combination

Qualifications/Requirements: GCE, A/AS: 18-20 points. Maths at A level. AGNVQ, Individual consideration. ND/C, Individual consideration. SQAH, CSYS required. SQAV, Individual consideration. IB, Individual consideration.

BSc Mathematics and Computing Science (i)

UCAS Code: GG15 • Mode: 4 Years Equal Combination

Qualifications/Requirements: GCE, A/AS: 18-20 points. Maths at A level. AGNVQ, Individual consideration. ND/C, Individual consideration. SQAH, CSYS required. SQAV, Individual consideration. IB, Individual consideration.

BSc Mathematics and Computing Science with a European Language (k)

UCAS Code: GG1N • Mode: 4 Years Equal Combination

Qualifications/Requirements: GCE, A/AS: 18-20 points. Maths and a modern foreign language at A level. AGNVQ, Individual consideration. ND/C, Individual consideration. SQAH, CSYS required. SQAV, Individual consideration. IB, Individual consideration.

MEng Computing and Information Technology

UCAS Code: G562 • Mode: 4 Years Single Subject

Qualifications/Requirements: GCE, A/AS: 24 points. AGNVQ, Distinction (in specific programmes) with A/AS. ND/C, Merit overall and 4 Distinctions. SQAH, Individual consideration. SQAV, Individual consideration. IB, Individual consideration.

MEng Computing and Information Technology

UCAS Code: G563 • Mode: 5 Years Single Subject

Qualifications/Requirements: GCE, A/AS: 24 points. AGNVQ, Distinction (in specific programmes) with A/AS. ND/C, Merit overall and 4 Distinctions. SQAH, Individual consideration. SQAV, Individual consideration. IB, Individual consideration.

UNIVERSITY OF SUSSEX S90

BA Economics and Applied Quantitative Methods

UCAS Code: L1G5 • Mode: 3 Years Major/Minor

Qualifications/Requirements: GCE, A level grades: BBB. AGNVQ, Merit with 6 additional units. ND/C, Merit overall (in specific programmes). SQAH, Individual consideration. SQAV, Individual consideration. IB, Individual consideration.

SWANSEA INSTITUTE OF HIGHER EDUCATION S96

BA 3D Computer Animation

UCAS Code: E541 • Mode: 3 Years Single Subject

Qualifications/Requirements: Please refer to prospectus.

BSc Computing and Information Systems

UCAS Code: G520 • Mode: 3 Years Single Subject

Qualifications/Requirements: GCE, A/AS: 4 points. AGNVQ, Merit. ND/C, 2 Merits. SQAH, Grades CCCC. SQAV, Individual consideration. IB, 24 points.

Mod 3D Computer Animation

UCAS Code: E540 • Mode: 3 Years Single Subject

Qualifications/Requirements: Please refer to prospectus.

UNIVERSITY OF TEESSIDE T20

BSc Business Informatics (HND top-up)

UCAS Code: NG14 • Mode: 1 Years Equal Combination

Qualifications/Requirements: ND/C, Higher National. SQAH, Not normally sufficient.

BSc Chemical Systems Engineering with Information Technology

UCAS Code: H8G5 • Mode: 3 Years Major/Minor

Qualifications/Requirements: GCE, A/AS: 12-14 points. Two Sciences. AGNVQ, Merit. ND/C, 3 Merits (in specific programmes). SQAH, Grades CCC including specific subject(s). SQAV, National including specific subject(s). IB, Individual consideration.

BSc Computing and Mathematics

UCAS Code: GG15 • Mode: 3/4 Years Equal Combination

Qualifications/Requirements: GCE, A/AS: 12-16 points. Maths at A level. AGNVQ, Merit. ND/C, Individual consideration. SQAH, Grades CCCC. SQAV, Individual consideration. IB, Individual consideration.

BSc Digital Visual Effects

UCAS Code: GWM2 • Mode: 3/4 Years Equal Combination

Qualifications/Requirements: GCE, A/AS: 16-20 points. AGNVQ, Merit. ND/C, Individual consideration. SQAH, Grades CCCC. SQAV, Individual consideration. IB, Individual consideration.

BSc Virtual Reality

UCAS Code: GG57 • Mode: 3/4 Years Equal Combination

Qualifications/Requirements: GCE, A/AS: 16-20 points. AGNVQ, Merit. ND/C, Individual consideration. SQAH, Grades CCCC. SQAV, Individual consideration. IB, Individual consideration.

BSc Visualisation

UCAS Code: G5W2 • Mode: 3/4 Years Major/Minor

Qualifications/Requirements: GCE, A/AS: 16-20 points. AGNVQ, Merit. ND/C, Individual consideration. SQAH, Grades CCCC. SQAV, Individual consideration. IB, Individual consideration.

HND Visualisation

UCAS Code: 2W5G • Mode: 2 Years Major/Minor

Qualifications/Requirements: GCE, A/AS: 6 points. AGNVQ, Pass. ND/C, Individual consideration. SQAH, Grades CC. SQAV, Individual consideration. IB, Individual consideration.

THAMES VALLEY UNIVERSITY T40

BA Digital Arts with Information Systems

UCAS Code: W9G5 • Mode: 3 Years Major/Minor

Qualifications/Requirements: GCE, A/AS: 8-12 points. AGNVQ, Merit. ND/C, Merit overall. SQAH, Grades CCC.

BA Digital Arts with Multimedia Computing

UCAS Code: W9GM • Mode: 2/3 Years Major/Minor

Qualifications/Requirements: GCE, A/AS: 8 points. AGNVQ, Merit. ND/C, Merit overall. SQAH, Grades CCCC. IB, 26 points.

BSc Information & Knowledge Management with Multimedia Computing

UCAS Code: P2GM • Mode: 2/3 Years Major/Minor

Qualifications/Requirements: GCE, A/AS: 8 points. AGNVQ, Merit. ND/C, Merit overall. SQAH, Grades CCCC. IB, 26 points.

UNIVERSITY OF ULSTER U20

BSc Mathematics, Statistics and Computing

UCAS Code: G921 • Mode: 4 Years Single Subject

Qualifications/Requirements: GCE, A level grades: CDD. Maths. AGNVQ, Distinction with A level. ND/C, Merit overall and 1 Distinction. SQAH, Grades CCCD. SQAV, Individual consideration. IB, 26 points.

BSc Mathematics, Statistics and Computing (Hons)

UCAS Code: G920 • Mode: 4 Years Single Subject

Qualifications/Requirements: GCE, A level grades: BCC. Maths. AGNVQ, Distinction with A level. ND/C, Merit overall and 4 Distinctions. SQAH, Grades BBBC. SQAV, Individual consideration. IB, 30 points.

UNIVERSITY COLLEGE LONDON (UNIVERSITY OF LONDON) U80

BSc Statistics, Computing, Operational Research and Economics

UCAS Code: Y624 • Mode: 3 Years Single Subject

Qualifications/Requirements: GCE, A level grades: ABC. Maths at A level. AGNVQ, Individual consideration. ND/C, Merit overall (in specific programmes). SQAH, Grades BBCCC including specific subject(s). SQAV, National including specific subject(s). IB, 30 points.

UNIVERSITY OF WESTMINSTER W50

BA Management of Business Information

UCAS Code: NG17 • Mode: 3 Years Equal Combination

Qualifications/Requirements: GCE, A level grades – BC. AGNVQ, Distinction. ND/C, Merit overall and 2 Distinctions. SQAH, Grades BBBB. IB, 28 points.

UNIVERSITY OF WOLVERHAMPTON W75

BSc Virtual Reality Design

UCAS Code: G7K1 • Mode: 3/4 Years Major/Minor

Qualifications/Requirements: GCE, A/AS: 16 points. AGNVQ, Merit. ND/C, 3 Merits. SQAH, Grades BBBB. SQAV, Individual consideration. IB, 28 points.

BSc Virtual Reality Manufacturing

UCAS Code: G7H7 • Mode: 3/4 Years Major/Minor

Qualifications/Requirements: GCE, A level grades: DD. AGNVQ, Merit. ND/C, 3 Merits. SQAH, Grades CCCC. SQAV, Individual consideration. IB, 24 points.

UNIVERSITY COLLEGE WORCESTER W80

BA/BSc Heritage Studies/Information Technology

UCAS Code: GN59 • Mode: 3 Years Equal Combination

Qualifications/Requirements: GCE, A level grades: DD. AGNVQ, Merit. ND/C, Individual consideration. SQAH, Individual consideration. SQAV, Individual consideration. IB, Individual consideration.

Mod Drama/Information Technology

UCAS Code: WG45 • Mode: 3 Years Equal Combination

Qualifications/Requirements: GCE, A level grades: CD. AGNVQ, Merit. ND/C, Individual consideration. SQAH, Individual consideration. SQAV, Individual consideration. IB, Individual consideration.

Mod Education Studies/Information Technology

UCAS Code: XG95 • Mode: 3 Years Equal Combination

Qualifications/Requirements: GCE, A level grades: DD. AGNVQ, Merit. ND/C, Individual consideration. SQAH, Individual consideration. SQAV, Individual consideration. IB, Individual consideration.

Mod English and Literary Studies/Information Technology

UCAS Code: QG35 • Mode: 3 Years Equal Combination

Qualifications/Requirements: GCE, A level grades: CC. AGNVQ, Merit. ND/C, Individual consideration. SQAH, Individual consideration. SQAV, Individual consideration. IB, Individual consideration.

Mod Environmental Science/Information Technology

UCAS Code: FG95 • Mode: 3 Years Equal Combination

Qualifications/Requirements: GCE, A level grades – DD. AGNVQ, Merit. ND/C, Individual consideration. SQAH, Individual consideration. SQAV, Individual consideration. IB, Individual consideration.

Mod European Studies/Information Technology

UCAS Code: GT52 • Mode: 3 Years Equal Combination

Qualifications/Requirements: GCE, A level grades: DD. AGNVQ, Merit. ND/C, Individual consideration. SQAH, Individual consideration. SQAV, Individual consideration. IB, Individual consideration.

Mod Geography/Information Technology

UCAS Code: LG85 • Mode: 3 Years Equal Combination

Qualifications/Requirements: GCE, A level grades – DD. AGNVQ, Merit. ND/C, Individual consideration. SQAH, Individual consideration. SQAV, Individual consideration. IB, Individual consideration.

Mod Health Studies/Information Technology

UCAS Code: BG95 • Mode: 3 Years Equal Combination

Qualifications/Requirements: GCE, A level grades – DD. AGNVQ, Merit. ND/C, Individual consideration. SQAH, Individual consideration. SQAV, Individual consideration. IB, Individual consideration.

Mod History/Information Technology

UCAS Code: VG15 • Mode: 3 Years Equal Combination

Qualifications/Requirements: GCE, A level grades: DD. AGNVQ, Merit. ND/C, Individual consideration. SQAH, Individual consideration. SQAV, Individual consideration. IB, Individual consideration.

THE UNIVERSITY OF YORK Y50

MMath Mathematics/Computer Science (Equal)

UCAS Code: GG15 • Mode: 4 Years Equal Combination

Qualifications/Requirements: GCE, A/AS: 26-28 points. Maths at A level. AGNVQ, Distinction (in specific programmes) with A level. ND/C, Higher National (in specific programmes). SQAH, CSYS required. SQAV, Higher National including specific subject(s). IB, 34 points.

UNIVERSITY AND COLLEGE ADDRESSES

A20 The University of Aberdeen
Undergraduate Admissions
University Office
Regent Walk
Aberdeen AB24 8FX
Tel: 01224 272030
Fax: 01224 272031
admoff@admin.abdn.ac.uk
http://www.abdn.ac.uk/

A30 University of Abertay Dundee
Bell Street
Dundee DD1 1HG
Tel: 01382 308943
Fax: 01382 308081
iro@abertay-dundee.ac.uk
http://www.abertay-dundee.ac.uk

A40 The University of Wales, Aberystwyth
Admissions & Recruitment
Old College
King Street
Aberystwyth SY23 2AX
Tel: 01970 622018
Fax: 01970 627410
undergraduate-admissions@aber.ac.uk
http://www.aber.ac.uk/

A60 Anglia Polytechnic University
East Road
Cambridge CB1 1PT
Tel: 01223 363271 x2024
Fax: 01223 576156
degaalap@bridge.anglia.ac.uk
http://www.anglia.ac.uk/

A80 Aston University
Registry
Aston Triangle
Birmingham B4 7ET
Tel: 0121 359 3611 x4803
Fax: 0121 333 6350
prospectus@aston.ac.uk
http://www.aston.ac.uk/

B06 University of Wales, Bangor
Bangor LL57 2DG
Tel: 01248 382014
Fax: 01248 370451
Admissions@bangor.ac.uk
http://www.bangor.ac.uk/

B11 Barking College
Student Services
Dagenham Road
Romford
Essex RM7 OXU
Tel: 01708 766841
Fax: 01708 731067
hammond@barking-coll.ac.uk
http://www.barking-coll.ac.uk

B13 Barnsley College
Old Mill Lane Site
Church Street
Barnsley S70 2AX
Tel: 01226 730191 x229
Fax: 01226 216166
admissions@barnsley.ac.uk
http://www.barnsley.ac.uk/he

B16 University of Bath
Registry
Claverton Down
Bath BA2 7AY
Tel: 01225 826800
Fax: 01225 826366
admissions@bath.ac.uk
http://www.bath.ac.uk/

B20 Bath Spa University College
Newton Park
Bath BA2 9BN
Tel: 01225 875875
Fax: 01225 875444
http://www.bathspa.ac.uk

B26 Bell College of Technology
Almada Street
Hamilton
Lanarkshire
Scotland ML3 OJB
Tel: 01698 283100 x204
Fax: 01698 457524
registry@bell.ac.uk

B32 The University of Birmingham
Academic and Student Division
Edgbaston
Birmingham B15 2TT
Tel: 0121 414 4048
Fax: 0121 414 7926
prospectus@bham.ac.uk
http://www.birmingham.ac.uk

B40 Blackburn College
Art and Design Division
Faculty of Creative Arts
Feilden Street
Blackburn BB2 1LH
Tel: 01254 292938
Fax: 01254 678903

B41 Blackpool and The Fylde College
Ashfield Road
Bispham
Blackpool FY2 0HB
Tel: 01253 352352
Fax: 01253 356127
visitors@blackpool.ac.uk
http://www.blackpool.ac.uk

B44 Bolton Institute of Higher Education
Deane Road
Bolton BL3 5AB
Tel: 01204 528851
Fax: 01204 399074
enquiries@bolton.ac.uk
http://www.bolton.ac.uk

B50 Bournemouth University
Studland House
12 Christchurch Road
Bournemouth
Dorset BH1 3NA
Tel: 01202 503858
Fax: 01202 503869
postmaster@bournemouth.ac.uk
http://www.bournemouth.ac.uk/

B53 The Arts Institute at Bournemouth
Fern Barrow
Wallisdown
Poole BH12 5HH
Tel: 01202 363217
Fax: 01202 537729
registry@arts-inst-bournemouth.ac.uk
http://arts-inst-bournemouth.ac.uk

B56 The University of Bradford
Richmond Road
Bradford BD13 1HS
Tel: 01274 233033
Fax: 01274 385810
schools-liaison@bradford.ac.uk
http://www.brad.ac.uk/bradinfo/bradinfo.html

B60 Bradford College
Great Horton Road
Bradford BD7 1AY
Tel: 01274 753026
Fax: 01274 741060
admissions@bilk.ac.uk
http://www.bilk.ac.uk/

B72 University of Brighton
Academic Registry
Mithras House
Lewes Road
Brighton BN2 4AT
Tel: 01273 642813
Fax: 01273 642825
admissions@bton.ac.uk
http://www.brighton.ac.uk/

B78 University of Bristol
Senate House
Tyndall Avenue
Bristol BS8 1TH
Tel: 0117 9287678
Fax: 0117 9251424
admissions@bristol.ac.uk
http://www.bris.ac.uk/

B80 University of The West of England, Bristol
Admissions Office
Frenchay Campus
Coldharbour Lane
Bristol BS16 1QY
Tel: 0117 9750428
Fax: 0117 9763804
admissions@uwe.ac.uk
http://www.uwe.ac.uk

B84 Brunel University
Uxbridge UB8 3PH
Tel: 01895 203214
Fax: 01895 230167
courses@brunel.ac.uk
http://www.brunel.ac.uk/home.html

B90 The University of Buckingham
Hunter Street
Buckingham MK18 1EG
Tel: 01280 814080
Fax: 01280 822245
admissions@buck.ac.uk
http://www.buck.ac.uk

B94 Buckinghamshire Chilterns University College
Queen Alexandra Road
High Wycombe HP11 2JZ
Tel: 01494 603032
Fax: 01494 526032
http://www.buckscol.ac.uk/bchome.html

C05 Cambridge University
CIAO
Kellet Lodge
Tennis Court Road
Cambridge CB2 1QJ
Tel: 01223 333304
Fax: 01223 366383
ucam-undergraduate-admissions@lists.cam.ac.uk
http://www.cam.ac.uk/

C10 Canterbury Christ Church University College
Canterbury CT1 1QU
Tel: 01227 782420
admissions@cant.ac.uk
http://www.cant.ac.uk/

C15 Cardiff University
PO Box 494
Cardiff CF1 3YL
Tel: 029 20874404
Fax: 029 20874130
prospectus@cf.ac.uk/ADMISSIONS@cf.ac.uk
http://www.cf.ac.uk/

C20 University of Wales Institute, Cardiff
PO Box 377
Llandaff Centre
Western Avenue
Cardiff CF5 2SG
Tel: 01222 506012
Fax: 01222 506956
admissions@uwic.ac.uk
http://www.uwic.ac.uk/

C22 Carmarthenshire College
Job's Well Road
Carmarthen
Dyfed SA31 3HY
Tel: 01554 748000
Fax: 01554 756088
eirian.davies@ccta.ac.uk
http://www.ccta.ac.uk

C25 University of Central England in Birmingham
Perry Barr
Birmingham B42 2SU
Tel: 0121 331 6650
Fax: 0121 331 6358
susan.lewis@uce.ac.uk
http://www.uce.ac.uk/

C30 University of Central Lancashire
Admissions Office
Preston PR1 2HE
Tel: 01772 201201
c.enquiries@uclan.ac.uk
http://www.uclan.ac.uk/

C50 Cheltenham & Gloucester College of Higher Education
The Park
PO Box 220
Cheltenham GL50 2QF
Tel: 01242 532826
Fax: 01242 256759
admissions@chelt.ac.uk
http://www.chelt.ac.uk/

C55 Chester: *A College of The University of Liverpool*
Cheyney Road
Chester CH1 4BJ
Tel: 01244 375444
Fax: 01244 373379
s.cranny@chester.ac.uk
http://www.chester.ac.uk

C58 University College Chichester
Bishop Otter Campus
College Lane
Chichester
West Sussex PO19 4PE
Tel: 01243 816000 x6020
Fax: 01243 828351
admissions@chihe.ac.uk
http://www.chihe.ac.uk/

C60 City University
The Admissions Office
Northampton Square
London EC1V 0HB
Tel: 0171 477 8028
Fax: 0171 477 8559
ugadmissions@city.ac.uk
http://www.city.ac.uk/

C62 City College, Birmingham
47A George St
Birmingham B3 1QA

C63 City of Bristol College
Hartcliffe Centre
Bishport Avenue
Hartcliffe
Bristol BS13 ORJ
Tel: 0117 904 5610
Fax: 0117 904 5612

C66 City College Manchester
Admissions Unit
PO Box 40
Manchester M23 0GN
Tel: 0161 957 1794
Fax: 0161 945 3854
admissions@manchester-city-coll.ac.uk
http://www.manchester-city-coll.ac.uk/

C75 Colchester Institute
Sheepen Road
Colchester CO3 3LD
Tel: 01206 718693
Fax: 01206 763041

C78 Cornwall College with Duchy College
Pool
Redruth
Cornwall TR15 3RD
Tel: 01209 712911
Fax: 01209 718802
enquiries@cornwall.ac.uk

C85 Coventry University
Priory Street
Coventry CV1 5FB
Tel: 01203 838089
Fax: 01203 838793
http://www.coventry.ac.uk/

C90 Cranfield University
Royal Military College of Science
Shrivenham
Swindon SN6 8LA
Tel: 01793 785400
Fax: 01793 783966
laxon@rmcs.cranfield.ac.uk
http://www.cranfield.ac.uk/

C95 Cumbria College of Art and Design
Brampton Road
Carlisle CA3 9AY
Tel: 01228 400300
Fax: 01228 514491
aileenmc@cumbriacad.ac.uk

D13 Dartington College of Arts
(Registry & Examinations)
Totnes TQ9 6EJ
Tel: 01803 861621
Fax: 01803 863569
registry@dartington.ac.uk
http://www.dartington.ac.uk/

D26 De Montfort University
The Gateway
Leicester LE1 9BH
Tel: 0116 257 7300
Fax: 01162577515
http://www.dmu.ac.uk/

D39 University of Derby
Kedleston Road
Derby DE22 1GB
Tel: 01332 622222 x1056
Fax: 01332 294861
Admissions@derby.ac.uk
http://www.derby.ac.uk/

D52 Doncaster College
Waterdale
Doncaster DN1 3EX
Tel: 01302 553518
Fax: 01302 553559
william.russell@don.ac.uk
http://www.don.ac.uk/

D58 Dudley College of Technology
The Broadway
Dudley
West Midlands DY1 4AS
Tel: 01384 455433
Fax: 01384 454256
mark.ellerby@dudleycol.ac.uk
http://www.dudleycol.ac.uk/

D65 University of Dundee
Dundee DD1 4HN
Tel: 01382 344025
Fax: 01382 344107
srs@dundee.ac.uk
http://www.dundee.ac.uk/

D86 The University of Durham
Academic Office
Old Shire Hall
Old Elvet
Durham DH1 3HP
Tel: 0191 374 4685
Fax: 0191 374 3740
admissions@durham.ac.uk
http://www.dur.ac.uk/

E14 University of East Anglia
Admissions Office
The Registry
Norwich NR4 7TJ
Tel: 01603 592823
Fax: 01603 458596
admissions@uea.ac.uk
http://www.uea.ac.uk/

E28 University of East London
Longbridge Road
Dagenham RM8 2AS
Tel: 0181 849 3535
admiss@uel.ac.uk
http://www.uel.ac.uk/

E42 Edge Hill College of Higher Education
St Helens Road
Ormskirk L39 4QP
Tel: 01695 584312
Fax: 01695 579997
ibisona@staff.ehche.ac.uk
http://www.ehche.ac.uk/

E56 The University of Edinburgh
Schools Liaison Service
57 George Square
Edinburgh EH8 9JU
Tel: 0131 650 2129
Fax: 0131 650 6536
slo@ed.ac.uk
http://www.ed.ac.uk/

E70 The University of Essex
Wivenhoe Park
Colchester CO4 3SQ
Tel: 01206 872002
Fax: 01206 872808
admit@essex.ac.uk
http://www.essex.ac.uk/

E84 University of Exeter
Registrar & Secretary's Department
Northcote House
The Queen's Drive
Exeter EX4 4QJ
Tel: 01392 263027
Admissions@exeter.ac.uk/Prospectus@exeter.ac.uk
http://ex.ac.uk/

F66 Farnborough College of Technology
Boundary Road
Farnborough GU14 6SB
Tel: 01252 407002
admissions@farn-ct.ac.uk
http://www.farn-ct.ac.uk/

G14 University of Glamorgan
Pontypridd CF37 1DL
Tel: 01443 482684
Fax: 01443 482925
registry@glam.ac.uk
http://www.glam.ac.uk/home.html

G28 University of Glasgow
Glasgow G12 8QQ
Tel: 0141 330 4275
admissions@mis.gla.ac.uk
http://www.gla.ac.uk/admissions

G42 Glasgow Caledonian University
Cowcaddens Road
Glasgow G4 0BA
Tel: 0141 331 3270
Fax: 0141 331 3449
D.Black@gcal.ac.uk
http://www.gcal.ac.uk

G45 Gloucestershire College of Arts and Technology
The Registry
Brunswick Campus
Brunswick Road
Gloucester GL1 1HU
Tel: 01452 426557
Fax: 01452 426531

G56 Goldsmiths College (University of London)
Registry (Undergraduate/PGCE)
Lewisham Way
New Cross
London SE14 6NW
Tel: 0171 919 7576
Fax: 0171 919 7509
admissions@gold.ac.uk
http://www.gold.ac.uk/

G70 University of Greenwich
Avery Hill Campus
Mansion Site
Bexley Road
Eltham
London SE9 2PQ
Tel: 0181 331 8598
courseinfo@greenwich.ac.uk
http://www.gre.ac.uk/

G74 Greenwich School of Management
Meridian House
Royal Hill
Greenwich
London SE10 8RD

H06 Halton College
Kingsway
Widnes WA8 7QQ
Tel: 0151 495 3315
Fax: 0151 420 2408
halton.college@cityscape.co.uk
http://www.cityscape.co.uk/users/aj75/index.html

H16 Herefordshire College of Technology
Folly Lane
Hereford HR1 1LS
Tel: 01432 352235 x235
Fax: 01432 353449
http://www.herefordshire.com/hct/

H24 Heriot-Watt University, Edinburgh
Riccarton
Edinburgh EH14 4AS
Tel: 0131 451 3377
Fax: 0131 451 3630
admissions@hw.ac.uk
http://www.hw.ac.uk/

H36 University of Hertfordshire
(Registration & Admissions)
Mercer Building
College Lane
Hatfield AL10 9AB
Tel: 01707 286056
Fax: 01707 284870
http://www.herts.ac.uk/

H49 UHIp
Caledonian House
63 Academy Street
Inverness IV1 1BB
http://www.uhi.ac.uk

H60 The University of Huddersfield
Queensgate
Huddersfield HD1 3DH
Tel: 01484 422288 x2258
Fax: 01484 516151
prospectus@hud.ac.uk
http://www.hud.ac.uk/

H72 The University of Hull
Admissions Office
Cottingham Road
Hull HU6 7RX
Tel: 01482 465328
Fax: 01482 442290
admissions@admin.hull.ac.uk
http://hull.ac.uk

I50 Imperial College of Science, Technology and Medicine (University of London)
Registrar's Division
London SW7 2AZ
Tel: 0171 594 8014
Fax: 0171 594 8004
admissions@ic.ac.uk
http://www.ic.ac.uk/

K12 Keele University
Admissions Office
Keele ST5 5BG
Tel: 01782 584003
Fax: 01782 632343
aaa20@admin.keele.ac.uk
http://www.keele.ac.uk/depts/aa/homepage.htm

K24 The University of Kent at Canterbury
Office for Undergraduate Admissions
The Registry
Canterbury CT22 7NZ
Tel: 01227 782 7272
admissions@ukc.ac.uk
http://www.ukc.ac.uk/

K60 King's College London (University of London)
Central Registry
Cornwall House
Waterloo Road
London SE1 8TX
Tel: 0171 872 3391
ucas.enquiries@kcl.ac.uk
http://www.kcl.ac.uk/

K84 Kingston University
Cooper House
40-46 Surbiton Road
Kingston-upon-Thames KT1 2HX
Tel: 0181 547 2000 x3514
Fax: 0181 547 7080
admissions-info@kingston.ac.uk
http://www.kingston.ac.uk/

L07 The University of Wales, Lampeter
The Registry
Lampeter SA48 8HX
Tel: 01570 423530
Fax: 01570 423530
recruit@lampeter.ac.uk
http://www.lampeter.ac.uk

L14 Lancaster University
Division of External Affairs
University House
Lancaster LA1 4YW
Tel: 01524 592108
Fax: 01524 846243
ugadmissions@lancaster.ac.uk
http://www.lancs.ac.uk/

L23 University of Leeds
Taught Courses Office
Leeds LS2 9JT
Tel: 0113 233 4024
prospectus@leeds.ac.uk
http://www.leeds.ac.uk/

L24 Leeds, Trinity and All Saints College
Brownberrie Lane
Horsforth
Leeds LS18 5HD
Tel: 0113 283 7117
http://www.tasc.ac.uk/

L27 Leeds Metropolitan University
Calverley Street
Leeds LS1 3HE
Tel: 0113 283 5946
Fax: 0113 283 3114
course-enquiries@lmu.ac.uk
http://www.lmu.ac.uk/

L28 Leeds College of Art & Design
Jacob Kramer Building
Blenheim Walk
Leeds LS2 9AQ
Tel: 0113 202 8000
Fax: 0113 202 8001

L34 University of Leicester
University Road
Leicester LE1 7RH
Tel: 0116 252 2414
Fax: 0116 252 2447
admissions@le.ac.uk
http://www.le.ac.uk/

L39 University of Lincolnshire and Humberside
Admissions Processing Unit
Milner Hall
Cottingham Road
Hull HU6 7RT
Tel: 01482 463354
Fax: 01482 463052
tjohnson@humber.ac.uk
http://www.ulh.ac.uk/

L41 The University of Liverpool
SCILAS
Student Services Centre
Liverpool L69 3BX
Tel: 0151 794 2069
Fax: 0151 794 2060
scilas@liv.ac.uk
http://www.liv.ac.uk/

L43 Liverpool Community College
Bankfield Centre
Bankfield Road
Liverpool L13 0BQ
Tel: 0151 252 3847
Fax: 0151 252 3848

L46 Liverpool Hope
PO Box 6
Stand Park Road
Liverpool L16 9JD
Tel: 0151 291 3194
Fax: 0151 291 3048
http://www.livhope.ac.uk/

L51 Liverpool John Moores University
Roscoe Court
4 Rodney Street
Liverpool L1 2TZ
Tel: 0151 231 3293
Fax: 0151 231 3293
recruitment@livjm.ac.uk
http://www.livjm.ac.uk/

L53 Llandrillo College, North Wales
Llandudno Road
Colwyn Bay
North Wales LL28 4HZ
Tel: 01492 542338
admissions@llandrillo.ac.uk

L55 London Guildhall University
Course Enquiries Unit
84 Moorgate
London EC2M 6SQ
Tel: 0171 320 3444
Fax: 0171 320 3462
enqs@lgu.ac.uk
http://www.lgu.ac.uk

L79 Loughborough University
Ashby Road
Loughborough LE11 3TU
Tel: 01509 263171 x4009
Fax: 01509 265687
w.j.clarke@lboro.ac.uk
http://www.lboro.ac.uk/

L93 University of Luton
Park Square
Luton LU1 3JU
Tel: 01582 489286
pat.herber@luton.ac.uk
http://www.luton.ac.uk/

M20 The University of Manchester
University Admissions Office
Registrar's Department
Manchester M13 9PL
Tel: 0161 275 2074
Fax: 0161 275 2407
ug.admissions@man.ac.uk
http://www.man.ac.uk/

M25 The University of Manchester Institute of Science and Technology (UMIST)
Registrar's Department
PO Box 88
Manchester M60 1QD
Tel: 0161 200 4033
Fax: 0161 200 8765
ug.prospectus@umist.ac.uk
http://www.umist.ac.uk/

M40 The Manchester Metropolitan University
Academic Division
All Saints
Manchester M15 6BH
Tel: 0161 247 1035
Fax: 0161 247 6311
prospectus@mmu.ac.uk/ enquiries@mmu.ac.uk
http://www.mmu.ac.uk/

M77 Mid-Cheshire College
Hartford Campus
Northwich
Cheshire CW8 1LJ
Tel: 01606 74444
Fax: 01606 75101

M80 Middlesex University
White Hart Lane
London N17 8HR
Tel: 0181 362 5955
Fax: 0181 362 6076
admissions@mdx.ac.uk
http://www.mdx.ac.uk/

N07 Napier University
219 Colinton Road
Edinburgh EH14 1DJ
Tel: 0131 455 4277
Fax: 0131 455 4329
info@napier.ac.uk
http://www.napier.ac.uk/

N14 University College Northampton
Park Campus
Moulton Park
Northampton NN2 7AL
Tel: 01604 735500
Fax: 01604 720636
admissions@northampton.ac.uk
http://www.northampton.ac.uk

N21 University of Newcastle upon Tyne
Admissions Office
6 Kensington Terrace
Newcastle upon Tyne NE1 7RU
Tel: 0191 222 7251
admissions-enquiries@ncl.ac.uk
http://www.ncl.ac.uk/

N23 Newcastle College
Ryehill Campus
Scotswood Road
Newcastle upon Tyne NE4 7SA
Tel: 0191 200 4110
Fax: 0191 272 4297
sdoughty@ncl-coll.ac.uk
http://www.ncl-coll.ac.uk/

N28 New College Durham
Framwellgate Moor
Durham DH1 5ES
Tel: 0191 375 4027
Fax: 0191 375 4222
admissions@newdur.ac.uk
http://www.newdur.ac.uk/

N36 Newman College of Higher Education
Genners Lane
Bartley Green
Birmingham B32 3NT
Tel: 0121 476 1181 x225
Fax: 0121 478 1250
registry@newman.ac.uk
http://www.newman.ac.uk/

N37 University of Wales College, Newport
University Information Centre
Caerleon Campus
PO Box 101
Newport NP6 1YH
Tel: 01633 432432
Fax: 01633 432850
uic@newport.ac.uk
http://www.newport.ac.uk/

N41 Northbrook College Sussex
Littlehampton Road
Goring by Sea
Worthing
West Sussex BN12 6NU
Tel: 01903 606001
Fax: 01903 606007
admissions@MBCOL.ac.uk
http://www.NBCOL.ac.uk

N49 NESCOT
Reigate Road
Ewell
Epsom KT17 3DS
Tel: 0181 394 3078
Fax: 0181 394 3030
rwood@nescot.ac.uk mkristensen@nescot.ac.uk
http://www.nescot.ac.uk/

N56 The North East Wales Institute of Higher Education
Plas Coch
Mold Road
Wrexham LL11 2AW
Tel: 01978 293045
Fax: 01978 290008
k.mitchell@newi.ac.uk
http://www.newi.ac.uk/

N58 North East Worcestershire College
Blackwood Road
Bromsgrove
Worcs B90 1PQ
Tel: 01527 572822
Fax: 01527 572901

N63 University of North London
Academic Registry
Grants/Loans Office (Room M19)
166-220 Holloway Road
London N7 8DB
Tel: 0171 753 3242
admissions@unl.ac.uk
http://www.unl.ac.uk/

N72 North Tyneside College
Embleton Avenue
Wallsend
Tyne & Wear NE28 9NJ
Tel: 0191 229 5203
admissions@ntyneside.ac.uk
http://www.ntyneside.ac.uk

N77 University of Northumbria at Newcastle
Ellison Building
Ellison Place
Newcastle upon Tyne NE1 8ST
Tel: 0191 227 4472
Fax: 0191 227 4017
rg.admissions@unn.ac.uk
http://www.unn.ac.uk/

N78 Northumberland College
College Road
Ashington
Northumberland NE63 9RG
Tel: 01670 841203
Fax: 01670 841201
http://www.northland.ac.uk/

N82 Norwich: City College
Ipswich Road
Norwich NR2 2LJ
Tel: 01603 773131
Fax: 01603 773301
registry@ccn.ac.uk
http://www.ccn.ac.uk/

N84 The University of Nottingham
University Park
Nottingham NG7 2RD
Tel: 0115 951 5753
Fax: 0115 951 6566
undergraduate-enquiries@nottingham.ac.uk
http://www.nottingham.ac.uk

N91 The Nottingham Trent University
Academic Registry
Burton Street
Nottingham NG1 4BU
Tel: 0115 948 6402
Fax: 0115 948 6063
ann-marie.cancemi@ntu.ac.uk
http://www.ntu.ac.uk/

O33 Oxford University
College Admissions Office
University Offices
Wellington Square
Oxford OX1 2JD
Tel: 01865 270209
Fax: 01865 270708
undergraduate.admissions@admin.ox.ac.uk
http://www.ox.ac.uk/

O66 Oxford Brookes University
Gipsy Lane Campus
Headington
Oxford OX3 0BP
Tel: 01865 741111
Fax: 01865 483983
http://www.brookes.ac.uk/

O80 Oxfordshire School of Art and Design
Broughton Road
Banbury
Oxon OX16 9QA
Tel: 01295 257979
enquiries@northox.ac.uk
http://www.northox.ac.uk/Banbury/default.htm

P20 University of Paisley
High Street
Paisley
Renfrewshire PA1 2BE
Tel: 0141 848 3859
Fax: 0141 848 3623
fras-ap0@paisley.ac.uk
http://www.paisley.ac.uk/

P60 University of Plymouth
Registry
Drake Circus
Plymouth PL4 8AA
Tel: 01752 232147
Fax: 01752 232141
admissions@plymouth.ac.uk
http://www.plym.ac.uk/

P65 Plymouth College of Art and Design
Tavistock Place
Plymouth PL4 8AT
Tel: 01752 203428
enquiries@pcad.ply.ac.uk
http://www.soc.plym.ac.uk/soc/mla/pcad.html

P80 University of Portsmouth
University House
Winston Churchill Avenue
Portsmouth PO1 2UP
Tel: 01705 843295
Fax: 01705 843082
admissions@port.ac.uk
http://www.port.ac.uk/

Q50 Queen Mary and Westfield College (University of London)
Mile End Road
London E1 4NS
Tel: 0171 775 3353
Fax: 0171 975 5588
Admissions@qmw.ac.uk
http://www.qmw.ac.uk/

Q75 The Queen's University of Belfast
Belfast BT7 1NN
Tel: 01232 245133 x3079
Fax: 01232 247895
admissions@qub.ac.uk
http://www.qub.ac.uk/

R06 Ravensbourne College of Design and Communication
Walden Road
Chislehurst
Kent BR7 5SN
Tel: 0181 289 4900
Fax: 0181 325 8320
info@rave.ac.uk
http://www.rave.ac.uk/

R10 Reading College and School of Arts and Design
King's Road
Reading
Berkshire RG1 4HJ
Tel: 0118 967 5013
enquiries@reading-college.ac.uk
http://reading-college.ac.uk

R12 The University of Reading
Faculty of Agriculture & Food
No 2 Earley Gate
Whiteknights Road
PO Box 239
Reading RG6 6AU
Tel: 0118 931 6273
Fax: 0118 931 6248
ug-prospectus@reading.ac.uk
http://www.rdg.ac.uk/UG

R36 The Robert Gordon University
Customer Services Department (Admissions)
Schoolhill
Aberdeen AB10 1FR
Tel: 01224 262105
Fax: 01224 262133
admissions@rgu.ac.uk
http://www.rgu.ac.uk/

R48 Roehampton Institute London
Roehampton Lane
London SW15 5PU
Tel: 0181 392 3165
Fax: 0181 392 3220
admissions@roehampton.ac.uk
http://www.roehampton.ac.uk/

R72 Royal Holloway, University of London
The Registry
Egham Hill
Egham TW20 0EX
Tel: 01784 443343
Fax: 01784 473662
liaison-office@rhbnc.ac.uk
http://www.rhbnc.ac.uk/

S03 The University of Salford
Salford M5 4WT
Tel: 0161 295 5509
Fax: 0161 295 3126
a.l.farrell@university-management.salford.ac.uk
http://www.salford.ac.uk/homepage.html

S05 SAE Technology College
United House
North Road
London N7 9DP
http://www.sae.edu

S08 Sandwell College
Wednesbury Campus
Woden Road South
Wednesbury
West Midlands WS10 0PE
Tel: 0121 556 6000/6626
Fax: 0121 253 6661

S10 University College Scarborough
Filey Road
Scarborough YO11 3AZ
Tel: 01723 362392
registry@ucscarb.ac.uk
http://www.ucscarb.ac.uk/

S18 The University of Sheffield
Undergraduate Admissions Office
14 Favell Road
Sheffield S3 7QX
Tel: 0114 222 4833
ug.admissions@sheffield.ac.uk
http://www.shef.ac.uk/uni/admin/admit/

S21 Sheffield Hallam University
Admissions Office
Surrey Building
Sheffield S1 1WB
Tel: 0114 225 2022
Fax: 0114 253 4023
c.arnold@shu.ac.uk
http://www.shu.ac.uk/

S22 Sheffield College
Head Office
PO Box 345
Sheffield S2 2YY
Tel: 0114 2602400
Fax: 0114 2602401

S23 Shrewsbury College of Arts and Technology
London Road
Shrewsbury
Shropshire SY2 7PR
Tel: 01743 342397
Fax: 01743 342534
mail@s-cat.ac.uk

S24 St Martin's College, Lancaster: Ambleside: Carlisle
Bowerham Road
Lancaster LA1 3JD
Tel: 01524 384600
admissions@ucsm.ac.uk
http://www.ucsm.ac.uk/

S26 Solihull College
Blossomfield Road
Solihull
West Midlands B91 1SB
Tel: 0121 678 7046
Fax: 0121 678 7200
enquiries@staff.solihull.ac.uk
http://www.solihull.ac.uk/

S27 University of Southampton
Academic Registrar's Department
Southampton SO17 1BJ
Tel: 01703 594392
Fax: 01703 593037
prospenq@soton.ac.uk
http://www.soton.ac.uk

S28 Somerset College of Arts and Technology
Wellington Road
Taunton
Somerset TA1 5AX
Tel: 01823 366366
Fax: 01823 366357
http://www.zynet.co.uk/scat1/

S30 Southampton Institute
East Park Terrace
Southampton SO14 0YN
Tel: 01703 319348
Fax: 01703 319852
MS@Solent.ac.uk
http://www.solent.ac.uk/

S32 South Devon College
Newton Road
Torquay
Devon TQ2 5BY
Tel: 01803 386404
Fax: 01803 386403
courses@s-devon.ac.uk

S33 South Bank University
Central Admissions & International Office
103 Borough Road
London SE1 0AA
Tel: 0171 815 6116
enrol@sbu.ac.uk
http://www.sbu.ac.uk/

S35 Southport College
Mornington Road
Southport
Merseyside PR9 0TT
Tel: 0151 9342627
Fax: 01704 546240

S36 University of St Andrews
Old Union Building
St Andrews KY16 9AJ
Tel: 01334 462150
Fax: 01334 463388
admissions@st.andrews.ac.uk
http://www.st-and.ac.uk/

S38 Southwark College
Surrey Docks Centre
Drummond Road
London SE16 4EE
Tel: 0171 815 1504
Fax: 0171 815 1525
ucas@southwark.ac.uk
http://www.southwark.ac.uk

S51 St Helens College
Brook Street
St Helens
Merseyside WA10 1PZ
Tel: 01744 623202
Fax: 01744 623402
http://www.sthelens.mernet.org.uk/

S59 The College of St Mark and St John
Derriford Road
Plymouth PL19 9AL
Tel: 01752 636700
Fax: 01752 636819
kearnh@marjon.ac.uk
http://194.80.168.100/

S72 Staffordshire University
College Road
Stoke-on-Trent ST4 2DE
Tel: 01782 292749
Fax: 01782 745422
admissions@staffs.ac.uk
http://www.staffs.ac.uk/

S75 The University of Stirling
Admissions Office
Stirling FK9 4LA
Tel: 01786 467043
Fax: 01786 466800
Admissions@stir.ac.uk
http://www.stir.ac.uk/

S76 Stockport College of Further & Higher Education
Wellington Road South
Stockport SK1 3UQ
Tel: 0161 958 3421
Fax: 0161 958 3305
stockcoll@cs.stockport.ac.uk
http://www.stockport.ac.uk/

S78 The University of Strathclyde
Registry-General
McCance Building
16 Richmond Street
Glasgow G1 1XQ
Tel: 0141 763 1354
Fax: 0141 552 5860
j.foulds@mis.strath.ac.uk
http://www.strath.ac.uk/Campus/prospect/info/index3.htm

S79 Stranmillis University College: A College of The Queen's University of Belfast
Stranmillis
Belfast BT9 5DY
http://www.stran-ni.ac.uk

S81 Suffolk College
Rope Walk
Ipswich IP4 1LT
Tel: 01473 296448
http://www.suffolk.ac.uk/

S84 University of Sunderland
St Mary's Building
Chester Road
Sunderland SR1 3SD
Tel: 0191 515 2433
student-helpline@sunderland.ac.uk
http://www.sunderland.ac.uk/

S87 The University of Surrey
The Registry
Guildford GU2 5XH
Tel: 01483 259996
Fax: 01483 300803
http://www.surrey.ac.uk/

S90 University of Sussex
Undergraduate Admissions Office
Sussex House
Falmer
Brighton BN1 9RH
Tel: 01273 877092
Fax: 01273 678335
ug.admissions@sussex.ac.uk
http://www.sussex.ac.uk/

S93 University of Wales Swansea
Singleton Park
Swansea SA2 8PP
Tel: 01792 295357
Fax: 01792 295110
admissions@swan.ac.uk
http://www.swan.ac.uk/

S96 Swansea Institute of Higher Education
Mount Pleasant
Swansea SA1 6ED
Tel: 01792 481094
Fax: 01792 481085
enquiry@sihe.ac.uk
http://www.sihe.ac.uk/home.html

S98 Swindon College
Regent Circus
Swindon
Wilts SN1 1PT
Tel: 01793 498466

T10 Tameside College
Ashton Centre
Beaufort Road
Aston-under-Lyne OL6 6NX
Tel: 0161 908 6789
Fax: 0161 908 6611
info@tamesidecollege.ac.uk

T20 University of Teesside
Borough Road
Middlesbrough TS1 3BA
Tel: 01642 384221
Fax: 01642 342071
H.Cummins@tees.ac.uk
http://www.tees.ac.uk/

T40 Thames Valley University
St Mary's Road
Ealing
London W5 5RF
Tel: 0181 231 2072
Fax: 0181 231 2056
christine.marchant@tvu.ac.uk
http://www.tvu.ac.uk/

T80 Trinity College Carmarthen
College Road
Carmarthenshire SA31 3EP
Tel: 01267 676720
Fax: 01267 230933

U20 University of Ulster
University House
Cromore Road
Coleraine BT52 1SA
Tel: 01265 44141
Fax: 01265 40902
sda.barnhill@ulster.ac.uk
http://www.ulst.ac.uk/

U80 University College London (University of London)
Registrar's Division
Gower Street
London WC1E 6BT
Tel: 0171 380 7387
Fax: 0171 380 7387
degree-info@ucl.ac.uk
http://www.ucl.ac.uk/

W17 University College Warrington
Padgate Campus
Fearnhead
Warrington WA2 0DB
Tel: 01925 494494
Fax: 01925 816077
registry.he@warr.ac.uk
http://www.warr.ac.uk/unicoll/

W20 The University of Warwick
Academic Office
Coventry CV4 7AL
Tel: 01203 523760
Fax: 01203 524170
ugadmissions@admin.warwick.ac.uk
http://www.csv.warwick.ac.uk/

W25 Warwickshire College, Royal Leamington Spa and Moreton Morrell
Warwick New Road
Leamington Spa
Warwickshire CV32 5JE
Tel: 01926 318000
Fax: 01926 318111
Enquiries@warkscol.ac.uk

W40 West Herts College, Watford
Marketing Department
Hempstead Road
Watford WD1 3EZ
Tel: 01923 812565
Fax: 01923 257540
admis.cas@westherts.ac.uk

W50 University of Westminster
Metford House
15-18 Clipstone Street
London W1M 8JS
Tel: 0171 911 5720
Fax: 0171 911 5858
http://www.wmin.ac.uk/

W52 Westminster College
Vincent Square
London SW1P 2PD
Tel: 0171 828 1222 x271
Fax: 0171 931 0347

W65 West Thames College
London Road
Isleworth
Middx TW7 4HS
Tel: 0181 568 0244 x306
Fax: 0181 569 7787
admissions@west-thames.ac.uk
http://www.west-thames.ac.uk/

W67 Wigan and Leigh College
PO Box 53
Parson's Walk
Wigan
Lancs WN1 1RR
Tel: 01942 761600

W73 Wirral Metropolitan College
Carlett Park Campus
Eastham
Wirral
Merseyside L62 0AY
Tel: 0151 551 7472
Fax: 0151 551 7401
h.e.enquiries@wmc.ac.uk
http://www.wmc.ac.uk

W75 University of Wolverhampton
Administration & Records Unit
Central Registry
Compton Park Campus
Wolverhampton WV3 9DX
Tel: 01902 323728
Fax: 01902 322528
admissions@wlv.ac.uk
http://www.wlv.ac.uk

W80 University College Worcester
Henwick Grove
Worcester WR2 6AJ
Tel: 01905 855009
Fax: 01905 748162
http://www.worc.ac.uk/worcs.html

W81 Worcester College of Technology
Deansway
Worcester WR1 2JF
Tel: 01905 725563
Fax: 01905 28906

W85 Writtle College
Writtle
Chelmsford CM1 3RR
Tel: 01245 424200
Fax: 01245 420456
postmaster@writtle.ac.uk
http://www.writtle.ac.uk/

Y50 The University of York
Heslington
York YO1 5DD
Tel: 01904 433535
Fax: 01904 433538
admissions@york.ac.uk
http://www.york.ac.uk

Y70 York College of Further and Higher Education
Tadcaster Road
York YO2 1UA
Tel: 01904 770398
Fax: 01904 770499

Y80 Yorkshire Coast College of Further and Higher Education
Lady Edith's Drive
Scalby Road
Scarborough
N Yorks YO12 5RN
Tel: 01723 372105
Fax: 01723 501918
admissions@ycoastco.ac.uk
http://www.ycoastco.ac.uk

USEFUL SOURCES OF INFORMATION

In addition to looking at the following publications, you will find that the UCAS website contains a large amount of useful information, including links to many other websites. The UCAS website can be accessed at: www.ucas.ac.uk. If you wish to go directly to the linked websites, these can be accessed through: www.ucas.ac.uk/links/index.html.

USEFUL READING

Course and institution choice

IT NOW!
GTI

GTI IT Business
GTI

Computing Careers Yearbook 1999
VNU Publications

Computer Users' Yearbook
VNU Publications

COSHEP/UCAS Entrance Guide to Higher Education in Scotland
John Smith & Son and Sheed and Ward – annual

Degree Course Offers
Brian Heap, Trotman & Co – annual

How to Complete Your UCAS Form
Tony Higgins, Trotman & Co – annual

The Potter Guide to Higher Education
Dalebank Books – annual

The Sixth Formers' Guide to Visiting Universities and Colleges
ISCO publications – annual

The Student Book
Trotman & Co – annual

Survey of HND Courses
Trotman & Co

UCAS Handbook
UCAS – annual

University and College Entrance: The Official Guide
UCAS. Available from Sheed and Ward – annual

Open Days, Pre-Taster Courses & Education Conventions
UCAS – annual

USEFUL CONTACTS

British Computer Society
1 Sanford Street
Swindon
SN1 1HJ
Tel: 01793 417417

Business in the Community
44 Baker Street
London
W1M 1DH
Tel: 0171 224 1600

Business Software Alliance
79 Knightsbridge
London
SW1X 7RB
Tel: 0171 245 0304

Design and Technology Association
Wellesbourne House
Walton Road
Wellesbourne
Warks
CV35 9JB
Tel: 01789 470007

Institution of Electrical Engineers
Savoy Place
London
WC2R 0BL
Tel: 0171 240 1871

Institute of Measurement and Control
87 Gower Street
London
WC1E 6AA
Tel: 0171 387 4949

Information Technology Industry Training Organisation
16 Berners Street
London
W1P 3DD
Tel: 0171 580 6677

Skill
Chapter House
18-20 Crucifix Lane
London
SE1 3JW
Tel: 0171 450 0620